# MUSCULOSKELETAL ULTRASOUND

# Musculoskeletal Ultrasound

**Marnix Van Holsbeeck, M.D.**
Department of Radiology
Henry Ford Hospital
Detroit, Michigan

**Joseph H. Introcaso, M.D., D.M.D.**
Department of Radiology
University of Michigan Medical School
Ann Arbor, Michigan

Mosby
Year Book

St. Louis Baltimore Boston Chicago London Philadelphia Sydney Toronto

Mosby
Year Book

Dedicated to Publishing Excellence

Sponsoring Editor: James D. Ryan/Anne S. Patterson
Associate Managing Editor, Manuscript Services: Deborah Thorp
Production Project Coordinator: Karen Halm
Proofroom Supervisor: Barbara Kelly



Mosby-Year Book, Inc.
11830 Westline Industrial Drive
St. Louis, MO 63146



4 5 6 7 8 9 0 Y C 95

**Library of Congress Cataloging-in-Publication Data**

Van Holsbeeck, Marnix.
Musculoskeletal ultrasound/Marnix van Holsbeeck, Joseph H. Introcaso.
p. cm.
Includes bibliographical references.
Includes index.
ISBN 0-8151-8975-3
1. Musculoskeletal system—Ultrasonic imaging. I. Introcaso, Joseph H. II. Title.
[DNLM: 1. Bone Diseases—diagnosis. 2. Muscular Diseases—-diagnosis. 3. Musculoskeletal System—pathology. 4. Ultrasonic Diagnosis. WE 141 V256m]
RC925.7.V36 1991 91-13223
616.7'07543—dc20 CIP
DLC
for Library of Congress

*To Beatrice, Lodewijk, Elise*
*Annabel, and Hendrik*

*MvH*

*To Marian, Joseph, William,*
*and Mary*

*JHI*

# FOREWORD

The capability for accurate diagnosis of musculoskeletal conditions has dramatically increased in recent years, as technological advances, including imaging studies of superior and exciting quality, have added to the information provided by physical examination and plain radiography. Ultrasound has clearly taken its place as one of the imaging modalities that has expanded our knowledge of disease states, their evaluation, and treatment.

The merits of ultrasonography as a diagnostic study are obvious. It is noninvasive and has no known risk. It can be done quickly, with little discomfort to the patient, and the capacity for bilateral imaging makes comparison with the asymptomatic contralateral limb possible. In most centers, it is much less expensive than other imaging studies. Disadvantages, such as learning curve, ease of interpretation, and variable image quality, seem to be diminishing as technology and experience increases.

The organization of *Musculoskeletal Ultrasound* is well suited to expand our knowledge in the application of this imaging technique. The first two chapters outline the basic science of ultrasound, the physical principles related to its clinical application, and describe artifacts that may interfere with image quality and lead to potential misinterpretation.

The subsequent four chapters describe in detail the imaging technique, normal anatomic features, and pathologic sonographic findings in a variety of soft tissue lesions, focusing specifically on muscle, tendon, bursa, and ligaments. The application of ultrasonography to soft tissue inflammation, and its diagnostic value in disruption and healing of muscles, tendons, ligaments, and musculotendinous junctions, are carefully and completely addressed with illustrations and multiple image reproductions.

Chapters 7 and 8 outline the application of ultrasonography in joint pathology. In Chapter 7, articular pathology, synovitis, joint effusions, meniscal lesions, aneurysms, and loose bodies are among those clinical problems in which musculoskeletal ultrasound has application. Chapter 8 is given entirely to an exhaustive evaluation of the state of the art use of ultrasound as it applies to pediatric problems, focusing particularly on congenital dislocation of the hip. In addition, Chapter 8 outlines the potential use of ultrasound in identifying injuries before ossification has occurred, and its use in distinguishing epiphyseal separations from true joint dislocations in the pediatric population.

The middle section of the book outlines four intriguing, but less common, uses of ultrasonography, including evaluation of conditions in the skin and subcutaneous tissue, evaluation of rheumatoid diseases, foreign bodies, and pain syndromes following arthroscopy.

The final three chapters of this book describe the current status of ultrasound as it applies to the shoulder, wrist, elbow, hand, hip, and knee. Not only is the normal sonographic anatomy well outlined, but conditions specific to these joints are enumerated, and the sonographic features are described and illustrated.

Clearly, well performed and interpreted ultrasonography is a valuable asset to the diagnostic armamentarium of musculoskeletal lesions. This exhaustive text of the varied, common, and not-so-common applications of musculoskeletal ultrasound is a valuable reference for the clinician and ultrasonographer alike.

Edward V. Craig, M.D.
*Associate Professor*
*Department of Orthopaedic Surgery*
*University of Minnesota*
*Minneapolis, Minnesota*

# FOREWORD

I am pleased to have been asked to contribute a foreword to Drs. van Holsbeeck and Introcaso's book on *Musculoskeletal Ultrasound.* In the present atmosphere of rapidly advancing techniques in imaging, as well as the increasing pressure for cost containment, this is an extremely timely work.

Vast areas of investigation have been opened as a result of the development of ultrasound as an imaging tool. In the past 15 years, there have been dramatic advances in imaging in the form of computerized tomography and magnetic resonance. These have developed, however, as highly expensive procedures with fixed facilities. By contrast, ultrasound provides economic, noninvasive imaging of tissues in static or dynamic states, and in serial studies where indicated. Its capacity to be easily transported adds to its value.

These unique aspects of imaging have made ultrasound especially valuable in the field of sports medicine. Ultrasound meets two very specific needs in sports medicine created by the pressure in professional athletics to return the athlete to competition as quickly as possible. First, it allows immediate identification of problems, thereby avoiding many diagnostic surgical procedures. Secondly, it allows the study of problems which are only manifest during physical activity. These same attributes also enhance the study of inflammatory conditions including infections, traumatic conditions including fractures, and dislocation and neoplastic conditions.

Drs. van Holsbeeck and Introcaso, two of the premier authorities on ultrasound, have provided a text equally valuable to the clinician and to the radiologist. In a concise manner, the book describes the clinical indications for ultrasound imaging. The authors also provide information about theory and technique for the benefit of the practicing radiologist or technician. The book is a timely and valuable text in an important and expanding field. I compliment Drs. van Holsbeeck and Introcaso for an excellent job and a fine accomplishment.

David J. Collon, M.D.
***Chief, Orthopaedic Section***
***Bone and Joint Center***
***Henry Ford Hospital***
***Detroit, Michigan***

# FOREWORD

The practice of medicine has always been a balance of art and science. Examination of the soft tissues of the musculoskeletal system has been the privilege of the clinician for centuries. Those clinicians gifted with exceptionally skilled hands and other senses of clinical observation are fortunate. This clinical information formed the basis for their therapeutic decisions and professional renown. However, these data are subjective and not comparable among various examiners. The development of high resolution real-time ultrasound imaging tips the balance toward more objective and quantitative measures.

Development of the sonographic evaluation of the musculoskeletal system has in large part been limited by technology. Current state of the art equipment has broken down these technological barriers. This text demonstrates the broad spectrum of applications of diagnostic ultrasound to evaluate the musculoskeletal system. It will have a profound impact on the practice of medicine in rheumatology, orthopedics, sports medicine, and traumatology. I have personally had the pleasure of seeing these techniques develop and grow in our rheumatology-orthopedic unit while Dr. van Holsbeeck was with us.

Ultrasound is the ideal modality for the examination of soft tissues because of its multiplanar and real-time capabilities. Sonography yields anatomic information during active and passive mobilization that is unattainable with other modalities. In addition, synovial and cartilage thickness can be accurately quantitated, providing an objective means of following patients with inflammatory arthritides. Ultrasound examination of deep seated joints such as the hip and shoulder is especially valuable. Joint effusions, loose bodies, tendonitis, and tendon and muscle ruptures can all be demonstrated sonographically. The noninvasive nature of the examination and lack of ionizing radiation make it very well accepted by patients, especially children.

This book is a necessity for rheumatologists, orthopedists, traumatologists, sports medicine, physical medicine and rehabilitation physicians. These physicians will quickly learn to utilize musculoskeletal ultrasound as an extension of the clinical diagnostic skills with which they are already so familiar.

Jan Dequeker, M.D., Ph.D.
*Professor of Rheumatology*
*Head, Division of Rheumatology*
*Department of Internal Medicine*
*Arthritis and Metabolic Bone Disease Research Unit*
*Katholieke University Leuven*
*Belgium*

# PREFACE

We have divided the topics discussed in this text into four categories: physical principles, tissue types (i.e., muscle, tendon, ligament), specific joints and areas of special interest. Conventions we use in the display of images are as follows. Longitudinal images are displayed with superior at the observer's left. Transverse images are oriented in the same manner as CT images; the patient's right at the observer's left. Exceptions to this format are indicated in the specific figure caption.

Correlation of imaging findings with surgical findings is usually a difficult task. We are greatly indebted to Dr. Mark Martens for his outstanding cooperation in providing surgical follow-up. His enthusiasm and energy are greatly appreciated.

We have greatly enjoyed working on this project and hope that you will enjoy the result.

MARNIX VAN HOLSBEECK, M.D.
JOSEPH H. INTROCASO, M.D., D.M.D.

# ACKNOWLEDGMENTS

We would like to thank all who have contributed to this project. Special thanks are due to our parents and families, whose patience, understanding and countless years of tuition payments have made this work possible. Dr. Michael DiPietro's chapter on Pediatric Applications of Musculoskeletal Ultrasound significantly enhances this work and is greatly appreciated. Dr. Mark Martens provided superb clinical and surgical correlation for many of the cases presented here. This assistance was invaluable in confirming our sonographic impressions. Dr. Christian Hessler contributed greatly in the constructive review of our work and this manuscript. Dr. Karen Stuck assisted in the review of a portion of the manuscript. Mr. James D. Ryan, our executive editor, provided an extraordinary amount of freedom to us in the preparation of this manuscript, as well as valuable guidance. Mr. Kristo van Holsbeeck performed the statistical analysis for a number of studies presented in this text. Dr. Roeland Lysens performed clinical examinations of the patients who were included in our control groups. Finally, but not least, our teachers who have shaped us over the years.

We thank you for your support!

Marnix van Holsbeeck, M.D.
Joseph H. Introcaso, M.D., D.M.D.

# CONTENTS

# 1

# Physical Principles of Ultrasound Imaging

Ultrasound is defined as sound having a frequency greater than that which is audible by humans. The human ear functions over a frequency range of 15 to 20,000 cycles per second (hertz). Thus, any sound having a frequency greater than 20 kilohertz (kHz) falls into the category of ultrasound. Medical imaging today most commonly uses frequencies from 2 to 12 megahertz (MHz).

The practical use of ultrasound has evolved slowly, largely due to limitations imposed by equipment. In 1912 the first significant attempt at practical application of ultrasound was made in the unsuccessful search for the wreck of the Titanic. Technological advances made during the second World War lead to the development of SONAR (*S*ound *N*avigation *A*nd *R*anging), which played an important role in the war at sea. Following the war, Dr. Douglas Howry applied this technology to medical applications with limited success. It was not until the development of B-mode imaging that ultrasound began to take a prominent position in medical diagnosis. Another leap forward occurred with the advent of gray scale imaging in 1972. The application of digital computers to sonographic imaging in the late 1970s led to the development of real time gray scale imaging as we know it today.

## FUNDAMENTAL PRINCIPLES

Sound waves and x-ray photons both are forms of energy transmission. However, that is where the similarity between the two ends. Their interactions with matter are quite dissimilar. The ways in which these two forms of energy interact with matter determines how they can be used in medical imaging. Unlike radiography, ultrasound imaging most commonly utilizes energy reflected back to the source to produce an image. This is referred to as *pulse-echo imaging.*

X-rays are best transmitted through a vacuum, but sound requires matter for its transmission. The speed at which x-ray photons travel is constant. However, the speed of sound varies with the type of matter through which it passes. Table 1–1 lists the speed of sound through a variety of substances. The factors that determine the speed of sound through a substance are density and compressibility. Materials with the greatest density and least compressibility will transmit sound at the highest velocity.

Sound is reflected at interfaces between materials. Two factors influence reflectivity: the acoustic impedance of the two materials and the angle of incidence of the sound beam. Acoustic impedance is the product of a material's density and the speed of sound within that substance. Reflectivity is greatest at interfaces between materials with dissimilar acoustic impedance, as defined by the following equation:

$$R = \left(\frac{Z_2 - Z_1}{Z_2 + Z_1}\right)^2 \times 100$$

where R is the percentage of sound beam reflected, and $Z_1$ and $Z_2$ are the acoustic impedances, assuming an angle of incidence of 90 degrees.

The data in Table 1–2 indicate that interfaces between soft tissue and air should be highly reflective,

**TABLE 1–1.**
Speed of Sound Through Various Substances

| Transmitting Substance | Speed of Sound (m/sec) |
|---|---|
| Air | 331 |
| Fat | 1,450 |
| Water | 1,540 |
| Liver | 1,549 |
| Blood | 1,570 |
| Muscle | 1,585 |
| Cortical bone | 4,080 |

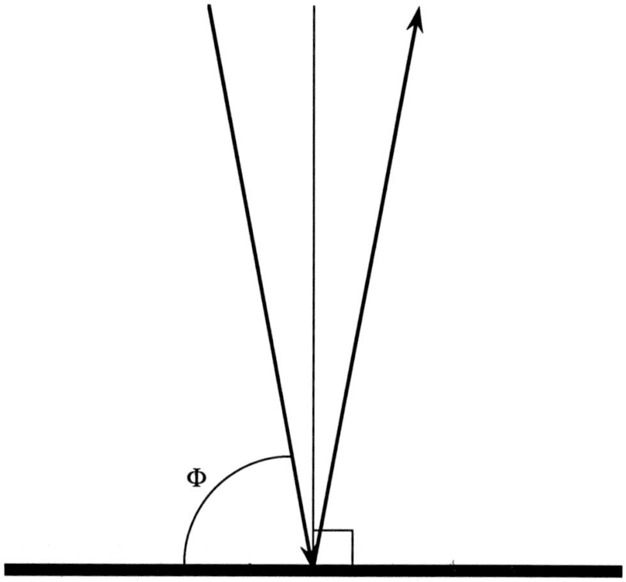


**FIG 1–1.**
Reflection of sound and angle of incidence (Φ).

which is what we observe clinically. With this in mind, it is clear why a coupling gel must be used to assure contact between the ultrasound transducer and the patient's skin. The previous equation tells us that 99.9% of a sound beam is reflected at any tissue-air interface. In areas of poor contact with the skin, where an air gap exists, essentially no energy is available for imaging.

Reflection of a sound beam varies greatly with the angle of incidence. The least reflection occurs with the sound beam perpendicular to the reflecting interface; in other words, an angle of incidence of 90 degrees. As the angle of incidence is decreased, the percentage of the sound beam reflected increases. Beyond a certain angle the entire sound beam is reflected. In addition, the direction of the reflected beam is determined by the angle of incidence (Fig 1–1). This is important to keep in mind when imaging an object with a curved surface, such as the femoral condyles or the diaphysis of a long bone. As the angle of incidence is decreased, the beam will be reflected away from the transducer and not contribute to the image (Fig 1–2). Therefore, the optimal scanning pattern for a curved surface is an arc that keeps the beam perpendicular to the surface of the object to be imaged (Fig 1–3).

Spatial localization will also be impaired when the

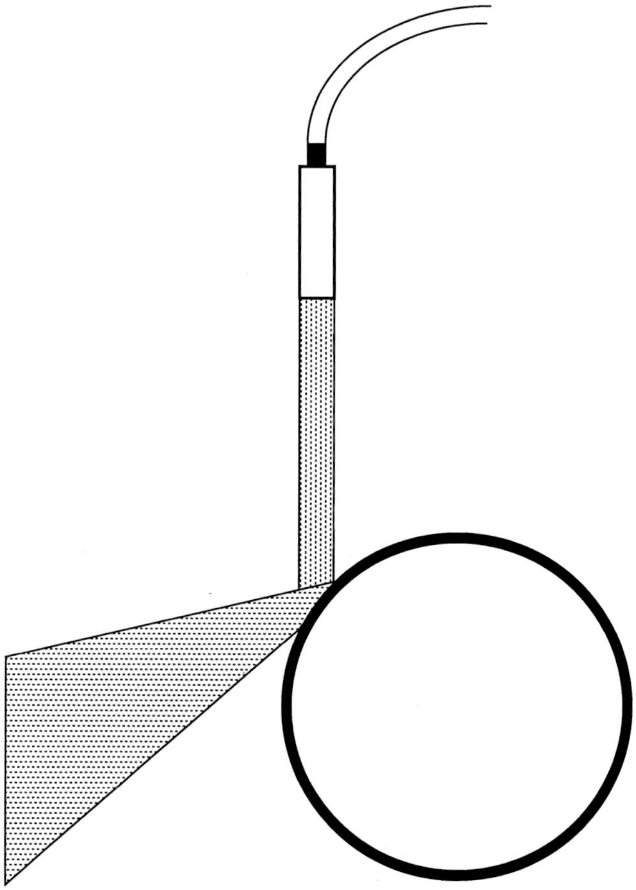

**FIG 1–2.**
Improper imaging of a curved object.

**TABLE 1–2.**
Acoustic Impedance of Various Materials

| Material | Acoustic Impedance (gm/cm² sec × 10⁻⁵) |
|---|---|
| Air | 0.0004 |
| Fat | 1.38 |
| Water | 1.54 |
| Blood | 1.61 |
| Muscle | 1.70 |
| Cortical bone | 7.8 |

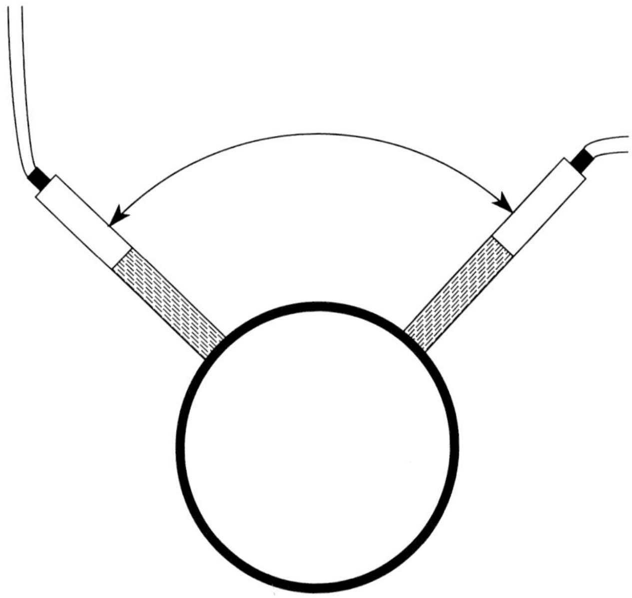

**FIG 1–3.**
Optimal imaging of a curved object.

angle of incidence is less than 90 degrees. This phenomenon is known as *refraction*. Refraction is a change in direction of a sound beam occurring at an interface between two dissimilar materials when the beam is not perpendicular to the interface. A change in wavelength of the sound beam occurs in accommodation to the different speed of sound in the new material. This acts to "bend" the beam. The magnitude of the change in direction of the sound beam is proportional to the difference of the speed of sound within the two materials and inversely proportional to the angle of incidence. In most circumstances, the error introduced by refraction is not significant. However, under certain conditions, the true location of an object will differ significantly from its imaged position.

As a sound beam passes through a material, a portion of its energy is absorbed by frictional forces. The energy is converted to heat and no longer contributes to the imaging process. Viscosity, relaxation time, and temperature of the material, along with the frequency of the the sound beam all effect absorption. Of these factors, the one that can be modified in the clinical setting is the frequency of the transducer. The degree of absorption of a sound beam in soft tissue is directly proportional to frequency. If the frequency of the sound beam is doubled, absorption will double. This becomes important in selecting a transducer, because spatial resolution is also proportional to frequency. The best transducer for a specific examination is one having the highest frequency that can penetrate to the desired depth within the soft tissues being examined. This will result in images with the greatest possible spatial resolution.

## EQUIPMENT

Four types of ultrasound transducers are currently available: (1) sector scanners, (2) annular arrays, (3) radial arrays, and (4) linear arrays. The basic principles governing their function are identical for the four types. They differ only in how some of the components are arranged (Fig 1–4). Annular arrays are the least familiar, because they are used predominantly in cardiac and ophthalmalogical examinations.

The heart of an ultrasound transducer is a piezoelectric crystal. These crystals, discovered by Pierre and Jacques Curie in 1880, have unique physical and electrical properties. When a voltage pulse is applied to a piezoelectric crystal, it vibrates and can produce sound at a specific resonant frequency. In addition, if a mechanical force is applied to a piezoelectric crystal, an electrical potential will result. These properties make piezoelectric crystals ideal for ultrasound transducers because the same element serves as both transmitter and receiver of the ultrasound beam. The original materials described by the Curies were quartz and Rochelle salts, but these have been replaced in modern medical imaging equipment by lead zirconate titanate.

How these crystals are incorporated into an imaging transducer is demonstrated schematically in Figures 1–4,A through D. Two opposing sides of the crystal are plated with a metallic conductor, usually gold, to serve as electrical contacts. Electrical leads attached to the plated surfaces deliver the electrical pulses to the crystal and conduct the potential generated by the crystal when it detects the reflected sound beam. The crystal surface facing the patient may be in contact with an acoustic lens that helps to focus the sound beam. These lenses are usually formed of polystyrene and use the principle of refraction to focus the sound beam. This unit is separated from the patient by an insulating layer that serves to protect the patient from possible electrical shock and protects the crystal from contamination. The other side of the piezoelectric crystal is in contact with a backing block that dampens the vibration of the crystal and absorbs sound moving away from the patient. This assembly is protected from noise by a layer of acoustic insulating material and enclosed within a plastic housing.

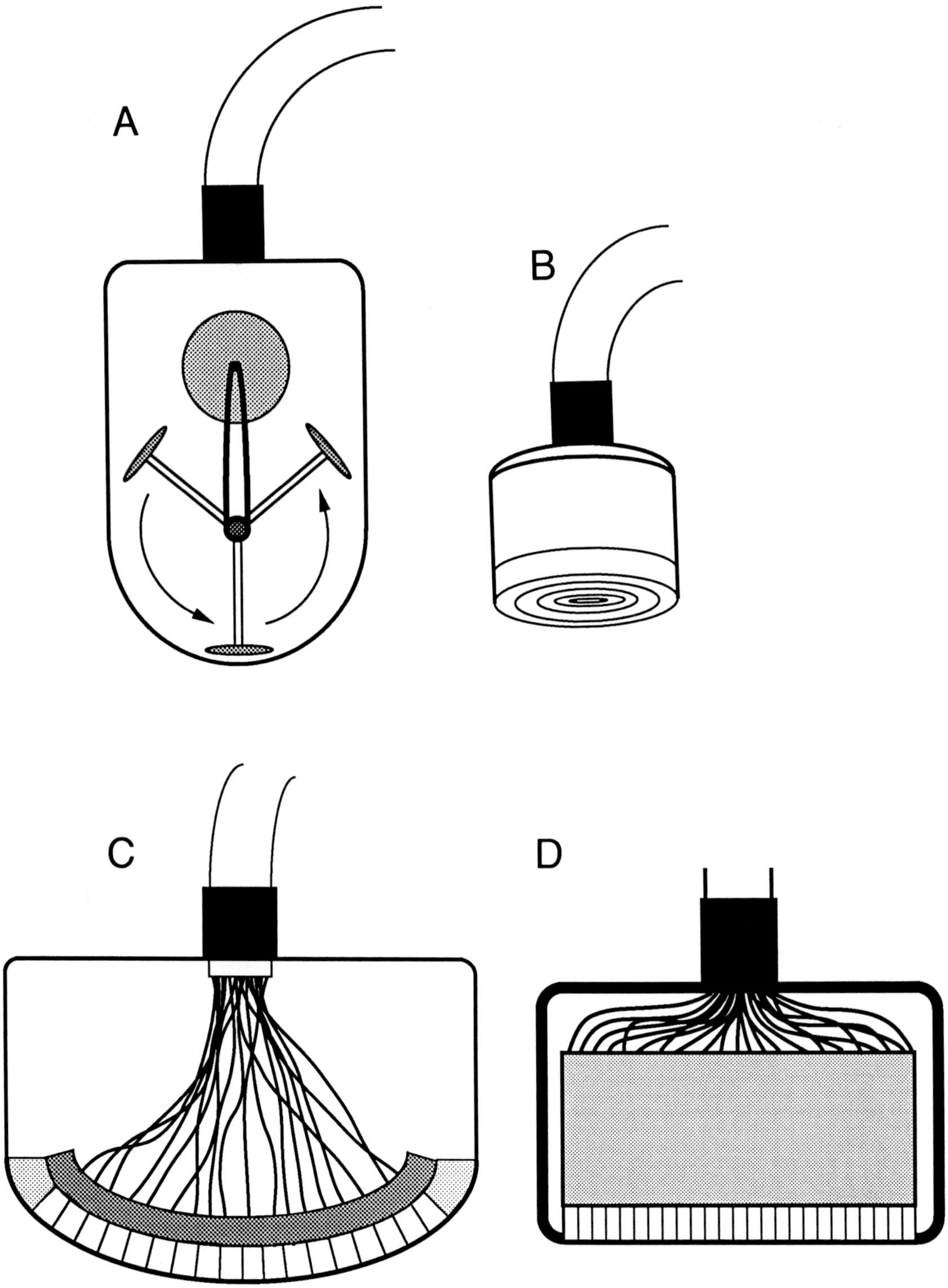


**FIG 1–4.**
**A,** sector scanning transducer. **B,** annular array transducer. **C,** radial array transducer. **D,** linear array transducer.

Operating characteristics of a transducer are described by its resonant frequency and the Q factor. Two properties of a transducer are described by the Q factor: the purity of sound produced and the "ring down" time. *Ring down time* refers to the time required for the crystal to stop vibrating. Piezoelectric crystals with a high Q factor produce pure sound but have a long ring down time. Likewise, crystals with a low Q factor produce a sound beam containing a relatively wide variety of frequencies but have a short ring down time. Crystals with a relatively low Q factor are most desirable for use in medical imaging because they are sensitive to a wider frequency range of sound returning to the transducer, and the short ring down time reduces interference with the returning signal. Careful selection of the backing block utilized in the transducer can enhance the operating characteristics by further reducing the ring down time.

The sound beam emitted by the transducer has a shape that changes with distance from the transducer (Fig 1–5). In the near field, the borders of the sound beam are almost parallel, referred to as the *Fresnel zone.* At a certain distance from the transducer, the beam diverges, referred to as the *Fraunhofer zone.* The point of transition varies with the frequency and width of the sound beam. As the frequency and width of the beam increase, the Fresnel zone (parallel borders) becomes longer. Side lobes are also present, extending laterally from the main beam, called the *Twilight zone.* (No! Just kidding.) These side lobes are usually less than 1% of the intensity of the main beam and thus very rarely clinically significant. When they do contribute to an image, they have the same effect as beam width artifact, described in chapter 2. Acoustic lenses and firing sequence of an array of crystals can be used to modify the shape of the ultrasound beam.

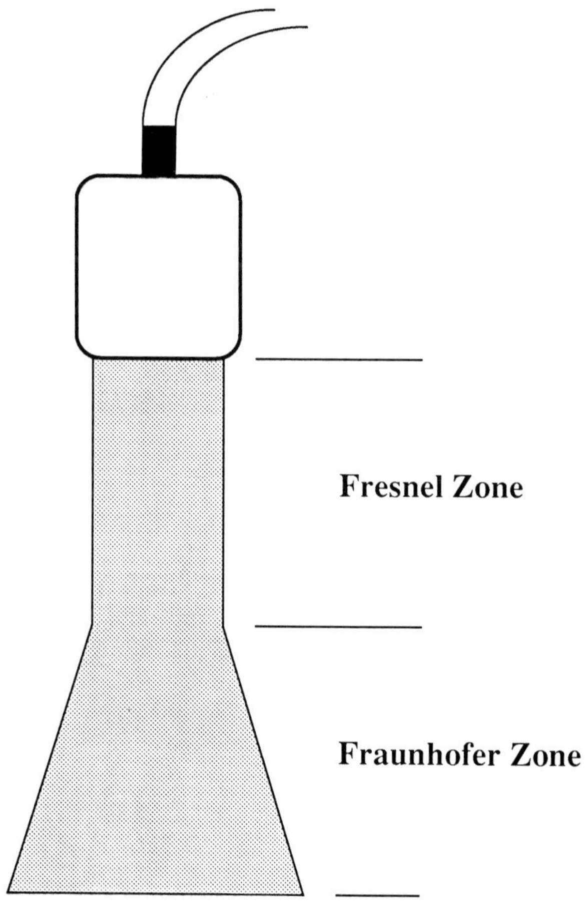


**FIG 1–5.**
Fresnel and Fraunhofer zones of a sound beam.

## IMAGING

Ultrasound images are composed of a matrix of picture elements. Gray scale images are produced by the display of echos returning to the transducer as picture elements (pixels) varying in brightness in proportion to the intensity of the echo. The location of the echo to be displayed is determined by the location on the transducer receiving the echo and the time of flight. Time of flight is the time elapsed between initiation of the sound pulse and its return to the transducer.

In ultrasound imaging we speak of two types of resolution: axial and horizontal. *Axial resolution* is the ability to distinguish two objects as being separate when they lie directly over each other; that is, aligned sequentially along the length of the beam. Frequency of the transducer and Q factor determine axial resolution. A higher-frequency transducer will provide greater axial resolution. Since Q factor of the transducer is determined by the manufacturer, selection of transducer frequency is the only control the examiner has over axial resolution.

*Horizontal resolution* is the ability to distinguish two objects as separate when they are located side by side at the same distance from the transducer. Beam width is the determining factor in horizontal resolution. Factors determining beam width relate to focusing of the ultrasound beam, principally whether imaging is performed in the Fresnel or Fraunhofer zones. The examiner has little control over beam width, and therefore horizontal resolution, because it is largely a function of transducer design. Therefore, careful attention should be paid to transducer specifications when an ultrasound machine is purchased.

Divergence of the sound beam, absorption, and scattering all play a role in attenuation of the ultrasound beam within the body. A sound beam undergoes an exponential decrease in intensity as it passes through tissue. If it is uncorrected, this would result in images that are markedly decreased in diagnostic information with increasing distance from the transducer. Correction for attenuation is made by amplifying echos returning to the transducer using an exponential function based on the time of flight. This type of correction is called *time gain compensation.* The examiner may modify the correction function using controls on the ultrasound unit to optimize the information displayed.

## TRANSMISSION ULTRASOUND

Recent developments in equipment and image processing techniques have made the production of high-quality images using transmitted ultrasound possible, in much the same way computed tomography images are produced. Two transducers are utilized, positioned facing each other. One serves as the signal source, and the other is the receiver. The most recent examples have demonstrated exquisite contrast resolution and anatomical detail. Imaging is currently limited to the extremities, due to the requirement of a water path. Hopefully, further development will continue, resulting in a clinical transmission ultrasound machine.

## CONCLUSION

The purpose of this chapter is not to provide a thorough review of ultrasound physics that will enable the reader to pass their specialty board examination. Its goal is to present technical information about ultrasound imaging that will allow the reader to improve the diagnostic quality of images produced daily in clinical practice. Its success can be determined only by the individual reader, but if you have endured reading it to this point, perhaps that too can be considered a success.

## BIBLIOGRAPHY

Bronzino J: *Technology for Patient Care.* St Louis, CV Mosby, 1977.

Carpenter DA: Ultrasonic transducers. *Clin Diagn Ultrasound* 1980; 5:31.

Curry TS, Dowdey JE, Murry RC: *Christensen's Introduction to the Physics of Diagnostic Radiology,* ed 3. Philadelphia, Lea & Febiger, 1984.

Goodsitt M, in Taveras J, Ferrucci JT (eds): *Radiology: Diagnosis-Imaging-Intervention.* Philadelphia, JB Lippincott Co, 1986, vol 1.

Hentz VR, Green PS, Arditi M: Imaging studies of the cadaver hand using transmission ultrasound. *Skeletal Radiology* 1987; 16:474–480.

Kremkau F: *Diagnostic Ultrasound: Physical Principles and Exercises.* New York, Grune & Stratton, 1980.

Rose JL, Goldberg BB: *Basic Physics in Diagnostic Ultrasound.* New York, Wiley & Sons, 1979.

Sarti D: *Diagnostic Ultrasound,* ed 2. Chicago, Year Book Medical Publishers, 1987.

Wells PNT: *Biomedical Ultrasonics.* New York, Academic Press, 1977.

Wells PNT: Real-time scanning systems. *Clin Diagn Ultrasound* 1980; 5:69.

# 2

# Artifacts in Musculoskeletal Ultrasound

All imaging modalities are subject to artifacts that are unique to that system. In radiography systems, artifacts degrade images and reduce their diagnostic value. Sonographic imaging differs in that some artifacts may facilitate making the correct diagnosis. Artifacts in ultrasound can be categorized in three ways, much like an old Clint Eastwood movie. There are the good, the bad, and the ugly. Therefore, it is especially important to be aware of the various types of artifacts and the circumstances in which they may be encountered. We must recognize the good artifacts to optimize our diagnostic accuracy. The bad and ugly artifacts must be identified so that they may be corrected or ignored.

## THE GOOD . . .

### Shadowing

At highly reflective interfaces almost all of the energy of a sound beam incident on that interface will be reflected. A minimal amount of energy will pass deep to this interface and be available for imaging. The result is a signal void deep to the hyperreflective object. On sonographic images this signal void appears similar to the shadow cast by a building on a sunny day.

In clinical imaging, shadowing is seen at interfaces of materials that differ significantly in acoustic impedance. Classic examples of materials that produce shadowing in vivo are bone, air, calcifications, and biliary and renal calculi. "Dirty shadowing" is a characteristic exhibited by gas within the soft tissues. It is a form of reverberation artifact that produces false echos deep to the highly reflective soft tissue–gas interface. If dirty shadowing is not observed, the differential diagnosis can usually be narrowed by consideration of the region being examined. For example, shadowing from an object within muscle is usually caused by myositis ossificans, arterial calcification, or a foreign body (Fig 2–1). Clinical history will further narrow the list, usually allowing a definitive diagnosis to be made.

Refractile shadowing or critical angle shadowing is observed when objects with highly curved surfaces, such as the diaphysis of a long bone, are imaged. However, the interface need not be as highly reflective as a soft tissue–bone interface. Large arteries and the gallbladder can demonstrate refractile shadowing. The shadow is observed at the lateral margins of the object, where the sound beam contacts the interface at a very oblique angle. Due to both refraction and reflection, essentially none of the incident sound beam returns to the transducer from that region. The result is an acoustic shadow.

On occasion, shadowing will not be observed deep to objects that usually shadow. This can be seen when the object expected to shadow is small relative to the ultrasound beam width or the axial resolution of the transducer. Inappropriate positioning of the focal zone will also minimize acoustic shadows. In addition, high levels of background noise will obscure shadowing. These problems will be discussed further under the sections on bad and ugly artifacts.

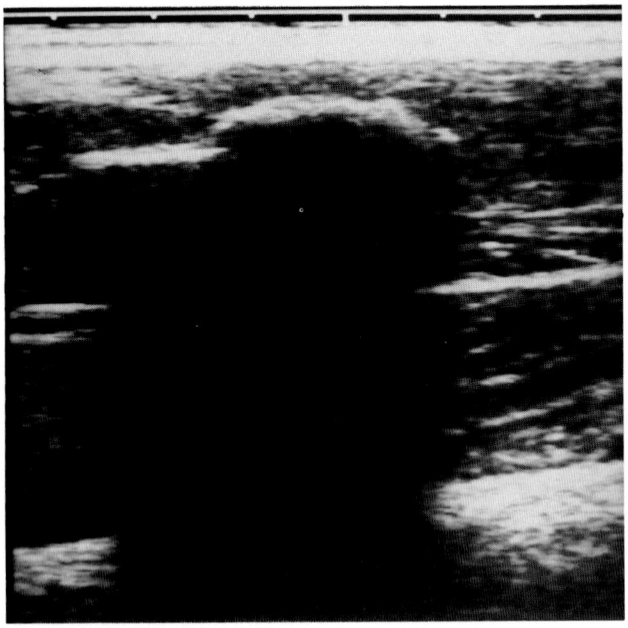

**FIG 2–1.**
Acoustic shadowing deep to myositis ossificans.

## Enhanced Through-Transmission

The intensity of echos returning to the transducer decreases exponentially with increasing depth in the tissues being examined. Superficial echos may be $10^6$ times greater in amplitude than those from the deeper tissues. If the discrepancy is uncorrected, the result would be a rapid decline in image definition with increasing depth. Time gain compensation is the primary means of correcting for this in the image to be displayed. Echos returning to the transducer from deeper structures are amplified based on a curve that increases exponentially with time. This method of correction results in the artifact we know as increased through-transmission.

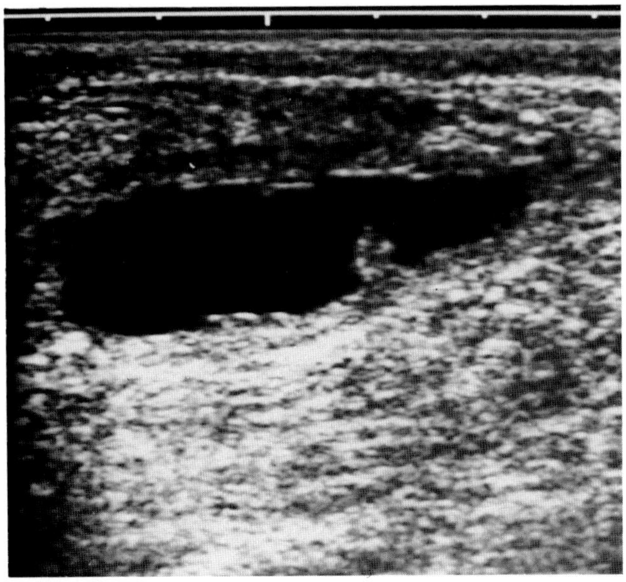

**FIG 2–2.**
Enhanced through-transmission deep to a muscle cyst.

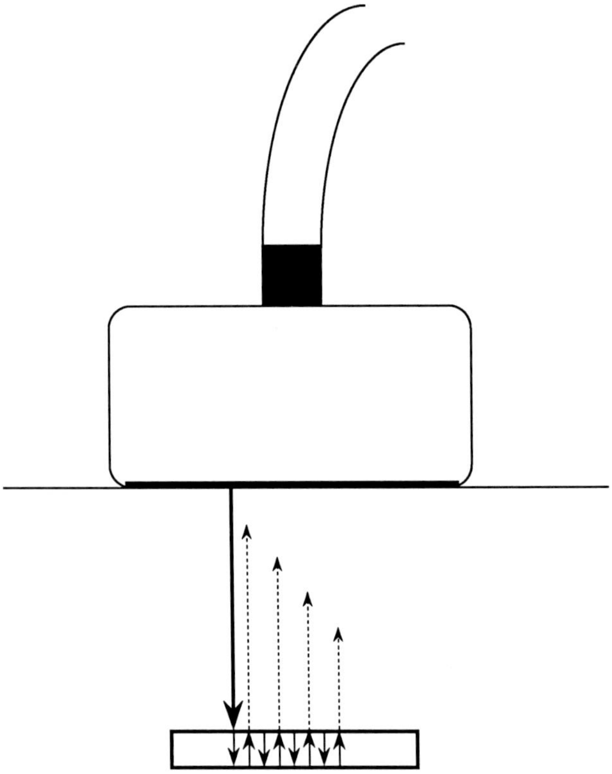

**FIG 2–3.**
Schematic representation of reverberation in a comet tail artifact.

This method of image processing assumes that all materials attenuate sound equally—a false assumption. When a structure that does not attenuate the sound beam as much as surrounding tissues is encountered, more sound is available to image the structures at deeper levels. Therefore, echos returning to the transducer have greater amplitude. These echos are further amplified by the time gain compensation. The result is a false impression of increased echogenicity of the deeper structures. Enhanced through-transmission is most commonly seen deep to anechoic structures, usually simple fluid. A fluid-filled bursa and the urinary bladder are common examples (Fig 2–2). With this in mind, it is clear why enhanced through-transmission is one of the criteria used to make the diagnosis of a simple cyst.

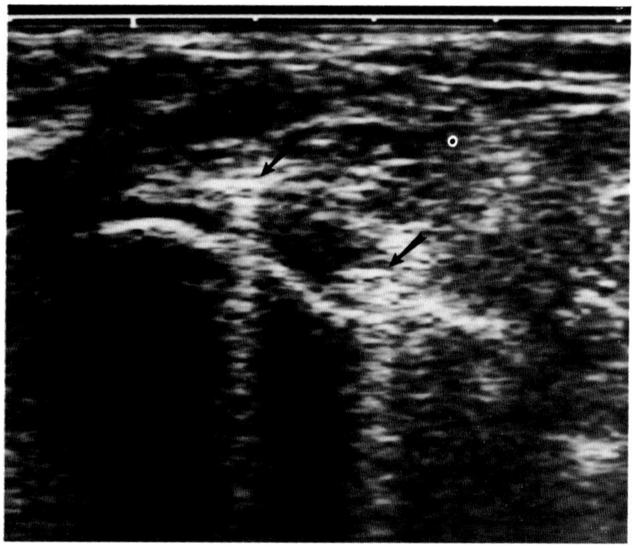

**FIG 2–4.**
Comet tail artifact associated with a metallic orthopedic staple. *Arrows* indicate the two legs of the metallic orthopedic staple.

### Comet Tail Artifact

Metal and glass will produce characteristic bands of increased echogenicity deep to the object. These bands cross tissue boundaries, including those tissues that produce shadowing. The intensity of these echogenic bands decreases with distance from the object, giving the appearance of the tail of a comet.

This artifact has been shown to be the result of reverberation occurring within the metallic or glass object. Figure 2–3 demonstrates diagrammatically how the sound beam is repeatedly reflected between the highly reflective anterior and posterior surfaces of the object. The periodicity of the bands within the comet tail is equal to the thickness of the object.

Recognition of this artifact allows the examiner to quickly diagnose metallic and glass foreign bodies (Fig 2–4). The position of the object can be accurately established, however, the size of relatively small objects cannot be measured accurately. This is especially true of cylindrical metal or glass objects and is a by-product of the reverberation.

## THE BAD . . .

### Refraction

Artifact resulting from refraction is the depiction of real structures in a false location. Refraction occurs at interfaces between substances that transmit sound at different velocities, such as fat (1,450 m/sec) and muscle (1,585 m/sec). The sound beam is "bent" at these interfaces in proportion to the difference in velocity of sound transmission within the two materials and the angle of incidence of the sound beam. Bending of the sound beam results in the depiction of structures deep to the interface in an incorrect location (Fig 2–5). Since we cannot control the speed of sound in various tissues, refraction artifact must be minimized by having the angle of incidence as close to 90 degrees as possible. The example in Figure 2–5 demonstrates schematically what can occur

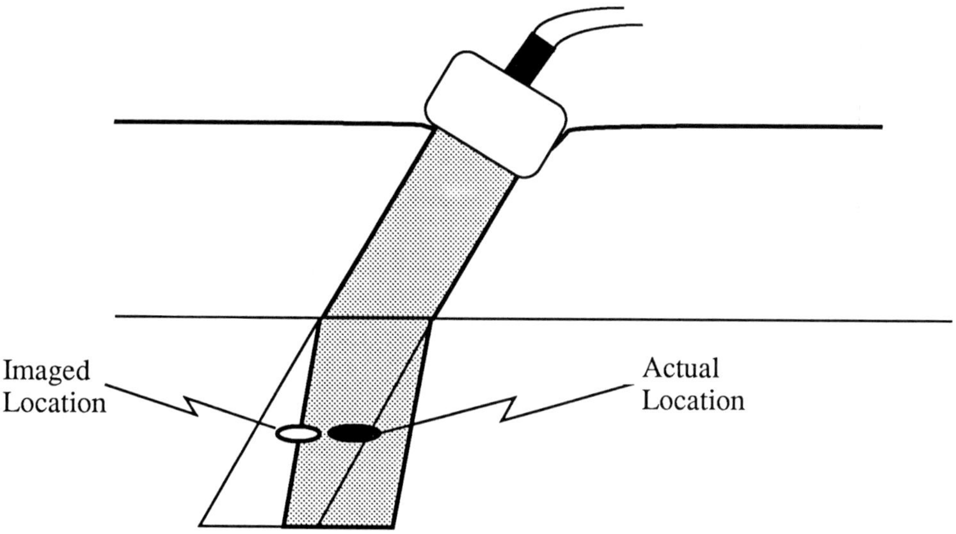


**FIG 2–5.**
Refraction artifact resulting in a real lesion depicted in an incorrect location.

when one is scanning obliquely. A lesion is identified but depicted in an incorrect location. If needle aspiration were attempted using a needle guide, the lesion would be missed.

## Reverberation

Earlier we discussed how a reverberation artifact, the comet tail, can be beneficial by helping to characterize foreign bodies. However, other types of reverberation artifacts can be misleading and, if unrecognized, will result in an incorrect diagnosis. Fortunately, situations in which these are encountered in musculoskeletal ultrasound are rare.

Reverberation occurs at highly reflective interfaces, such as the diaphragm. The sound beam is reflected back and forth within the body, resulting in phantom structures, sometimes mirror images. Intensity of the echos decreases exponentially with the increasing number of iterations. The sites at which these artifacts are most frequently encountered are the pelvis and around the diaphragm. Although we have not encountered a case while imaging in the extremities, the possibility of phantom echos should always be kept in mind.

## Anisotropic Reflectors

An anisotropic substance is one that displays different properties depending on the direction of measurement. In abdominal imaging the kidney is the only anisotropic reflector. Its anisotropic reflectivity is often subtle in vivo and therefore often unappreciated. Musculoskeletal ultrasound involves imaging of strongly anisotropic reflectors, such as tendons. Muscle demonstrates anisotropy, but it is much less pronounced than tendon. Figure 2–6 demonstrates the anisotropic character of the flexor tendons of the wrist. The image on the right was obtained with the sound beam perpendicular to the surface of the tendons, and they appear rather echogenic. Immediately after obtaining this image, the transducer was obliqued slightly without changing the position of the face of the transducer. The resulting image is displayed on the left. Now the same tendons appear markedly hypoechoic. Examination of tendons should always be performed with the sound beam perpendicular to their surface. During real-time examination, this is not as difficult as it may appear.

## Speed of Sound Artifact

Ultrasound equipment determines the distance of an object from the transducer by measuring the time elapsed between the origination of the sound pulse and its return to the transducer. This is referred to as the *time of flight.* In calculating distance, the machine assumes a constant speed of sound. However, as we discussed in the previous chapter (see Table 1–1), this is a false assumption. Although the speed of sound within human tissues does not vary greatly, it may still produce a significant artifact. The image in Figure 2–7 was obtained while scanning the liver of a patient with retroperitoneal lipomatosis. The apparent rent in the diaphragm is an artifact created by the difference in the speed of sound passing through

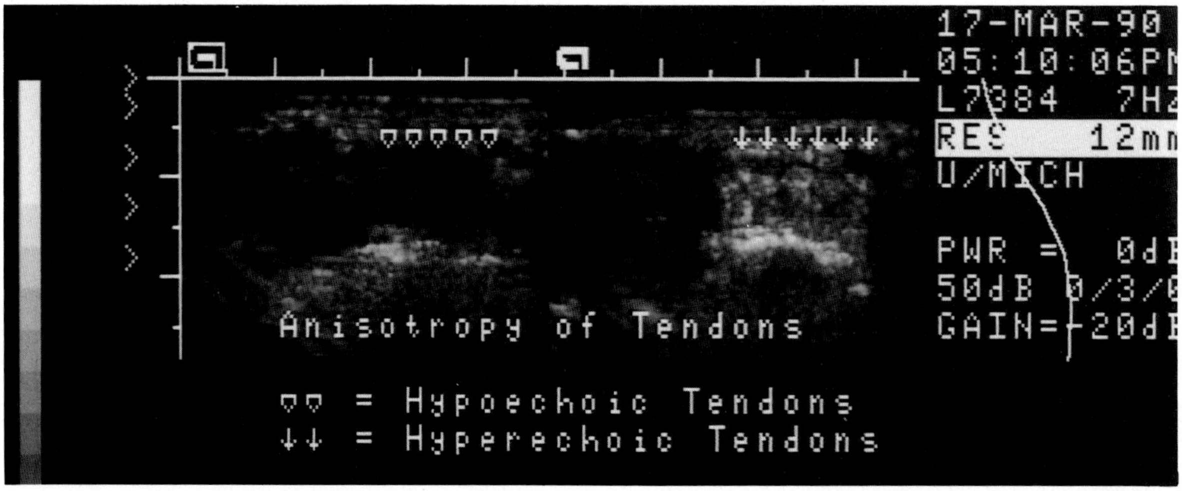


**FIG 2–6.**
Anisotropy of the flexor tendons of the wrist.

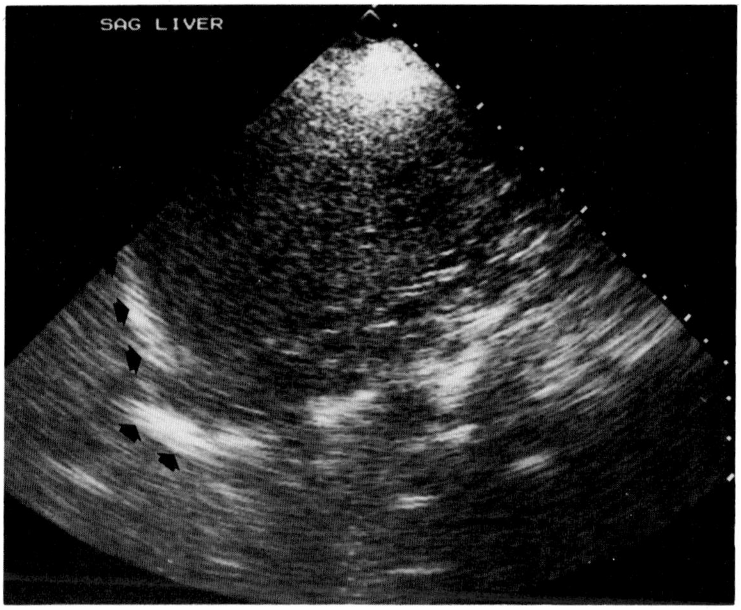


**FIG 2–7.**
Speed of sound artifact. A false discontinuity in the diaphragm *(arrows)* is imaged due to speed of sound artifact and refraction in this patient with retroperitoneal lipomatosis. (Courtesy of Dr. Jonathan Rubin, University of Michigan Hospitals, Ann Arbor, Michigan.)

two different tissues (liver and fat). In this example, refraction also plays a role. This type of artifact may also be encountered at muscle-fat interfaces in the extremities of obese patients.

### Beam Width Artifact

An ultrasound beam has a width that varies according to the design characteristics of the transducer. Therefore, like other cross-sectional imaging modalities, ultrasound images a volume of tissue. When an object is smaller than the width of the ultrasound beam, echos depicted at that location are a combination of those from the object and the surrounding tissues. This volume averaging, as it is called in computed tomography and magnetic resonance imaging (MRI), can give the appearance of echos within simple cysts, as well as eliminating shadowing deep to small calcifications. Normally, this does not interfere with the diagnosis. However, in musculoskeletal ultrasound, we are often dealing with very small structures, so this artifact is seen much more frequently than in abdominal imaging. The examiner must always remember that a highly echogenic focus may be a calcification despite the lack of acoustic shadowing (Fig 2–8).

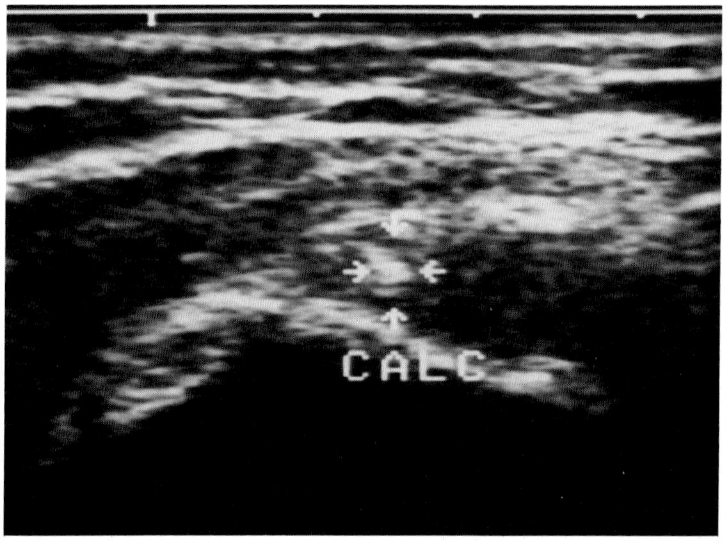


**FIG 2–8.**
Beam width artifact resulting in lack of acoustic shadowing deep to a calcification in the rotator cuff.

## THE UGLY. . .

### Motion Artifact

Patient motion can degrade ultrasound images as well as radiographs. The displayed image on an ultrasound machine is an average of several data acquisitions. When movement occurs, the image is blurred, sometimes severely limiting the diagnostic value. State of the art machines provide cine loop functions that can help in some cases. On other machines, the degree of persistence (i.e. the number of acquisitions averaged) can be reduced.

### Electrical Noise

Ultrasound machines are generally well insulated from electronic noise. However, some circumstances may arise where electromagnetic interference from high-voltage transformers or other equipment degrades images (Fig 2–9). If you are ever requested to perform a portable examination on a patient in the MRI holding area, you may be confronted with a bizarrely distorted image. Strong magnetic fields don't like cathode ray tubes. These types of artifacts are never diagnostic problems but, rather, minor annoyances.

### Frame Buffer Drop Out

All modern ultrasound machines use digital image processing. The displayed image is stored in a matrix of computer memory. When a memory chip fails, a number of pixels that compose the image will be missing. The pattern of loss will be dependent on the design of the display circuitry. Once again, it is not a difficult artifact to recognize but merely a sign that a call for maintenance is required.

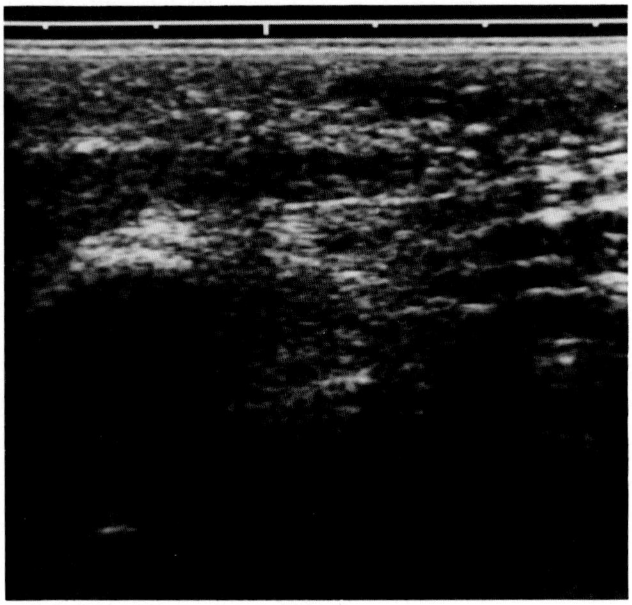

**FIG 2–9.**
Electrical noise artifact caused by electrocautery equipment. The *dotted lines* running obliquely downward from left to right are the result of electrocautery equipment used during this intraoperative ultrasound.

## BIBLIOGRAPHY

Fornage BD: Achilles tendon. US examination. *Radiology;* 1986, 159:759–764.

Fornage BD: The hypoechoic normal tendon—a pitfall. *J Ultrasound Med* 1987, 6:19–22.

Laing FC: Commonly encountered artifacts in clinical ultrasound. *Semin Ultrasound* 1983; 4:27–43.

Metreweli C: Practical abdominal ultrasound. Chicago, Year Book Medical Publishers, 1978.

Rubin JM, Carson PL, Meyer CR: Anisotropic ultrasonic backscatter from the renal cortex. *Ultrasound Med Biol* 1988; 14:507–511.

Thickman DI, Ziskin MC, Goldenberg NJ, et al: Clinical manifestations of the comet-tail artifact. *J Ultrasound Med* 1983; 2:225–230.

Ziskin MC, Thickman DI, Goldenberg NJ, et al: The comet-tail artifact. *J Ultrasound Med* 1982; 1:1–7.

# 3

# Sonography of Muscle

Ultrasound was the first imaging modality available for the evaluation of muscle pathology. Xeroradiography and low-kilovoltage radiography lack sufficient contrast resolution for the evaluation of muscle injury. Even computed tomography (CT) cannot define muscle structure sufficiently to detect the most common types of muscle pathology. Intravenous (IV) contrast enhancement does not improve the performance of CT in this area. In addition, the transverse axial plane of section provided by CT is usually unsuitable for evaluation of muscle injury. Muscle retracts along its long axis when injured, making it difficult to visualize on transverse axial images.

The multiplanar capability and improved tissue characterization provided by magnetic resonance imaging (MRI) have made it quite suitable for the evaluation of muscle pathology. Traumatic and ischemic injury can be well demonstrated using T1-, proton density–, and T2-weighted images. In addition, MR $^{31}$P spectroscopy allows detailed analysis of the metabolic state of muscle (Zochodne et al., 1978; Lenkenski et al., 1988). However, MRI lacks the ability to perform real-time dynamic examination. Cost and availability factors further limit the utility of MRI in the diagnosis of muscle pathology.

Sonography can provide all of the information available with MRI and more with regard to muscle pathology. Its spatial resolution and definition of muscle structure is usually superior to that provided by MRI. Real-time examination available only with ultrasound elucidates some types of muscle lesions, which are occult on static examinations (Graf and Schuler, 1988). The degree of functional deficit resulting from fibrosis, muscular cysts, or myositis ossificans can be fully assessed using real-time sonography. This information may have a major impact on decision making in competitive athletics, disability, and medicolegal cases.

The availability, ease of examination, and low cost relative to MRI make follow-up of healing lesions practical. The majority of patients referred for evaluation of muscle lesions are athletes. Several studies have demonstrated that approximately 30% of all sports injuries are muscular in origin (Peterson and Renström, 1986). In these patients the decision of when to return to training or competition is extremely important. Reinjury resulting from premature resumption of activity can be costly for both the individual athlete and the team. Serial sonographic examinations can accurately evaluate the rate and stage of healing, significantly decreasing the likelihood of reinjury.

## EXAMINATION TECHNIQUE

Muscles can be relatively large structures, some extending over two joints. The longest can measure more than 50 cm. A muscle tear tends to be a long fusiform lesion. In cases of complete rupture, the severed ends of the muscle may retract more than 10 cm. With these factors in mind, it is clear that we must have the ability to examine a long anatomic segment. Therefore, long linear array transducers are the most desirable for the examination of muscles. The majority of cases can be performed using a long 7.5-MHz linear array. This will provide excellent resolution. Obese patients may require a 5-MHz trans-

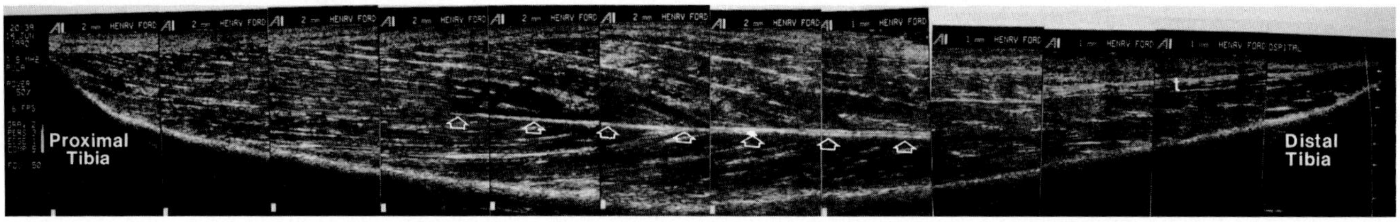


**FIG 3–1.**
Normal muscular anatomy of the anterior tibial compartment: composite longitudinal image. This image demonstrates the normal muscular anatomy of the anterior tibial compartment in a normal young athlete. Serial longitudinal images were obtained and then juxtaposed to produce a composite image demonstrating the entire length of the tibialis anterior muscle. This is a circumpennate muscle that functions in dorsiflexion of the foot. Note the featherlike appearance of the muscle created by the convergence of the fibroadipose septa upon the aponeurosis (*open arrows*). Abbreviations: tibialis anterior tendon (t).

ducer for better penetration. In addition to a long linear transducer, a machine that has the capability of displaying two images side by side is essential. This capability allows the examiner to provide a composite image demonstrating a very long segment, even an entire muscle (Fig 3–1).

A standoff cushion is required for the majority of examinations. This allows optimal visualization of the superficial fascia and musculotendonous junction. Fascial defects, muscle hernias, and superficial muscle tears will be missed if a standoff cushion is not utilized. Standoff cushions will also facilitate the production of composite images by conforming to the irregular contours of the skin surface. If a radial array or sector scanner must be used because a high-frequency linear array is not available, a standoff will help by reducing the obliquity of the beam in the region of interest. Regardless of the transducer utilized, a standoff will help to place the region of interest within the optimal focal zone of the transducer.

Pain resulting from muscle injury is generally well localized, unlike chest or abdominal pain. Therefore, we direct our examination toward finding the point of maximal tenderness. This technique is called *sonographic palpation.* The patient is directed to point with one finger to the symptomatic region, or the referring physician may mark a specific region of interest on the skin. Systematic examination of the indicated area is then performed by gentle compression with the ultrasound transducer. During the examination, the degree of compression should be as consistent as possible. Fascia and fibroadipose septa compose the most echogenic elements of muscle structure. When these structures are more closely approximated by increased compression, the overall echogenicity of the muscle appears increased. If no abnormality is clearly evident at the point of maximal tenderness, comparison with the corresponding area on the asymptomatic side is mandatory.

Muscles are dynamic structures; therefore, they cannot be properly evaluated with static images alone. Real-time sonography allows examination of muscle structure under dynamic conditions. State of the art ultrasound machines with cine loop capability are very helpful during dynamic examination. A videocassette recorder may also be valuable during this portion of the examination. Identification of muscles is made by location, origin, insertion, and function. This can easily be established under sonographic observation. Initial observation is performed without stress, followed by increasing graded isometric contraction. Small muscle tears may be occult on images obtained during relaxation but clearly visible during isometric contraction.

The examination is started with the transducer oriented in the long axis of the muscle using sonographic palpation. After the area of abnormality is established, images are obtained during relaxation and isometric contraction. Then the transducer is rotated 90 degrees, and the process repeated for transverse imaging. Comparable images of the asymptomatic side are obtained to facilitate detection of subtle abnormalities.

## NORMAL SONOGRAPHIC ANATOMY

Individual skeletal muscle fibers are enveloped by the endomysium, which consists of an extensive network of capillaries and nerves (Moore, 1980). These muscle fibers are grouped into bundles surrounded by the perimysium. The perimysium is composed of connective tissue, blood vessels, nerves, and adipose tissue (Clemente, 1988). Some authors refer to the perimysium as *fibroadipose septa* (Graf and Schuler,

1988; Fornage, 1989). A sheath of dense connective tissue called the *epimysium* surrounds the entire muscle. Single muscles or groups of muscles may be separated by a fascial layer.

The internal architecture of skeletal muscle varies depending on its designated function. Muscles having their fibers arranged parallel to the long axis of the muscle are best suited for movement over a long distance. Unipennate, bipennate, and circumpennate configurations are better suited for lifting greater weight over shorter distances (Fig 3–2). Attachment of muscles to the bones is accomplished by tendons and the fibro-osseous junction (Sharpey's fibers). Each muscle has at least one belly, or venter, and two tendons. However, a muscle may have more than one belly with fibrous intersections separating them, such as the rectus abdominus. Multiple origins of a muscle with a single belly is another possible configuration, seen in the biceps, triceps, and quadriceps.

These different arrangements can easily be observed sonographically. Variations from these systematic arrangements are easily detected with a thorough ultrasound examination. Muscle bundles have a hypoechoic appearance. The fibroadipose septa of the perimysium are seen as hyperechoic lines separating the muscle bundles (Figs 3–3 through 3–6). Epimysium, nerves, investing fascia, tendons, and fat also appear hyperechoic relative to the muscle bundles. These features allow easy recognition ot the pennate structure of muscle. Fat planes between muscles will aid in differentiation of separate muscles.

Identification of the pennate structure is most easily made in longitudinal images. In transverse images, muscle will have a speckled appearance. Images obtained during isometric contraction will demonstrate an apparent increase in muscle mass. This is due to the thickening of muscle bundles during contraction. Since these muscle bundles appear hypoechoic, the overall echogenicity of the muscle decreases during isometric contraction. Hypertrophy of muscle bundles in a well-conditioned athlete will also result in decreased echogenicity. This may prove to be a valuable tool in evaluating the level of condi-

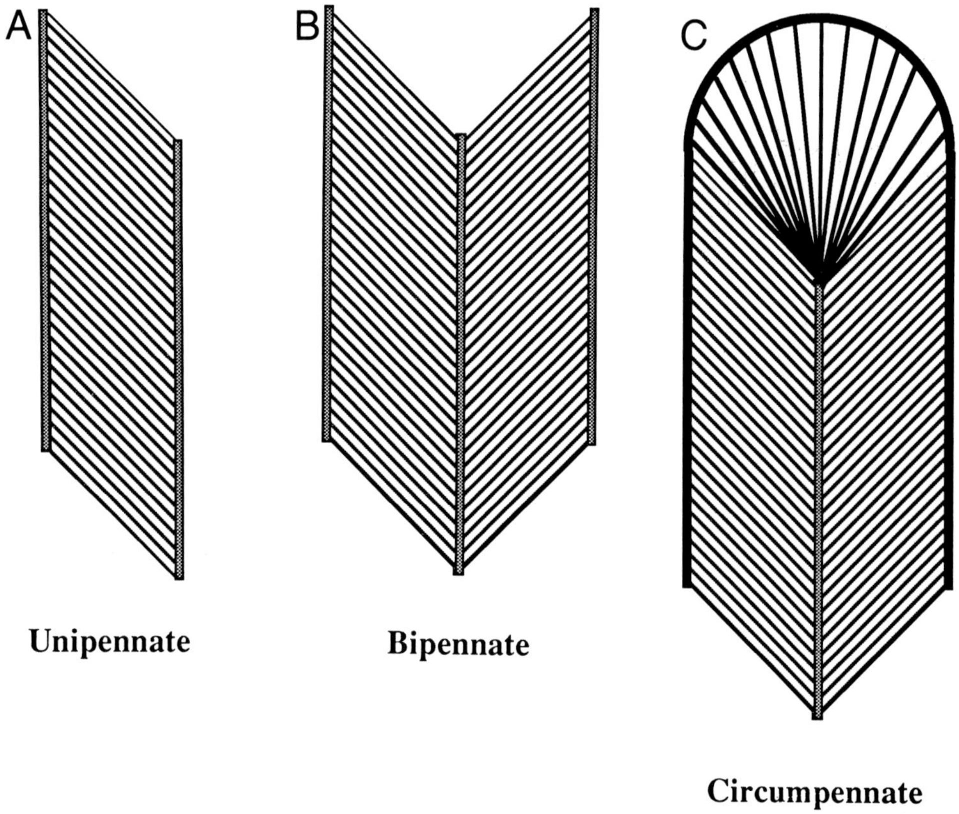


**FIG 3–2.**
Pennate structure of muscles. The diagrams depict the most common arrangements of muscle fibers in the extremities. These arrangements allow greater weight to be lifted over a shorter distance. Parallel arrangement of muscle fibers, seen more commonly in the abdomen, head, and neck, are more suitable for lifting less weight over long distances.

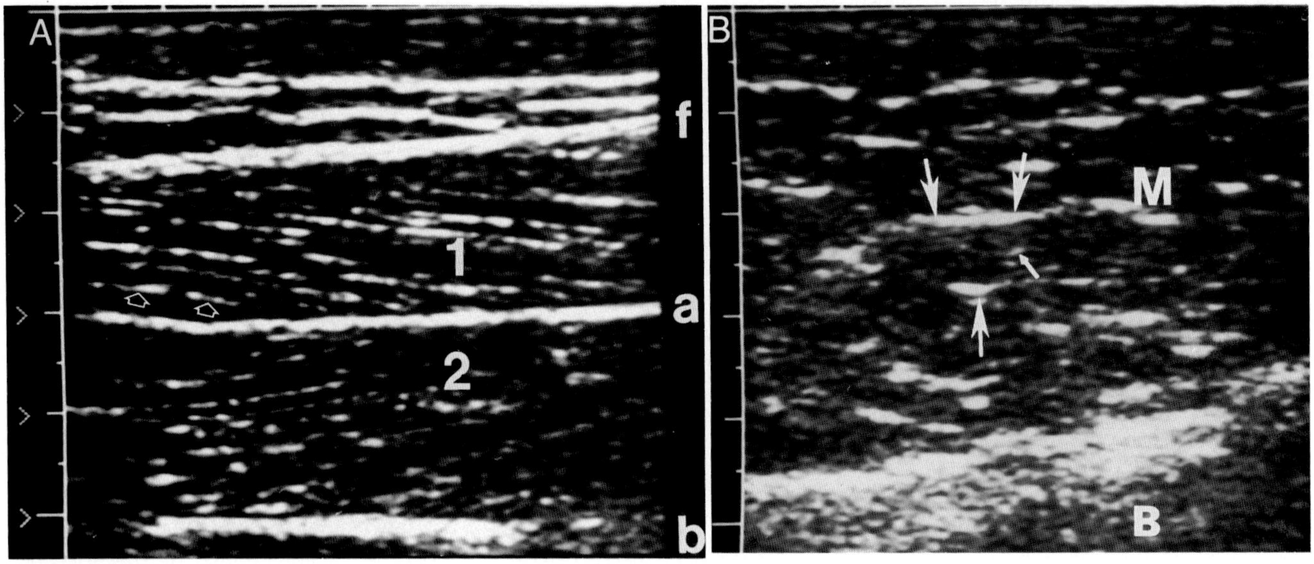


**FIG 3–3.**
**A,** normal anterior thigh muscles: longitudinal sonogram. This image was obtained in a normal 27-year-old man. The fibroadipose septa *(open arrows)* are seen as hyperechoic lines separating the hypoechoic muscle bundles. These septa converge on the highly reflective aponeurosis *(a)*, giving the appearance of a feather. The aponeurosis divides the rectus femoris *(1)* and the vastus intermedius *(2)* muscles. Two highly reflective lines bound the muscles: the fascia *(f)* superficially and the bone *(b)* deep to the muscles. Subcutaneous fat lies superficial to the fascia. **B,** normal anterior thigh muscles: transverse sonogram. Same patient as in **A.** In transverse images the fibroadipose septa *(arrows)* appear as hyperechoic lines of varying lengths. The pennate arrangement seen in longitudinal images is not evident on oblique or transverse images. Muscle *(M)*, femur *(B)*.

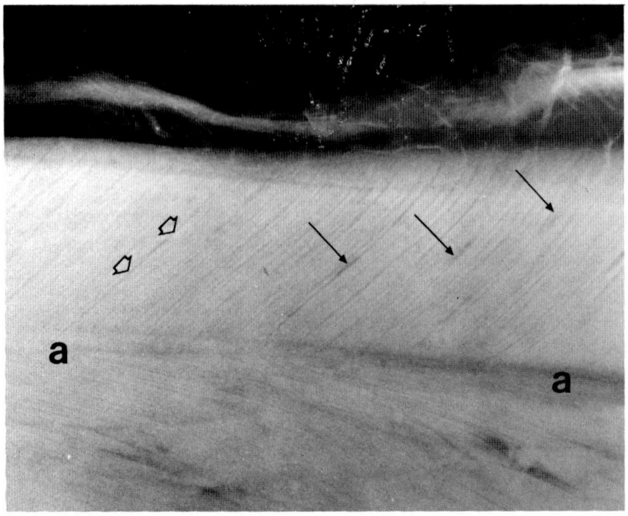


**FIG 3–4.**
Specimen radiograph of the rectus femoris muscle: longitudinal anatomical section. This radiograph was obtained using mammographic technique on a specimen 1 cm thick. Note the pennate structure of the muscle defined by the radiolucent fibroadipose septa *(open arrows)* converging on the thick fibrous aponeurosis *(a)* or intermuscular septum. These septa contain collagen, nerves, and blood vessels in addition to fat. A periodicity can be seen in the presence of thicker septa *(long arrows)*.

tioning in athletes. Conversely, during relaxation, firm compression with the transducer will result in a more echogenic appearance of the muscle.

Another aspect of muscle physiology that must be kept in mind is blood flow. Although muscle composes approximately 40% of body mass, it receives only 15% of total body blood flow at rest (Guyton, 1982). However, exercise can increase blood flow by 20-fold, resulting in a volumetric increase in muscle size by 10% to 15% (see Fig 3–48). For example, tiptoe exercise performed over several minutes will increase the volume of the posterior compartment of the calf sufficiently to cause anterior and lateral bowing of the interosseous membrane. Volume of the compartment will return to normal within approximately 10 to 15 minutes following exercise.

## MUSCLE PATHOLOGY

The overwhelming majority of muscle pathology is traumatic in origin, either occupational or sports related. Lesions can be further categorized into "intramuscular" lesions and muscle boundary lesions. We consider the intramuscular lesions to involve the muscle belly or venter. Muscle boundary lesions

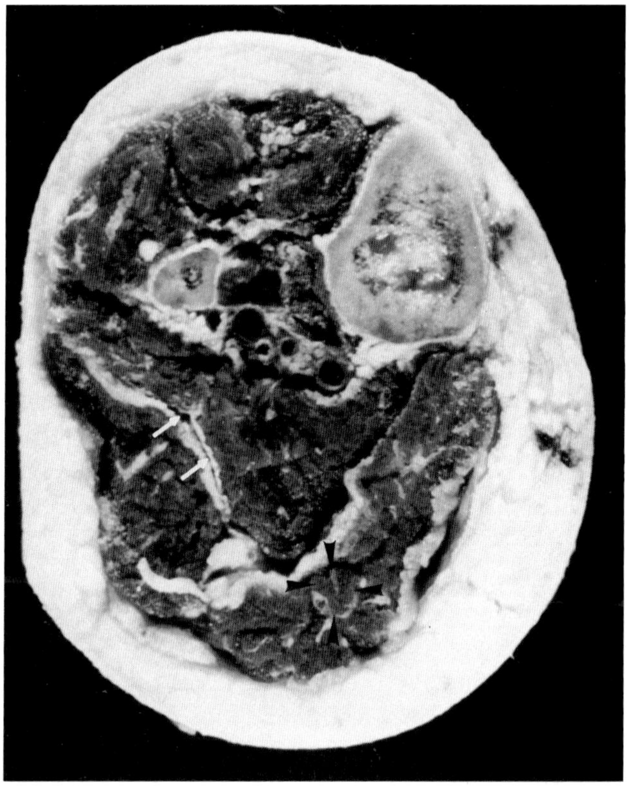

**FIG 3–5.**
Cross-sectional anatomical specimen of the lower leg: photograph. Muscles are divided into compartments by fibrous septa, which contain vessels *(white arrows)*. Compare the appearance of the fibroadipose septa cut transversely with the sonographic image in Figure 3–3,B. The *arrowheads* indicate a muscle bundle surrounded by septa.

are those involving the muscle-fascia/tendon interface.

## INTRAMUSCULAR PATHOLOGY

There are many etiologies of muscle pain, and not all of these are associated with an identifiable lesion. Delayed muscle soreness and muscle cramps are two well-recognized entities that are frequently encountered clinically. In exceptional cases it may be difficult to distinguish these causes of muscle pain from the sequellae of muscle rupture. Sonography is extremely valuable in these cases to exclude anatomical lesions.

Delayed muscle soreness is typically characterized by pain, tenderness, and frequently swelling of the involved muscle groups. Onset of symptoms is usually observed approximately 12 to 24 hours following strenuous exercise. The symptoms slowly resolve over several days. Everyone has experienced delayed muscle soreness at one time or another. Usually this ensues after playing three sets of tennis on the first day of a vacation. The remainder of the vacation is spent in a sedentary manner recuperating from the injury. Studies conducted with athletes and military recruits have demonstrated acute elevation of muscle enzymes measured in serum (Peterson and Renström, 1986; Nicholas and Hershman, 1986). These changes are highly suggestive of microruptures and

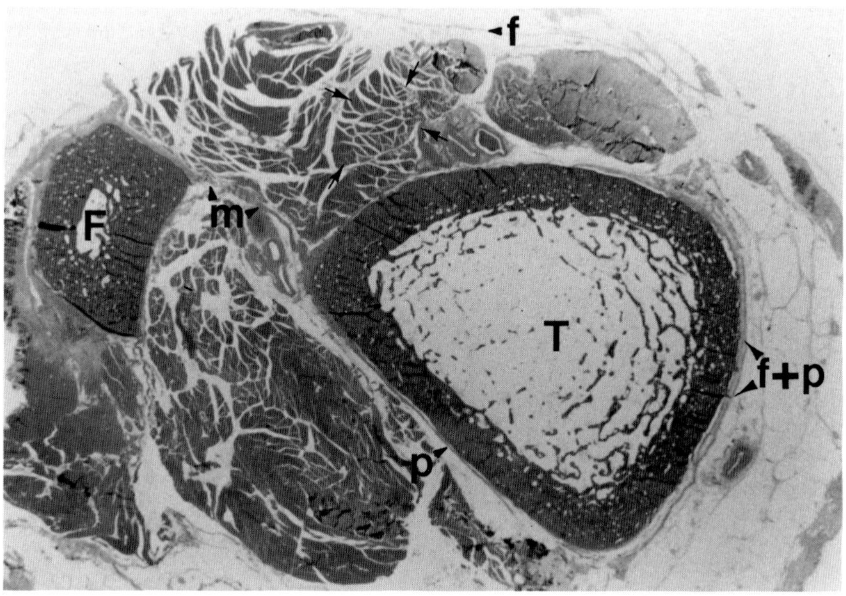


**FIG 3–6.**
Normal muscle architecture: transverse histological section (10 μm thick). This hematoxylin and eosin preparation was made from a specimen proximal to the one depicted in Figure 3–5. The fibroadipose septa are seen as clefts due to histological preparation. A muscle bundle similar to the one in Figure 3–5 is indicated by *arrows*. Note the further division within the bundle by finer septa. The interosseous membrane *(m)*, fascia *(f)*, and periosteum *(p)* are composed of dense fibrous tissue, seen as hyperechoic lines sonographically. Fascia of the lower leg is intimately related to the periosteum covering the anterior medial surface of the tibia *(f+p)*. Tibia *(T)*, fibula *(F)*.

myofibrilar lysis. Histologic evidence of this injury has been observed. However, MRI and sonography have failed to demonstrate the microscopic lesions in these cases.

Muscle cramps affect a broad spectrum of people, ranging from well-conditioned athletes to the elderly. They are initiated by a focus of spontaneous muscle activity that results in muscle spasm and pain (Nicholas and Hershman, 1986). The pathogenesis of muscle cramps remains unknown, but multiple predisposing factors have been identified. Prolonged exercise, dehydration, extremes in temperature, lactic acidosis, chronic hemodialysis, atherosclerotic disease, and varicose veins are among the factors cited as predisposing to muscle cramps. Muscle rupture may also result in muscle cramps, often leading the clinician away from the correct diagnosis. In the past, the correct diagnosis came to light only when the symptoms did not resolve after 3 to 4 days. The role of ultrasound in the evaluation of prolonged muscle cramp is to diagnose underlying muscle rupture so that appropriate therapy may be instituted. Therapy for muscle cramps often includes passive stretching of the muscle. In simple cases this maneuver will result in alleviation of pain and resolution of the cramp. However, this diagnostic/therapeutic maneuver will result in increased pain and progression of the lesion in cases where there is underlying muscle or tendon injury.

### Muscle Rupture

A muscle rupture may be caused by either compression (direct trauma) or distraction (indirect trauma). In compression injuries, the muscle is crushed against the underlying bone by an external force. This type of injury is frequent in contact sports and motor vehicle accidents. The collision of two football players with the impact of one's helmet into the quadriceps muscle of the other is an example of how this injury may result. Muscle fibers involved are macerated along with the associated vessels, usually resulting in hematoma formation. The investing fascia may also be torn with extension of the hematoma beyond the muscle belly, possibly involving multiple compartments. Healing of these lesions

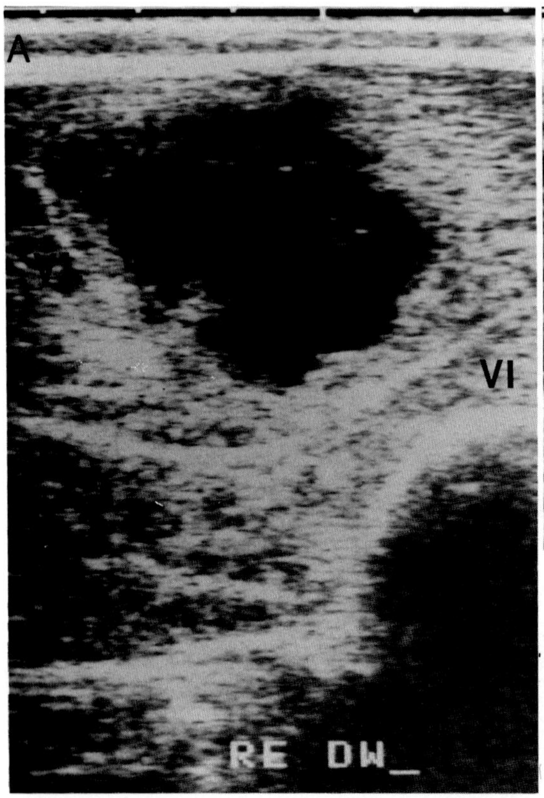


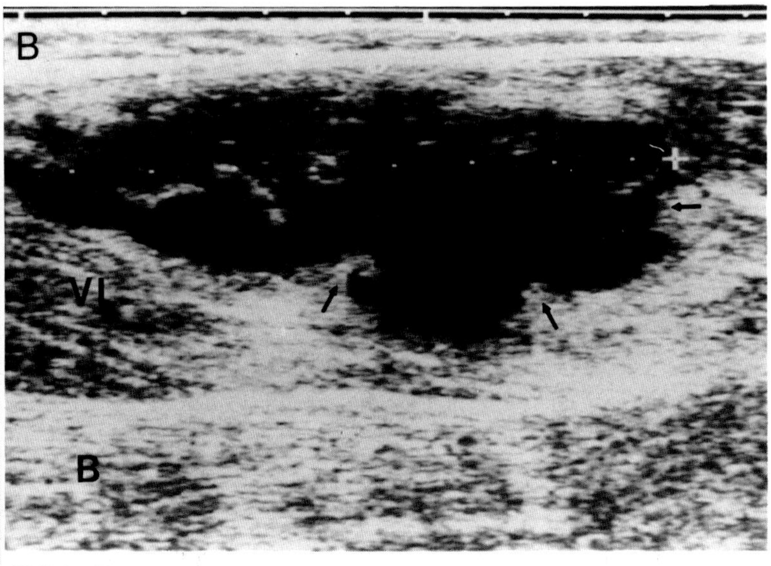


**FIG 3–7.**
**A,** compression rupture of the vastus lateralis: transverse sonogram. The patient is a 23-year-old football player who was kicked in the lateral thigh 6 days ago. This transverse image was obtained over the lateral aspect of the midthigh. A large hypoechoic fluid collection with very irregular borders is identified within the substance of the vastus lateralis muscle *(VL)*. The irregular border of the lesion is characteristic of compressive muscle rupture. Approximately 20% of significant compression ruptures progress to myositis ossificans. Biceps muscle *(B)*, vastus intermedius muscle *(VI)*. **B,** compression rupture of the vastus lateralis: longitudinal sonogram. Same patient as in **A.** The compressive injury is again identified within the substance of the vastus lateralis muscle *(VL)*. Compressive force applied to muscle results in diffuse injury with irregular borders in the region of rupture *(arrows)*. Measurement of the rupture site is important for comparison with follow-up examinations. Biceps muscle *(B)*.

takes place slowly with considerable reparative tissue, which evolves into extensive scar tissue. This invariably results in considerable long-term functional deficit.

Sonography is well suited for the evaluation of compressive muscle rupture (Lehto and Alanen, 1987). These lesions are characterized by an irregular cavity with shaggy borders (Fig 3–7). When imaged acutely, the echogenicity of the hematoma may limit evaluation of the extent of the lesion, resulting in an underestimation of the size. Approximately 48 to 72 hours following injury, the collection will become essentially anechoic, allowing measurement of the true extent of the injury. Follow-up examinations to evaluate healing will demonstrate progressive filling in of the lesion from the periphery with echogenic tissue. Scar tissue will appear hyperechoic, and acoustic shadowing will be seen deep to regions of myositis ossificans. Areas of myositis ossificans will have a pseudotumor appearance, which is detected significantly earlier with sonography than with conventional radiographs.

Distraction muscle ruptures differ from compression ruptures in both their pathogenesis and morphology. They are the result of the intrinsic force generated by sudden forceful contraction of a muscle (Peterson and Renström, 1986). Sports that have a high incidence of distraction muscle ruptures are track and field, weight lifting, football, and gymnastics. Muscles of the lower extremity are most frequently affected, particularly those muscles that span two joints (Nicholas and Hershman, 1986; Bouvier et al., 1982; Genety et al., 1980). The most susceptible are the hamstrings, rectus femoris, and triceps surae muscles. Distraction injury in the upper extremities and trunk is seen far less frequently (see Fig 3–18).

At the time of injury, the patient experiences a sharp stabbing pain in the region of the injured muscle. Pain dissipates slightly with the muscle at rest but can be reproduced by contraction of the involved muscle. Loss of function is proportional to the extent of injury. A complete rupture results in total loss of function in the region served by that muscle. The extremity is swollen secondary to both the inflammatory reaction and hemorrhage. In cases of superficial muscle rupture, skin discoloration may be observed approximately 24 hours following injury (Peterson and Renström, 1986). The area of skin discoloration should not be mistaken for the site of injury because

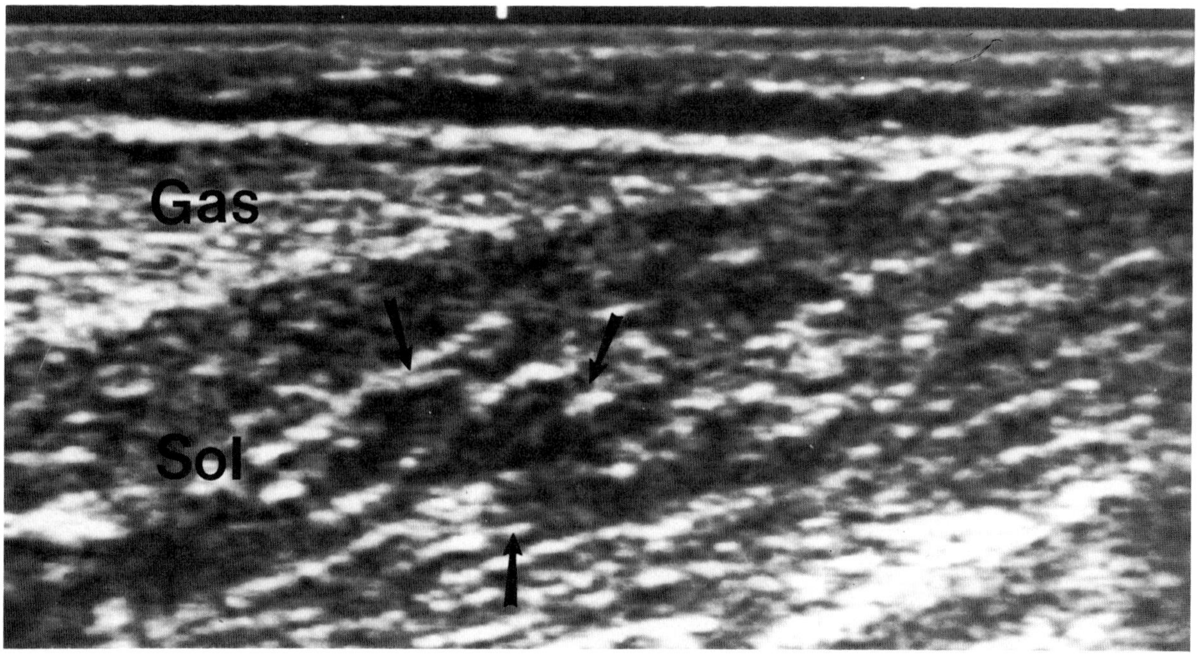


**FIG 3–8.**
Elongation injury of the soleus muscle: longitudinal sonogram. This 32-year-old tennis player tripped during a match and experienced a sudden sharp pain in the posterior calf. He presented with diffuse swelling of the calf 5 days following injury. Multiple flame shaped hypoechoic areas *(arrows)* interrupt the normal pennate structure of the soleus muscle *(Sol)*. This appearance is pathognomonic for elongation injury but is not seen in all cases. These lesions resolve rapidly, resulting in restoration of normal muscle architecture. Gastrocnemius muscle *(Gas)*.

blood will track inferiorly as it makes its way toward the subcutaneous tissues. Therefore, the site of rupture invariably lies superior to the region of skin discoloration.

In contrast to compressive injury, distraction ruptures are subdivided into three groups based on sonographic findings: elongation injury (grade I), partial rupture (grade II), and complete rupture (grade III) (Peetrons, 1987). Elongation injury occurs when a muscle is stretched to its elastic limit. Patients report severe pain throughout the muscle, without point tenderness. This entity is clinically indistinguishable from muscle cramp. Anatomically, the lesions are predominantly microscopic, with injury involving less than 5% of muscle substance. On gross pathologic examination, small cavities containing serosanguinous fluid are identified (Figs 3–8 and 3–9). The cavities range from 3 to 7 cm and have a cross-sectional diameter from 2 mm up to 1 cm. These areas are believed to represent small fluid collections that fill voids created by myofibrilar retraction following rupture (Peterson and Renström, 1986; Zuinen et al., 1982). Ultrasound demonstrates these cavities as flame-shaped (see Fig 3–8) hypoechoic areas in the muscle belly (see Fig 3–9). Examination in multiple planes is necessary in some cases to differentiate subtle lesions from artifact. Follow-up examination performed 2 weeks following injury will demonstrate restoration of normal muscle architecture. Therapy for elongation injury is rest of the involved muscle and palliation of pain.

Partial muscle rupture (grade II) is a more exten-

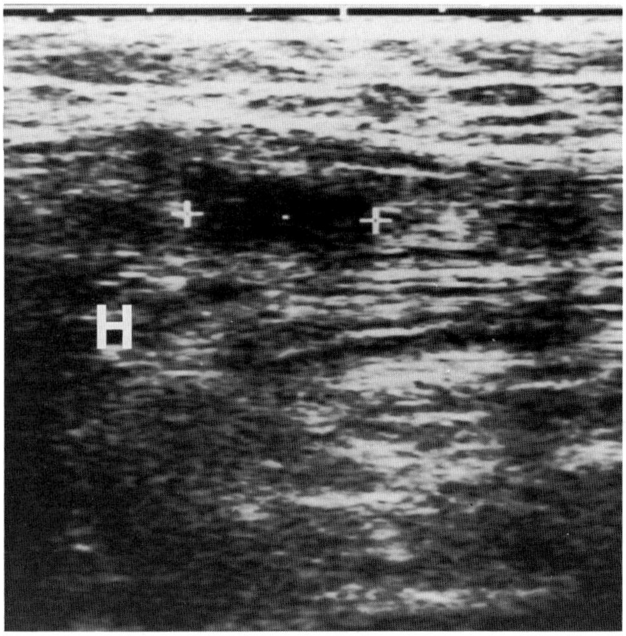


**FIG 3–9.**
Elongation injury of the hamstrings: longitudinal sonogram. A 21-year-old female ballet dancer presented with persistent pain in the posterior thigh on the right after overstretching the leg during a performance. The transducer is positioned along the length of the hamstrings at the posterolateral aspect of the distal thigh. A small ovoid essentially anechoic fluid collection is noted within the substance of the hamstrings *(H)*, consistent with elongation injury. Small focal ruptures, 2 cm or less, are considered elongation injuries.

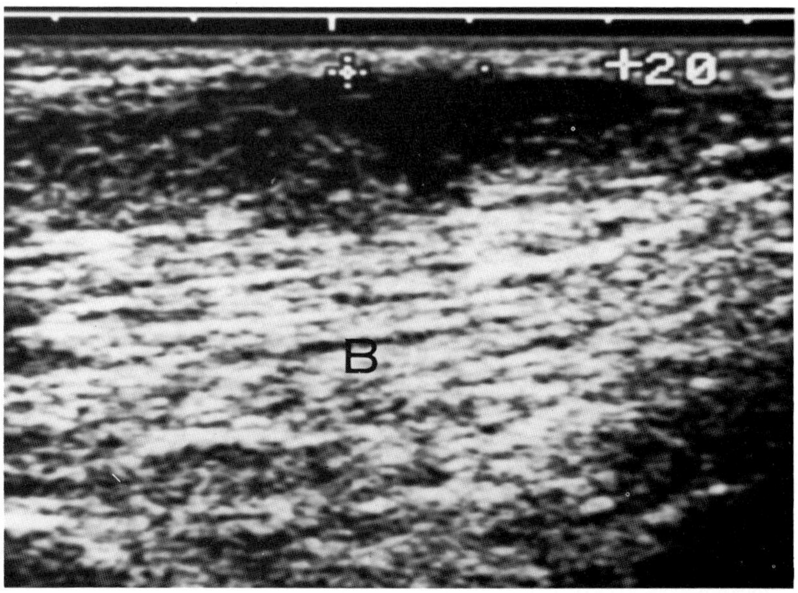


**FIG 3–10.**
Partial rupture of the superficial biceps femoris muscle: longitudinal sonogram. The patient is a 19-year-old female sprinter who experienced sudden onset of stabbing pain in the posterior thigh during acceleration. She presented 10 days after injury with ecchymosis and a limp. The transducer is positioned in the length of the biceps muscle over the area of ecchymosis. A superficial anechoic cleft *(+)* 2.5 cm long is present within the biceps muscle *(B)*. Retracted muscle bundles at the proximal and distal margins of the cavity appear irregularly hypoechoic. Greater than 5% of the cross-sectional area of the long head of the biceps is involved, making this a type II partial rupture.

sive injury, with the muscle stretched well beyond the limit of its elasticity. It involves greater than 5% of muscle substance but less than the entire muscle cross section. At the time of injury, the patient often experiences a "snap," accompanied by the sudden onset of sharp focal pain. Acutely, muscle function is completely absent. Partial function will return slowly over several days. Unlike elongation injury, point tenderness and focal swelling are present. Ecchymosis may develop if the muscle is located superficially. This usually develops 2 to 24 hours following injury and is found distal to the site of rupture. Rarely, the rupture can be detected clinically as a palpable defect.

Sonography clearly demonstrates the discontinuity in the muscle with disruption of the fibroadipose septa. A hypoechoic gap is identified within the muscle substance (Figs 3–10 and 3–11). Fragments of torn muscle originating from the walls of the cavity can be identified within the hematoma, which fills the gap. Gentle pressure applied with the transducer will demonstrate these torn muscle fragments to be free floating, referred to as the *bell clapper sign* (Fig 3–12) (Fornage et al., 1983; Harcke et al., 1988; Wagner et al., 1980). The proportion of intact muscle can be accurately evaluated by imaging in multiple planes. Mass effect displacing adjacent perimysium, fascia, and tendons may also be observed. Follow-up examinations will show a hyperechoic wall (Fig 3–13) that corresponds to granulation tissue and regenerating muscle fibers. This progressively thickens and fills in the cavity. The sonographic triad of a hypoechoic cavity within the muscle substance, a thick hyperechoic wall of the cavity, and the bell clapper sign are pathognomonic for muscle rupture. In extensive injury, therapy may include surgical repair of the ruptured segment. Management of the accompanying hematoma will be discussed later.

Complete muscle rupture (grade III) is seen less frequently than either elongation injury or partial rupture. Initial clinical presentation is quite similar to that of partial rupture, although associated syncope has been reported (Peetrons, 1987). In complete rupture, total functional impairment persists. When the muscle is superficially located, the gap between the retracted ends of the ruptured muscle can usually be palpated. Ecchymosis is also more common in cases of complete rupture.

Complete separation and retraction of the severed

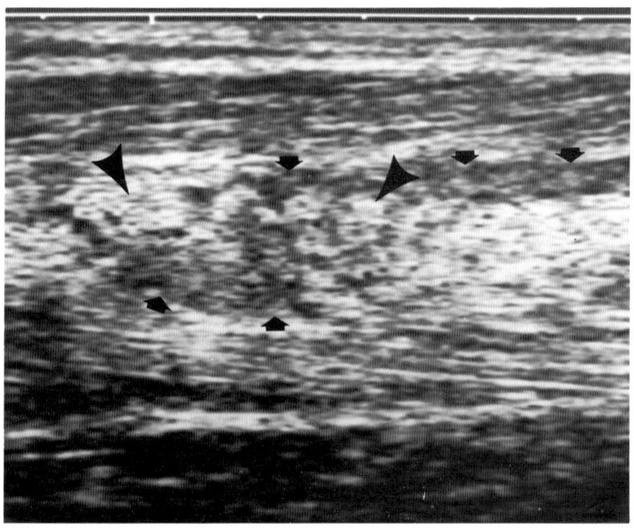

**FIG 3–11.**
Fresh partial hamstrings rupture: longitudinal sonogram. This 24-year-old sprinter had acute onset of sharp pain in the posterior right leg during a 100-m race. Pain progressively increased throughout the day. This examination was performed the same day, with the transducer aligned along the length of the biceps. The normal pennate structure of the muscle is disrupted with slightly rounded ends of the ruptured muscle bundles *(large arrowheads)* separated by slightly hypoechoic hemorrhage *(small arrows)*.

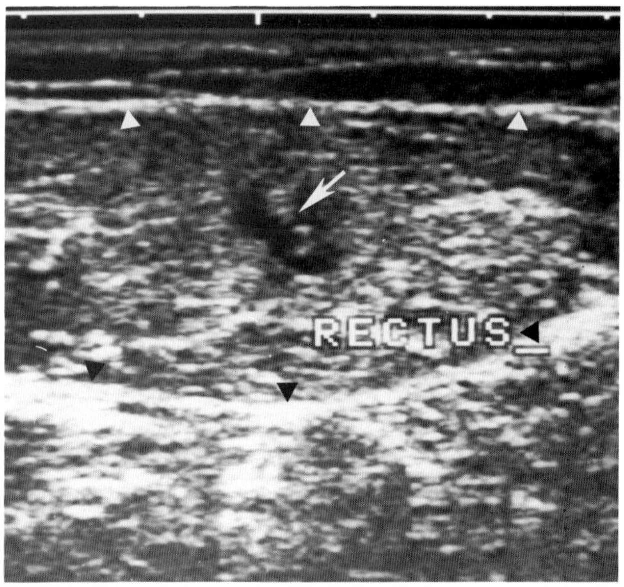


**FIG 3–12.**
Partial rectus muscle tear: transverse sonogram. This 20-year-old football player presented 2 days after experiencing sharp pain in the anterior right thigh. A flap of torn muscle tissue *(white arrow)* with one end freely floating is seen in the rupture defect (a bell clapper sign). The torn end of the muscle flap is clearly identified as moving freely within the hematoma on real-time imaging. Fascia *(white arrowheads)*, aponeurosis *(black arrowheads)*.

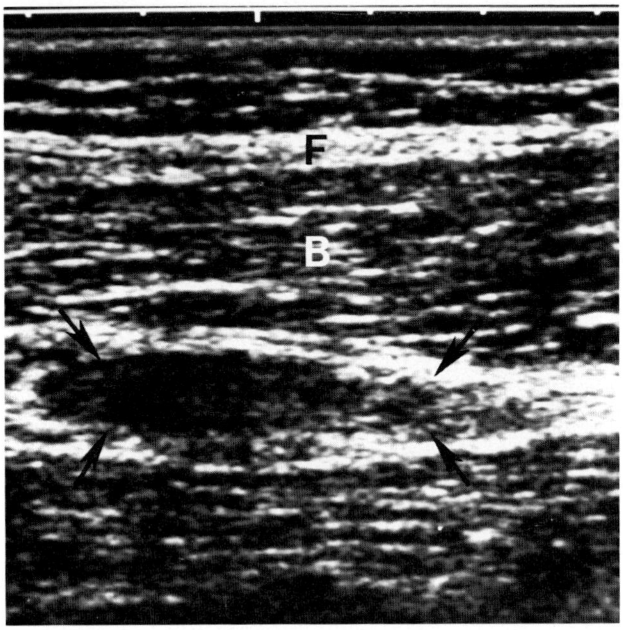


**FIG 3–13.**
Healing partial hamstrings tear: longitudinal sonogram. Three weeks ago, this 25-year-old professional soccer player experienced sharp pain in the posterior right thigh during a game. He has experienced persistent pain and has been unable to exercise since the injury. The transducer is positioned over the midposterior thigh along the length of the biceps *(B)*. An ovoid hypoechoic lesion with thick echogenic borders *(arrows)* is identified within the substance of the hamstrings, consistent with a healing partial muscle rupture. Abbreviations: fascia *(F)*.

muscle are apparent on sonographic examination (Figs 3–14 through 3–17). The distal portion of the retracted muscle may "bunch up," resembling a soft tissue mass (Peterson and Renström, 1986). Hematoma fills the void created by the retracted muscle ends. The investing fascia may be intact. However, extension of hematoma through a rent in the investing fascia is observed in the majority of cases. Several days following injury, enhanced through transmission will be noted deep to the hematoma. Nerve involvement may also be noted sonographically. However, electromyographic evaluation is indicated when neurological deficit is evident clinically, despite identification of an intact nerve on sonographic images.

### Hematoma

Hematoma formation is a hallmark of muscle rupture (Fig. 3–18). The magnitude of the hematoma is generally an indicator of the extent of the underlying injury. However, muscle hematoma may be out of proportion to the associated injury in patients with hemophilia or on anticoagulant therapy. Direct trauma will result in contusion of the highly vascular fibroadipose septa. These septa will be markedly thickened on sonographic images (Figs 3–19 through 3–21).

Elongation injury may also cause contusion of the fibroadipose septa. In these cases, disruption of vessels in the epimysium may also occur, producing intermuscular hematoma. Intermuscular hematoma is characterized by blood dissecting within the fascial planes between muscles (see Fig 3–19). More extensive injury will result in formation of intramuscular fluid collections easily identified on ultrasound images. In cases of complete rupture, these collections may exceed 100 mL and extend beyond the investing fascia. Large hematomas may exhibit considerable mass effect (Fig 3–22). The resulting compartment syndrome will further compromise the condition of surrounding muscles and nerves. Prompt evacuation of large hematomas is therefore highly recommended. Aspiration of large hematomas is also desirable in cases of partial rupture. This will allow faster healing and reduce scar formation by allowing closer approximation of wound margins.

The evolution and resolution of intramuscular hematoma does not differ from hematomas elsewhere in the body. Their initial sonographic appearance is that of a homogeneous hypoechoic fluid collection. Over the course of several hours, serum, cellular and fibrin elements will separate (see Fig 3–22) out, resulting in an identifiable fluid-fluid level. Further changes occurring over several days will result in the collection becoming uniformly anechoic (Fig 3–23). Fluid aspirated from the collection at this stage in evolution will have the characteristic crank case oil appearance. These collections will be resorbed slowly, over a period of weeks, without intervention.

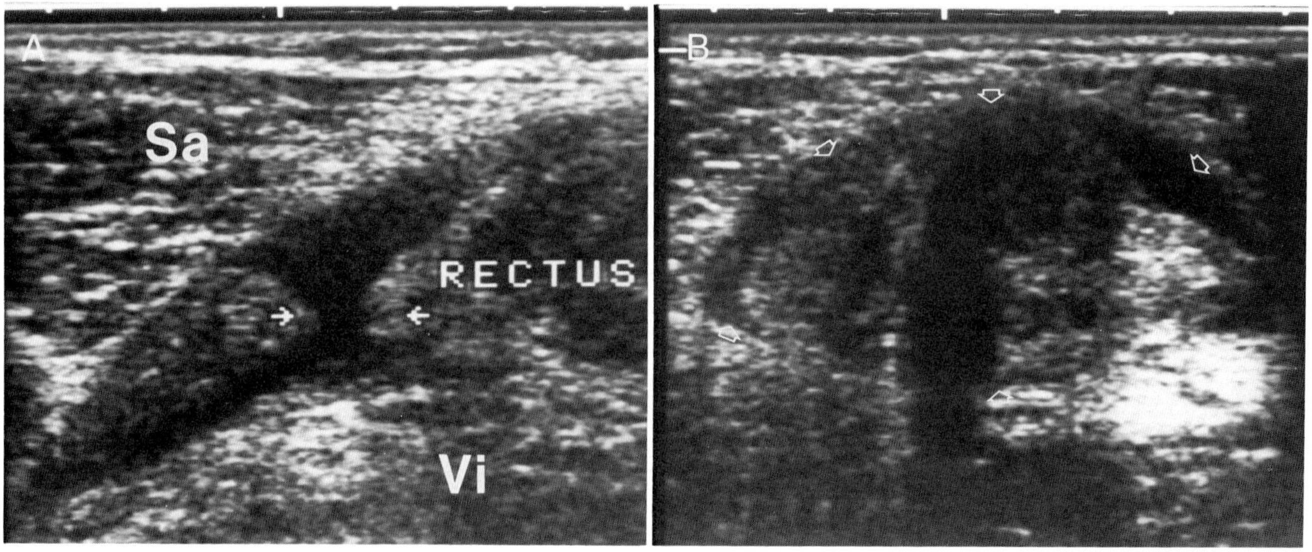


**FIG 3–14.**
**A,** complete rectus muscle rupture by distraction: longitudinal sonogram. The patient is a 21-year-old soccer player who felt a "pop" and sharp stabbing pain while running. This longitudinal image of the proximal anterior thigh was obtained 5 days after injury. The rounded proximal and distal ends of the ruptured muscle *(arrows)* are separated by hematoma throughout the cross-section of the muscle. Distraction injury of the rectus femoris is the most commonly observed muscle rupture. Abbreviations: sartorius muscle *(Sa),* vastus intermedius muscle *(Vi).* **B,** complete rectus muscle rupture by distraction: transverse sonogram. Same patient as in **A.** The transducer is positioned transversely over the point of maximal tenderness, just proximal to the site of rupture. A finding characteristic of complete rupture is fluid contained within the epimysium *(open arrows)* surrounding the entire circumference of the muscle.

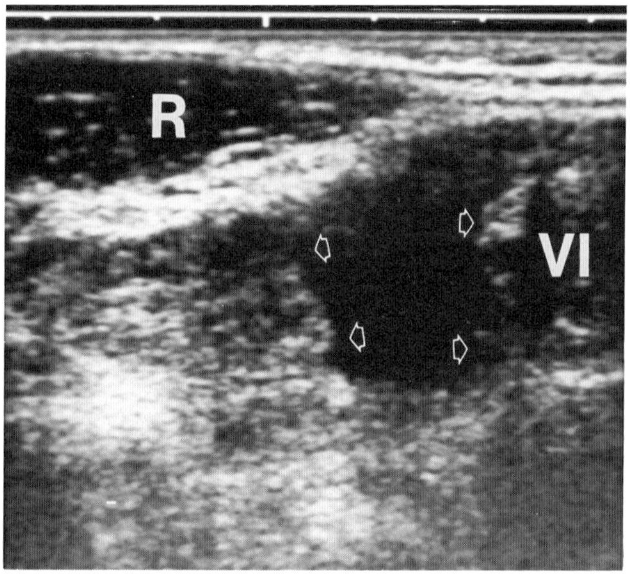


**FIG 3–15.**
Complete vastus intermedius rupture by distraction: longitudinal sonogram. A 27-year-old male soccer player presented with a complaint of stabbing pain in the anterior right thigh, which developed suddenly while playing. This image was obtained with the transducer placed along the length of the rectus *(R)* at the point of maximal tenderness. A 2-cm gap *(open arrows)* is present in the vastus intermedius *(VI)* throughout its cross section. Compare the smooth distinct margins of this distraction injury with the rough margins of compressive injury seen in Figure 3–7.

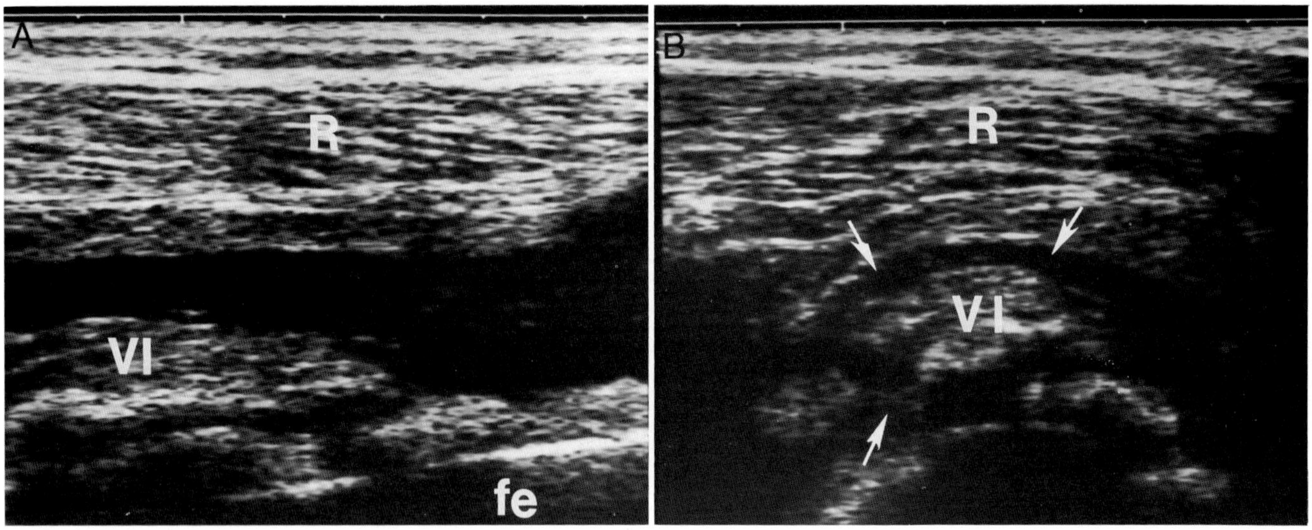


**FIG 3–16.**
**A,** bell clapper sign: longitudinal sonogram. The patient is a 20-year-old football player who experienced sudden onset of sharp pain in the anterior right midthigh. No definite clinical signs of complete rupture were present. Transducer position is just superior to that in Figure 3–15. A complete tear of the vastus intermedius *(VI)* is present. Pressure applied with the transducer demonstrated free movement of the torn muscle ends. The bell clapper sign is present in both partial and complete ruptures. Abbreviations: rectus muscle *(R)*, femur *(fe)*. **B,** bell clapper sign: transverse sonogram. Same patient as in **A.** The transducer is positioned transversely just proximal to the site of rupture. Hemorrhage *(arrows)* completely surrounds the vastus intermedius *(VI)*, immediately beneath the rectus *(R)*.

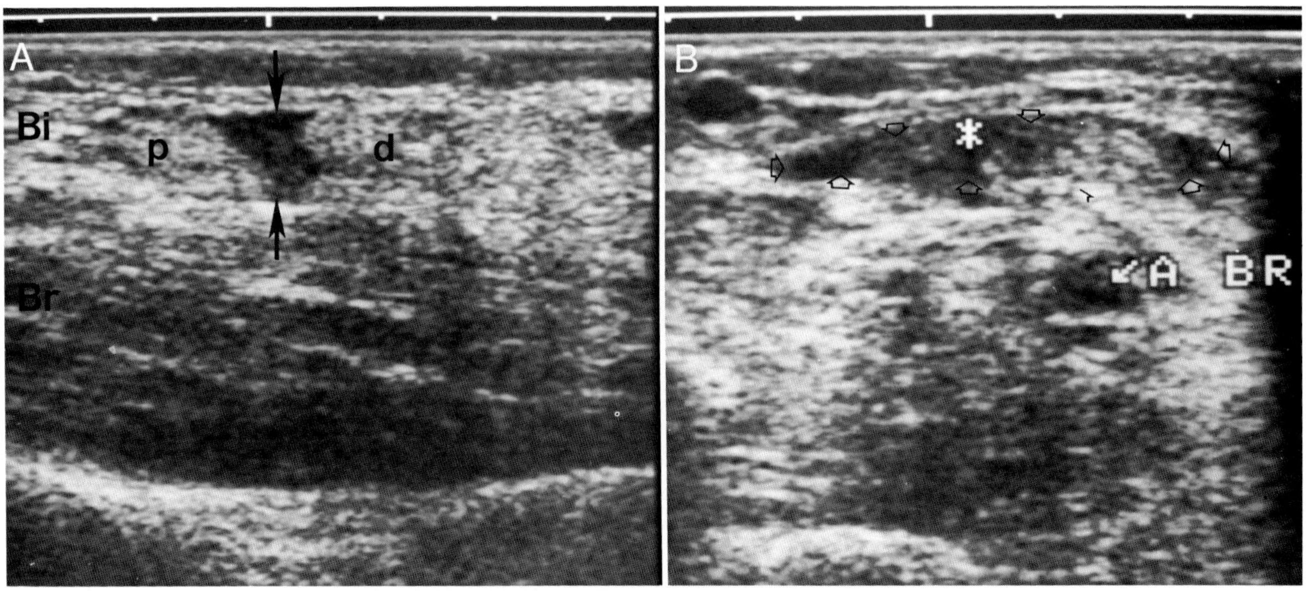


**FIG 3–17.**
**A,** complete biceps brachii rupture: longitudinal sonogram. While lifting a package, this 60-year-old diabetic man felt a sharp pain at the proximal aspect of the right elbow. The transducer is positioned over the medial aspect of the distal humerus along its length. A gap *(arrows)* is identified through the entire cross-section of the biceps muscle *(Bi)*. The retracted ends of the ruptured muscle are rounded and increased in echogenicity. Echogenicity of the injured muscle must be compared with the normal brachial muscle *(Br)* deep to the biceps. Hemorrhage separates the proximal *(p)* and distal *(d)* ends of the muscle. **B,** biceps brachii rupture: transverse sonogram. Same patient as in **A.** This transverse image was obtained just proximal to the point of maximal tenderness. Hemorrhage contained within the epimysium of the biceps *(open arrows)* outlines the retracted proximal portion of the muscle *(asterisk)*. The brachial artery *(white arrow)* is seen deep to the biceps.

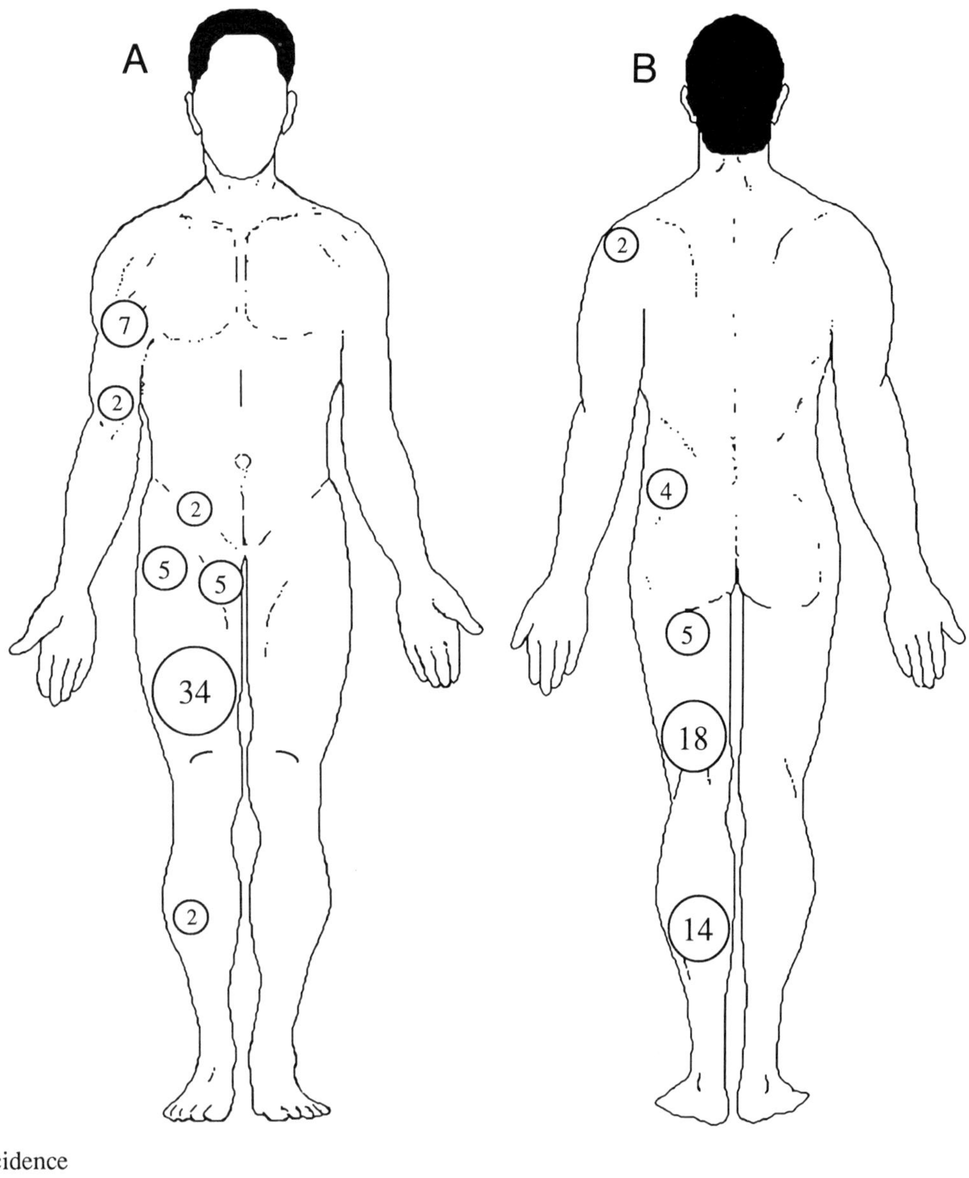


**FIG 3–18.**
Frequency of muscle ruptures by site.

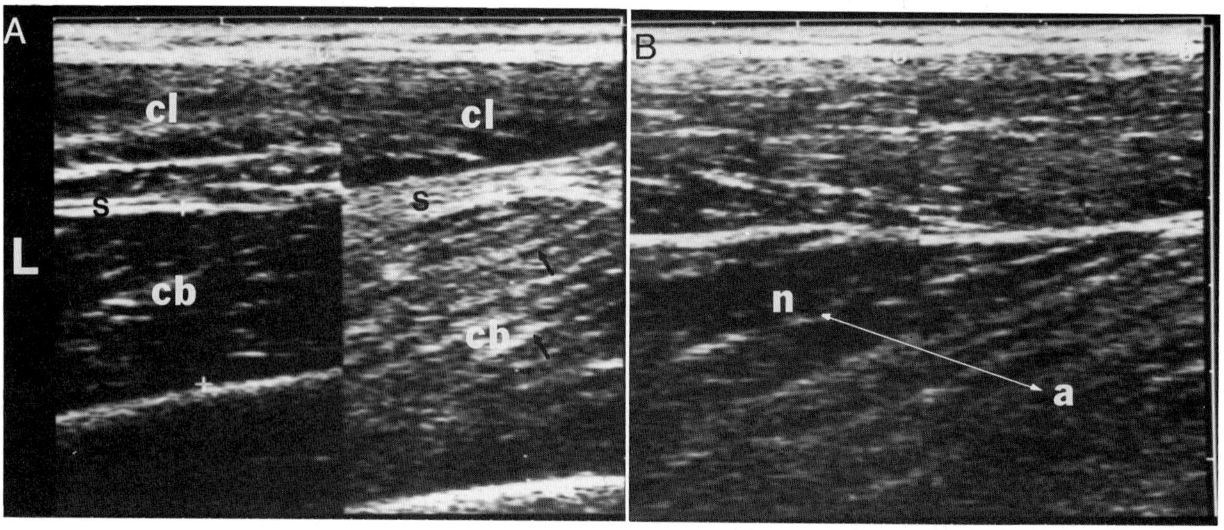


**FIG 3–19.**
**A,** muscular contusion: longitudinal sonogram. Two hours after being hit in the right posterior thigh with the helmet of another player, this 20-year-old football player presented for sonographic examination. The transducer is positioned over the midportion of the hamstrings, along its length. In the image on the *right,* the short head of the biceps muscle (caput breve, *cb*) on the symptomatic leg is markedly swollen. Thickening and increased echogenicity is noted in the fibroadipose septa *(arrows)* and intermuscular septum *(s)*. Compare this with the appearance of the asymptomatic side, shown on the *left*. Long head of the biceps muscle (caput longum, *cl*). **B,** The transducer is moved slightly superiorly to demonstrate the transition point from normal to contused muscle. Normal *(n),* abnormal *(a)*.

**FIG 3–20.**
Recent iliopsoas contusion: longitudinal sonogram. At a local carnival, this 20-year-old man was injured by the steering wheel while driving one of the "bump cars." He complained of pain just medial to the left anterior-superior iliac spine and walked with a limp. The transducer is positioned along the length of the iliacus muscle with the proximal side of the transducer at the anterior superior iliac spine. Comparison with the asymptomatic right side *(r)* is provided. The left *(l)* iliacus is markedly swollen and diffusely increased in echogenicity. Normal fibroadipose septa can no longer be identified at the site of injury, but no gap in the muscle substance was identified. Aspiration yielded approximately 20 mL of blood. Abbreviations: iliac bone *(b)*.

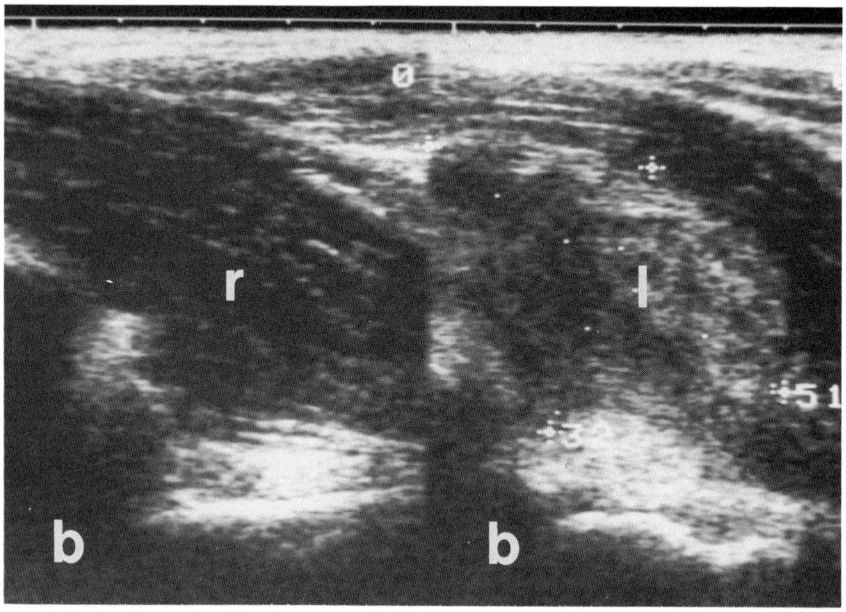

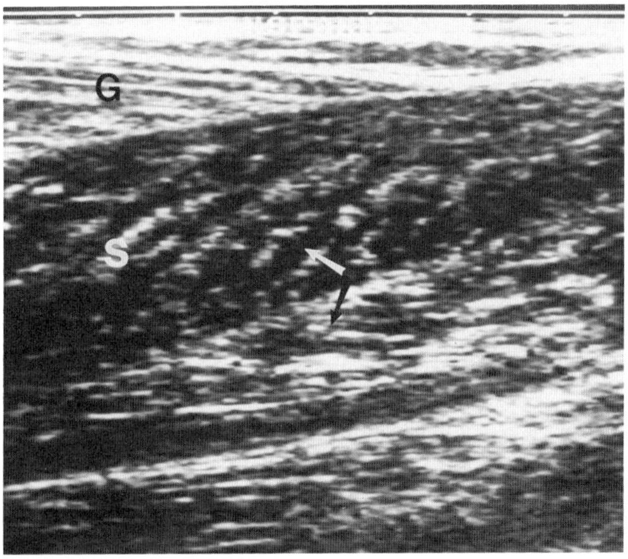


**FIG 3–21.**
Resolving muscular contusion of the calf muscles: longitudinal sonogram. Two weeks ago, this patient sustained a blow to the posterior calf. Muscle bundles *(white arrow)* and fibroadipose septa of the soleus *(s)* are markedly thickened and decreased in echogenicity. Compare this appearance with flexor hallucis longus *(black arrow)* deep to the soleus. Abbreviations: normal gastrocnemius *(G)*.

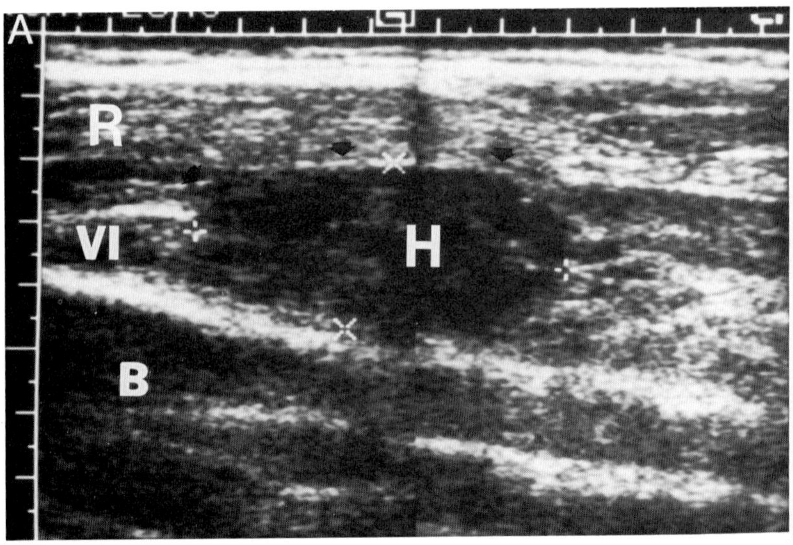


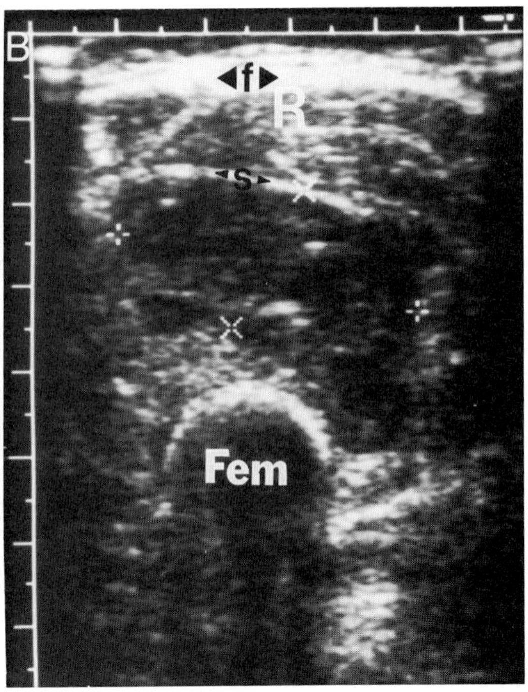


**FIG 3–22.**
**A,** quadriceps hematoma: longitudinal sonogram. This 30-year-old man developed a distraction rupture of the quadriceps during a soccer game 3 days ago. A hypoechoic fluid collection *(H)* is identified within the vastus intermedius *(VI)*, dorsal to the rectus *(R)*. The intramuscular hematoma elevates the intermuscular septum *(small arrows)*. Fibrin clots are seen floating within the hematoma. Femur *(B)*. **B,** quadriceps hematoma: transverse sonogram. Same patient as in **A.** Anterior bowing of the intermuscular septum *(<s>)* is better appreciated on this transverse image through the hematoma *(+)*. Note the three hyperechoic arcs identified in this image: the fascia cruris *(<f>)*, intermuscular septum *(<s>)*, and the femoral cortex *(Fem)*.

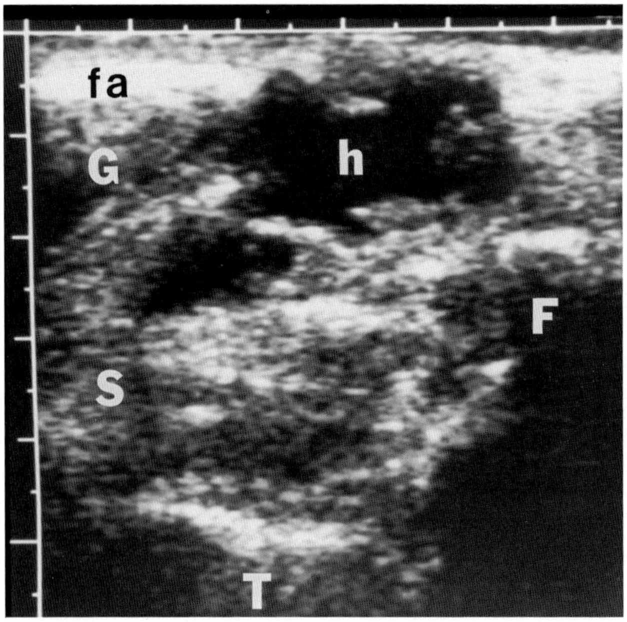


**FIG 3–23.**
Intramuscular hematoma after compression injury: transverse sonogram. An ecchymosis developed at the posterior superior aspect of the left calf of this 40-year-old man after he was kicked by a horse. At the time of presentation, 12 days following injury, he was unable to walk on his left leg. The transducer is positioned transversely over the calf muscles just below the knee. A characteristic compression rupture is identified. An anechoic hematoma *(h)* with markedly irregular borders is present, involving both the gastrocnemius *(G)* and the soleus *(S)*. Note that the superficial fascia *(fa)* and the intermuscular septum are ruptured. This type of multiple compartment involvement is frequently associated with compressive injury but is unheard of in distraction injuries. Abbreviations: fibula *(F)*, tibia *(T)*.

## Healing Muscle Rupture

Healing of muscle rupture occurs slowly, taking from 3 to 16 weeks to progress to completion. Time required for healing is proportional to the extent of injury and related to the location of the lesion. Lesions in the calf tend to heal more slowly than others. Surprisingly, muscle has a tremendous capacity for regeneration. If the sarcolemmal sheaths are intact, complete regeneration of normal muscle architecture will occur by regeneration of myocytes (Bullough and Vigorita, 1984). This is typically seen in elongation injury and following infection. More extensive injury requires two different types of repair: regeneration and scar formation. These processes are in competition with each other. The objective of therapy is to ensure that regeneration predominates. In gross muscle rupture, regeneration occurs by growth of undamaged muscle fibers at the wound margin and formation of new fibers. These new fibers are formed by recruiting reserve cells found in the endomysium (Bullough and Vigorita, 1984). Formation of scar tissue will inhibit the process of regeneration.

The role of sonography in the evaluation of healing muscle rupture lies in three areas. First is the assessment of the extent of injury and measurement of the separation of wound margins. These two factors are the best predictors of the proportion of scar formation that can be measured sonographically. The greater the percentage of muscle substance involved and the greater distraction gap results in a greater proportion of scar tissue. This information will guide the clinician in determining what intervention, if any, is indicated. Ultrasound-guided aspiration of a hematoma may be desirable at the time of examination to reduce the distraction gap.

After initial evaluation, the role of sonography is to determine the stage of healing. The earliest detectable change toward healing is increased echogenicity of the margins of the wound (Figs 3–24 through 3–26). This hyperechoic band will progressively increase in thickness as healing proceeds. Echogenicity of the inner aspect of this band will be slightly less than the periphery. Eventually, the entire cavity will be filled in (Fig 3–27). Over a period of weeks this region will demonstrate further organization, and more normal muscle architecture with fibroadipose septa can be identified (Fig 3–28). These serial follow-up examinations are extremely valuable in the process of determining when it is safe for the patient to resume limited athletic activity. In our experience, resumption of training at the time when the lesion has filled in but further organization is not evident will lead to a high incidence of recurrent injury. Premature resumption of athletic activity will therefore prolong the recovery period and increase scar formation to the detriment of both the individual athlete and his or her team.

The final role of ultrasound in the evaluation of healing muscle rupture is assessment of the magnitude of scar formation (Fig 3–29). Fibrosis is evident sonographically as hyperechoic tissue (Figs 3–30 and 3–31). Following distraction rupture, the fibrous scar is usually linear in contrast to compressive ruptures where the scar is usually nodular or triangular (Fig 3–32). Fibrous scar from superficial tears will result in focal retraction of the fascia or intermuscular septum (Fig 3–33) toward the center of the lesion (see Fig 3–32). Evaluation of the extent of fibrosis is a direct indicator of the loss of a muscle's ability to produce tension. In addition, the risk of recurrent injury is proportional to the amount of residual fibrous scar (Nicholas and Hershman, 1986; Peterson and Renström, 1986) (Fig 3–34).

An infrequent complication seen in cases of healing muscle rupture is the formation of a muscle cyst (Fig 3–35). In our experience, the calf is the most common location of these cysts. The appearance of these cysts may vary slightly from the classic sonographic characteristics of a simple cyst. A perceptible

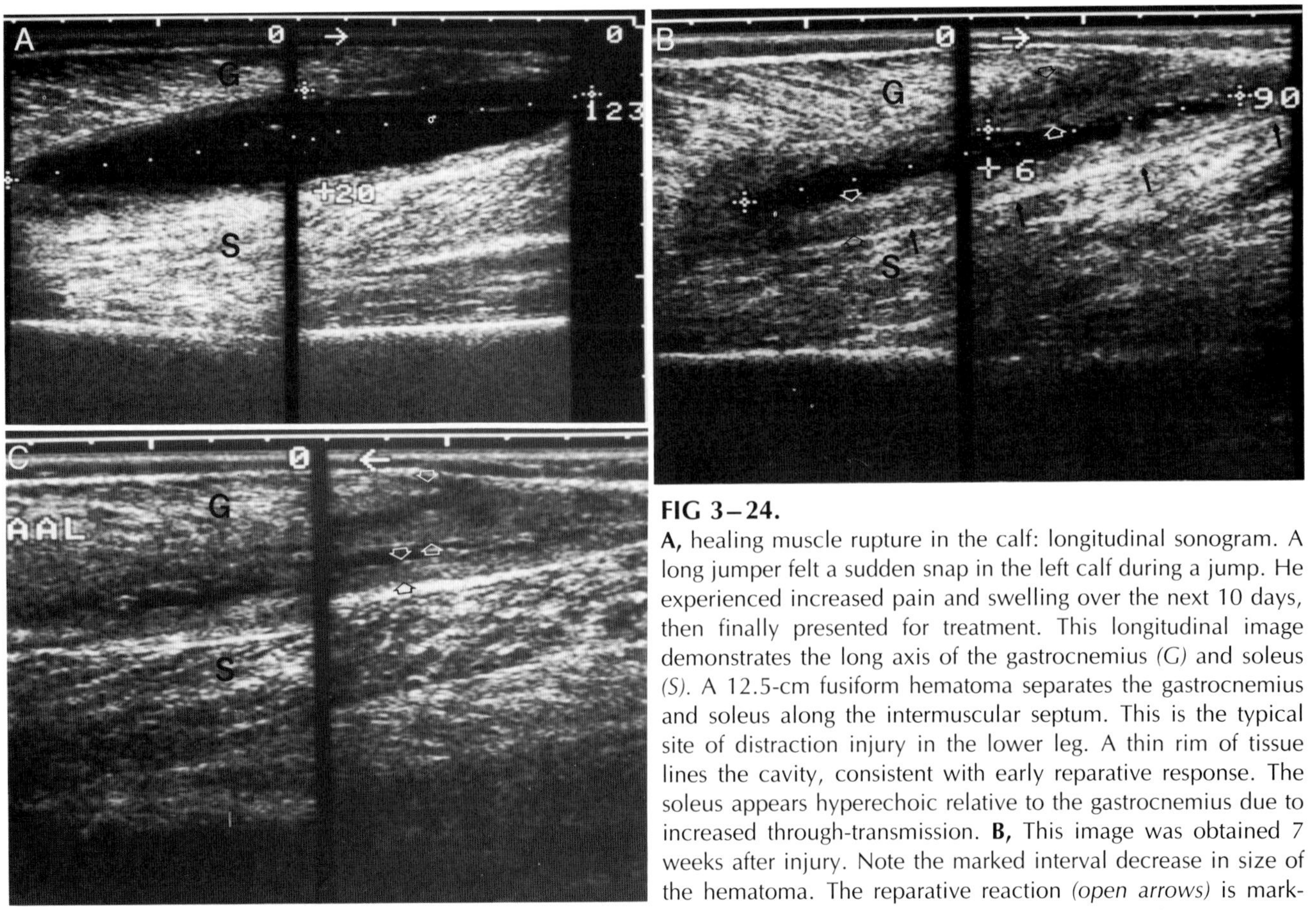


**FIG 3–24.**
**A,** healing muscle rupture in the calf: longitudinal sonogram. A long jumper felt a sudden snap in the left calf during a jump. He experienced increased pain and swelling over the next 10 days, then finally presented for treatment. This longitudinal image demonstrates the long axis of the gastrocnemius *(G)* and soleus *(S)*. A 12.5-cm fusiform hematoma separates the gastrocnemius and soleus along the intermuscular septum. This is the typical site of distraction injury in the lower leg. A thin rim of tissue lines the cavity, consistent with early reparative response. The soleus appears hyperechoic relative to the gastrocnemius due to increased through-transmission. **B,** This image was obtained 7 weeks after injury. Note the marked interval decrease in size of the hematoma. The reparative reaction *(open arrows)* is markedly increased, and the intermuscular septum is better visualized *(black arrows)*. Medial gastrocnemius *(G)*, soleus *(S)*. **C,** 12 weeks have now elapsed since the injury. Healing continues normally as demonstrated by further filling in of the lesion by reparative tissue *(open arrows)*. Normal muscle architecture begins to be seen at the periphery of the lesion. Abbreviations: medial gastrocnemius *(G)*, soleus *(S)*.

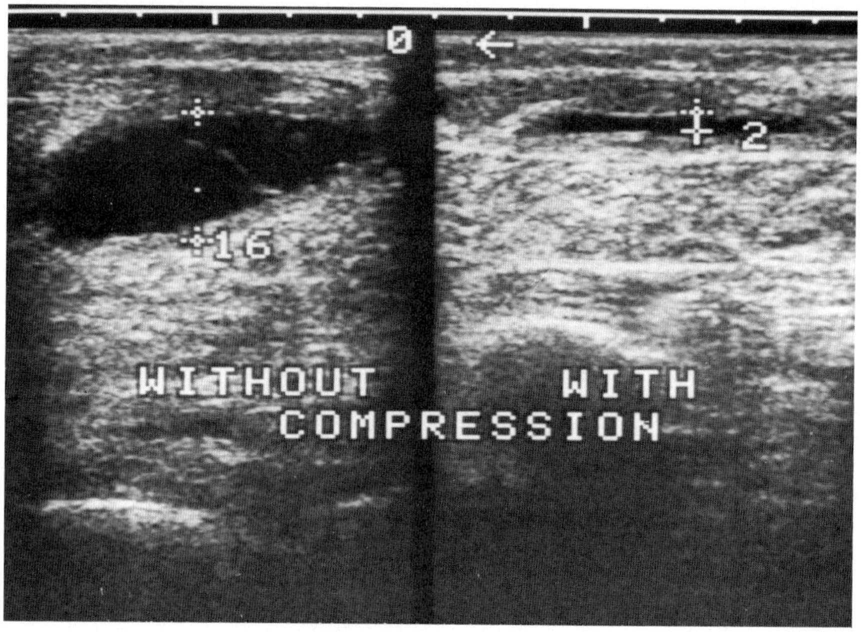


**FIG 3–25.**
Healing muscle rupture: transverse sonograms. This transverse sonogram of the proximal left calf was obtained during the first examination of the patient in Figure 3–24. The compression image demonstrates without question that this is a fluid collection. Reparative tissue along the walls of the cavity does not change in dimension.

thin wall may be identified with focal areas of nodular thickening (Fig 3–36). Septations may also be present (Zuinen et al., 1980, 1982). Little, if any, change occurs in the size of these lesions over observation periods as long as 3 years. Muscle cysts may cause chronic muscle pain, and surgical excision is indicated in such cases. At surgery these lesions are demonstrated to be cysts with synovial lining filled with synovial fluid. These likely arise from pleuripotential reserve cells found in the epimysium.

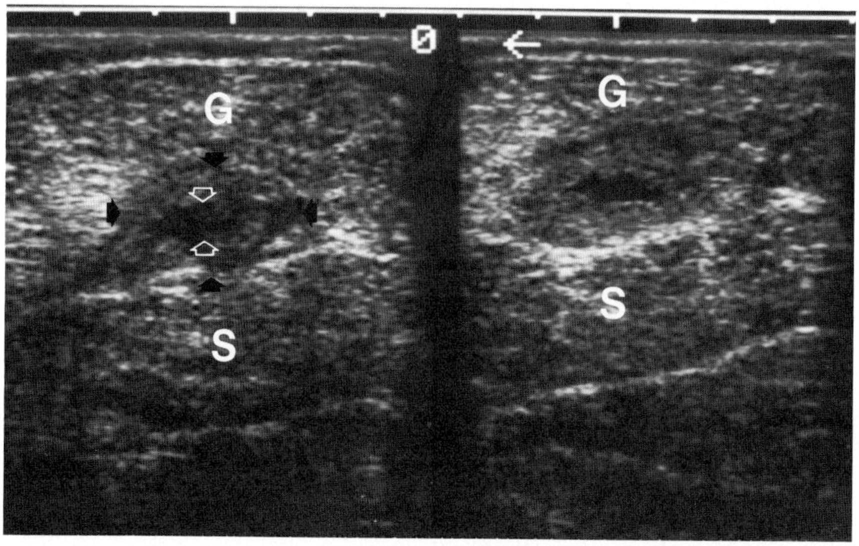


**FIG 3–26.**
Healing muscle rupture in the calf: transverse sonograms. Same patient as in Figures 3–24 and 3–25, 12 weeks after trauma. Note the concentric appearance of the reparative tissue that fills in the lesion *(black arrows)*. Residual hematoma *(white arrows)* is seen centrally within the lesion. The compression image on the right demonstrates that soft tissue of the reparative reaction is not compressible, unlike fluid (see Fig 3–25). Abbreviations: medial gastrocnemius *(G)*, soleus *(S)*.

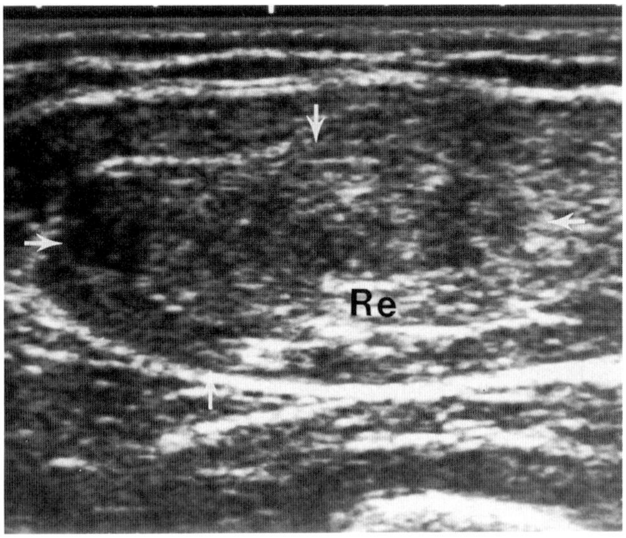


**FIG 3–27.**
Healing rectus muscle tear: transverse sonogram. This patient has persistent weakness of the right leg following a rectus femoris tear 3 weeks ago. A hypoechoic area *(arrows)* is seen involving half the cross-sectional area of the rectus muscle *(Re)*. Normal echogenic fibroadipose septa are not present in this region.

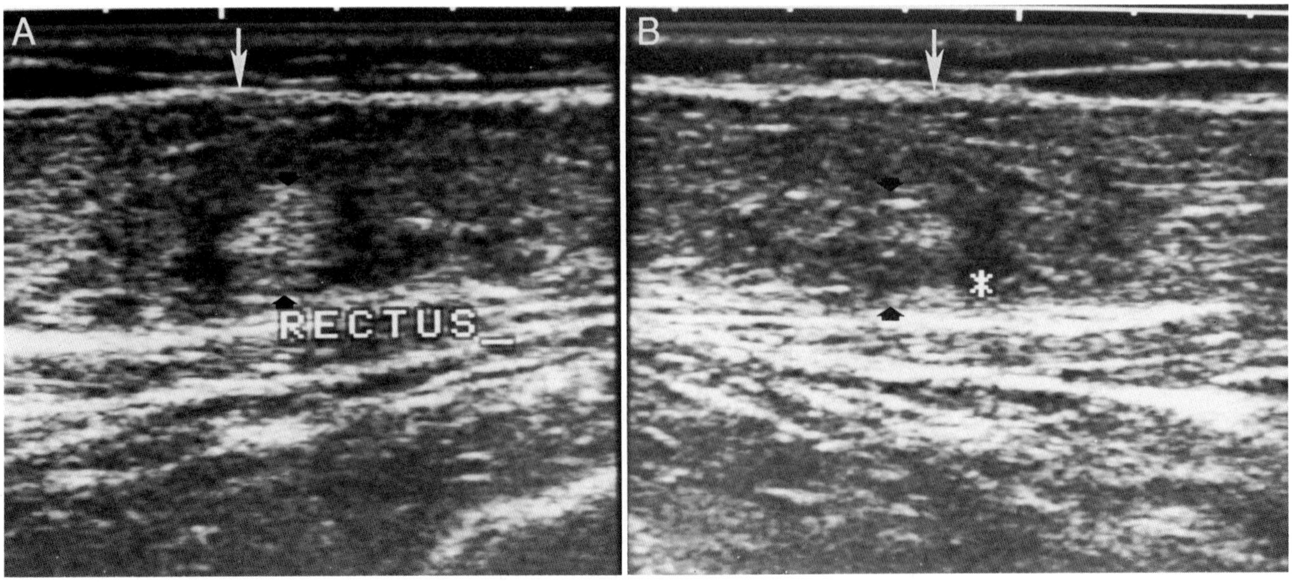


**FIG 3–28.**
**A,** Healing rectus muscle tear: longitudinal sonogram. Two gaps with an intervening segment of retracted muscle *(black arrows)* are identified within the rectus femoris of this 25-year-old basketball player 4 weeks after injury. The overlying fascia is slightly elevated *(white arrow)*. **B,** 6 weeks after injury, regeneration of muscle bundles is observed in the proximal gap, restoring continuity with the middle segment *(black arrows)*. The distal gap *(asterisk)* remains visible. Elevation of the fascia *(white arrow)* is less prominent.

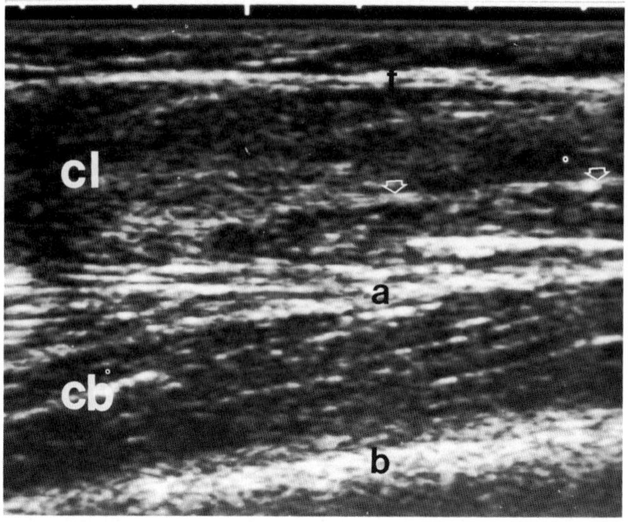


**FIG 3–29.**
Reparative tissue in the hamstrings: longitudinal sonogram. Seven weeks after a complete hamstrings rupture, the gap is filled in with homogeneous hypoechoic tissue. A long hyperechoic line *(open arrows)* is identified running almost perpendicular to the axis of normal fibroadipose septa of the long head of the biceps (caput longum, *cl*). This is consistent with a fibrous scar. Note the normal pennate structure of the short head of the biceps (caput breve, *cb*). Abbreviations: aponeurosis *(a)*, fascia *(f)*, femur *(b)*.

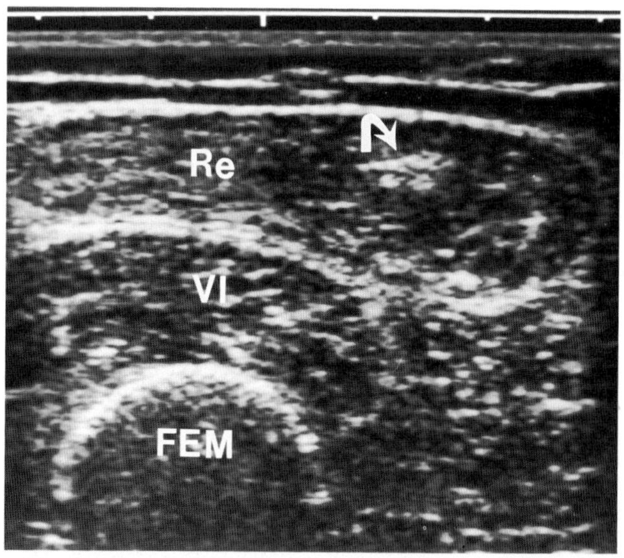


**FIG 3–30.**
Rectus scar after distraction injury: transverse sonogram. The patient experienced continued weakness of the right leg after distraction injury 3 months ago. The transducer is positioned transversely over the anterior midthigh demonstrating two divisions of the quadriceps: the rectus femoris *(Re)* and the vastus intermedius *(VI)*. An echogenic ovoid area *(curved arrow)* within the substance of the rectus represents residual scar from prior rupture. Abbreviations: femur *(FEM)*.

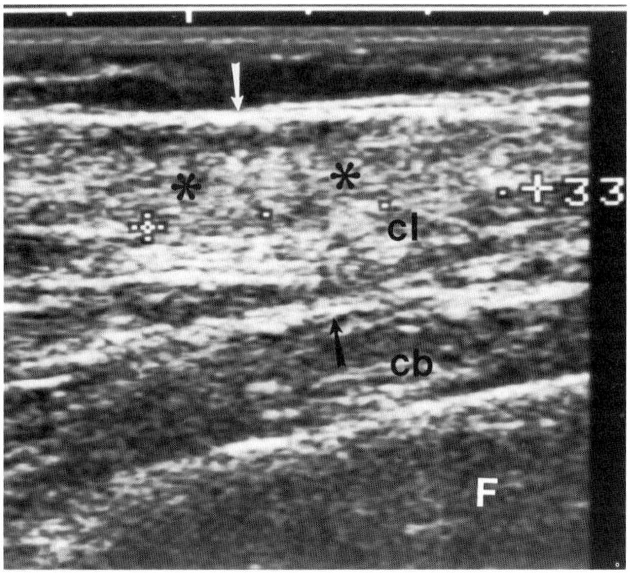


**FIG 3–31.**
Fibrosis in the hamstrings: longitudinal sonogram. One year following hamstrings injury this soccer player continued to experience weakness on flexion of the left knee. This longitudinal image of the posterior distal thigh demonstrates homogeneous hyperechoic scar *(asterisk)* within the substance of the long head of the biceps *(cl)*. Note the concave contour of the fascia *(white arrow)* and aponeurosis *(black arrow)*, indicative of muscle volume loss and scar retraction. Abbreviations: femur *(F)*, short head of the biceps *(cb)*.

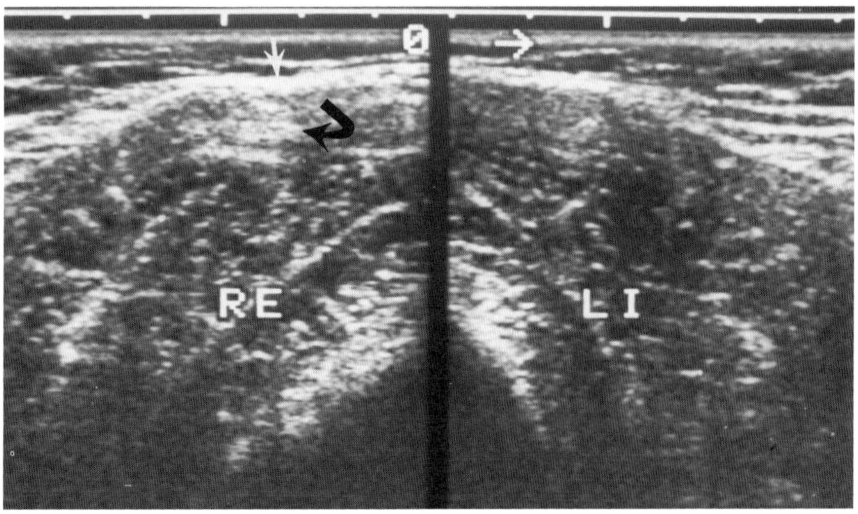


**FIG 3–32.**
Scar in the gluteus muscle: transverse sonogram. This 16-year-old female victim of a motor vehicle accident was struck on the right gluteal region 3 months ago. She complained of continued pain in the right gluteal region *(RE)* during normal activity. Side by side comparison demonstrates a region of increased echogenicity *(curved arrow)* in the right gluteus maximus deep to the fascia *(white arrow)*, which appears retracted. Increased echogenicity and focal retraction are hallmarks of scar formation. Abbreviations: left gluteus maximus *(LI)*.

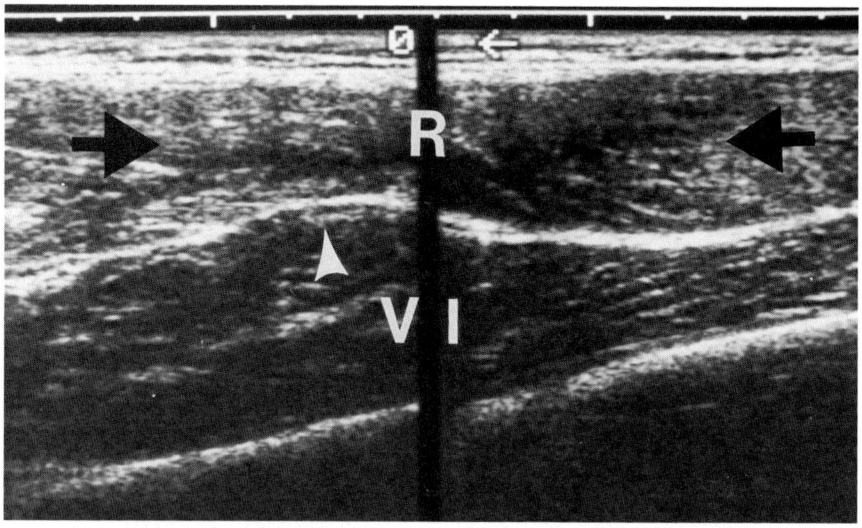


**FIG 3–33.**
Volume loss after rectus tear: longitudinal sonogram. One and one half years ago this 24-year-old male football player suffered a complete rectus tear. He presented for sonographic evaluation after being unable to attain his previous level of performance. The normal pennate structure of the rectus *(R)* is replaced with inhomogeneous tissue *(large black arrows)*. Marked retraction due to scarring of the healing rectus results in anterior bowing of the aponeurosis *(arrowhead)* and bulging of the vastus intermedius *(VI)*.

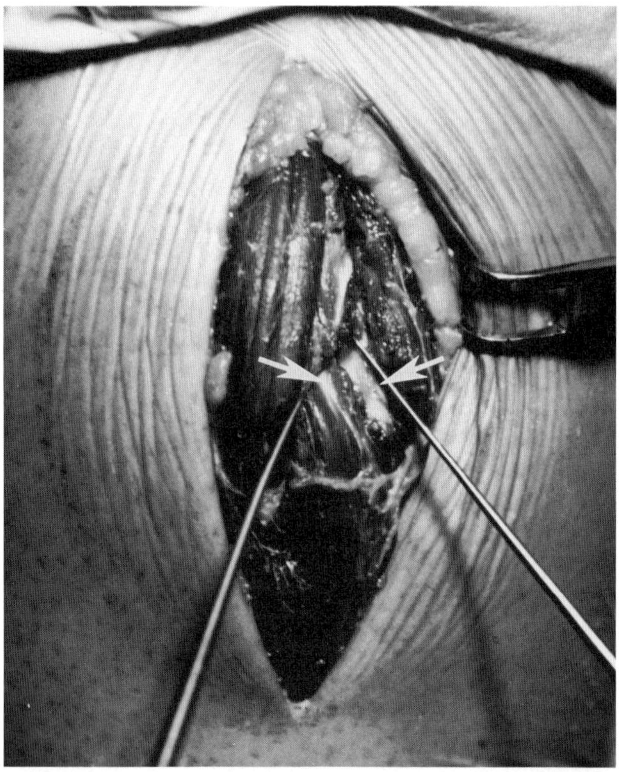

**FIG 3–34.**
Rectus fibrosis: surgical findings. In severe cases of fibrosis *(arrows)* or myositis ossificans, surgical excision is indicated to improve muscle function. This procedure is indicated primarily for those athletes who wish to continue high-level competition. However, it may also be indicated for alleviation of chronic pain. Ultrasound is extremely valuable in the diagnosis and localization of these lesions, facilitating surgical planning.

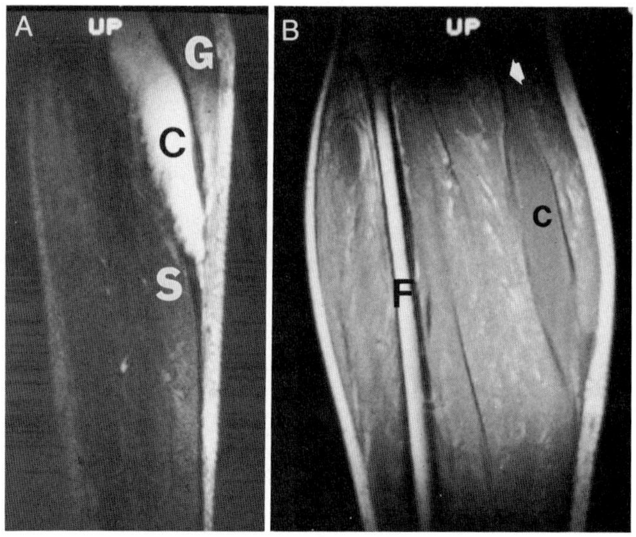


**FIG 3–35.**
**A,** Muscle cyst in the calf: sagittal MR image. This middle-aged man experienced blunt trauma to the right calf 4 months ago. In this sagittal T2-weighted spin-echo image, a large fluid collection of high-signal intensity *(C)* is identified between the gastrocnemius *(G)* and soleus *(S)*. The margin of the cyst adjacent to the soleus appears frayed. The location suggests a muscle-aponeurosis lesion. Duration of the lesion indicates markedly delayed healing, often seen in calf muscle rupture. **B,** muscle cyst in the calf: coronal MR image. Same patient as in **A.** The cyst is identified as a fusiform collection of slightly decreased signal intensity on this coronal T1-weighted spin-echo image. Location of the lesion *(c)* is confirmed, between the gastrocnemius and the soleus along the aponeurosis *(arrow)*. Abbreviations: fibula *(F)*.

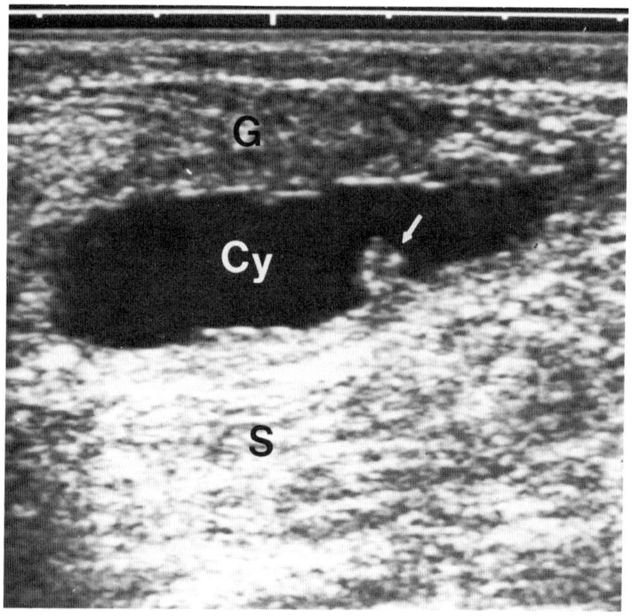


**FIG 3–36.**
Muscle cyst in the calf: transverse sonogram. Same patient as in Figure 3–35. The transducer is positioned transversely over the middle of the cyst. Increased through transmission is noted deep to the cyst *(Cy)*. This image demonstrates a torn muscle bundle *(arrow)* along the deep margin of the lesion, accounting for the frayed appearance of this border on the MRI examination. Abbreviations: gastrocnemius *(G)*, soleus *(S)*.

### Myositis Ossificans

Muscular contusion with intramuscular hematoma may calcify and then ossify. This lesion, myositis ossificans, is a frequent finding in athletes involved in contact sports. Often the inciting trauma is not extensive. In 40% of cases the lesion cannot be linked to a specific traumatic event. These lesions are most common in the thigh and pelvis. Identification of myositis ossificans is important because these lesions can cause chronic pain as they enlarge. Surgical removal may be required for palliative reasons. In addition, documentation of these lesions may prevent misdiagnosis at a later date. A young person may present with a painful extremity, and radiographs demonstrate an ossified mass in the diaphyseal region adjacent to the cortex. If the radiograph is misinterpreted as a tumor, the pathologist will likely interpret the histologic specimen as a low-grade periosteal or parosteal sarcoma. Previous sonographic documentation of the size and location of myositis ossificans will prevent this unfortunate occurrence.

The evolution of myositis ossificans is easily followed with sonography. Maturation of these lesions takes approximately 5 to 6 months. Initially, within 3 weeks of injury, the lesion is identified as a soft tissue mass with disorganized inhomogeneous internal architecture. At this stage, the lesion is indistinguishable sonographically from a soft tissue neoplasm. Clinicians may refer to this lesion as *gelosis*, a palpable firm mass within muscle. This term comes from the Latin gelare, to freeze (Fig 3–37).

Three to 4 weeks following injury, the first calcifications appear (Fig 3–38). These calcifications are identifiable sonographically well before they are evident on conventional radiographs. The predominantly peripheral distribution of these calcifications is characteristic of myositis ossificans. This is soon followed by ossification (Kramer et al., 1979). During this process of maturation, acoustic shadowing will become evident. At this stage it is important to note that no abnormal soft tissue extends beyond the borders defined by the shadowing calcifications (Fig 3–39). This helps to differentiate myositis ossificans from parosteal sarcoma. Further evidence to exclude parosteal sarcoma can be obtained by examination of the tissues deep to the calcifications. This requires careful angling of the transducer to direct the sound beam between shadowing calcifications. If normal

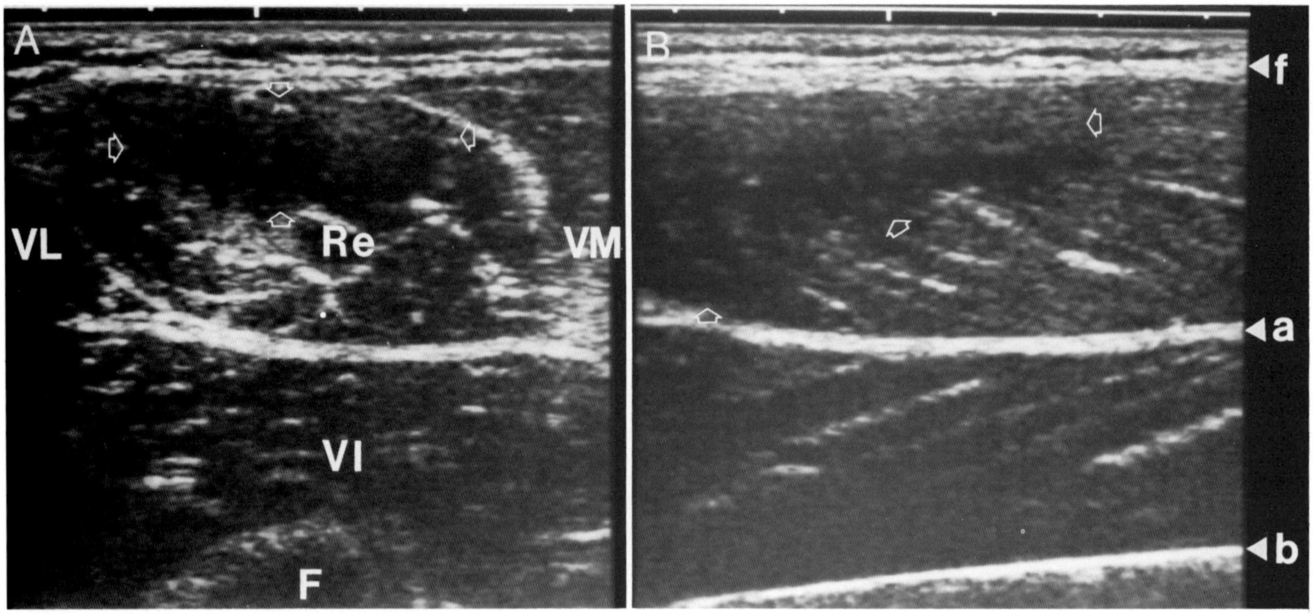


**FIG 3–37.**
**A,** gelosis of the rectus muscle: longitudinal sonogram. A basketball player presented for sonographic examination 3 weeks after being hit in the anterior thigh by another player. The clinician palpated a firm mass in the rectus muscle. This transverse image was obtained over the rectus approximately 20 cm superior to the patella. A hypoechoic area *(open arrows)* composed of soft tissue is identified within the rectus *(Re)* corresponding to the clinically palpable abnormality. No fibroadipose septa are seen in this region. Abbreviations: vastus lateralis *(VL)*, vastus medialis *(VM)*, vastus intermedius *(VI)*, femur *(F)*. **B,** gelosis of the rectus muscle: transverse sonogram. Same patient as in **A.** The transducer has been rotated to lie along the length of the quadriceps. A region of hypoechoic soft tissue replaces the normal muscle architecture. Note that this appearance differs from that of a healing muscle rupture. Abbreviations: aponeurosis *(a)*, bone *(b)*, fascia *(f)*.

periosteum adjacent to the cortex can be demonstrated, parosteal sarcoma has been excluded.

Heterotopic bone formation in patients following total joint replacement, burns, or neurological injury will demonstrate sonographic features similar to those of myositis ossificans. In these disorders, the distribution of ossification is predominantly periarticular (Fig 3–40), and the clinical history will clinch the diagnosis. Sonographic follow-up of these patients can be quite valuable, especially in those patients with ankylosing spondylitis or Forestier's disease undergoing total hip arthroplasty, due to their predilection for heterotopic bone formation. Although satisfactory therapy is not yet available, clinical trials with diphosphonates have shown promise in reducing formation of heterotopic bone.

Congenital myositis ossificans is an extremely rare disorder, also referred to as *myositis ossificans progressiva.* In this idiopathic disorder, muscle ossification is symmetrical and bilateral. Systemic diseases such as dermatomyositis and scleroderma are also characterized by muscle calcifications. However, calcifications will also be identified in the subcutaneous tissues of patients with these diseases.

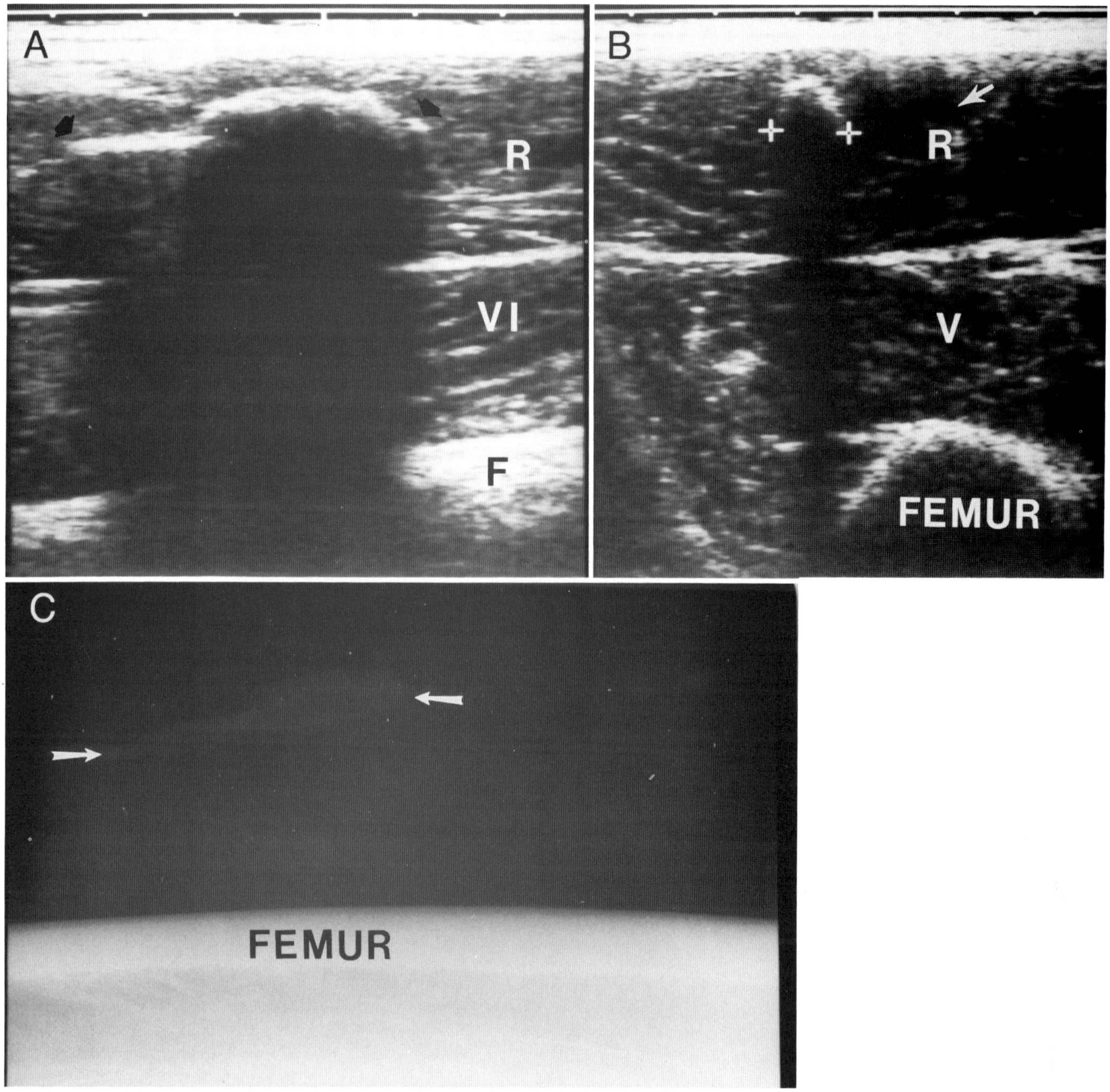


**FIG 3–38.**

**A,** gelosis progressing to myositis ossificans in the rectus muscle: longitudinal sonogram. Same patient and transducer position as in Figure 3–37,B, 3 months following injury. A linear-curvilinear hyperechoic line *(arrows)* with acoustic shadowing is identified within the rectus in the same location as the previously noted abnormality. Incomplete shadowing is noted deep to the linear segment due to volume averaging. Shadowing makes evaluation of the depth of the lesion difficult. Scanning obliquely will help to demonstrate the deep extent of myositis ossificans. Abbreviations: femur *(F),* rectus *(R),* vastus intermedius *(VI).* **B,** myositis ossificans in the rectus muscle: transverse sonogram. Same patient as in Figures 3–37 and 3–38,A. The region of ossification demonstrated to be much narrower than it is long. A hypoechoic region *(arrow)* adjacent to the ossification represents reparative tissue. Abbreviations: rectus *(R),* vastus intermedius *(V).* **C,** Myositis ossificans in the rectus muscle: plain film. Same patient as in Figures 3–37 and 3–38,A and B. This low-kilovoltage radiograph demonstrates a subtle area of myositis ossificans *(arrows).* Two small round areas of ossification that were outside the sonographic field of view are also noted. Ultrasound more clearly demonstrates lesions of myositis ossificans. Sonographic identification of ossification is much less dependent on technique, unlike radiographs, which may be overpenetrated. Localization of the ossification to a specific muscle is also far more precise with ultrasound than conventional radiography. As demonstrated in this case, ultrasound will demonstrate the soft tissue abnormality that will lead to myositis ossificans.

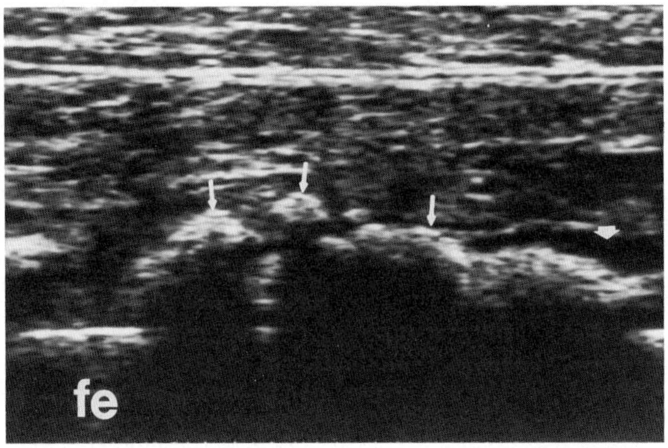


**FIG 3–39.**
Myositis ossificans of the anterior thigh: longitudinal sonogram. The patient is a 40-year-old construction worker 5 weeks following injury. He was hit with a metal reinforcing rod in the right thigh. An area of myositis ossificans *(small arrows)* is seen within the vastus intermedius just superficial to the femoral cortex *(fe).* No soft tissue mass is identified extending beyond the ossification, thus excluding parosteal sarcoma. A fusiform collection of residual hematoma *(small arrow)* is seen at the caudal margin of the ossification. Myositis ossificans has two different presentations: completely intramuscular or periosteal extending into muscle. Figure 3–38 clearly depicted completely intramuscular ossification. This case demonstrates ossification with periosteal base.

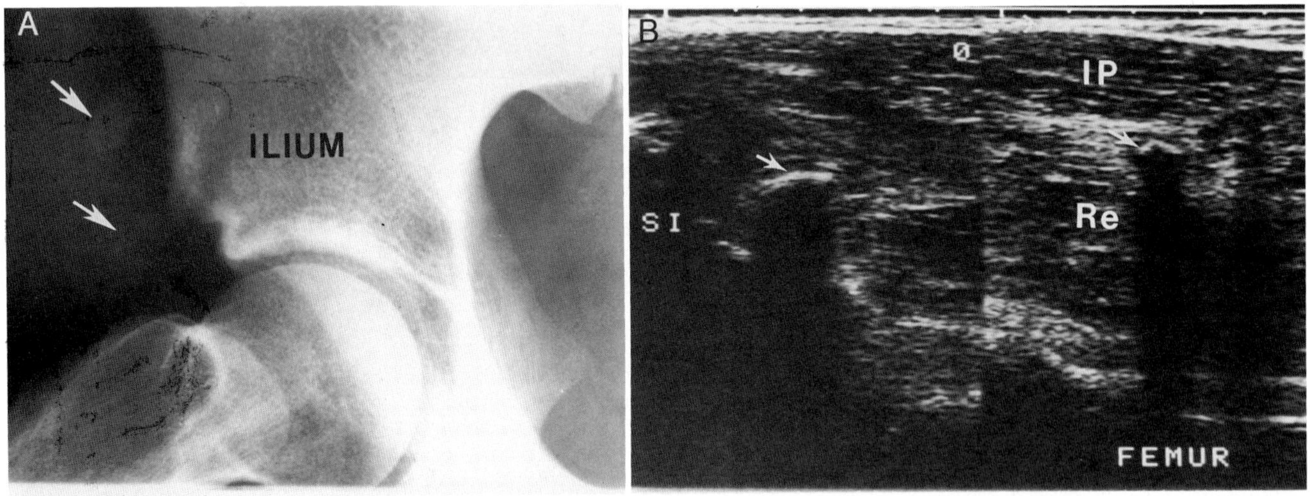


**FIG 3–40.**
**A,** Localization of myositis ossificans: conventional radiograph of the hip. This soccer player was struck in the upper thigh by another player 7 months ago. He presented with a complaint of significantly decreased kicking power. Two sites of ossification *(white arrows)* are identified lateral to the ilium. On radiographs it cannot be determined if the ossification is muscular, capsular, or intra-articular. **B,** Localization of myositis ossificans: longitudinal sonogram. Same patient as in **A.** The transducer is aligned in the length of the lower extremity with its cranial extent touching the anterior inferior iliac spine *(SI)* at the observer's left. This longitudinal image clearly demonstrates the lesion *(arrows)* to be located within the rectus muscle *(Re),* deep to the iliopsoas *(IP).*

## Myositis

Myositis is a general term used to denote inflammation of muscle. It can occur as a result of trauma, infection, or systemic disease. Viral infection is frequently accompanied by muscle involvement characterized by myalgia. A classic example of this type of muscle pain is epidemic pleurodynia, Bornholm disease. This is associated with Coxsackie virus infection. Viral inclusions in muscle have been demonstrated by electron microscopy in association with viral infection (Resnick and Niwayama, 1988), but we have been unable to identify any sonographic changes.

Bacterial infection involving muscle is a serious condition that is most common in children and young adults. Although rare in temperate climates, pyogenic myositis is quite common in tropical regions (Resnick and Niwayama, 1988). It accounts for 4% of all surgical admissions in East Africa. Muscles of the lower extremities are more commonly involved. However, any muscle may become infected. In the majority of cases, the infection remains contained within a single fascial compartment. The involved muscle initially feels hard to palpation, and induration of the skin makes it resemble the texture of wood (Resnick and Niwayama, 1988). Subsequently, the mass becomes fluctuant, and regional adenopathy is palpable.

Ultrasound is particularly valuable in the early diagnosis of bacterial myositis, when the clinical picture is still nonspecific. The sonographic appearance of the involved muscle is a negative of normal muscle. Muscle fibers become relatively hyperechoic. The fibroadipose septa are distended with inflammatory exudate and appear relatively hypoechoic (Fig 3–41). Comparison with the asymptomatic side will demonstrate an increase in diameter of the involved muscle (Fig 3–42). Given time, this will evolve into a frank abscess with central necrosis and formation of a collection of purulent material. Once the infection has reached this point, its appearance may be quite similar to that of an organizing hematoma. A hypoechoic fluid collection containing echogenic debris is identified. In some cases a fluid-fluid level may be observed (Harcke et al., 1988). The clinical presentation of high fever, chills, leukocytosis, and elevated erythrocyte sedimentation rate will ensure the diagnosis. Echogenic foci with characteristic dirty shadowing can be seen if gas-forming organisms are present.

An abscess resulting from pyogenic myositis can usually be distinguished from one secondary to osteomyelitis. In cases of pyogenic myositis the abscess is clearly centered in muscle. However, in muscle abscess resulting from osteomyelitis, purulent material will be seen tracking along the bony contours. Elevation of the periosteum with fluid separating the periosteum from the cortex will also be noted (van Sonnenberg et al., 1987; Abiri et al., 1988).

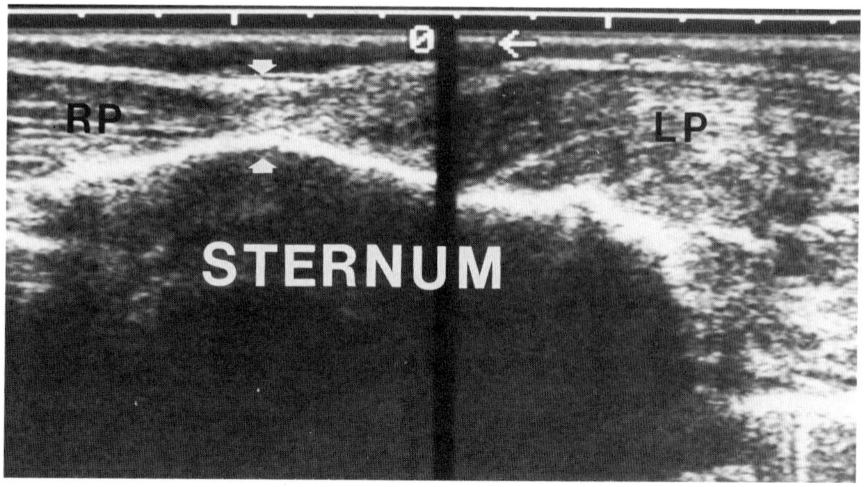


**FIG 3–41.**
Pyogenic myositis of the pectoral muscle: longitudinal sonogram. A 50-year-old man was admitted with fever and pain in the left upper thorax. A posteroanterior chest x-ray film demonstrated marked ill-defined opacity in the left suprahilar region. This opacity could not be localized on the lateral projection. Ultrasound was performed to exclude chest wall pathology. The transducer was positioned along the length of the pectoralis muscles in the midline of the chest. Orientation is similar to a CT section with the patient's left at the observer's right. The origin of the pectoralis muscles at the midline of the sternum is indicated by the small *white arrows.* Note the marked anterior bulge of the left pectoralis muscle *(LP)* compared with the right *(RP).* Fibroadipose septa are less well defined in the left pectoralis muscle.

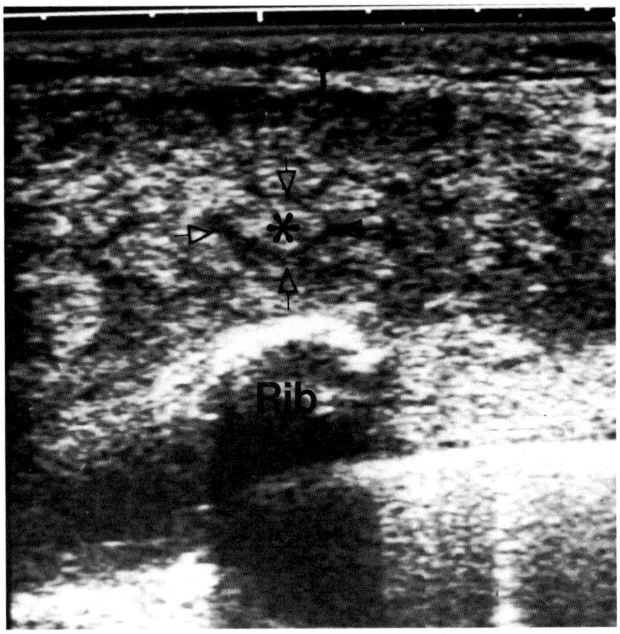


**FIG 3–42.**
Pyogenic myositis: transverse sonogram. Same patient as in Figure 3–41. The transducer has been rotated 90 degrees, parallel to the sternum, and moved laterally to yield a cross-sectional image of the pectoralis muscle. The pectoralis muscle is inhomogeneous and a reversal in echo texture of the muscle is seen. Fibroadipose septa *(arrows)* appear hypoechoic relative to the involved muscle bundles *(asterisk)*. Edema fluid is noted beneath the fascia *(f)*. Aspiration and culture revealed *Staphylococcus aureus* to be the causative organism.

## Compartment Syndrome

Muscle ischemia and ischemic necrosis are often difficult to diagnose clinically when they are not the result of atherosclerotic disease. A compartment syndrome is the most frequent cause of muscle ischemia in patients less than 50 years old. Compartment syndromes are characterized by elevated pressure within a muscle group surrounded by investing fascia (Anderson, 1983). Elevated interstitial pressure in an anatomically confined compartment results in impaired capillary perfusion. Although blood flow at the capillary level is severely compromised, major arteries remain patent. Therefore, normal pulses are observed distal to the involved compartment in the vast majority of cases. This frequently misleads the clinician.

Muscle edema is a frequent cause of increased pressure within a compartment. Due to the limited elasticity of the investing fascia, intracompartmental pressure may be significantly increased with slight edema. This edema is often seen as a result of strenuous exercise or blunt trauma. A tightly fitting plaster cast can have the same effect.

### *Acute Compartment Syndrome*

Compartment syndromes are divided into two categories: acute and chronic. Acute compartment syndrome is the result of a single insult, either exercise-induced muscular overload or trauma. It is characterized clinically by increasing pain associated with movement, particularly passive stretching. If exercise induced, the pain usually starts during exercise, but onset may be delayed as much as 12 hours following exercise. Swelling of the involved compartment may be clinically detectable. Parethesia in the sensory distribution of nerves running through the compartment will also be observed (Nicholas and Hershman, 1986). Most frequently affected are the anterior, posterior, and lateral compartments of the lower leg. A high priority must be placed on confirming the diagnosis of compartment syndrome because delay in treatment will result in permanent

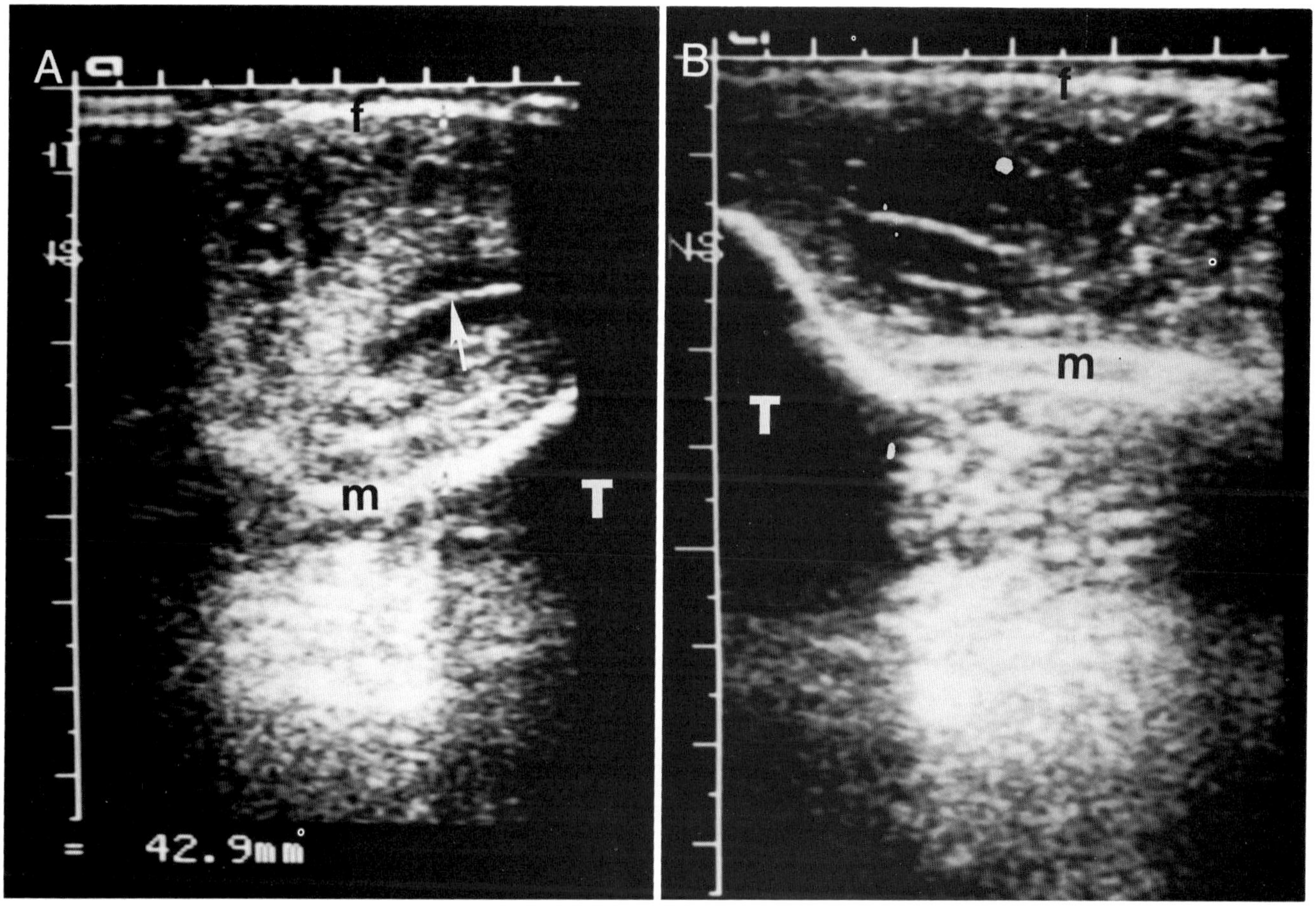


**FIG 3–43.**
**A,** acute anterior compartment syndrome of the lower leg: transverse sonogram. The patient is a 55-year-old male tennis player who usually plays one set each week. This past weekend he played in a tournament. One match lasted 4 hours. Several hours after the match, he experienced a gradual onset of pain in the anterior lateral aspect of the right lower leg, which reached a peak 2 days later. After 7 days he could no longer tolerate the pain and presented to the emergency room. He was admitted with a clinical diagnosis of deep venous thrombosis. The right lower leg was edematous, erythematous, and warm. Distal pulses were intact. Contrast venography was normal. Ultrasound examination confirmed the presence of normal veins and excluded a ruptured Baker cyst. This image was obtained at the level of the midtibia *(T).* The anterior tibial compartment of the right leg was markedly distended and increased in echogenicity. Superficial fascia *(f)* and the interosseous membrane *(m)* were bulging due to mass effect. Note the relative sparing of muscle architecture immediately adjacent to the larger fibroadipose septa *(arrow).* The anterior tibial compartment on the right measures 42.9 mm, 30% greater than the asymptomatic side. Computed tomography confirmed the presence of edema in the anterior compartment by demonstrating decreased attenuation within the compartment and absence of enhancement with bolus contrast administration. **B,** acute anterior compartment syndrome of the lower leg: transverse sonogram. Same patient as in **A.** This image depicts the anterior tibial compartment of the asymptomatic left leg at the same level. Compare this normal appearance of the compartment with that seen in **A.** Note the straight appearance of the superficial fascia *(f)* and interosseous membrane *(m).*

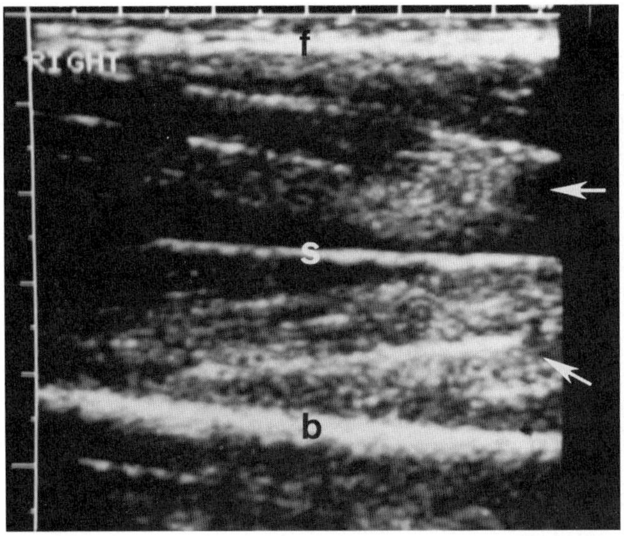


**FIG 3–44.**
Acute anterior compartment syndrome of the lower leg: longitudinal sonogram. Same patient as in Figure 3–43. This image is a longitudinal section obtained over the symptomatic anterior compartment. The muscles of the anterior compartment are markedly increased in echogenicity *(arrows).* Note that the region adjacent to the intermuscular septum *(s)* is relatively spared, likely due to the proximity to larger vessels in the septum. Abbreviations: fascia *(f)*, tibia *(b)*.

neurological and functional deficit. Surgical fasciotomy is the only treatment available. Fascial release adversely affects the strength of a muscle group, making it more than a benign procedure. This further increases the importance of accurate diagnosis (Garfin, 1981).

Until now the only way to confirm the diagnosis of a compartment syndrome was direct pressure measurement. A catheter was inserted into the suspect compartment using sterile technique and local anesthesia, and pressure measurements were obtained. Evaluation of chronic compartment syndrome required pressure measurement at rest, during exercise, and following recovery (Mubarak, 1981). A pressure difference of 30 to 40 mm Hg between the compartment and the subcutaneous tissues will result in significant tissue damage. "Noise" introduced by muscle contraction and limb motion significantly reduce the accuracy of this test. In addition, the invasive nature of the test makes it less desirable.

Computed tomography provides little assistance in making the diagnosis of compartment syndrome. Edema attributable to compartment syndrome may appear slightly lower in attenuation (20–30 Hounsfield units) following bolus contrast administration (Weissman and Sledge, 1986). However, this is subject to all of the variables that are inherent to IV contrast administration. This technique has not demonstrated a high degree of sensitivity or specificity.

The noninvasive nature of ultrasound makes it an attractive alternative to direct pressure measurement. Sonography also has the ability to exclude other entities that must be considered in the differential diagnosis of compartment syndrome. Posttraumatic hematoma, abscess, deep venous thrombosis, and ruptured Baker's cyst are all well evaluated with ultrasound (Magnussen et al., 1988; Gompels and Darlington, 1979; Lawson and Mittler, 1978).

Acute compartment syndrome is characterized sonographically by diffusely increased echogenicity of the muscles within the compartment (Figs 3–43 through 3–47). Unlike pyogenic myositis, the fibroadipose septa remain hyperechoic. In some cases, sparing of the periseptal muscle bundles may be observed. These spared muscle bundles will maintain their normal hypoechoic appearance, accentuating the feather-like structure of the muscles (Graf and Schuler, 1988). The explanation for this pattern of in-

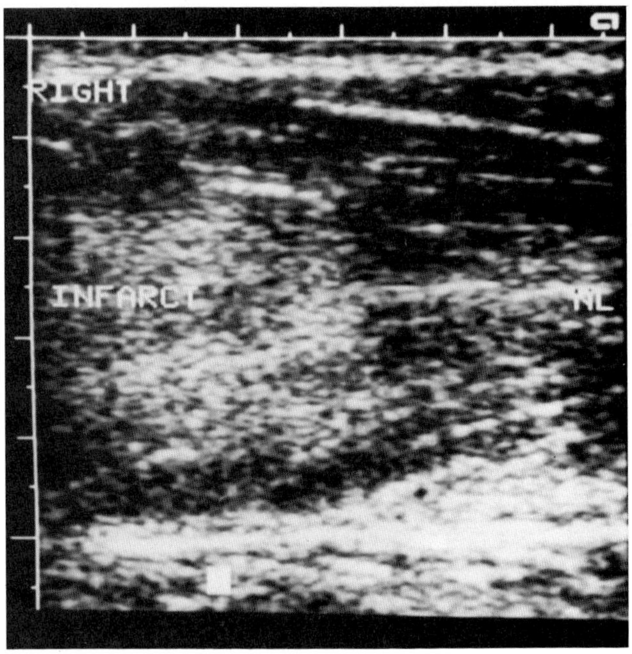


**FIG 3–45.**
Infarcted muscle: longitudinal sonogram. Same patient as in Figures 3–43 and 3–44. A gradual transition is seen from the abnormal infarcted muscle at the left and the viable muscle *(NL)* on the right. Unfortunately, the patient refused surgery for 7 days. When he finally consented to fasciotomy, irreversible damage to the deep peroneal nerve had already occurred. Considerable muscle necrosis was also present. As a result of the delay in treatment, this patient had significant permanent motor and neurological deficit.

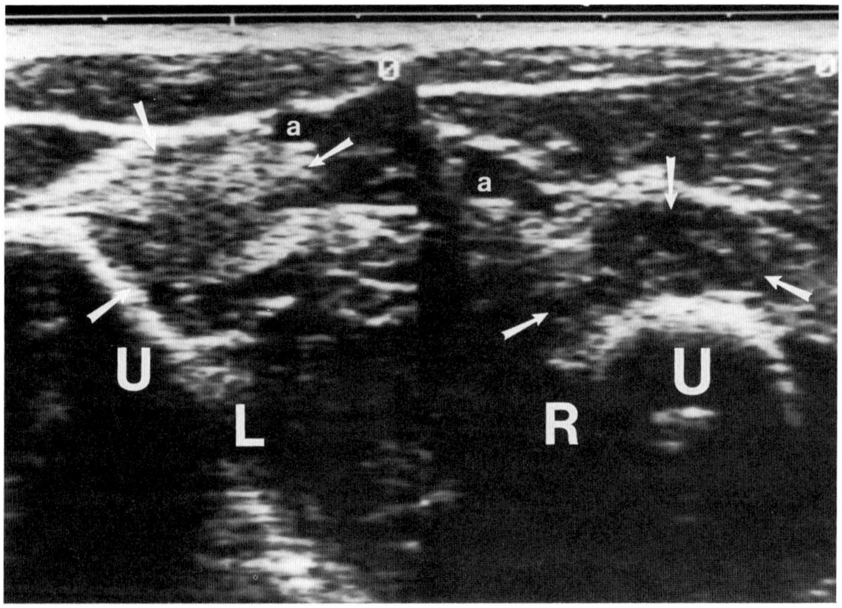


**FIG 3–46.**
Ischemia of the flexor digitorum profundus muscle: transverse sonogram. A 17-year-old boy spent the entire day doing carpentry with his father while on vacation. Six hours later, he began to experience severe pain in the left forearm and hand weakness. Physical examination revealed a swollen erythematous tender left forearm. Distal pulses were normal. The transducer is positioned transversely over the anterior forearm at the junction of the proximal and middle thirds. In this split screen image, the symptomatic left forearm is shown on the left and the normal right forearm to the right. The flexor digitorum profundus muscle *(arrows)*, located between the ulna *(U)* and ulnar artery *(a)*, is diffusely increased in echogenicity on the left. The right flexor digitorum profundus muscle is normal. Abbreviations: left *(L)*, right *(R)*.

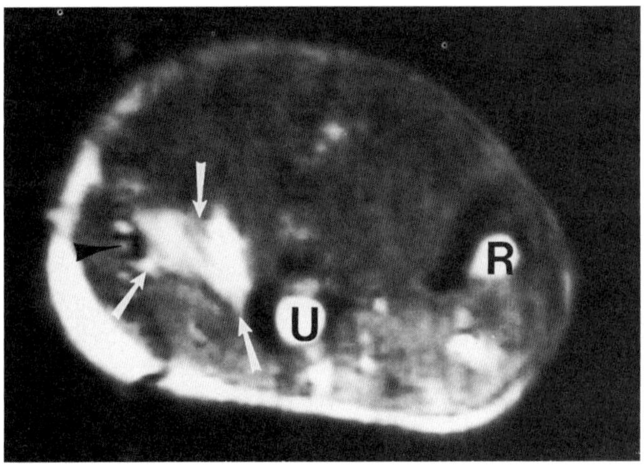


**FIG 3–47.**
Ischemia of the flexor digitorum profundus: T2-weighted axial spin-echo image. Same patient as in Figure 3–46 one day following sonographic examination. The patient's symptoms were unchanged. Note the markedly increased signal intensity of the left flexor digitorum profundus muscle *(arrows)*, indicative of edema. Flow void is seen in the ulnar artery *(arrowhead)*. The clinical service elected to follow this patient with serial sonographic examinations. An increase in diameter of the muscle would have indicated the need for fasciotomy. However, follow-up examinations demonstrated slow but steady resolution of the edema. Abbreviations: radius *(R)*, ulna *(U)*.

jury is the distribution of the vascular supply, which enters along the fibroadipose septa. Another sign of compartment syndrome is increased dimension of the compartment compared with the asymptomatic side. Bowing and displacement of the investing fascia will be evident.

Some investigators have reported observing an irregular sonographic pattern in compartment syndrome, having both solid and cystic components (Auerbach and Bowen, 1981; Peterson and Renström, 1986). This represents a later stage in the evolution of compartment syndrome. The loss of recognizable muscle architecture is consistent with progression beyond ischemia to frank infarction. Hypoechoic cystic areas will progress and may contain some material of higher echogenicity. The sonographic appearance may resemble that of extensive rhabdomyolysis. At this stage, the diagnosis comes too late. Extensive tissue damage has already occurred. Foot drop, nerve palsy, or Volkmann's contracture are common sequelae of compartment syndrome when it is not treated promptly. Muscle necrosis will heal with extensive fibrosis and ossification, which will also involve the surrounding fascia (Resnick and Niwayama, 1988).

### *Chronic Compartment Syndrome*

Chronic compartment syndrome is associated with recurrent pain during exercise. It usually occurs as a result of rapidly increased muscle mass following rigorous athletic conditioning. Muscle mass is increased significantly in a short period of time, too rapidly for the nonelastic fascia to adapt. The result is a pathological increase in intracompartmental pressure during exercise because it cannot expand to accommodate the transient increase in volume required secondary to increased blood flow (Nicholas and Hershman, 1986). The diagnosis of chronic compartment syndrome is more difficult than acute compartment syndrome. The pain is intermittent, predominantly during exercise. It may be characterized as sharp, cramplike, or burning pain. The level of exercise that brings on symptoms may vary considerably, from running a 4-minute mile to walking several blocks. Often the patient can continue to exercise but at a reduced level due to the discomfort. At the end of exercise, the pain usually resolves quickly but may persist for hours or even days. Symptoms are bilateral in 70% of patients (Mubarak, 1981; Martens et al., 1984).

Clinical examination during an active episode of chronic compartment syndrome reveals pain on movement of the extremity. Most commonly pain associated with plantar and dorsiflexion of the foot, as the anterior and posterior compartments of the lower leg are most often involved. Sensation in the distal portion of the extremity may be altered, but this is not a constant finding. Distal arterial pulses are always normal (Brahim and Zaccardelli, 1986). The differential diagnosis includes shin splints, stress fractures, periostitis, popliteal entrapment, and true claudication. It is easy to see why the diagnosis of chronic compartment syndrome is so difficult.

Pathological examination has demonstrated an increased water content within the involved compartment. This would explain the increased intracompartmental pressure that results in decreased perfusion. The decrease in perfusion has been demonstrated by decreased xenon 133 clearance and increased accumulation of lactate (Nicholas and Hershman, 1986). Phosphorus MR spectroscopy has also been used to demonstrate a decrease in muscle adenosine triphosphate.

An increase in dimension of approximately 10% has been demonstrated by sonographic measurement of the anterior tibial compartment following dorsiflexion exercise of the foot (Brahim and Zaccardelli, 1986; Gershuni et al., 1982). These investigators postulated that the limitation of further expansion by inelastic fascia results in ischemia and chronic compartment syndrome.

We examined the compartments of the lower leg in a large group of patients, which included normal young people, normal athletes, and athletes with the clinical diagnosis of chronic compartment syndrome. The greatest dimension of the compartment was measured using a standardized technique. The transducer was positioned transversely at a point equidistant from the tibiofemoral joint space and the medial or lateral maleolus, depending on the compartment to be measured. A line perpendicular to the interosseous membrane was established to measure the distance from the membrane to the compartmental fascia. The position utilized for measurement was marked on the skin along with an identical position on the other leg. Measurements of the anterior compartment are most important in patients who present with pain lateral to the anterior border of the tibia. In patients with pain posterior to the tibia and fibula, both the superficial and deep posterior compartments must be examined. Peroneal compartment syndrome is quite rare. We did not encounter a patient with peroneal compartment syndrome in this series. Dimensions of the compartments were obtained at rest, immediately following exercise, and after a 10-minute recovery period.

In our patients, exercise consisted of plantar and dorsiflexion of the foot with a 4-kg load at a rate of 60 cycles/minute. This was performed for 15 minutes or until symptoms forced the patient to stop. In all of these patients, direct pressure measurements were obtained at the same time sonographic measurements were taken.

Normal nonathletes and athletes demonstrated normal compartmental pressures. In these patients, an increase in dimension of the compartments immediately following exercise ranging between 10% and 15% was observed. Appearance of the compartments was slightly less echogenic immediately following exercise when compared with the resting images (Fig 3–48). These changes can be attributed to increased blood flow during exercise. Sonographic measurement following the 10-minute recovery period demonstrated a return to normal dimension and echogenicity. These findings were observed bilaterally in normal individuals.

Patients who carried the clinical diagnosis of chronic compartment syndrome could be divided into two groups based on sonographic findings. All of these patients demonstrated elevated compartment pressures: peak pressure with exercise elevated greater than 30 mm Hg over extracompartmental

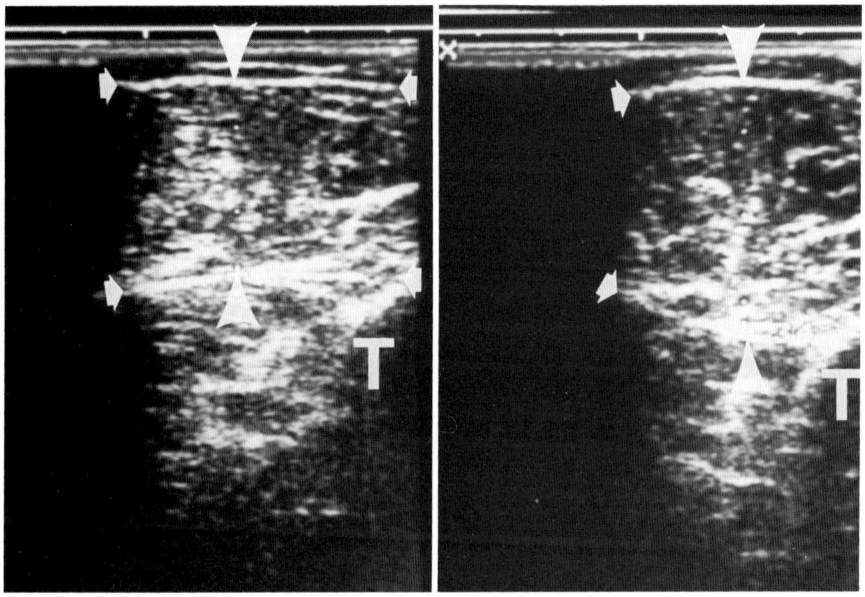


**FIG 3–48.**
Compartment measurement: transverse sonogram anterior tibial compartment. A normal 26-year-old male athlete. The image on the left was obtained of the anterior tibial compartment at rest and the image on the right obtained after 15 minutes of strenuous dorsiflexion of the foot. Note the increased dimension *(arrowheads)* of the compartment as demonstrated by displacement of the fascia and interosseous membrane *(arrows)*. In a normal exercise test, compartment volume should increase 10% to 15%. The muscles of the compartment are decreased in echogenicity following exercise. This is normal, likely the result of increased blood flow to the muscle. Abbreviations: tibia *(T)*.

pressure. The first group, which constituted approximately one third of the symptomatic patients, did not demonstrate any detectable increase in compartmental volume. Thickening and rigidity of the interosseous membrane were evident (Fig 3–49). In 5% of these patients herniation of muscle through the fascia cruris was observed.

The second group, comprising two thirds of the symptomatic patients, demonstrated normal expansion of the compartment, a 10% to 15% increase in volume. This finding was quite unexpected in light of the clearly abnormal intracompartmental pressures. However, return to normal compartmental dimension was markedly delayed. In some patients, the recovery period was as long as 1 hour, six times the upper limit of normal. One percent of patients demonstrated increased echogenicity of the muscle, similar to that seen in acute compartment syndrome (Fig 3–50).

Although unexpected, the discovery of two different sonographic patterns of presentation in chronic compartment syndrome is not inconsistent with the presumed etiology. Clearly, in the first group, fascial rigidity is responsible for increased intracompartmental pressure during exercise. However, the second group demonstrated a normal degree of compartmental expansion with exercise. It is possible that although the amount of expansion observed was within the "normal" range, it was insufficient for these athletes undertaking a rigorous training regimen. This would explain increased intracompartmental pressures at high levels of exercise and increased recovery times. More detailed data are needed to confirm this hypothesis.

## Rhabdomyolysis

Rhabdomyolysis is a general term denoting skeletal muscle necrosis. It is considered as an entity separate from compartment syndromes. Many etiologies of rhabdomyolysis have been documented. These have been divided into five groups: (1) primary muscle injury, (2) abnormalities of energy production (McArdle's disease, carnitine palmityl transferase deficiency), (3) infection, (4) hypoxia, and (5) miscellaneous causes (drugs and toxins) (Knochel, 1981). The most common causes seen today are primary muscle injury and hypoxia. Primary muscle injury is comprised largely of exertional rhabdomyolysis seen in athletes and military recruits, as well as crush injury secondary to trauma (Kaplan, 1980). Hypoxic injury is frequently seen as a result of peripheral vascular dis-

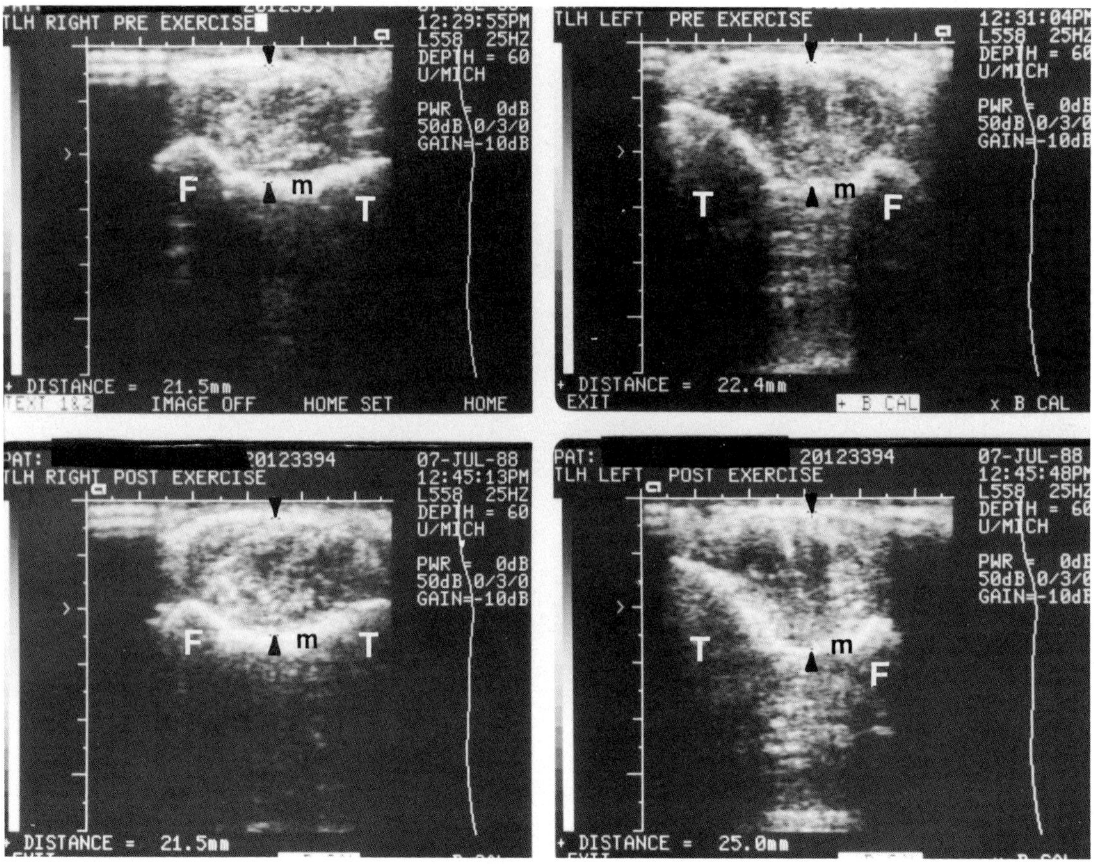


**FIG 3–49.**

Compartment measurement: transverse sonogram anterior tibial compartment. Each time after sprinting 50 yards, this football player experienced shooting pain in the anterior tibia. At the observer's left are images of the anterior tibial compartment of the patient's right leg. The left leg is at the observer's right; note preexercise and postexercise labels. Thickening of the interosseous membrane is present in the right leg. Dimensions of this compartment do not change with exercise, consistent with chronic compartment syndrome. A normal increase of slightly greater than 10% is noted in the left leg. Abbreviations: fibula *(F)*, tibia *(T)*.

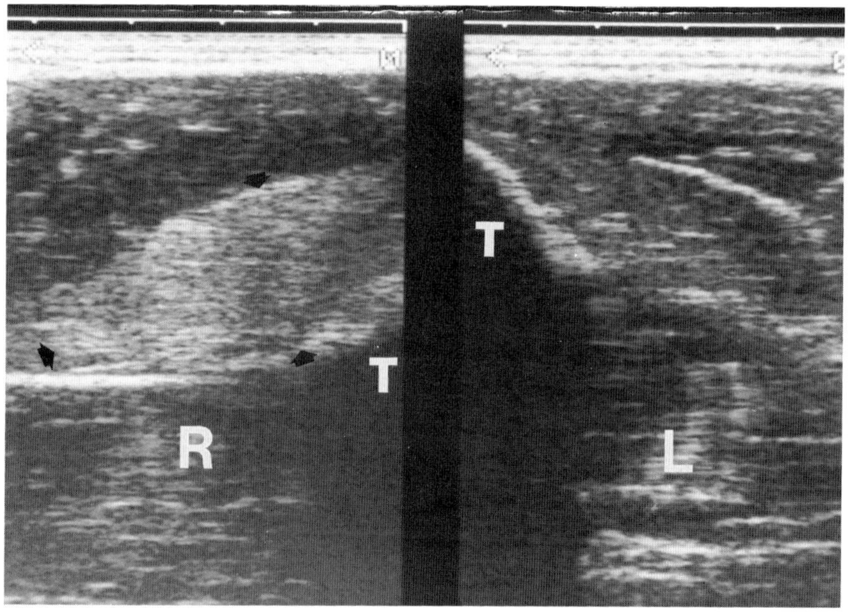


**FIG 3–50.**
Subacute anterior tibial compartment syndrome: transverse sonogram. One day following a tough soccer match, this 30-year-old man presented with intense pain in the right lower leg. He also experienced limitation of dorsiflexion of the foot and paresthesia. Anterior tibial compartment measurements were increased on the right compared with the asymptomatic left side. The right flexor hallucis muscle *(black arrows)* is increased in both dimension and echogenicity. This patient was followed sonographically. Over the course of 2 days, he improved both clinically and sonographically. An exercise stress test of the leg performed 2 weeks later demonstrated increased echogenicity of the flexor hallucis longus rather than the normal decrease in echogenicity. Abbreviations: left *(L)*, right *(R)*, tibia *(T)*.

ease. Unfortunately, an alarming increase in severe rhabdomyolysis has been observed in drug addicts, particularly those who abuse cocaine and heroin (Kaplan, 1980; Roth et al., 1988). Clinical signs indicative of rhabdomyolysis are myoglobinuria, hyperuricemia, hyperuricosuria, and elevated serum levels of muscle enzymes. Prompt diagnosis is extremely important because significant rhabdomyolysis will result in acute renal failure, secondary hyperkalemia, and disseminated intravascular coagulation (Kaplan, 1980; Nicholas and Hershman, 1986).

A region of rhabdomyolysis will appear as a hypoechoic area deep within the muscle (Kaplan, 1980; Vukanovic et al., 1983). Multiple lesions may be observed, and the muscle may be diffusely enlarged. The gluteal muscles are most involved in drug addicts and patients with epilepsy. Multiplicity of lesions and their deep location are rather characteristic of rhabdomyolysis (Kaplan, 1980). Sonographic features will differentiate rhabdomyolysis from distraction injury, but these lesions may strongly resemble abscesses. A clinical history of fever and leukocytosis will help make the diagnosis. However, noninfected regions of rhabdomyolysis can produce fever. Aspiration of the lesion is indicated when the diagnosis is not clear. Uncomplicated areas of rhabdomyolysis will yield clear serous fluid on aspiration. Hematomas may also have a similar appearance immediately following injury. However, hematomas are not usually associated with myoglobinuria and elevated serum levels of muscle enzymes. Several hours following injury the diagnosis of hematoma will be clear sonographically.

Contrast-enhanced CT shows nonspecific findings of circumscribed regions of decreased attenuation within the muscle (Kaplan, 1980; Vukanovic et al., 1983). Gross examination of the pathological specimen reveals a swollen firm pale muscle. Histopathological findings include muscle cell swelling, hyaline degeneration with loss of characteristic striation, and areas of interstitial inflammation (Kaplan, 1980; Geller, 1973). The inflammatory exudate surrounding necrotic muscle cells explains the hypoechoic appearance seen on sonographic images.

## MUSCLE BOUNDARY LESIONS

### Muscle-Aponeurosis Avulsion

We consider muscle boundary lesions to be pathology involving the investing fascia, an aponeurosis or the myotendinous junction. Muscle tears of the

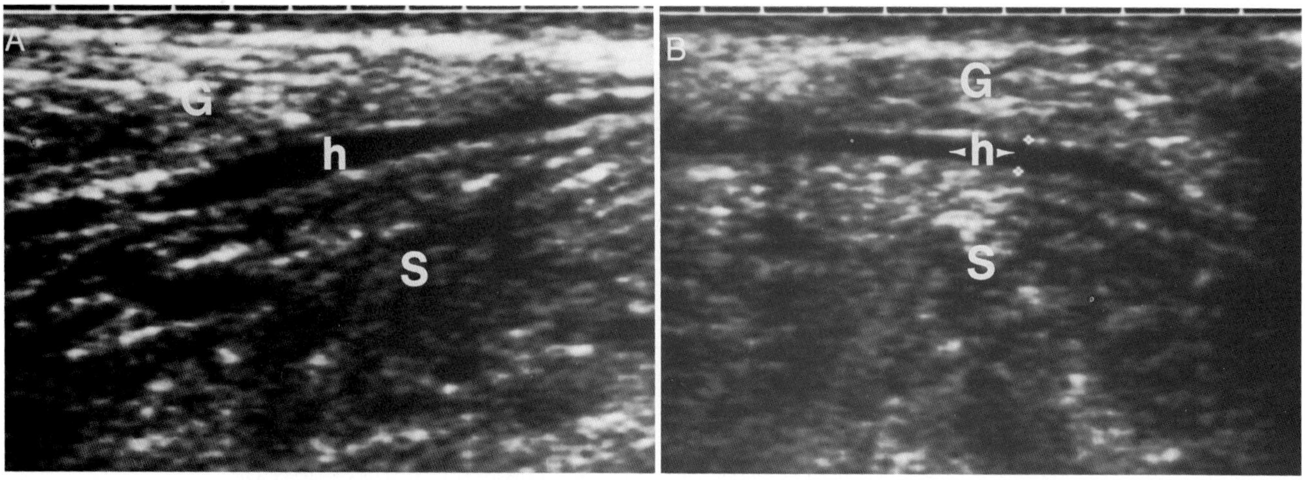


**FIG 3–51.**
Muscle aponeurosis avulsion: longitudinal **(A)** and transverse **(B)** sonograms of the calf. An anechoic cleft *(h)* along the aponeurosis between the gastrocnemius *(G)* and soleus *(S)* represents a resolving hematoma. This is the classic location and appearance of a muscle-aponeurosis avulsion in the calf.

calf frequently fall into this category. Distraction lesions involving the gastrocnemius and soleus typically occur along their attachment to the aponeurosis (Fig 3–51). Similar ruptures are seen along the soleus and flexor hallucis longus (Fig 3–52). These muscle-aponeurosis avulsions are commonly seen in athletes whose sports require speed: sprinting, soccer, football, baseball, and hockey (Fig 3–53). Sonography demonstrates a linear (Figs 3–52 and 3–54) rupture filled with blood that extends along the aponeurosis. A characteristic feature of muscle-aponeurosis avulsion is a change in orientation of the fibroadipose septa on either side of the aponeurosis in longitudinal images (Fig 3–55).

### Muscle Hernia

Fascial defects with associated herniation of muscle are uncommon lesions. Chronic compartment syndrome is the most frequent cause of muscle hernias, but they may also be found following trauma or surgery. In chronic compartment syndrome, hernia-

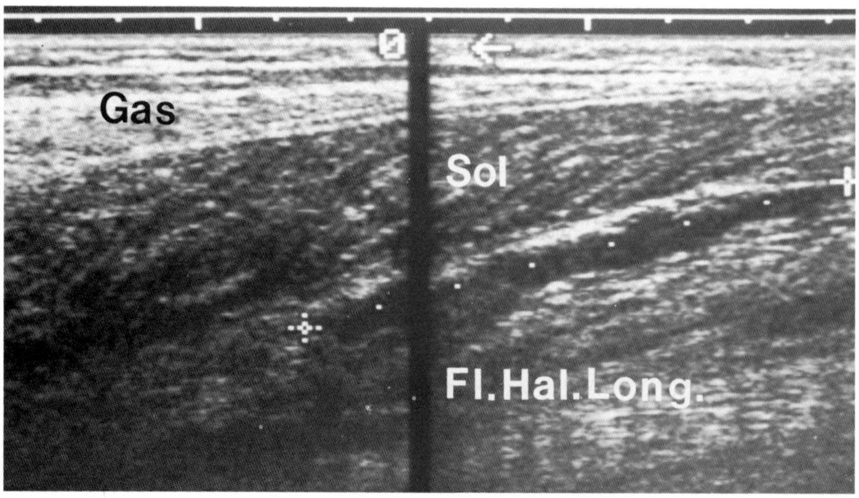


**FIG 3–52.**
Muscle aponeurosis avulsion: longitudinal sonogram of the calf. After 1 week of pain and swelling of the right calf, this 38-year-old man presented for sonographic evaluation. The transducer is aligned along the proximal portion of the flexor hallucis longus muscle. An avulsion of the flexor hallucis longus from the aponeurosis is present *(+)*. The aponeurosis is seen as a hyperechoic line superficial to the hematoma. Abbreviations: gastrocnemius *(Gas)*, soleus *(Sol)*.

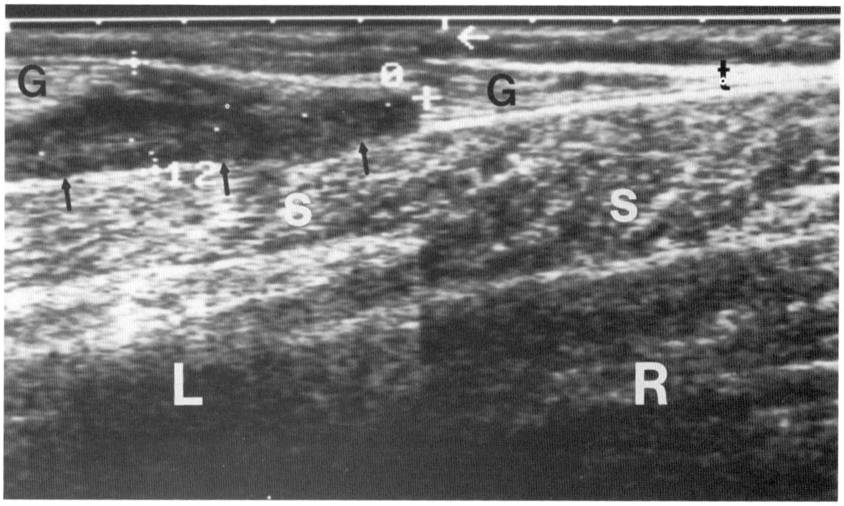


**FIG 3–53.**
Rupture of the gastrocnemius musculotendinous junction: longitudinal sonogram. A 25-year-old sprinter experienced sudden sharp pain in the distal left calf at the start of a race. He has limited plantar flexion of the foot, associated with severe pain. The transducer is positioned over the musculotendinous junction of the medial head of the gastrocnemius. Side to side comparison is provided in this split screen image. On the *left,* the tendon is retracted out of the field of view. A hematoma *(arrows)* is seen in the usual location of the musculotendinous junction. The retracted gastrocnemius is noted at the left corner of the image. A normal gastrocnemius musculotendinous junction is shown on the *right.* Abbreviations: gastrocnemius *(G),* proximal achilles tendon *(t),* soleus *(S),* left *(L),* right *(R).*

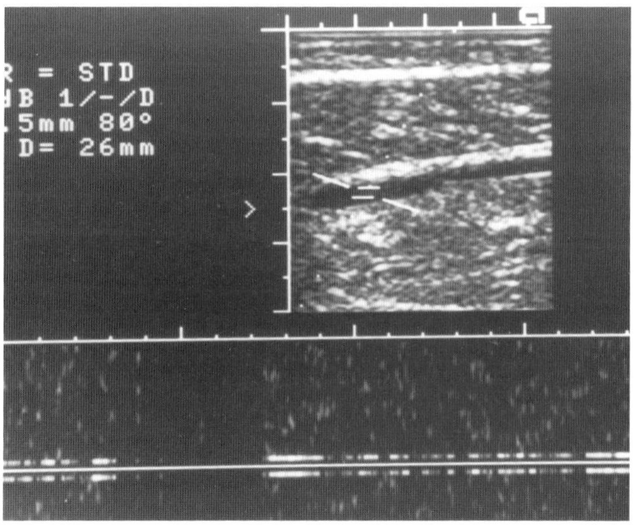


**FIG 3–54.**
Muscle-aponeurosis avulsion in the calf: Doppler study. Occasionally, a muscle-aponeurosis avulsion in the calf will appear similar to ectatic veins. In these cases, the diagnostic problem is easily solved with Doppler flow analysis. The avulsion of the soleus from the aponeurosis shown here demonstrates this problem. No Doppler signal is observed from the lesion.

tion occurs through weak areas in the fascia as a result of increased intracompartmental pressure. Often herniation occurs only during strenuous exercise and is reduced with rest. Therefore, examination should be performed immediately following strenuous exercise that reproduces the pain. Muscle hernias may be unilateral, bilateral, single, or multiple. They are frequently palpable clinically. Pain may be due to ischemia of the herniated muscle or irritation of an adjacent nerve. Neuroma formation is a rare complication.

Sonography will demonstrate the fascial defect as well as the extent of muscle herniation. The most common site of muscle hernia is the lower third of the calf overlying the anterior intermuscular septum between the anterior and lateral compartments. At this site, a branch of the superficial peroneal nerve exits through the fascia cruris (Nicholas and Hershman, 1986). Acutely, the herniated muscle will have a hyperechoic appearance due to crowding of the fibroadipose septa (Fig 3–56). However, if the hernia is not intermittent in nature, it will become hypoechoic as the involved muscle becomes edematous and eventually necrotic. The lesion will remain hypoechoic as the muscle is replaced by reparative tissue (Figs 3–57 and 3–58).

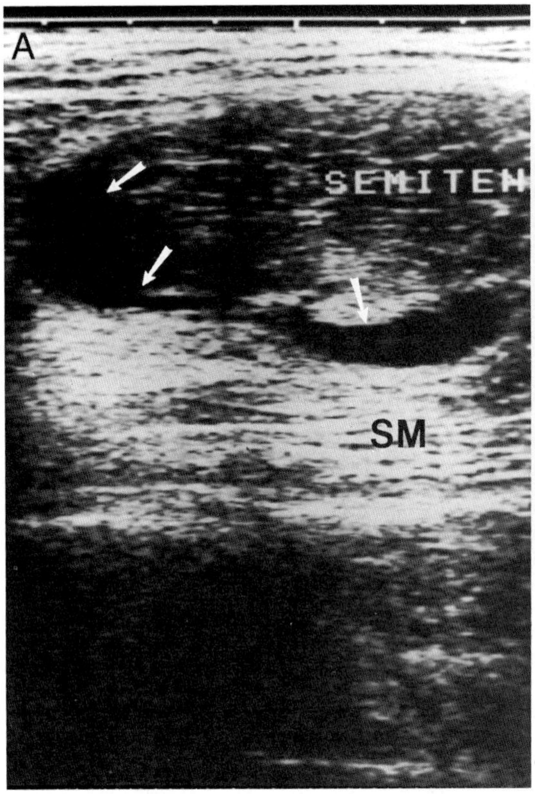


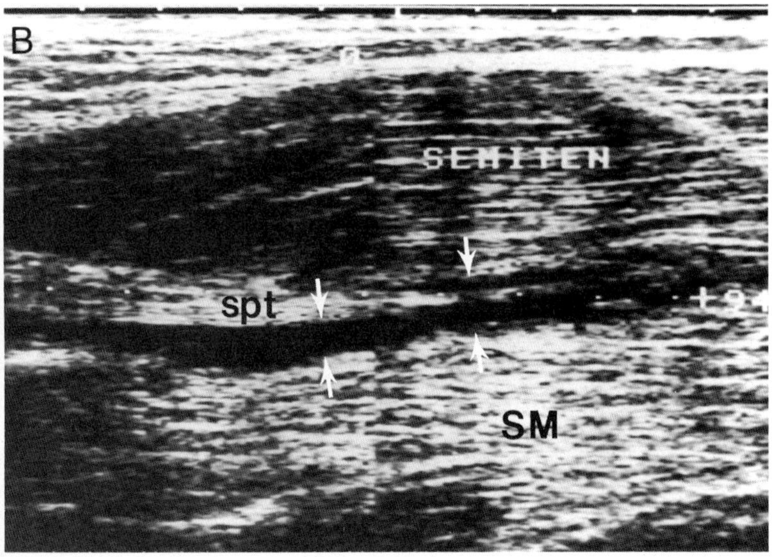


**FIG 3–55.**
**A,** semitendinosus muscle-aponeurosis avulsion: transverse sonogram. This 20-year-old soccer player was kicked in the back of the calf while flexing his knee to attempt a kick. He felt a snap in the posterior right thigh, which was accompanied by immediate function loss. At presentation 2 days later, the posterior thigh was markedly swollen. The transducer is positioned transversely over the posterior medial aspect of the upper thigh. Hematoma *(arrows)* surrounds the deep portion of the semitendinosus muscle. The avulsion has occurred between the semimembranosus *(SM)* and the semitendinosus muscles. **B,** semitendinosus muscle-aponeurosis avulsion: longitudinal sonogram. Same patient as in **A.** A platelike hematoma *(arrows)* is seen on both sides of the fibrous septum *(spt),* which separates the two muscles, the orientation of the fibroadipose septa changes on either side of the hematoma. No intramuscular rupture was noted.

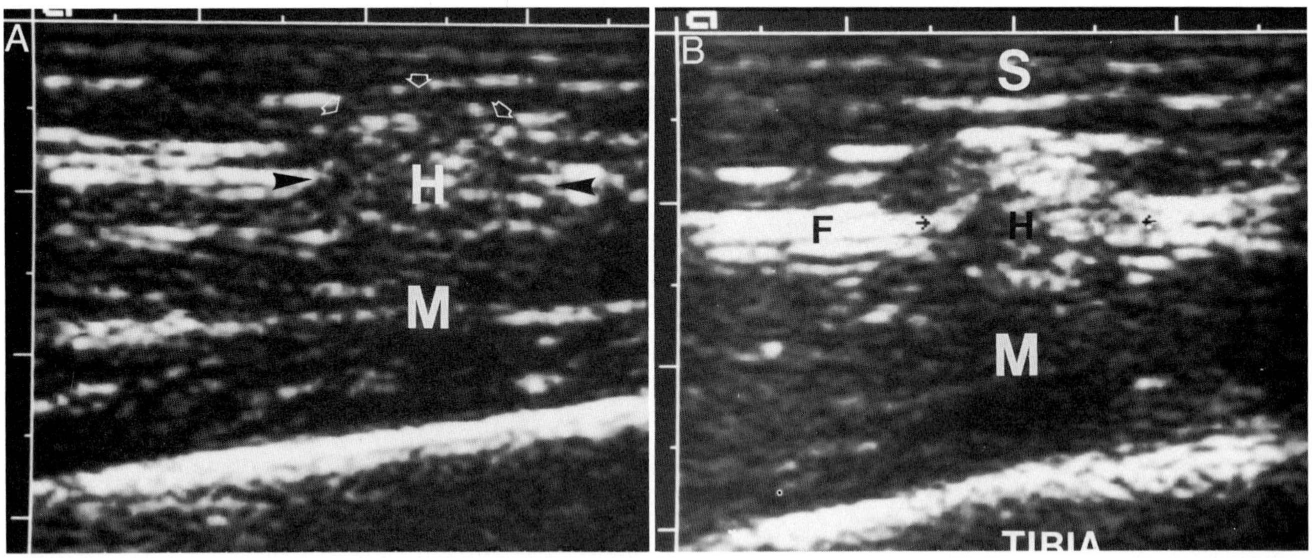


**FIG 3–56.**
**A,** small muscle hernia of the anterior tibial muscle: longitudinal sonogram. A 19-year-old soccer player experienced pain in both calves during play. He reports being able to palpate painful nodules lateral to the tibia at the end of exercise. Sonographic examination was performed immediately after the patient had run several miles. The transducer is positioned over the midportion of the anterior compartment along its length. A portion of the anterior tibial muscle *(M)* herniates through a rent in the fascia *(arrowheads)* of the anterior compartment, bulging into the subcutaneous tissues *(open arrows).* **B,** angling the transducer slightly provided better visualization of the fascial defect *(arrows).* The fascia *(F)* is seen as a hyperechoic layer 3 to 4 mm in thickness, separating the muscle compartment *(M)* from the subcutaneous tissues *(S).* Small hernias *(H)* tend to have a hyperechoic appearance compared with normal muscle.

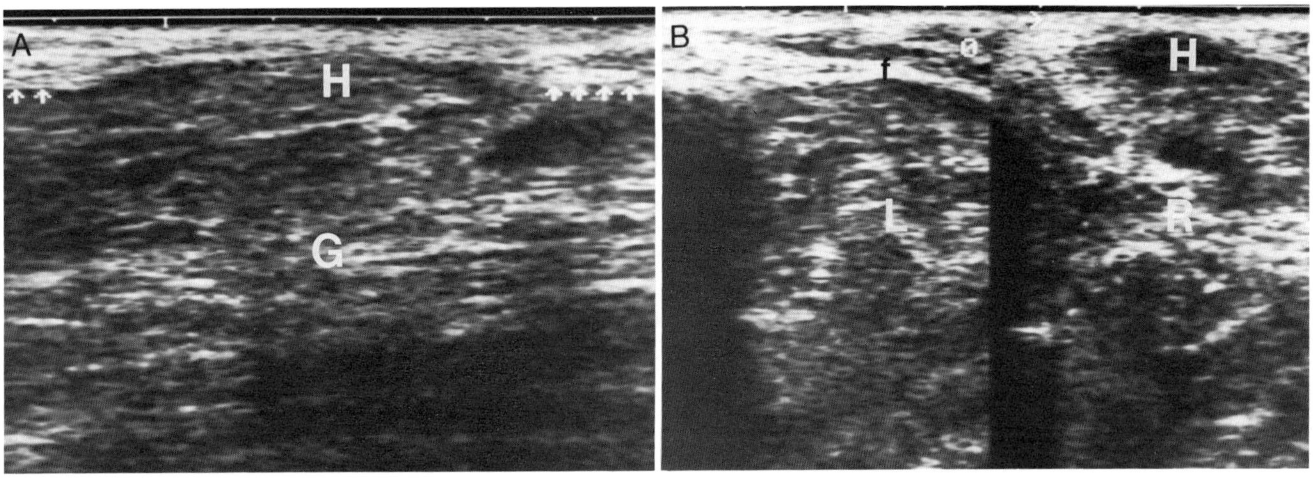


**FIG 3–57.**
**A,** large muscle hernia of the proximal gastrocnemius: longitudinal sonogram. The patient is a 25-year-old football player with chronic posterior compartment syndrome of the right leg. He experienced pain and swelling over the medial aspect of the proximal gastrocnemius, which was most severe following exercise. A 4.5-cm defect is noted in the fascia cruris with the gastrocnemius *(G)* herniating through the gap *(H)*. Normal hyperechoic fascia *(arrows)* is seen defining the extent of the lesion. **B,** large muscle hernia of the proximal gastrocnemius: transverse sonogram. Same patient as in **A.** In this split screen display, the image on the *left* demonstrates the normal left calf. Fascia *(f)* completely covers the gastrocnemius. On the *right,* the gastrocnemius protrudes through the large fascial defect. Abbreviations: left *(L)*, right *(R)*, hernia *(H)*.

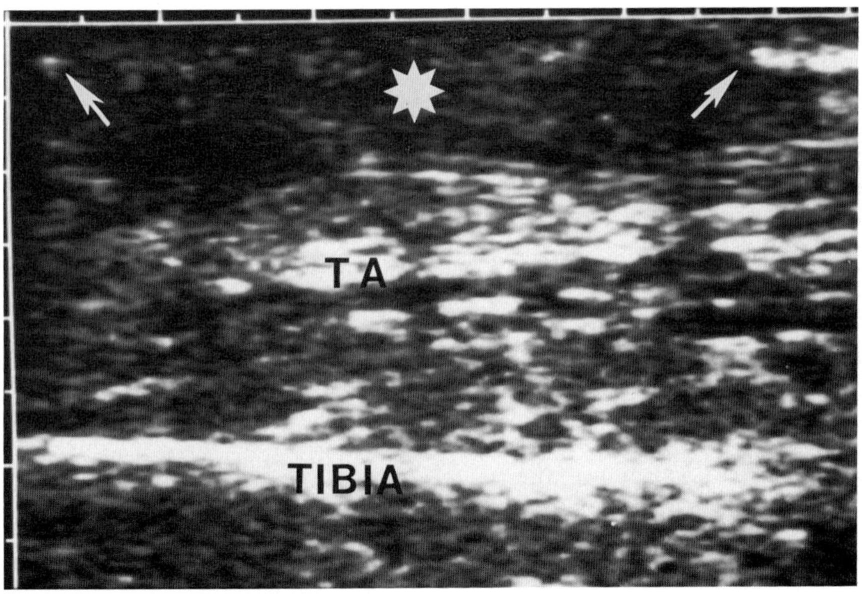


**FIG 3–58.**
Traumatic muscle hernia: longitudinal sonogram. A large fragment of glass ruptured the anterior tibial muscle after this 35-year-old woman dropped a vase. The fragment was surgically removed and the ruptured muscle repaired. Edema of the muscle prevented closure of the fascial defect. Two months following this injury, the patient presented with a complaint of limited dorsiflexion of the foot. The transducer is positioned over the middle segment of the anterior tibial muscle. A large hypoechoic segment of the muscle *(asterisk)* herniates through the remaining fascia *(arrows)*. Normal fibroadipose septa are not present in this region. During dorsiflexion of the foot, this segment was akinetic. A hypoechoic appearance of herniated muscle is rare. It can be seen in cases of muscle necrosis or, as in this case, may represent muscle undergoing regeneration following rupture. Abbreviations: anterior tibial muscle *(TA)*.

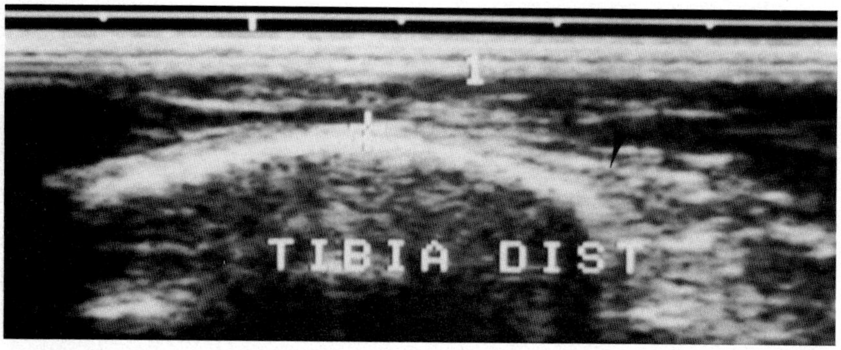


**FIG 3–59.**
Periosteum and fascia: transverse ultrasound of a specimen. This transverse image of the distal third of the tibia was obtained of an autopsy specimen. The fascia *(arrowhead)* joins the periosteum of the anterior tibia to form a hyperechoic layer *(+)* 1 mm thick overlying the anterior medial tibial cortex.

## Shin Splints

Chronic changes in fascia may also affect muscle function. Fascial thickening may result from infection of muscles, joints, or osteomyelitis. It may also be caused by trauma or exercise-induced overload. The American Medical Association standard nomenclature defines shin splints as "pain and discomfort in the legs from repetitive running on hard surfaces or forceful excessive use of the foot flexors." They suggest that the diagnosis be limited to musculotendonous inflammation, excluding fatigue fractures and ischemic disorders such as compartment syndrome. The patient typically presents with pain and tenderness lateral to the tibia and along the posteromedial border of the middle to lower tibia. They characterize the pain as a dull ache that varies in intensity. Onset of symptoms occurs after rhythmic repetitive exercise. Physical examination reveals pain on active movement of the ankle.

Fascial thickening over the facies medialis of the tibia (Figs 3–59 through 3–62) is observed on ultrasound examination. Thickening is also noted at the anchor points of the muscle compartment to the periosteum along the posteromedial and lateral borders of the tibia. Minimal focal thickening of the periosteum may be observed at these attachment sites. Although the etiology of pain in shin splints is not known, these findings suggest that the cause may be traumatic inflammation of the fascia and adjacent periosteum (Fig 3–63).

## Runner's Knee

Iliotibial band friction syndrome is another traumatic fascial lesion, also referred to as runner's knee. The syndrome is characterized by pain over the lateral aspect of the knee, particularly in the region of the lateral femoral condyle anterior to the origin of the lateral collateral ligament (Nicholas, 1986). Runners with excessive pronation of the feet and those who run on cambered roads are at increased risk. Patients with hip abduction contracture, genu varum,

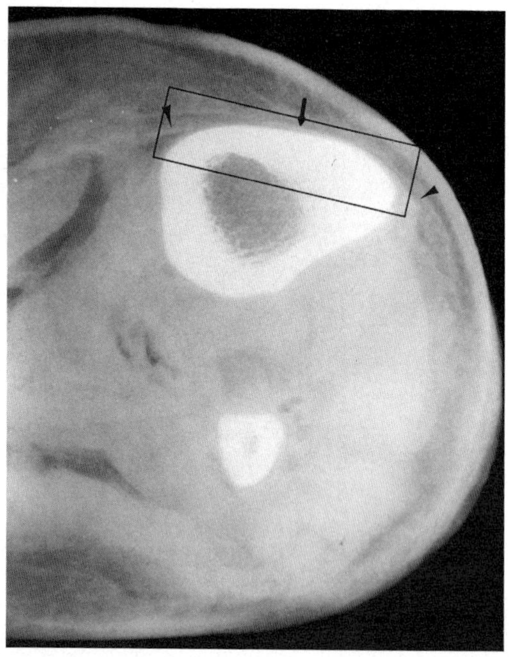

**FIG 3–60.**
Radiography of a specimen of the distal tibia: transverse section. A specimen radiograph was obtained through the same region shown in Figure 3–59. The specimen was 6 mm thick. Within the rectangular area is the region depicted in Figure 3–59. Fascia *(arrowheads)* is again seen fusing with the periosteum *(arrow)* of the anteromedial tibia, providing a firm point of anchorage.

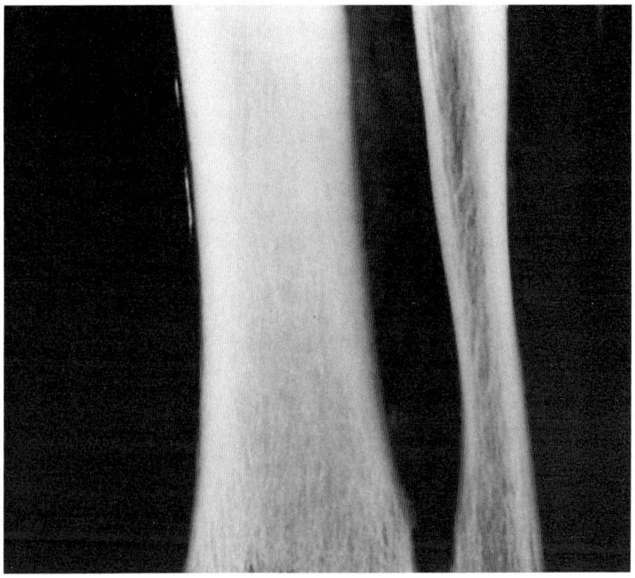

**FIG 3–61.**
Anteroposterior radiograph of a distal leg specimen. Under sonographic guidance a needle was placed under the hyperechoic fascial and periosteal layer covering the anterior medial tibia in Figure 3–59. Barium sulfate was then injected and this radiograph obtained. The barium is seen layering adjacent to the tibial cortex. Histological sections were then obtained.

tight heel cords, and internal tibial torsion are also predisposed to develop runner's knee. No clinical signs of internal derangement of the knee are present. Renne was first to recognize that this form of lateral knee pain was a distinct syndrome (Renne, 1975). He attributed it to inflammation of the synovium by movement of the iliotibial tract over a prominent lateral femoral condyle (Cravers, 1978; James et al., 1978; Noble, 1980; Sutker et al., 1981).

Fascia forming the iliotibial tract is seen as a hyperechoic lamellar structure adjacent to the lateral femoral condyle. Ultrasound examination is usually normal unless performed immediately following exercise that reproduces the pain. Following exercise the fascia is swollen and decreased in echogenicity (Fig 3–64). In some cases edema is so prominent it gives the appearance of a fluid-filled bursa (Fig 3–65). No synovial thickening is observed.

Therapy for iliotibial tract syndrome is conservative. Rest, cold compresses, and anti-inflammatory medication are the mainstays of treatment. These may be combined with therapeutic ultrasound. Local steroid injection may be of value in advanced cases. Stretching of the iliotibial tract and orthotics may be useful in preventing recurrence. A change in running habits is also beneficial. Patients should be advised to shorten their stride and alternate sides of the road while running. If these conservative measures fail, an effective release operation is available. This involves horizontal incision of the iliotibial band.

### Plantar Fascia Rupture

Long distance runners often suffer from chronic foot pain. Heel pain is frequently due to plantar fas-

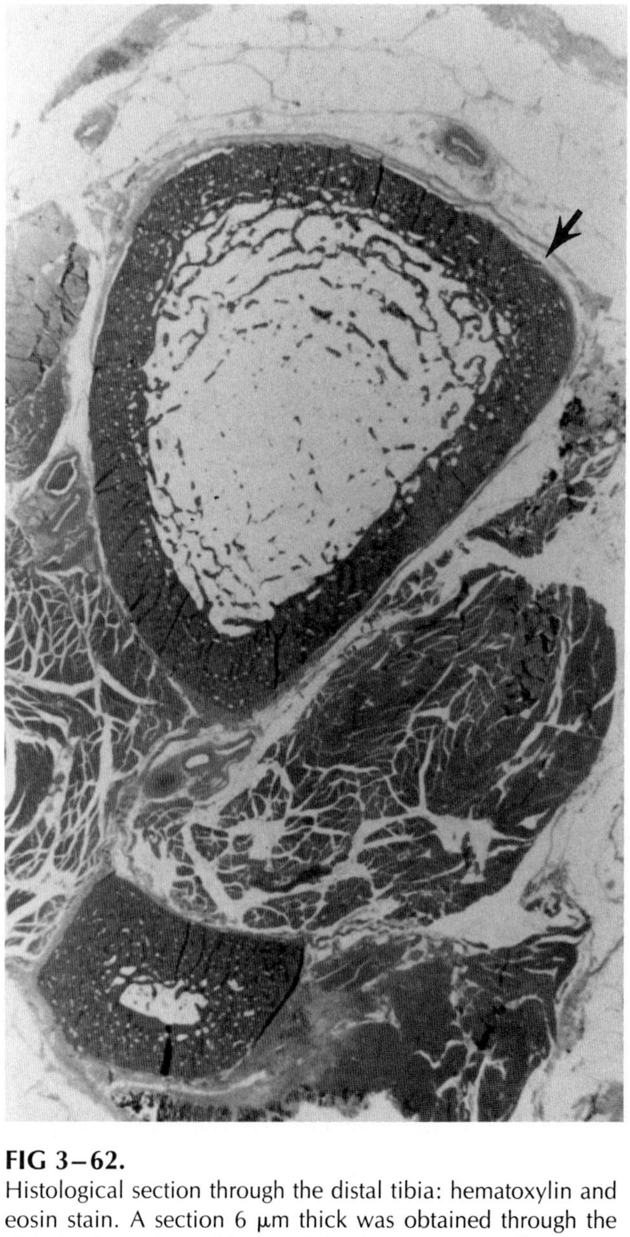

**FIG 3–62.**
Histological section through the distal tibia: hematoxylin and eosin stain. A section 6 μm thick was obtained through the tibia in the region of barium injection. Barium sulfate particles *(arrow)* are identified beneath the fascial-periosteal layer, which was dissected off of the cortex by the barium injection. This confirms the anatomical relationship demonstrated on sonographic images.

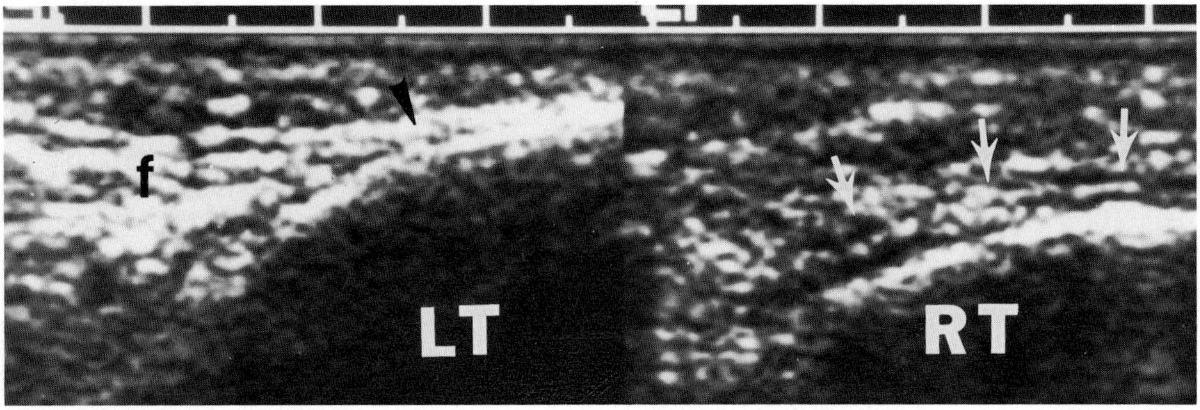


**FIG 3–63.**
Shin splint: transverse sonogram. This 20-year-old female volleyball player complained of pain over the lateral margin of the middle third of the right tibia during play. The point of maximal tenderness is established at the junction of the fascia cruris with the tibia. A transverse image over the point of maximal tenderness is shown side by side with a comparable image of the asymptomatic left leg. The image of the left side has been reversed to make these superimposable. On the right, the fascia-tibia junction *(white arrows)* is thickened and hypoechoic relative to the normal left side *(arrowhead)*. Also note that the subcutaneous tissues overlying the abnormal fascia-tibia junction are thickened as well. Abbreviations: normal fascia *(f)*.

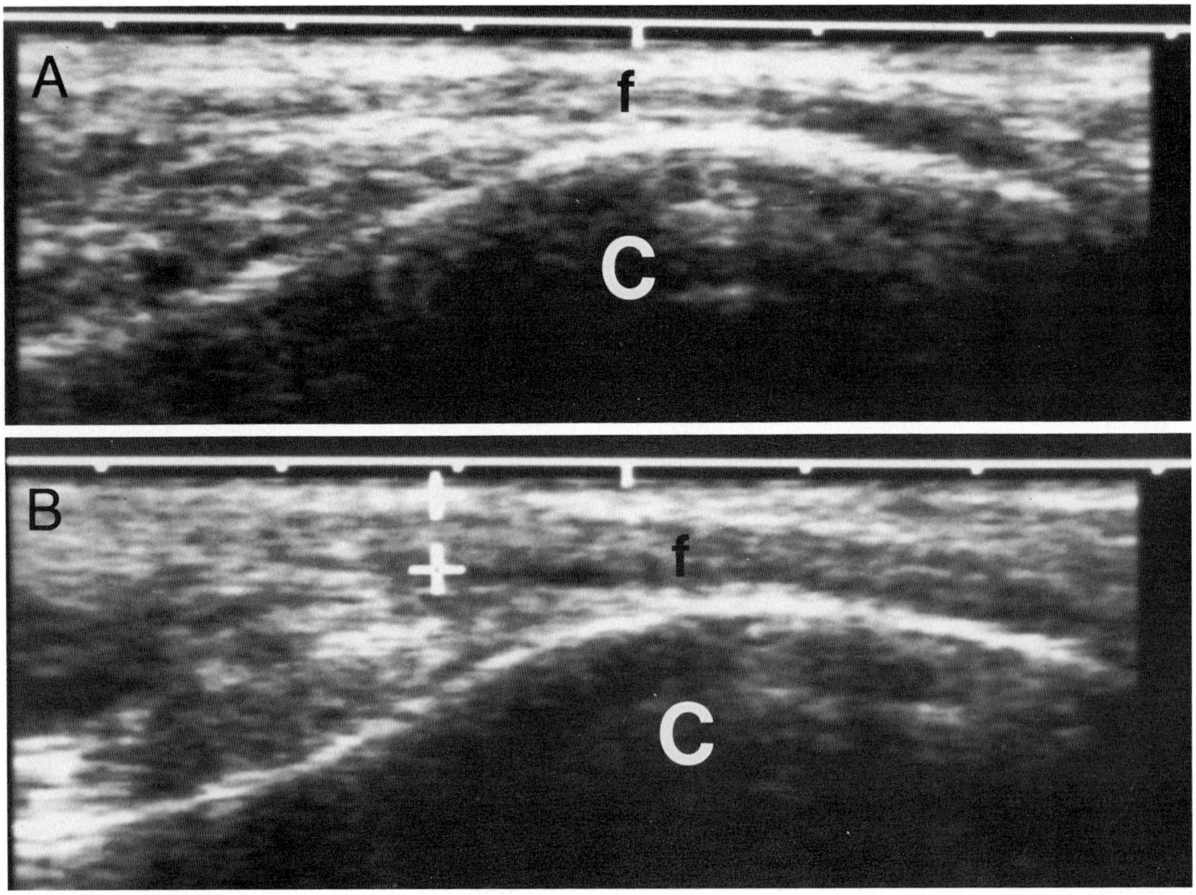


**FIG 3–64.**
Runner's knee: coronal sonogram. This 28-year-old jogger increased his exercise routine to 10 miles daily from 5 miles. After several weeks he developed pain over the lateral aspect of the knee, which forced him to stop after 9 miles. Pain recurred after 1 week of rest. This coronal image was obtained over the lateral femoral condyle anterior to the lateral collateral ligament prior to exercise. Fascia *(f)* overlying the lateral femoral condyle *(C)* appears hyperechoic, particularly the peripheral margins. **B,** runner's knee: coronal sonogram after exercise. Same patient and transducer position as in **A.** Immediately after running to his limit, the fascia *(f)* is thickened and hypoechoic. Abbreviations: lateral femoral condyle *(C)*.

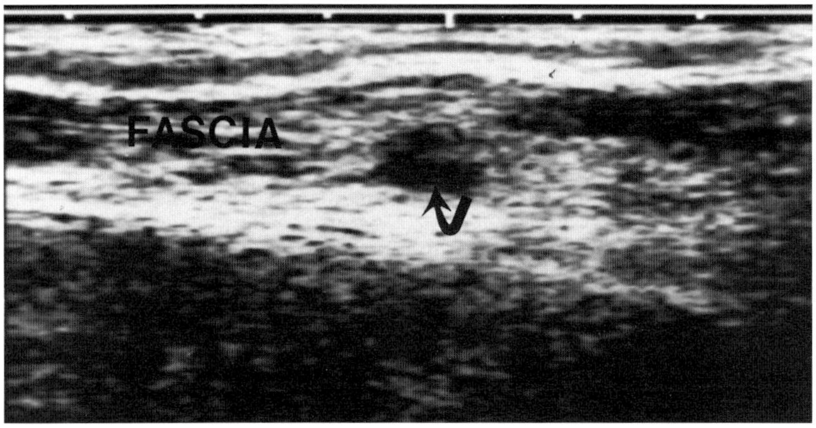


**FIG 3–65.**
Runner's knee: coronal sonogram. This 36-year-old male marathon runner was unable to finish his last two races due to pain over the lateral aspect of the left knee. At the end of a run, he had pain and swelling in this region. He presented for ultrasound examination after exercising until pain developed. The fascia is thickened and hypoechoic with a bursa-like structure seen deep to the fascia *(curved arrow)*.

ciitis, which is accompanied by a plantar spur (Resnick and Niwayama, 1988). This is clearly visible on conventional radiographs. Generalized and more severe pain may be the result of rupture of the plantar fascia (Peterson and Renström, 1986; Nicholas and Hershman, 1986). These lesions tend to heal poorly with abundant granulation tissue. Casts and special taping are usually not helpful. Surgical repair is often required; therefore, the diagnosis must be certain.

The majority of ruptures are found at the middle to posterior third of the fascia. Ultrasound usually demonstrates a hypoechoic thick fusiform lesion between the parallel fascicles of the plantar fascia or aponeurosis (Fig 3–66). Rarely discontinuity of the fascia will be observed because the patient usually presents well after rupture occurs. Sonography is essential to exclude bursitis of the bursa subcutanea calcanea (see Fig. 15–27).

## SUMMARY

Ultrasound is a reliable means of diagnosing intramuscular and muscle boundary lesions. Its real-time capability is unique in providing a means to evaluate structures under dynamic conditions. This allows diagnosis of lesions that would otherwise remain occult. Serial follow-up examinations provide valuable information about healing and prognosis for recurrent injury. These factors, as well as cost, make ultrasound the examination of choice for the evaluation of muscle pathology.

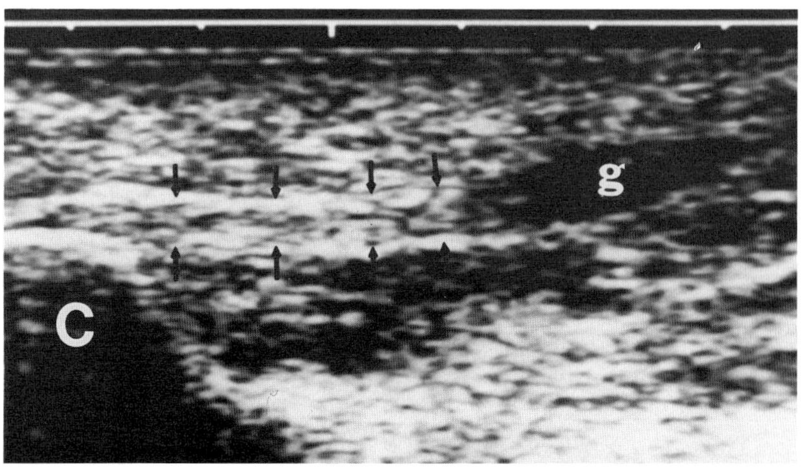


**FIG 3–66.**
Plantar fascia rupture: longitudinal sonogram. A marathon runner with intractable pain in the sole of the right foot presents for sonographic examination. The transducer is aligned along the length of the foot, touching the lateral tuberosity of the calcaneus *(C)* at the observer's left. The plantar fascia *(arrows)* is seen as a hyperechoic band attached to the lateral tuberosity. A gap *(g)* in the fascia is noted in its midportion. At surgery 3 weeks later, a discontinuity filled with granulation tissue was found.

## REFERENCES


Abiri MM, Kirpekar M, Ablow RC: Osteomyelitis: Detection with ultrasound. *Radiology* 1988; 169:795–797.

Anderson JE: *Grant's Atlas of Anatomy.* Baltimore, Williams & Wilkins Co, 1983.

Auerbach DN, Bowen A III: Sonography of the leg in posterior compartment syndrome. *AJR* 1981; 136:407–408.

Bouvier JF, Chassain AP, Veyriras E: Place de l'echotomographie dans les accident musculaires des membres inférieurs du footballeur. *Cinesiologie* 1982; 21:274–278.

Brahim F, Zaccardelli W: Ultrasound measurement of the anterior leg compartment. *Am J Sports Med* 1986; 14:300–302.

Bullough PG, Vigorita VJ: *Atlas of Orthopedic Pathology.* New York, Gower Medical Publishing, 1984.

Clemente CD: *Gray's Anatomy.* Philadelphia, Lea & Febiger, 1985.

Cravers S: Iliotibial tract friction syndrome in athletes. *Br J Sports Med* 1978; 12:69.

Fornage BD, Touche DH, Segal P, et al: Ultrasonography in the evaluation of muscular trauma. *J Ultrasound Med* 1983; 2:549–554.

Fornage BD: Ultrasonography of muscles and tendons, in *Examination Technique and Atlas of Normal Anatomy of the Extremities.* New York, Springer-Verlag, 1989.

Garfin SR, Tipton CM, Mubarak SJ, et al: Role of fascia in maintenance of muscle tension and pressure. *J Appl Physiol* 1981; 51:317.

Geller SA: Extreme exertion rhabdomyolysis. *Hum Pathol* 1973; 4:241–245.

Genety J, Brunet-Guedj E, Moyen B, et al: Pathologie musculaire du sportif. *La Vie Medicale* 1980; 1:1607–1612.

Gershuni DH, Gosink BB, Hargens AR, et al: Ultrasound evaluation of the anterior musculo-fascial compartment of the leg following exercise. *Clin Orthop* 1982; 167:185–190.

Gompels BM, Darlington LG: Gray scale ultrasonography and arthrography in the evaluation of popliteal cysts. *Clin Radiol* 1979; 30:539–545.

Graf R, Schuler P: *Sonographie am Stütz und Bewegungsapparat bei Erwachsenen und Kindern.* Weinheim, Edition Medizin VCH, 1988.

Guyton AC: *Textbook of Medical Physiology,* ed 6. Philadelphia, WB Saunders Co, 1982.

Harcke HT, Grissom LE, Finkelstein MS: Evaluation of the musculoskeletal system with sonography. *AJR* 1988; 150:1253–1261.

James SL, Bates BT, Osternig LR, et al: Injuries to runners. *Am J Sports Med* 1978; 6:40.

Kaplan GN: Ultrasonic appearance of rhabdomyolysis. *AJR* 1980; 134:375–377.

Knochel JP: Rhabdomyolysis and myoglobinuria. *Semin Nephrol* 1981; 1:75.

Kramer FL, Kuttz AB, Rubin C, et al: Ultrasound appearance of myositis ossificans. *Skeletal Radiol* 1979; 4:19–20.

Lawson TL, Mittler S: Ultrasonic evaluation of extremity soft tissue lesions with arthrographic correlation. *J Assoc Can Radiol* 1978; 29:58–61.

Lehto M, Alanen A: Healing of muscle trauma. Correlation of sonographical and histologic findings in an experimental study in rats. *J Ultrasound Med* 1987; 6:425–429.

Lenkinski RE, Holland GA, Allman T, et al: Integrated MR imaging and spectroscopy with chemical shift imaging of P-31 at 1.5T. *Radiology* 1988; 169:201–206.

Magnussen PA, Crozier AE, Gregg PJ: Detecting hematomas by ultrasound. *J Bone Joint Surg* 1988; 70(B):150.

Martens MA, Backaert M, Vermaut G, et al: Chronic leg pain in athletes due to recurrent compartment syndrome. *Am J Sports Med* 1984; 12:148.

Moore K: *Clinically Oriented Anatomy.* Baltimore, Williams & Wilkins Co, 1980.

Mubarak SJ: *Compartment Syndromes and Volkmann's Contracture.* Philadelphia, WB Saunders Co, 1981.

Nicholas JA, Hershman EB: *The Lower Extremity and Spine in Sports Medicine.* St Louis, CV Mosby Co, 1986.

Noble CA: Iliotibial band friction syndrome in runners. *Am J Sports Med* 1980; 8:232.

Peetrons P, Sintzoff S: Les accidents du membre inférieur chez les sportifs: Intégration des differents modes d'imagerie (abstract). *J Francophone Radiol,* November 1987.

Peterson L, Renström P: *Sports Injuries.* Chicago, Year Book Medical Publishers, 1986.

Renne JW: The iliotibial band friction syndrome. *J Bone Joint Surg* 1975; 57(A):1110.

Resnick D, Niwayama G: *Diagnosis of Bone and Joint Disorders.* Philadelphia, WB Saunders Co, 1988.

Roth D, et al: Acute rhabdomyolysis associated with cocaine intoxication. *N Engl J Med* 1988; 319:673–677.

Sutker AN, Barber FA, Jackson DW, et al: Iliotibial band friction syndrome in distance runners. *Physiol Sports Med* 1981; 9:69–73.

van Sonnenberg E, Wittich GR, Casola G, et al: Sonography of thigh abscess: Detection, diagnosis and drainage. *AJR* 1987; 149:769.

Vukanovic S, Hauser H, Curati WL: Myonecrosis induced by drug overdose: Pathogenesis, clinical aspects and radiological manifestations. *Eur J Radiol* 1983; 3:314–318.

Wagner P, Domergue A, Poirier C, et al: L'Echotomographie dans les accidents musculaires du sportif. *Ultrasons* 1980; 1:277–287.

Weissman BNW, Sledge CB: *Orthopedic Radiology.* Philadelphia, WB Saunders Co, 1986.

Zochmodne DW, Thompson RT, Driedger AA, et al: Metabolic changes in human muscle denervation: Topical $^{31}$P NMR spectroscopy studies. *Magn Reson Med* 1988; 7:373–383.

Zuinen C, Carlier L, Gaudissart JL: L'Echotomographie en Traumatologie Musculaire. *Med Sport* 1980; 54:379–382.

Zuinen C, Commandre F, Valdener V, et al: Echotomographie du Muscle. *Med Sport* 1982; 56:396–404.

# 4

# Sonography of Tendons

In this chapter, we will discuss the examination of tendons, with and without synovial sheaths, using ultrasound. Prior to the application of ultrasound to the evaluation of the musculoskeletal system, clinicians relied on low-kilovoltage radiography and xeroradiography to aid in the diagnosis of tenomuscular injury (Bock et al., 1981). These techniques provided little information beyond indicating the site of soft tissue swelling. Ultrasound now provides detailed information about the involved anatomy and nature of the pathology.

A synovial sheath is a tubular sac wrapped around the tendon (Warwick and Williams, 1980). The inner layer is tightly applied to the tendon and has a smooth inner surface juxtaposed with the outer layer. A minute amount of viscous mucoid fluid separates the two layers. Synovial sheaths are found in areas of stress in the hand, wrist, and ankle. These synovial sheaths facilitate the examination of individual tendons sonographically. In the shoulder, the long biceps tendon is an example of a tendon with a synovial sheath that plays an important role in shoulder disease.

Tendons without a synovial sheath are slightly more difficult to examine using ultrasound. They are enveloped by loose areolar connective tissue called the *paratenon* (Fornage et al., 1984; Fornage, 1986). It is possible to examine long tendons, especially if they are surrounded by fat. The Achilles, patellar, proximal gastrocnemius, and semimembranous tendons are a few examples. All tendinous insertions, including those hidden by bony structures and short tendons, may be examined by the experienced musculoskeletal ultrasonographer. In these cases, side to side comparison is essential (Fig 4–1).

## EXAMINATION TECHNIQUE

### Equipment

The evolution of musculoskeletal ultrasound was delayed, in part, by limitations imposed by equipment. Examination of tendons requires relatively high spatial and contrast resolution. This is achieved through the use of high-frequency transducers (5–10 MHz). Axial resolution increases with transducer frequency. However, a tradeoff between resolution and depth of field exists. Penetration of sound through soft tissues decreases as frequency increases. The depth of field may be limited to only 3 to 4 cm using a 12-MHz transducer. Selection of an appropriate frequency transducer is dependent on the region to be examined and the body habitus of the patient. When superficial tendons are examined, a standoff pad or a bag of intravenous (IV) fluid can be used to position the tendon of interest in optimal position within the transducer's field of view (Fornage et al., 1984).

Dynamic real-time imaging is needed to examine tendons throughout their physiological range of motion. Sector scanners have the disadvantages of a narrow near field of view, near field unsharpness, and obliquity of the sound beam when tendons are examined in their longitudinal axis. Linear array transducers provide optimal images, having their sound beam perpendicular to the tendon throughout the

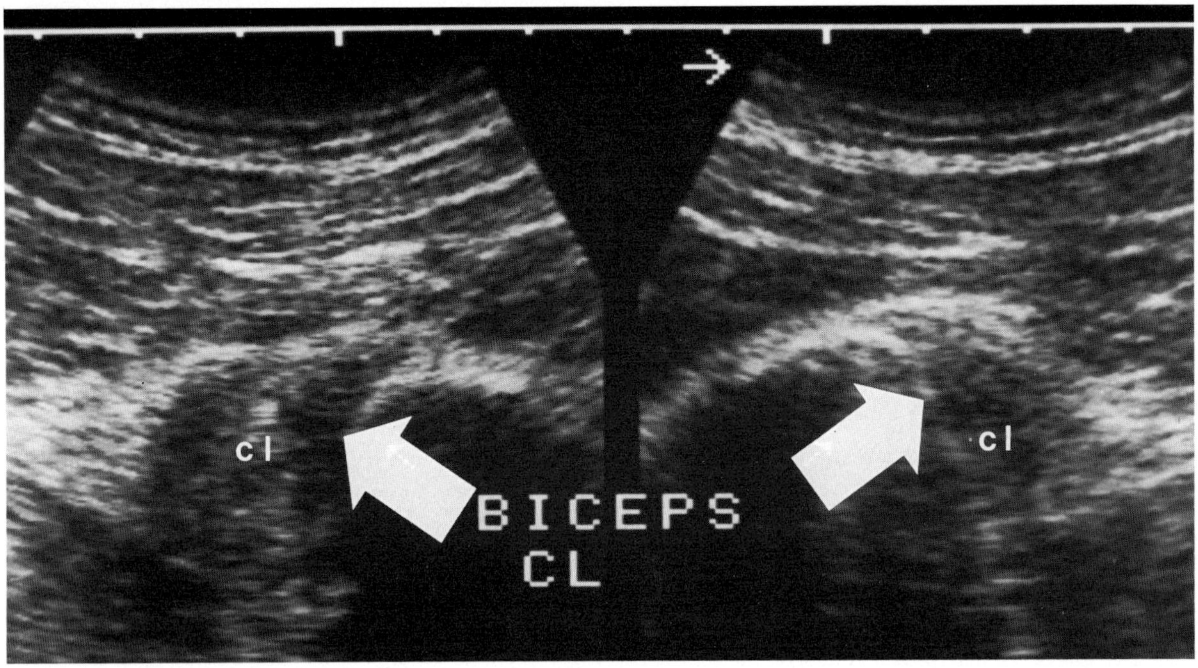


**FIG 4–1.**
Tendonitis of the hamstrings: transverse sonogram. The patient is a long distance runner with persistent pain in the right buttock. The patient experienced pain at the beginning of exercise, which then disappeared, only to recur later in the workout. Clinical examination revealed local tenderness over the ischial tuberosity. The ultrasound transducer is positioned transversely over the gluteal fold and the ischial tuberosity. Side to side comparison is made of the long head of the biceps *(CL)* at its insertion on the ischial tuberosity. The diameter of the right proximal hamstring insertion at the observer's left is nearly twice the diameter of the asymptomatic side *(white arrows)*. In addition, the right insertion is also hypoechoic relative to the normal side. Clearly, side to side comparison is necessary to detect the lesion.

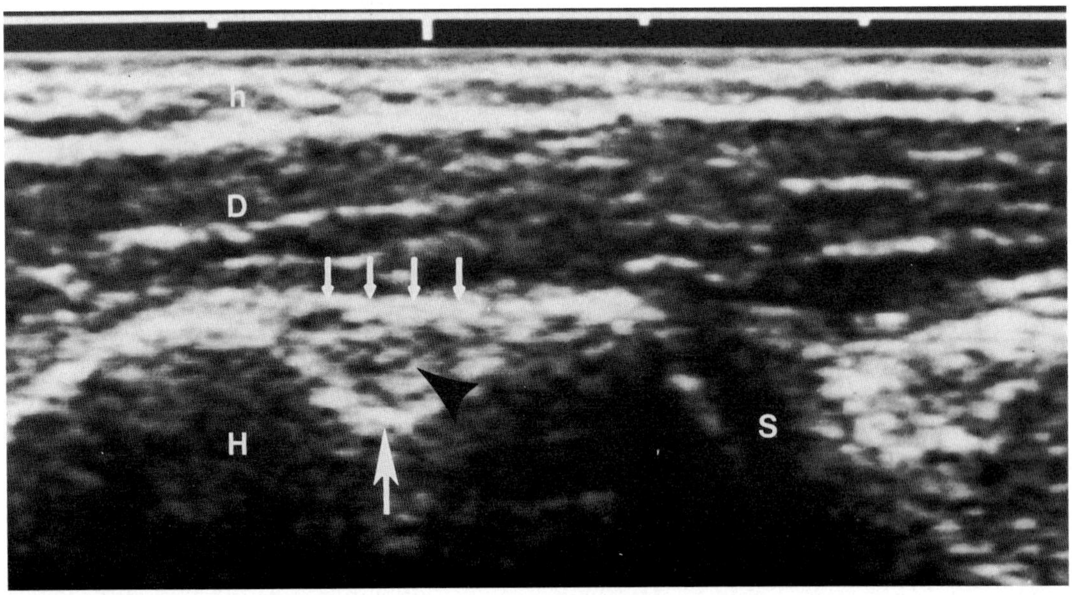


**FIG 4–2.**
Normal biceps tendon: transverse sonogram. Examination of an asymptomatic young male. This image was obtained with the right shoulder in the neutral position. The medial aspect is at the observers' right. The bicipital groove *(large arrow)* is identified at the anterior aspect of the humeral head. The transducer is positioned perpendicular to the long axis of the humerus over the proximal metaphysis. The long biceps tendon *(arrowhead)* is located in the middle of the bicipital sulcus and has a hyperechoic appearance. A small sonolucent area separates the tendon from the transverse ligament *(small arrows)*. This represents a normal amount of fluid in the synovial sheath. Abbreviations: humeral head *(H)*, hypoderm *(h)*, deltoid muscle *(D)*, subscapularis muscle *(S)*.

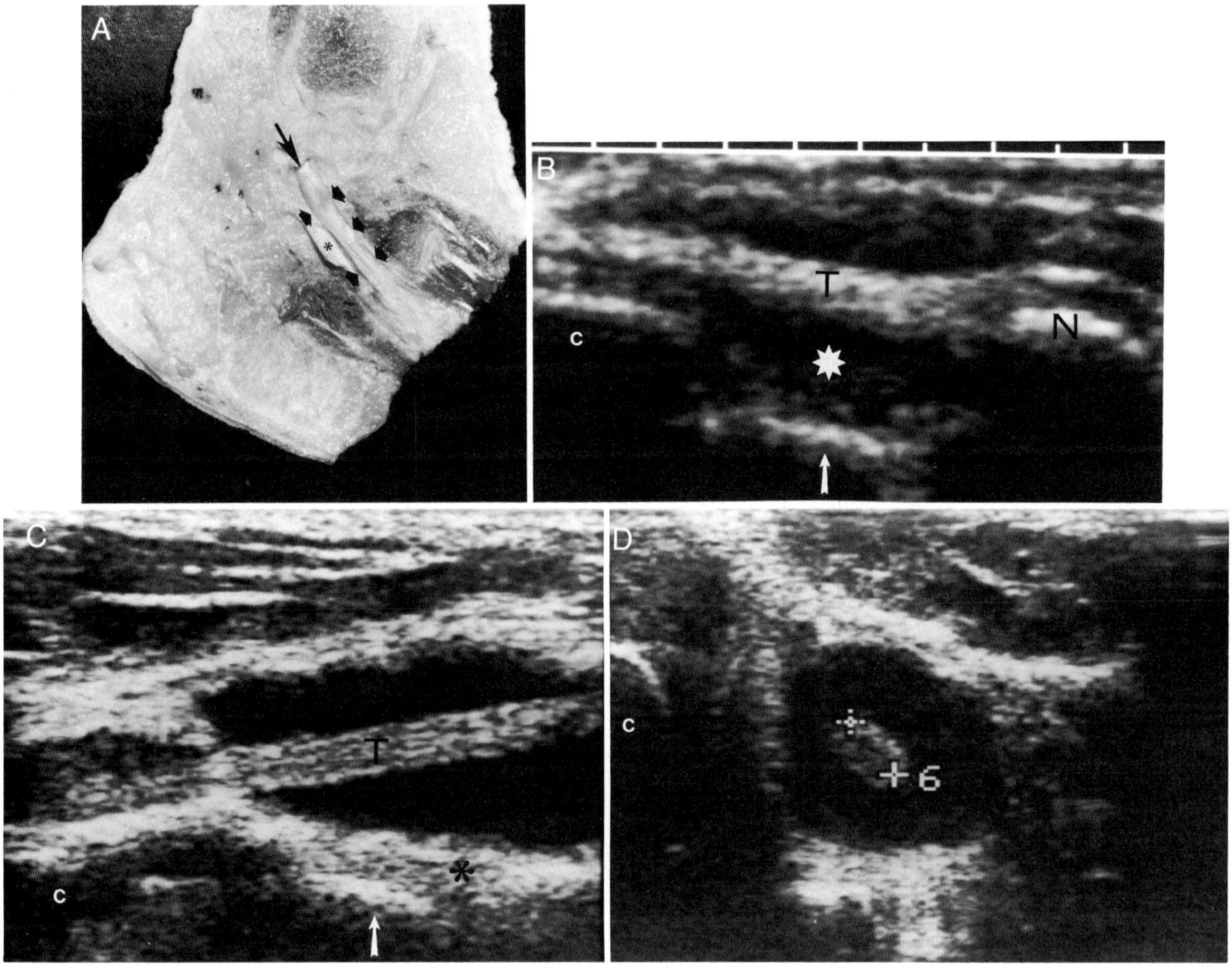


**FIG 4–3.**

**A,** anatomic section through the medial aspect of the ankle that bisects the tibialis posterior tendon sagittally. The tibialis posterior tendon *(arrow)* has a separate tendon sheath. A normal tendon sheath has a very small quantity of synovial fluid. This is seen as a membrane wrapped around the tendon *(small arrows)*. A small part of the flexor digitorum longus tendon is seen posterior to the tibialis posterior tendon *(asterisk)*. **B,** normal tibialis posterior tendon: longitudinal sonogram. The position of the transducer is at the medial plantar surface of the foot, overlying the talonavicular region. It is oriented along the long axis of the foot with the distal portion of the foot at the observer's right. The tibialis posterior tendon *(T)* is adjacent to the sustentaculum tali *(white arrow)* and inserts on the navicular bone *(N)*. Normal fat surrounding the tendon appears hypoechoic *(asterisk)*. Abbreviations: calcaneus *(c)*. **C,** longitudinal sonogram through an inflamed tendon sheath of the tibialis posterior tendon. The patient is a soccer player with pain and swelling over the medial aspect of the foot. Pain was localized to the region overlying the tibialis posterior tendon with maximal tenderness on palpation focused over the sustentaculum tali *(white arrow)*. Transducer position and orientation are identical to that in **B.** The tibialis posterior tendon is surrounded by a large amount of fluid in the tendon sheath. Individual fibers are visualized more clearly due to this pathological fluid collection. The surrounding fat is now hyperreflective *(asterisk)*. Abbreviations: calcaneus *(c)*. **D,** transverse image through the inflamed tibialis posterior tendon. Same patient as in **C.** The transducer is rotated 90 degrees and is now perpendicular to the long axis of the foot. The tendon sheath thickness exceeds the diameter of the tendon itself (greater than 6 mm). Abbreviations: calcaneus *(c)*.

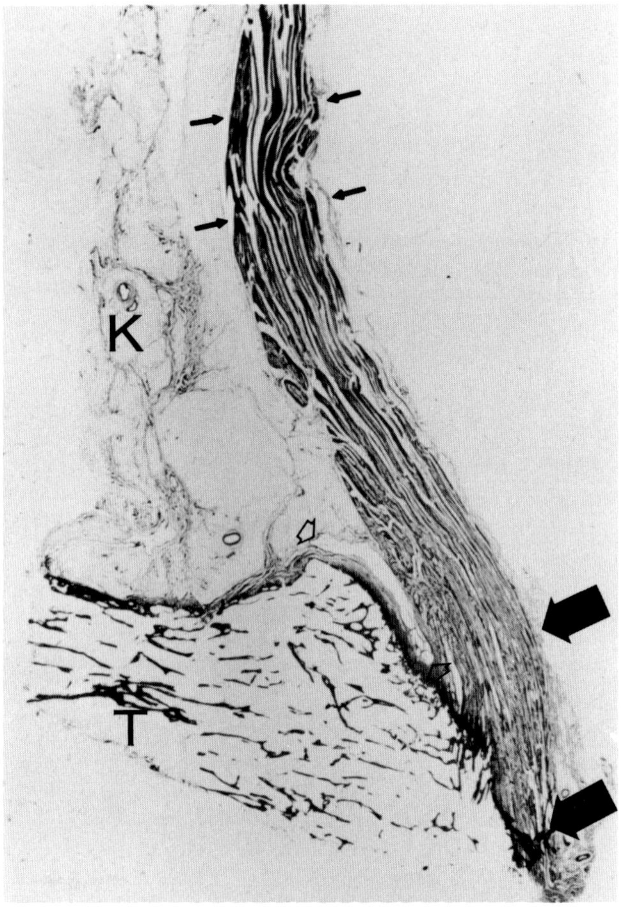


**FIG 4–4.**
Photomicrograph (×7) of an anatomical specimen of the Achilles tendon insertion: sagittal section. In the distal 1 cm of the tendon *(area between the black arrows)* the fibers are more densely arranged, and cartilage cells are found at higher magnification. Between the tendon and the superoposterior angle of the calcaneus, a small boomerang-shaped bursa is present *(open arrows).* The wall of the bursa is microscopically thin, but the lumen is relatively large. In vivo the bursa is filled with a small amount of serous fluid. Around the tendon there is a paper thin rim of connective tissue known as the peritenon *(small arrows).* The dorsal aspect of the peritenon is thicker than the ventral. Achilles tendon fat pad, Kager's fat pad *(K),* trabecular bone of the calcaneus *(T).* (Courtesy of Dr. E. Verbeken, K.U. Leuven, Belgium.)

imaging field. In our experience, state of the art ultrasound machines with high-frequency linear array transducers (5, 7.5, and 10 MHz) provide the best combination of field of view, resolution, and depth of field with electronic focusing of the sound beam. Using the transducer with the highest frequency that will penetrate to the desired depth obviates the need to change to a higher-frequency transducer in the middle of an examination, thus reducing the examination time.

### Scanning Technique

The examination is started by imaging in the transverse plane. This helps to orient the examiner and direct scanning in the longitudinal axis. The location of the tendon is established relative to easily recognized bony landmarks (Fig 4–2). In the majority of cases, scanning is done primarily in the longitudinal axis of the tendon. Longitudinal images allow the examiner to better distinguish separate tendons. Exceptions to this are the proximal insertion of the long biceps tendon in both the upper and lower extremities. These tendons are difficult to detect when scanning in the longitudinal axis. In these cases transverse imaging is primarily used, and careful side to side comparison is made.

## SONOGRAPHIC TENDON ANATOMY

Tendons are composed largely of parallel running fascicles of collagen fibers that interweave and interconnect (Warwick and Williams, 1980; Dillehay et al., 1984). These fascicles are large enough to be visible to the naked eye as well as sonographically (Fornage et al., 1984; Fornage, 1986). During postnatal growth, tendons increase in length interstitially (Warwick and Williams, 1980). At the musculotendinous junction, growth is especially rapid due to the increased cellularity in this region, which allows for rapid elaboration and maturation of collagen. Tendon growth is slowest at the attachment of the tendon to bone. Vascular networks within the tendon are sparse. Therefore, minimal bleeding is seen sonographically after tendon rupture (Zuinen et al., 1983).

Surrounding the tendon is either a synovial sheath or a dense connective tissue layer called the *epitendineum.* The synovial sheath, as previously described, contains a capillary film of fluid that serves as a lubricant. This fluid is seen as a dark halo surrounding the tendon on ultrasound images (Fig 4–3). The thickness of the sheath usually does not exceed a couple of millimeters. However, in acute tendon disease it may exceed the thickness of the tendon that it surrounds (see Fig 4–3,D).

In tendons without a synovial sheath (Fig 4–4), the epitendineum, a dense connective tissue layer, is tightly bound to the tendon. Connective tissue fibers permeate the fascicles, making the epitendineum adherent to the tendon. Blood vessels and nerves enter

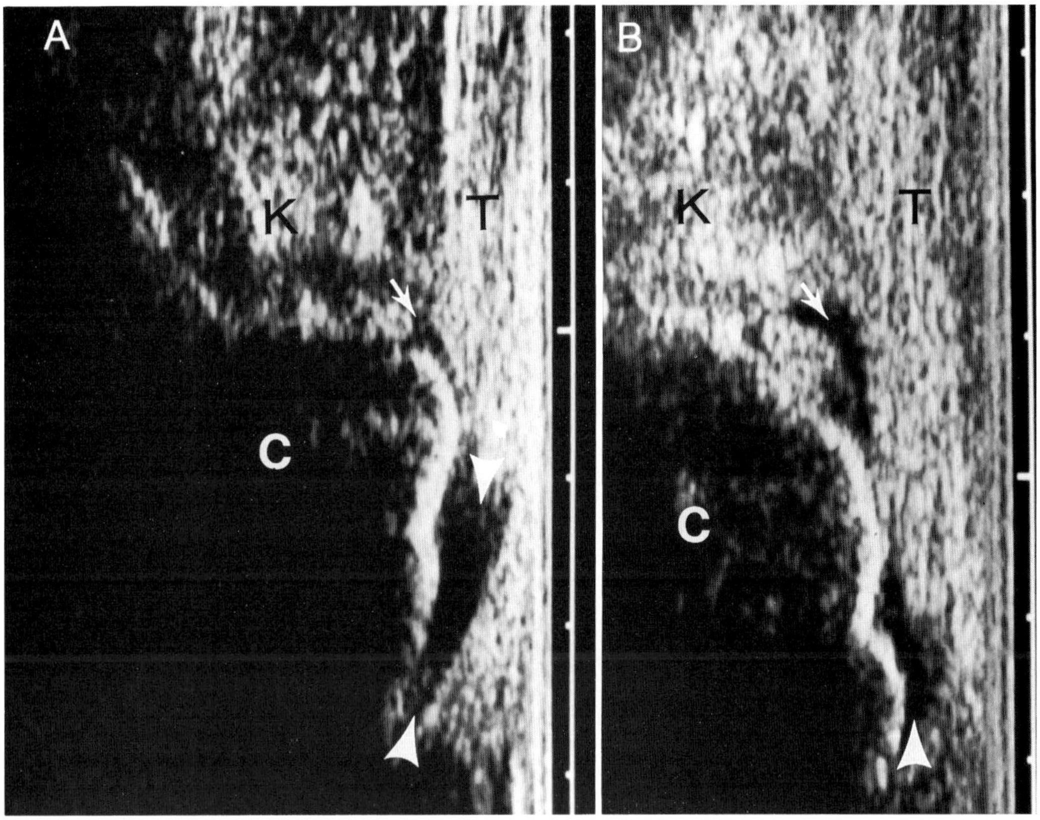


**FIG 4–5.**
**A,** Achilles tendon insertion. Longitudinal sonogram of the specimen in Figure 4–4 prior to sectioning with identical orientation. The sonographic architecture of the tendon *(T)* changes in the distal portion at its insertion *(white arrows)*. The insertion *(arrowheads)* is markedly hypoechoic, nearly anechoic. This region is approximately 1 cm long and consists of densely packed connective tissue with a structure similar to fibrocartilage (see Fig 4–4). This hypoechoic appearance is similar to that seen in cartilage elsewhere in the body. A normal retrocalcaneal bursa is seen as a hypoechoic boomerang-shaped cleft approximately 5 mm in craniocaudal dimension *(small white arrow)*. Abbreviations: Kager's fat pad *(K)*, calcaneus *(C)*, Achilles tendon *(T)*. **B,** retrocalcaneal bursitis: longitudinal sonogram. The orientation is identical to that in **A.** This is a patient with pain in the heel after a direct blow to the posterior surface of the Achilles tendon. There is tenderness localized over the area immediately proximal to the tendon insertion. The retrocalcaneal bursa *(arrow)* is enlarged. Bursitis, with production of nonviscous fluid, in an athlete is called *frictional bursitis*. The tendon insertion *(arrowhead)* is considerably shorter than the insertion seen in **A.** Variability in the length of the tendon insertion is great, but there is a strong correlation with the patient's height. Abbreviations: Kager's fat pad *(K)*, Achilles tendon *(T)*, calcaneus *(C)*.

the tendon along these fibers. Loose areolar connective tissue, called the *paratenon,* envelops the epitendineum. Sonographically, the epitendineum is seen as a reflective line surrounding the tendon (Fornage, 1986) (Fig 4–5; see Fig. 4–13).

The cross-sectional profile of a tendon is round (long biceps tendon; see Fig 4–2), oval (Achilles tendon; see Fig 5–6,B), or rectangular (patellar tendon; see Fig 4–18). This cross-sectional profile can change after athletic conditioning. A round Achilles tendon is found in nonconditioned subjects, and a more ovoid Achilles tendon is found in athletes.

A narrow band of fibrocartilage joins the tendon to bone (Cooper and Misol, 1970; Niepel and Sit AJ, 1979). This is referred to as the *tendon insertion.* The tendon insertion is an avascular structure. For example, the insertion of the Achilles tendon is approximately 1 cm long. The sonographic appearance is a relatively well-defined hypoechoic zone in the distal portion of the tendon (see Fig 4–5). It has a triangular shape when scanned longitudinally. The hypoechoic character of this fibrocartilaginous insertion is similar to the sonographic appearance of cartilage elsewhere in the body (Aisen et al., 1984). Thickening of the tendon insertion is seen in traumatic disease (Merkel et al., 1982).

## SONOGRAPHIC DIAGNOSIS OF TENDON PATHOLOGY

### Tendons With a Synovial Sheath

#### *Tendonitis*

Acute tendonitis is detected by increased fluid within the synovial sheath. The increased fluid within the sheath is seen as an anechoic halo around the tendon on transverse images (Figs 4–3,D and 4–6). This differs from the normal synovial sheath in which the capillary fluid film appears as a hypoechoic halo (Fig 4–7). On longitudinal views, the tendon is bounded on both sides by anechoic lines (Fig 4–8). Often the width of this fluid collection exceeds the diameter of the tendon that it surrounds.

In subacute cases of tendonitis, tendon thickening can be observed. We have seen this most commonly in the long biceps tendon of swimmers with a shallow bicipital groove (Fig 4–9). In these patients, tendon dislocation may be seen (see Fig 4–9,C).

Chronic tendonitis is often difficult to diagnose sonographically. Frequently there is not an increase in synovial fluid. The most common finding is thickening of the tendon itself (Middleton et al., 1985b, 1986a). Comparison with the asymptomatic side is essential to make the diagnosis of chronic tendonitis (Bruce et al., 1982; Crass et al., 1984, 1986; Demarais et al., 1984; Dillehay et al., 1984; Blei et al., 1986; Fornage, 1986).

The differential diagnosis in cases with these findings would include rheumatoid disease. However, this can usually be distinguished from other disease processes by irregular thickening of the synovial lining. This will result in echos within the abnormally increased synovial fluid collection (see Fig 10–15).

#### *Tendon Rupture*

The biceps tendon is a common site in which tendon rupture can be observed. Predisposing lesions are rotator cuff tendonitis and inflammation of the long biceps tendon (Middleton et al., 1986b). These lesions are frequently found in athletes involved in baseball, tennis, squash, and field events (javelin throwing, discus, etc.). Clinicians usually refer to lesions involving tendons of the shoulder with edema as *shoulder impingement* (Jobe and Jobe, 1983). Patients who suffer from shoulder impingement usually develop tears of the long biceps tendon and rotator cuff many years later (Neer, 1983).

Sonographic findings in rupture of the biceps tendon are an empty bicipital groove (Fig 4–1) and a fusiform muscle mass located distally (see Fig 9–25). It is important to recognize this pair of findings to avoid misinterpretation of the muscle mass as a soft tissue neoplasm (Chui, 1988).

The posterior tibial tendon in the ankle is also

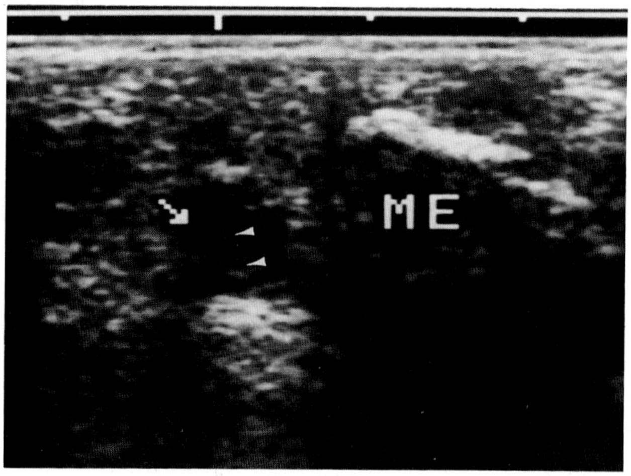


**FIG 4–6.**
Peroneus tendonitis: transverse sonogram. The patient is a soccer player with pain and swelling inferior to the lateral malleolus. The pain is worse during exercise on irregular surfaces, especially on grass and sand. The transducer is positioned inferior to the lateral malleolus *(ME)* with the axis of the transducer parallel to the long axis of the tibia. In most patients a standoff must be used for proper evaluation. The plantar surface is on the observer's left. The common peroneus tendon sheath *(white arrow)* surrounds both tendons *(arrowheads)* and is fluid filled. The thickness of the tendon sheath exceeds that of a normal tendon sheath, which should only be 1 to 2 mm.

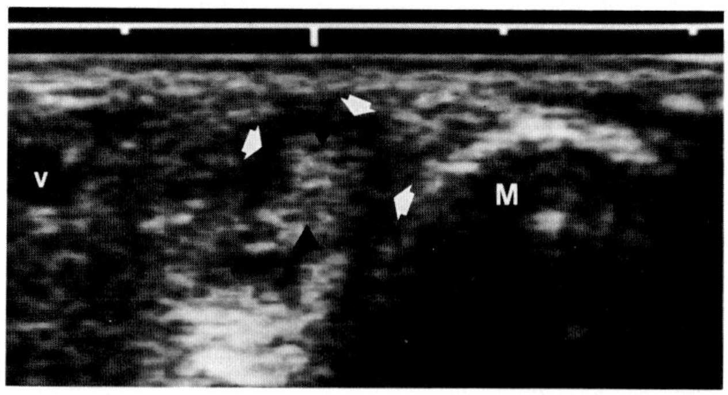


**FIG 4–7.**
Normal peroneus tendon sheath: transverse sonogram. A normal volunteer. The transducer is positioned inferior and anterior to the lateral malleolus, parallel to the long axis of the tibia. The plantar surface of the foot is on the observer's left. A fluid film 1 mm thick, contained within the tendon sheath *(arrows)*, surrounds the two peroneus tendons *(arrowheads)*. Acoustic enhancement is present adjacent to the tendon sheath. Abbreviations: lateral malleolus *(M)*, vein *(V)*.

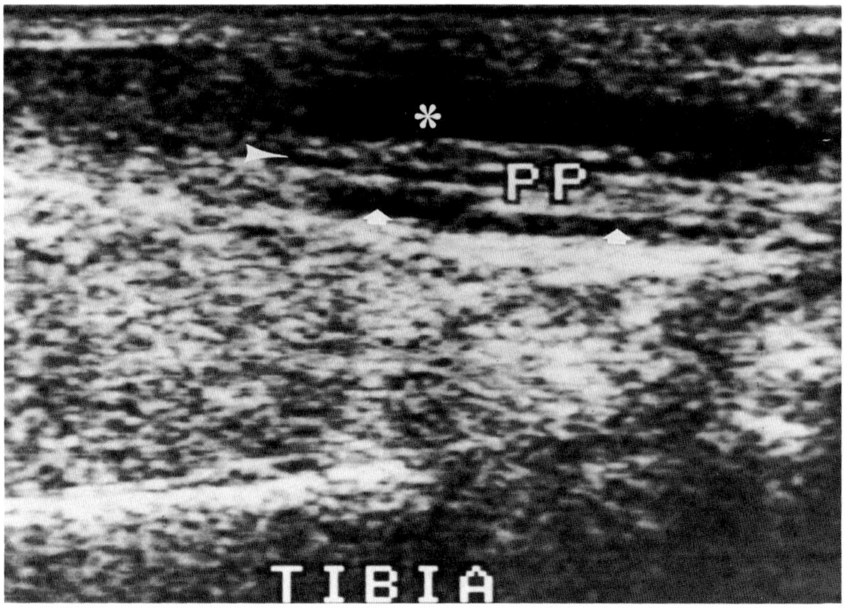


**FIG 4–8.**
Peroneus tendonitis: longitudinal sonogram. Same patient as in Figure 4–6. The transducer is positioned posterior to the lateral malleolus and slightly oblique parallel to the long axis of the common peroneus tendon. The superior aspect is at the observer's left. The two peroneus tendons *(PP)* are surrounded by a large fluid collection. A small amount of fluid separates the two tendons *(arrowhead)*. Some fluid is seen medial to the tendons *(arrows)*, but most of the fluid is pooling laterally *(white asterisk)*. The synovial sheath is curved at its superior extent.

prone to rupture (Fig. 4–1). Patients with this injury develop a painful flat foot. Posterior tibial tendon rupture can occur in athletes, but this type of tendon rupture is seen predominantly in patients with rheumatoid arthritis. The sonographic diagnosis is simple; a gap is identified within the tendon. A small fluid collection within the tendon sheath separates the retracted ends of the tendon.

An example of incomplete tendon rupture is seen in Figures 4–11, A and B. These images demonstrate a partial tear of the peroneus tendon (Hardaker et al., 1985). Anechoic clefts are seen within the tendon, and fluid is also present within the tendon sheath. Both findings are probably secondary to functional overload. It is unusual to see a large amount of hemorrhage within the sheath following tendon rupture due to the hypovascularity of tendons.

### *Tendon Luxation*

Luxation of the long biceps tendon is a rare finding but is easily diagnosed sonographically (O'Donoghue, 1982). Scanning is initiated in the transverse plane with the arm in internal rotation. The bicipital groove is identified on the anterior aspect of the humeral head. This serves as a landmark that helps to distinguish the subscapularis muscle from the supraspinatus. The diagnosis of luxation of the long biceps tendon can be made when the bicipital groove is empty and there is concavity of the transverse ligament. The long biceps brachii tendon can dislocate only if the attachment of the transverse ligament is torn (Petersson, 1986). Identification of the medially displaced long biceps tendon confirms the diagnosis. It is most frequently seen under the subscapularis tendon (see Fig 4–9,C) (Petersson, 1986).

Luxation of the peroneus tendon (Church, 1977; McConkey and Favero, 1987; Oden, 1987) is often attributable to chronic trauma (Eichelberger et al., 1982) seen in athletes involved in football, soccer, gymnastics, and dancing. A standoff or water bath is mandatory for sonographic examination of the lateral malleolar region. The examination is conducted with the foot everted against resistance to accentuate the dislocation. The transducer is placed just inferior to the lateral malleolus, scanning in the transverse plane. If luxation is present, the tendon will be seen overlying the lateral malleolus. Excess fluid may also be noted surrounding the tendon (Fig 4–12). Patients with this condition were treated surgically, and a torn superior peroneal retinaculum was found in all cases (Das et al., 1985).

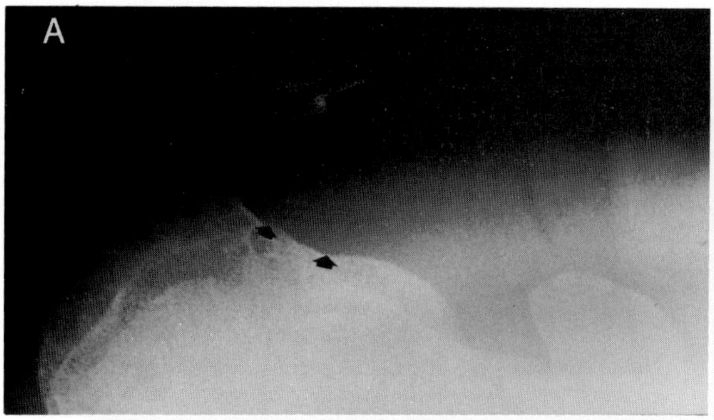


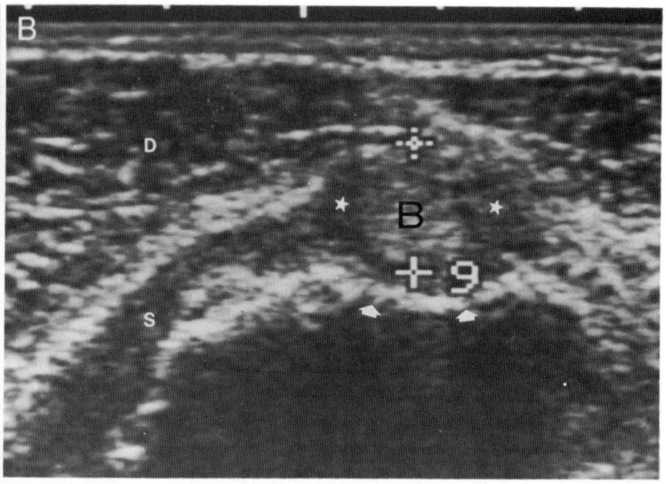


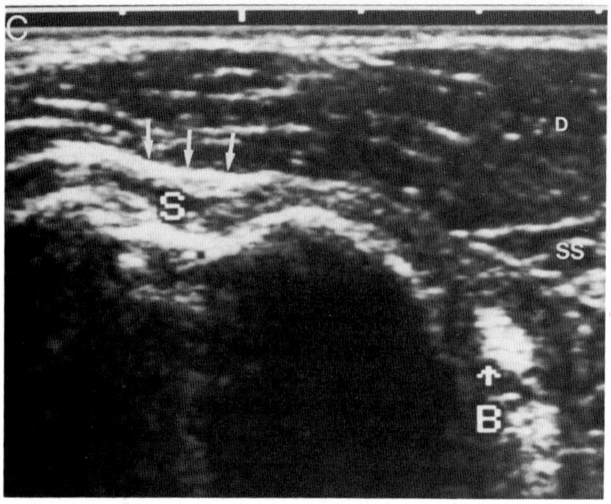


**FIG 4–9.**
**A,** investigation of a shallow bicipital groove. Conventional radiograph of the bicipital groove. A shallow bicipital groove *(arrows)* is a normal variant seen in about 8% of the population. The most frequent sequela of a shallow bicipital groove is tendonitis involving the long biceps tendon. **B,** ultrasound of the bicipital groove. The transducer is positioned perpendicular to the long axis of the humerus over the anterior aspect of the humeral epiphysis. The arm is in neutral position. The diameter of the tendon is significantly increased, 9 mm instead of 2 to 4 mm. Fluid within the synovial sheath is also increased *(asterisk).* Compare this image with Figure 4–2. Abbreviations: deltoid muscle *(D),* subscapularis muscle *(S),* bicipital groove *(small white arrows),* long biceps tendon *(B).* **C,** ultrasound of the bicipital groove: biceps tendon dislocation. The patient is a swimmer with right anterior shoulder pain. An infrequent finding in athletes with a shallow bicipital groove is dislocation of the long biceps tendon. Transducer position is identical to that in **B.** The bicipital groove or sulcus *(S)* is empty. The transverse ligament settles within the groove *(arrows),* and the long biceps tendon is seen medially beneath the subscapularis muscle *(ss).* Abbreviations: deltoid muscle *(D).*

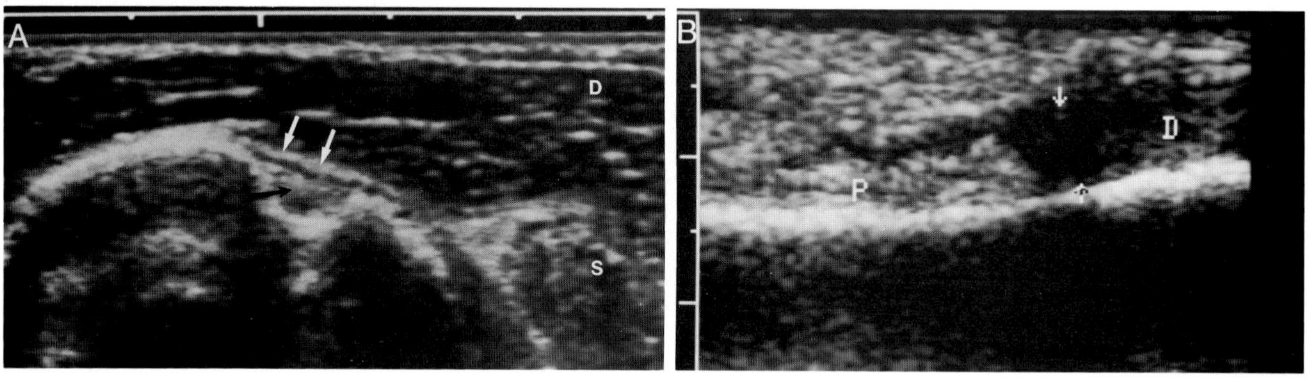


**FIG 4–10.**
**A,** biceps tendon rupture: transverse sonogram. A 50-year-old female tennis player presented with a soft tissue mass at the anterior aspect of the upper arm. A biceps tendon rupture was suspected, but the clinician requested an ultrasound examination to exclude a soft tissue tumor. The transducer is positioned transversely over the bicipital groove (compare with Figure 4–2). The transverse ligament is easily identified intact *(arrows).* However, the round appearance of the biceps tendon and the hypoechoic tendon sheath are absent from the groove. The biceps tendon cannot be visualized medially beneath the subscapularis muscle. Reflective material in the bicipital groove *(black arrow)* proved to be fibrosis. Abbreviations: deltoid muscle *(D),* subscapularis muscle *(S).* **B,** complete rupture of the posterior tibial tendon: longitudinal sonogram. The patient is a 75-year-old man who plays tennis once a week. Five months ago he felt a snap in a lateral aspect of his foot during accidental ankle eversion. He presents today after having sought treatment from many clinicians for painful swelling of the ankle, without relief. This image was obtained along the length of the posterior tibial tendon immediately posterior and superior to the medial malleolus. A discontinuity *(arrows)* is noted in the tendon. Fluid fills the gap created by the retracted ends of the tendon and tracks along the synovial sheath. Abbrreviations: proximal posterior tibial tendon *(P),* distal posterior tibial tendon *(D).*

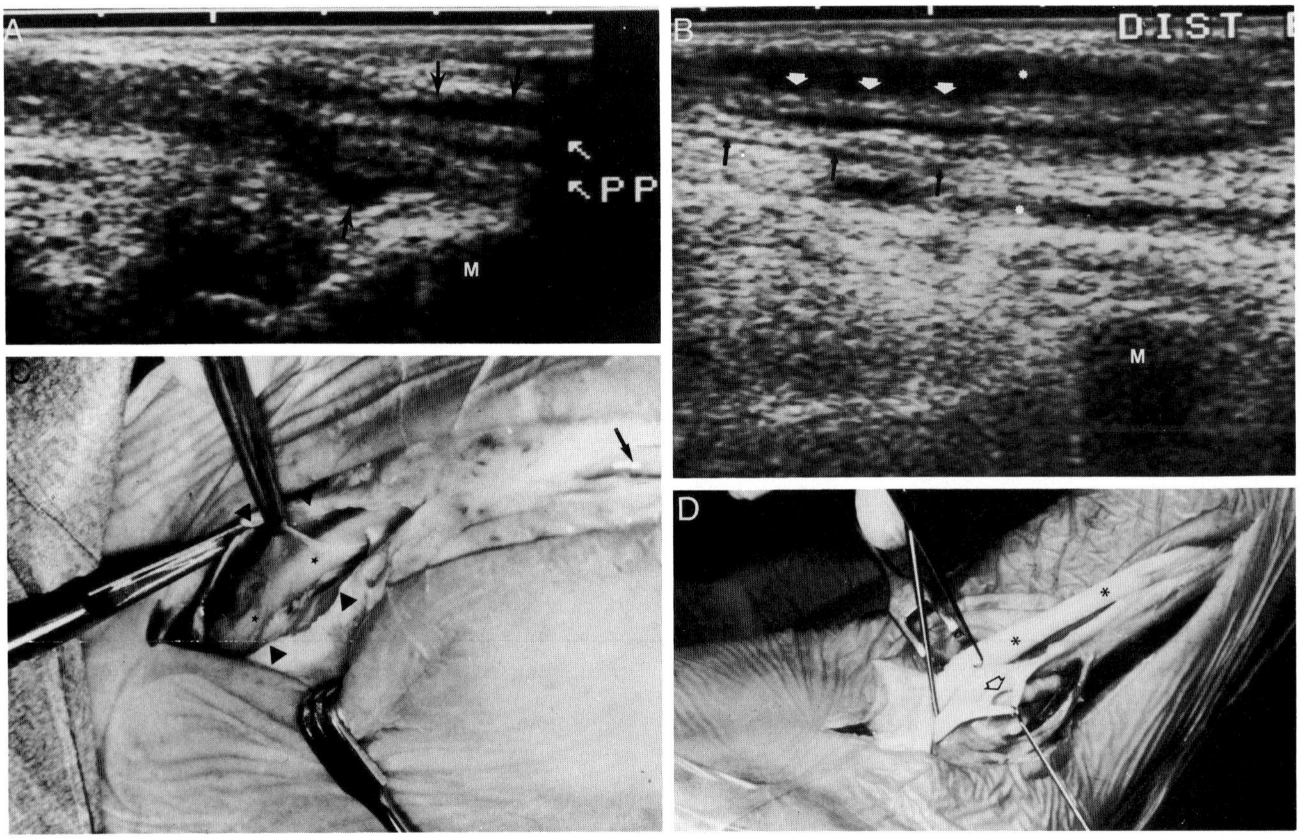


**FIG 4–11.**
**A,** tendonitis of the peroneus tendons progressing to incomplete rupture: sonogram. A ballet dancer with persistent ankle pain. The pain is localized to the lateral aspect of the ankle and is worsened by eversion of the foot. The transducer is positioned posterior to the lateral malleolus, parallel to the tibia. An increased amount of fluid *(black arrows)* is found in the sheath of the peroneus tendons *(PP).* Abbreviations: lateral malleolus *(M).* **B,** incomplete rupture of the peroneus tendons: longitudinal sonogram taken 5 months later. Position of the transducer is identical to that in **A.** Fluid within the tendon sheath has increased *(white asterisk).* A hypoechoic gap is seen centrally within the peroneus brevis tendon, which represents a fluid collection within the tendon *(black arrows).* Individual tendon fibers are visualized more clearly due to this pathological fluid collection. The peroneus longus tendon is frayed and a large defect is seen in the posterior aspect *(white arrows).* **C** and **D,** intraoperative photographs: peroneus tendonitis. Same patient as in **A** and **B.** The skin incision is made longitudinally, posterior to the lateral malleolus. The skin is retracted anteriorly revealing the opened tendon sheath *(black arrowheads)* and the peroneus tendon *(stars).* When the tendon sheath was incised, a pathological amount of synovial fluid was present, as seen sonographically. Its reflection is indicated by the *black arrow* at the right. **D,** the peroneus brevis tendon is seen anteriorly *(asterisk),* and the peroneus longus is exposed. Fine hook retractors reveal the defect in the tendon *(open arrow).*

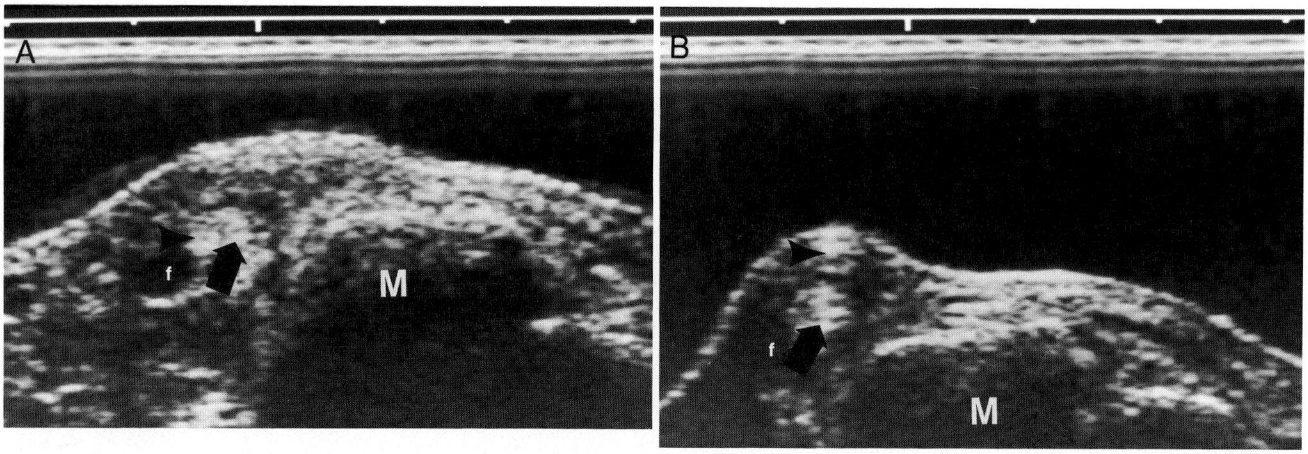


**FIG 4–12.**
**A,** peroneus tendon luxation: transverse sonogram through the common peroneus tendon sheath. A ballet dancer presented with pain and swelling over the distal fibula. She could reproduce the pain by dorsiflexion and eversion of the foot. Examination of the lateral malleolus required a standoff. The position of the transducer is perpendicular to the long axis of the tibia, posterior to the lateral malleolus. There is a slightly increased amount of fluid *(f)* in the peroneus tendon sheath. The peroneus brevis tendon *(black arrow)* is located anteriorly near the lateral malleolus *(M)*. The peroneus longus tendon *(arrowhead)* is located more posteriorly. **B,** peroneus tendon luxation: transverse sonogram with the foot in dorsiflexion and eversion. Same patient as in **A** with identical positioning of the transducer. With the patient's foot in dorsiflexion and everted, the peroneus longus tendon dislocates laterally and anteriorly, overriding the peroneus brevis tendon *(black arrow)*.

## Tendons Without a Synovial Sheath

### *Tendonitis*

Tendonitis is almost always attributable to chronic trauma and a high level of athletic activity (Roels and Martens, 1978; Feretti et al., 1985). Plain film findings can be detected only in advanced cases and only if you use low-kilovoltage technique (Fig 4–13). The sonographic findings are identical in all locations. Focal thickening of the tendon (van Holsbeeck, 1987) and widening of the distance between the longitudinal fascicles are invariably present (Fig 4–14). Focal hypoechoic areas within the tendon are also present. Fat surrounding the tendon has increased echogenicity as seen in Figure 4–15 (Hoffa's fat pad, Kager's triangle). The longitudinal tendon fibers are seen to be intact (Fig 4–13). In some cases, hypoechoic foci are seen around the tendon (Figs 4–16 and 4–17). These were shown to be fibromyxoid tissue on histological analysis of 53 cases. The tendon itself may have a nodular appearance (Fig 4–18). Bony enthesophytes (Figs 4–19 and 4–20) may be seen at the tendon insertion in cases of prolonged tendinitis. In some patients, tendonitis persists for years with episodic flare-ups during the sports season. Rarely, calcium deposits may be detected within the tendon. Calcium is deposited preferentially in the cartilaginous tendon insertion (Fig 4–21).

In those patients with the sonographic diagnosis of patellar tendonitis who underwent surgery, all were found to have the epitendineum intact. On longitudinal sectioning of the tendon, degenerating or necrotic tissue was identified just below the patellar apex. The superficial tendon fibers in this region were grossly normal. Histological examination revealed mucoid degeneration, fibrinous necrosis, clefts, and microtears. Areas of regeneration with reparative and granulation tissue were also seen (Roels and Martens, 1978). The widened distance between longitudinal fascicles seen sonographically proved to be ruptured collagen fibers under light microscopy, especially those cross-linking the fascicles. Focal thickening of the tendon was the result of tiny pockets of necrotic debris. Hypoechoic foci were attributable to edema and mucoid degeneration. The nodular appearance of tendons was due to collections of granulation and scar tissue.

Merkel et al. (1982) described the pathological findings in traumatic tendonitis using electron microscopy. They reported rupture of collagen bundles as well as a decrease in the mean diameter of collagen fibrils (from 450Å to 345Å). The fibrils also displayed disorder in their arrangement. Merkel et al. (1982) also noted an increase in quantity of acid mucopolysaccharide ground substance. They concluded that the increased ground substance and edema in-

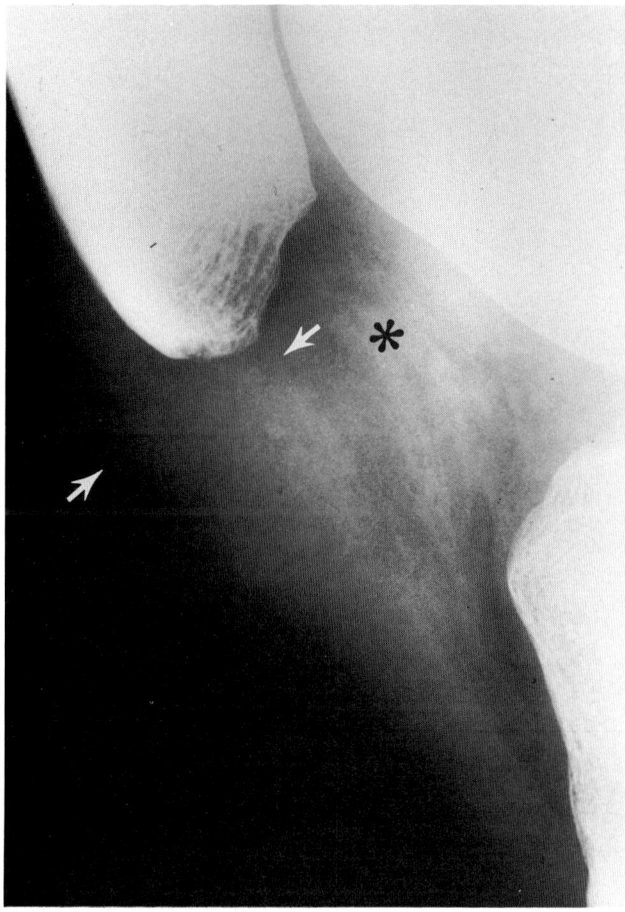

**FIG 4–13.**
Jumper's knee: conventional radiograph with soft tissue contrast. A soccer player presented with knee pain at the beginning and end of exercise. Pain and tenderness were localized under the apex of the patella. The image *above* is a lateral view of the knee produced using a Gevaert MRL cassette and mammography film. Thickening of the patellar tendon is seen in the infrapatellar region *(white arrows)*. The adjacent infrapatellar fat pad (Hoffa's fat pad) is infiltrated and is less radiolucent *(black asterisk)* when compared with the rest of the infrapatellar fat. A small calcification is seen within the posterior third of the patellar tendon at its point of maximal thickening.

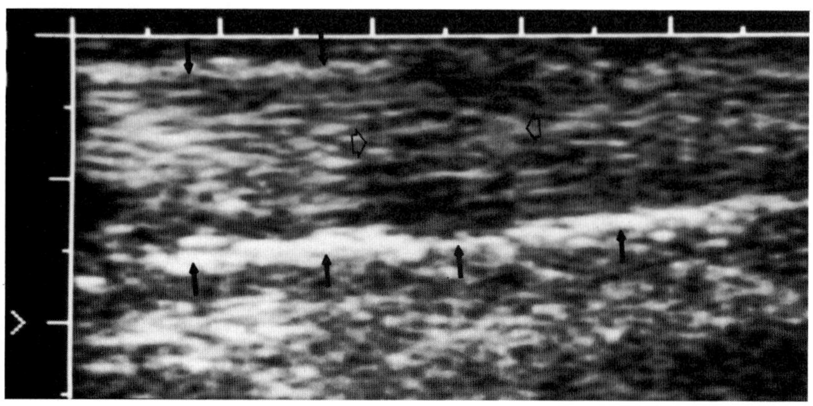

**FIG 4–14.**
Achilles tendonitis: longitudinal sonogram. A long distance runner with a 3-month history of pain in the Achilles tendon. The pain is located 5 cm above the tendon insertion. The transducer is positioned over the heel, dorsal to the Achilles tendon, oriented along the long axis of the tendon. The lesion *(open arrows)* is identified sonographically as a hypoechoic focal swelling of the tendon with increased distance between the individual longitudinal fibers. The longitudinally oriented tendon fibers of the Achilles tendon can easily be identified. There is no synovial sheath and no fluid surrounding the tendon. The hyperreflective area *(black arrows)* surrounding the tendon corresponds to the epitendineum.

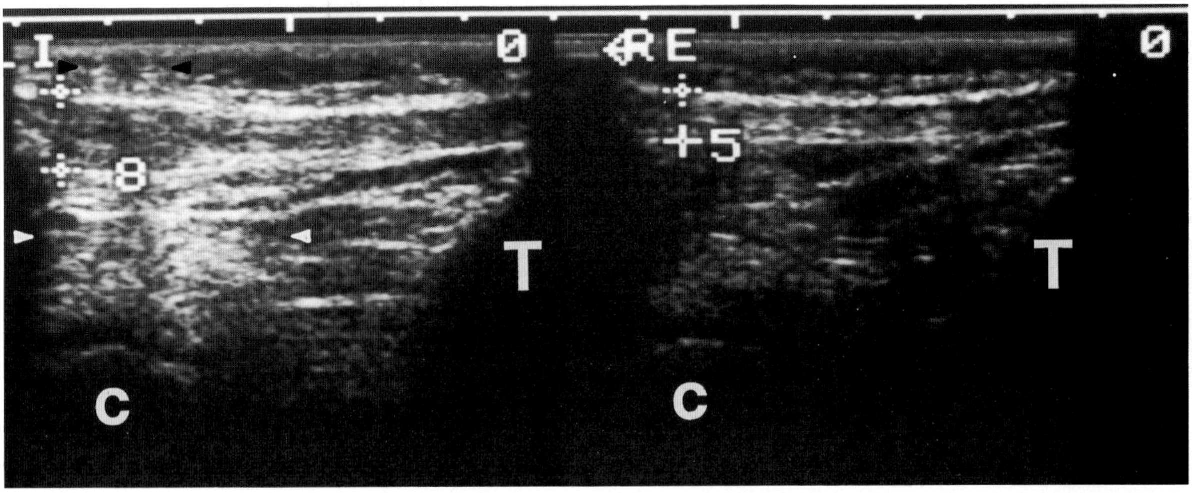


**FIG 4–15.**
Comparison of two longitudinal images through the patellar tendons: longitudinal sonograms. A basketball player complained of left knee pain at the end of the season. The pain was constantly present. At night he could not tolerate the light pressure of blankets on his knee. The transducer is positioned in the long axis directly overlying the patellar tendon. Both tendons are measured under the patellar apex. A normal sagittal diameter for an athlete is approximately 5 mm. The image on the *right* demonstrates a normal tendon. The infrapatellar portion of the tendon on the *left* is widened and appears markedly hypoechoic relative to the surrounding fat. The surrounding fat demonstrated increased echogenicity *(arrowheads)* relative to normal fat. Abbreviations: femoral condyle *(C)*, tibia *(T)*.

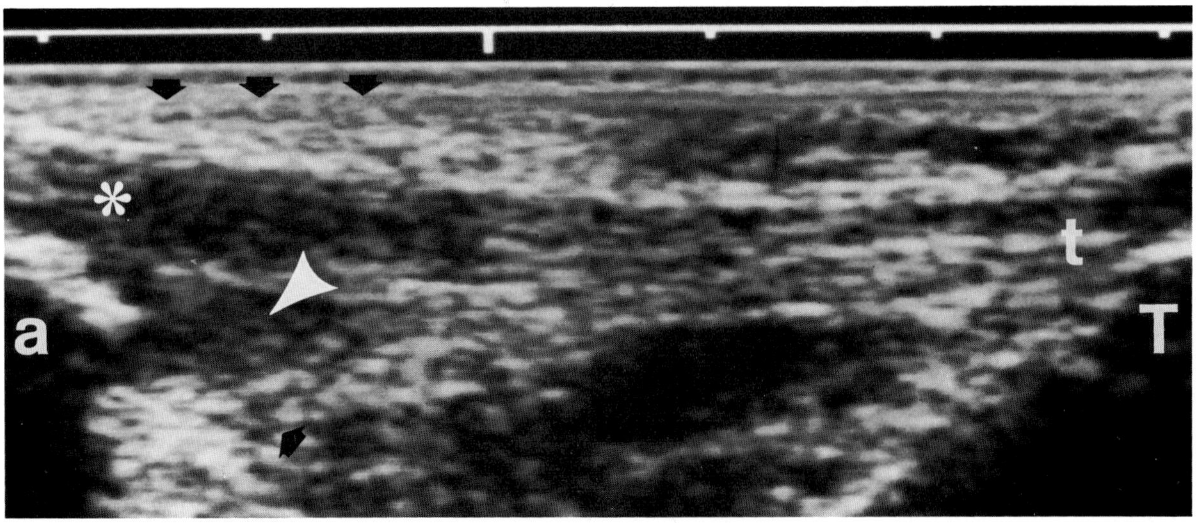


**FIG 4–16.**
Jumper's knee: longitudinal sonogram. A soccer player presented with knee pain after practice. Pain and point tenderness were located under the patellar apex. The transducer is positioned in the long axis of the right patellar tendon. The proximal part of the patellar tendon is swollen and relatively hypoechoic *(white asterisk)*. A collection of hypoechoic material *(white arrowhead)* is seen deep to the patellar tendon and inferior to the patellar apex *(a)*. Histological analysis has shown this to be fibromyxoid tissue. The hypoderm and the infrapatellar fat pad adjacent to the lesion are hyperechoic due to inflammation *(black arrows)*. Normal portion of the distal tendon *(t)* and insertion on the tibia *(T)*.

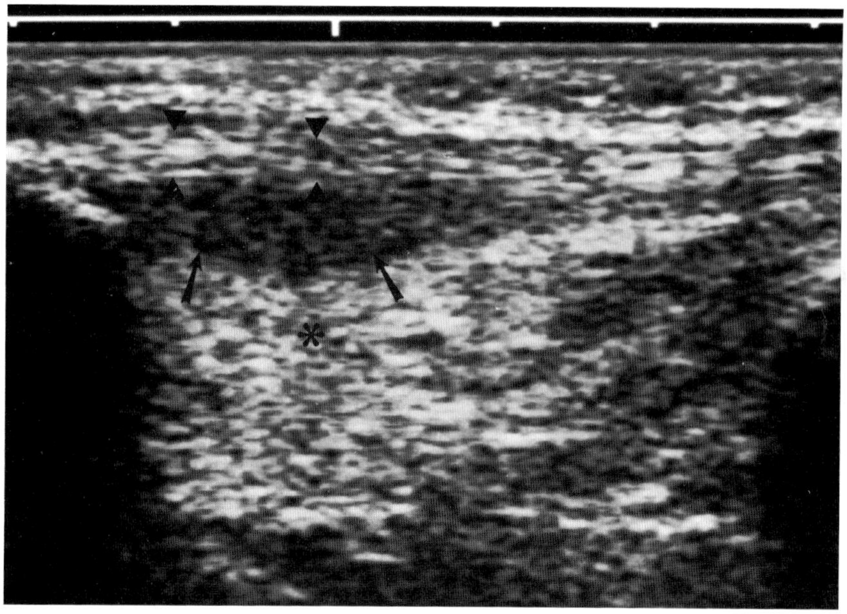

**FIG 4–17.**
Jumper's knee: longitudinal sonogram. A volleyball player experienced the same knee pain at the end of the season. The pain was worse during practice, especially on flexion. The pain disappeared after warming up. The transducer is positioned in the long axis of the right patellar tendon. The fibers of the tendon *(arrowheads)* are continuous and have a normal appearance throughout their length. Hypoechoic fibromyxoid tissue *(black arrows)* is present surrounding the proximal portion of the tendon. Hoffa's fat pad, deep to this hypoechoic fibromyxoid tissue, is hyperechoic *(black asterisk)*.

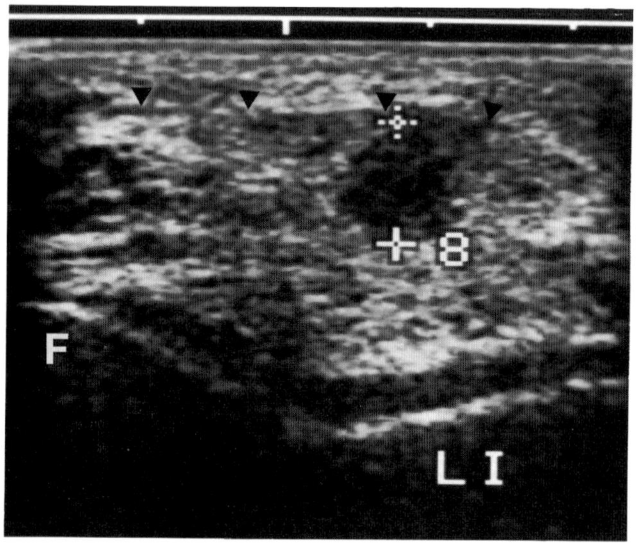


**FIG 4–18.**
Jumper's knee: transverse sonogram. A soccer player had chronic knee pain for almost 1 year. The exercise-induced pain prevented him from participating in professional sports. The left patellar tendon is scanned transversely with the transducer positioned perpendicular to its longitudinal axis. The patellar tendon is convex anteriorly *(black arrowheads)*. A nodule *(crosses)* is seen within the deep aspect of the tendon. The articular cartilage of the femoral condyles is seen deep to the tendon. Abbreviations: lateral femoral condyle *(F)*.

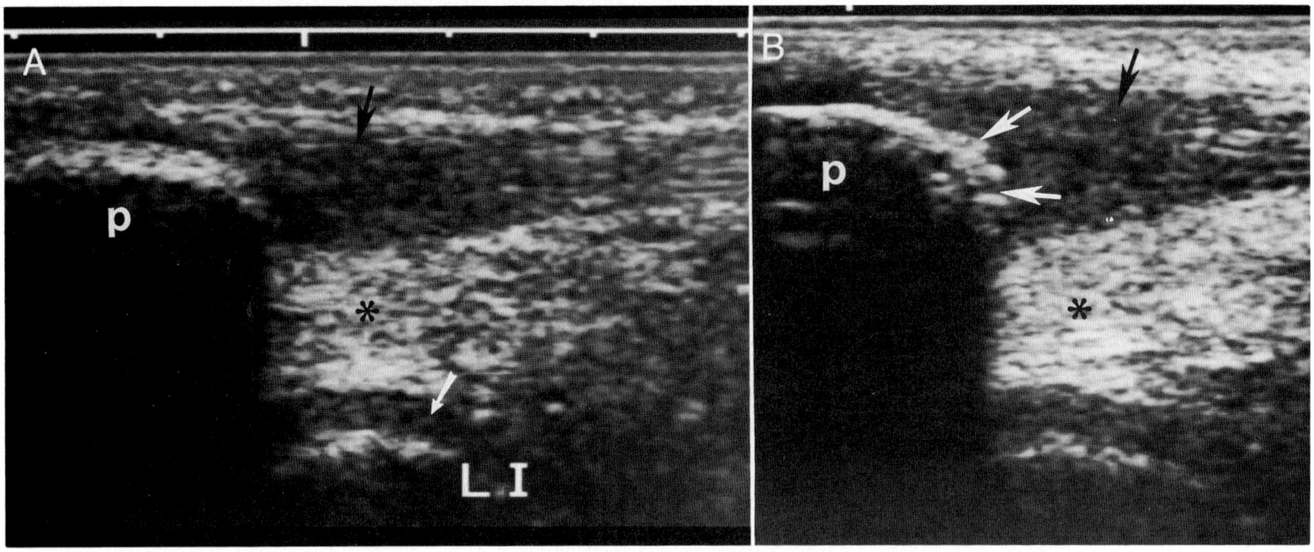


**FIG 4–19.**
**A,** progression of subacute patellar tendonitis to chronic tendonitis. Taken 7 months after the onset of symptoms, including infrapatellar pain during training. The transducer is positioned in the long axis of the tendon over the patellar apex. The hypoechoic lesion is located inferior to the patellar apex *(P)*. Only a few tendon fibers remain intact in this hypoechoic region *(black arrow)*. The portion of Hoffa's fat pad deep to the area of tendonitis is hyperechoic *(black asterisk)*. Cartilage of the femoral condyle *(white arrow)*. **B,** 2 years after the onset of symptoms. Transducer position is identical to that in **A.** There is increased infrapatellar swelling *(black arrow)*. A sign of the chronicity of patellar tendonitis is the irregular new bone formation *(white arrows)* identified as inhomogeneous hyperreflectivity. Symbols: hyperechoic fat *(asterisk)*.

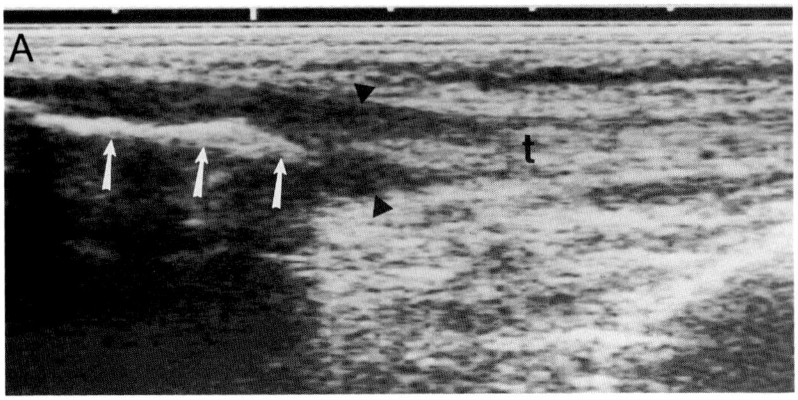


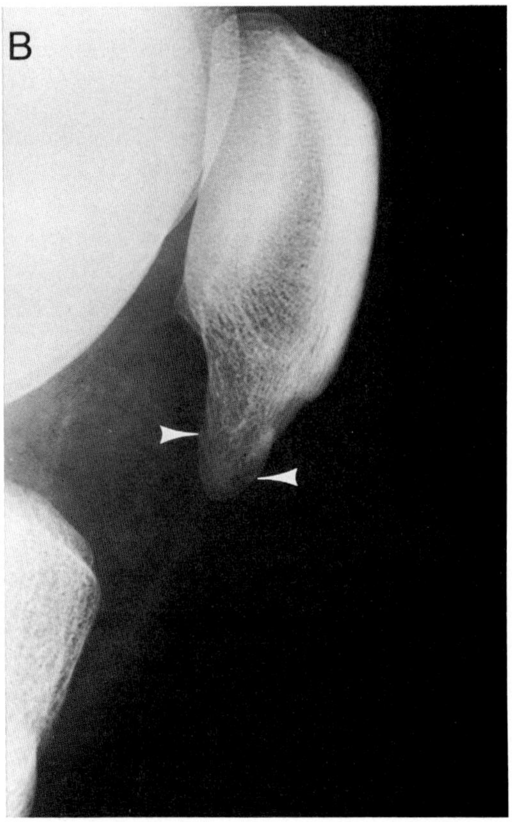


**FIG 4–20.**
**A,** chronic recurrent patellar tendonitis: longitudinal sonogram. In each of the past 4 years this patient experienced symptoms of patellar tendonitis at the end of the sporting season. The transducer is positioned along the length of the patellar tendon over the patellar apex. Lengthening of the patellar apex *(white arrows)* is seen. The patellar tendon *(t)* shows the characteristic lesion of patellar tendonitis *(black arrowheads)*. **B,** chronic recurrent patellar tendonitis: conventional radiograph. Same patient as in **A.** Patellar lengthening with well-organized new bone formation *(white arrowheads)* is a characteristic feature of chronic recurrent patellar tendonitis.

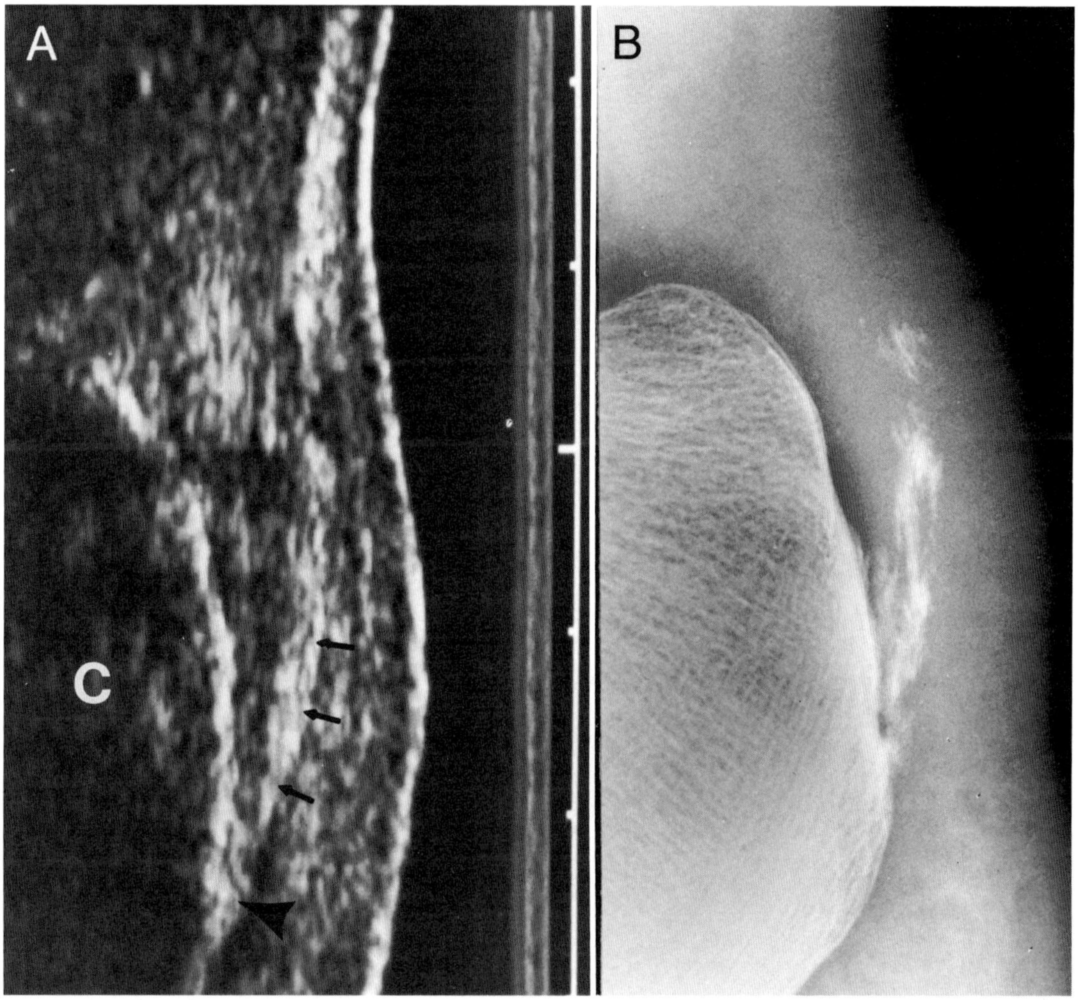


**FIG 4–21.**
**A,** calcifying tendonitis: sagittal sonogram. A long distance runner with pain and swelling over the Achilles tendon insertion. A few days prior to examination he experienced an acute exacerbation of symptoms. The area of tenderness is now markedly swollen and erythematous. The transducer is oriented in the long axis of the Achilles tendon overlying its insertion. The triangular fibrocartilagenous insertion demonstrates increased reflectivity. The most hyperechoic areas are the distal part of the insertion *(black arrowhead)* and the marginal areas of the insertion *(black arrows)*. Small areas of acoustic shadowing correspond to calcifications within the tendon. Abbreviations: calcaneus *(C)*. **B,** calcifying tendonitis: Gevaert MRL radiograph. Same patient as in **A.** Conventional radiograph of the heel (lateral view) demonstrates calcifications within the triangular tendon insertion surrounded by soft tissue swelling.

terfered with nutrient transport, compounding the metabolic disturbance caused by trauma.

Patellar tendonitis (jumper's knee) is frequently found in people who participate in sports that require repetitive activity of the extensor mechanism of the knee. Some examples are football, soccer, basketball, track, and field (Roels and Martens, 1978; Martens et al., 1982). However, focal patellar tendonitis can be seen in patients who do not participate in athletics. These patients usually have a history of blunt trauma to the patellar tendon. In rare cases, the cause may be iatrogenic. Lesions identical to those found in blunt trauma can be seen following arthroscopy. When a transtendonous approach is chosen, granulation tissue and tendonitis can be observed at the puncture site (see Fig 12–3).

Presenting symptoms include pain on climbing stairs, pain on long walks, and pain when sitting with knees bent. Patellar hypermobility, quadriceps wasting, and local tenderness are frequent findings on physical examination. Localized swelling over the apex of the patella was clinically noted in approximately 5% of our patients. The mean age of our patients with patellar tendonitis is 20 years old.

Unfortunately, many cases of patellar tendonitis are completely overlooked or misdiagnosed as bursitis or chondromalacia. Correct diagnosis is imperative for institution of adequate therapy. Sonography can be especially important in selecting those cases that are best suited for surgical treatment.

In patellar tendonitis, two patterns may be seen sonographically: focal or global tendon thickening. Focal tendon thickening was observed in 91.5% of our cases. In the majority of these (97%), the site of thickening was proximal (Martens et al., 1982), near the apex of the patella (Fig 4–22). The remaining 3% of cases of focal thickening demonstrated the lesion near the tibial tuberosity (Figs 4–23 and 4–24).

Global tendon thickening comprised 8.5% of our cases. This was the result of very long standing disease and is frequently a precursor to tendon rupture. Figures 4–25 and 4–26 demonstrates the sonographic appearance of global tendon thickening. Patients with global tendon thickening or foci of amorphous material within the tendon are candidates for surgical treatment.

The differential diagnosis of patellar tendonitis includes two other disease processes that can elicit identical clinical symptoms. Bursitis involving one of the peritendonous bursae is a frequent cause of localized tenderness over the patella or tibial tuberosity (Bywaters, 1979). This is easily differentiated from

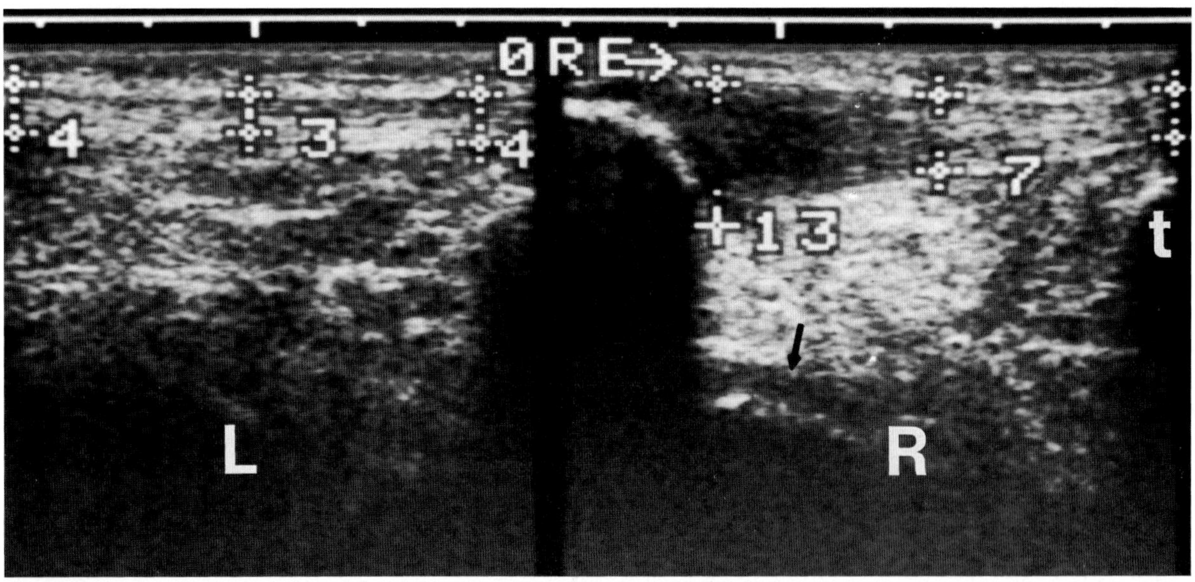


**FIG 4–22.**

Measurement in patellar tendonitis: sagittal sonograms. The patient's normal left knee is seen to the left *(L)*. The infrapatellar portion of the patellar tendon is on the left side of the image and the tibial insertion is on the right. The sagittal diameters are expressed in millimeters. Normal tendons have a sagittal diameter between 3 and 5 mm. Comparison with the asymptomatic side is very helpful. The patient's symptomatic right knee is seen on the right *(R)*. Traumatic lesions in the patellar tendon are located under the patellar apex in more than 90% of cases. The infrapatellar portion of the tendon is hypoechoic and enlarged. The sagittal diameter is more than three times the normal diameter. Hoffa's fat pad is increased in echogenicity, which allows better visualization of the condylar cartilage *(black arrow)*. Abbreviations: tibia *(t)*.

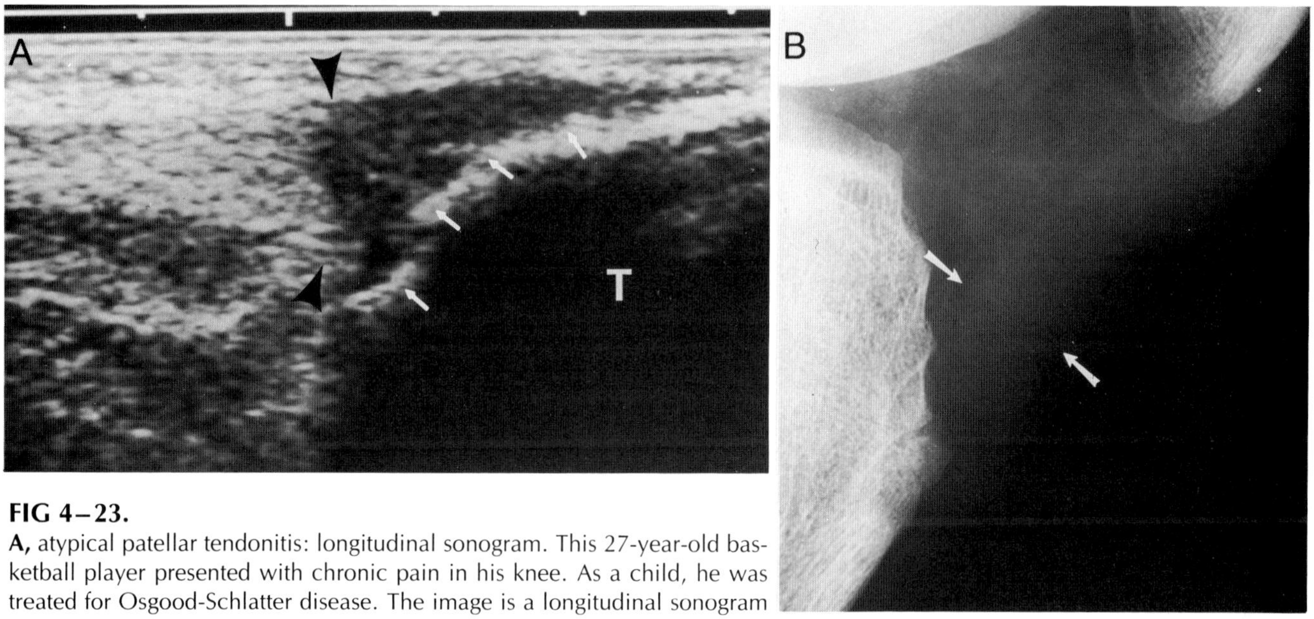


**FIG 4–23.**
**A,** atypical patellar tendonitis: longitudinal sonogram. This 27-year-old basketball player presented with chronic pain in his knee. As a child, he was treated for Osgood-Schlatter disease. The image is a longitudinal sonogram through the patellar tendon. The distal end of the patellar tendon and the insertion on the tibia have a hypoechoic appearance. The diameter of the tendon is markedly increased *(black arrowheads)*. The insertion on the tibia *(T)* has a highly irregular surface *(small arrows)*. **B,** atypical patellar tendonitis: conventional radiograph. A conventional radiograph depicting a lateral view of the knee. Swelling *(white arrows)* of the distal end of the patellar tendon and irregularity of the patellar tendon insertion are present.

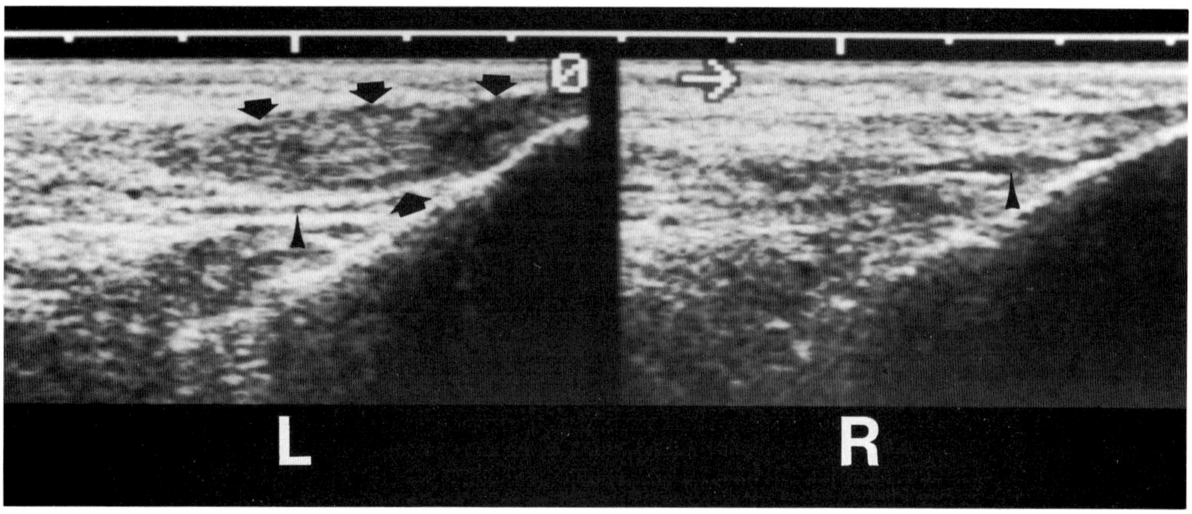


**FIG 4–24.**
Distal patellar tendonitis with incomplete rupture: longitudinal sonograms. A 40-year-old man with chronic pain in the left knee after blunt trauma to the anterior aspect of the knee. Longitudinal sonograms through the patellar tendons. The entire distal half of the left patellar tendon is swollen and hypoechoic *(arrows)*. Some of the fibers are interrupted in the most distal 2 cm of the tendon. The right patellar tendon is normal. Normal pretibial bursa *(arrowheads)*.

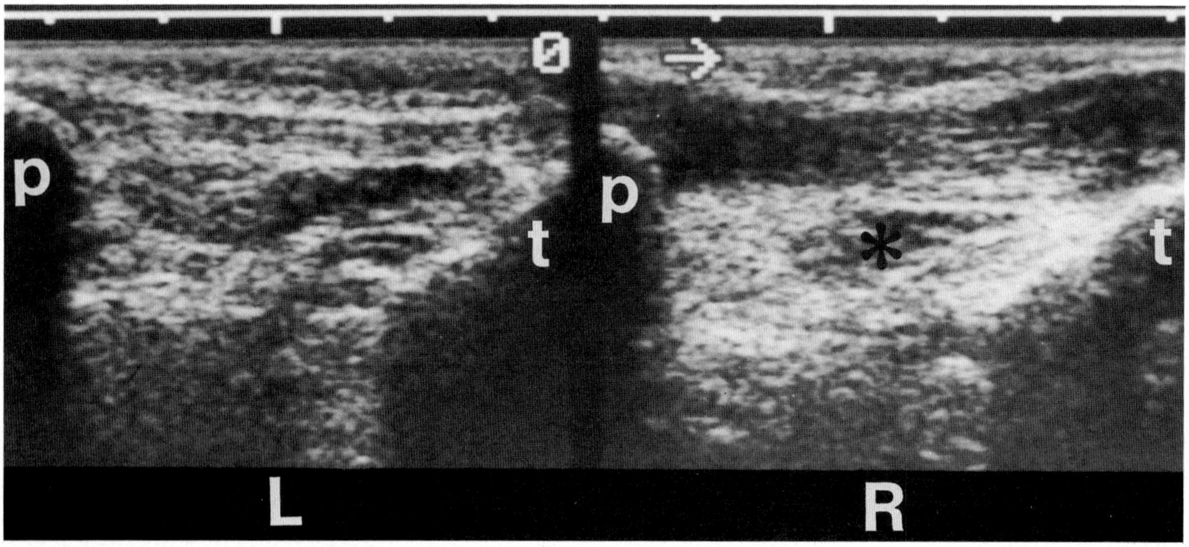


**FIG 4–25.**
Global patellar tendonitis: longitudinal sonograms. A patient with right patellar tendonitis for more than 4 years. He played volleyball, but due to constant pain, he stopped all sports activity. The transducer is positioned along the length of the patellar between the apex of the patella *(p)* at the *left* and the insertion on the tibia *(t)*. The entire tendon is thickened and hypoechoic in appearance. Hoffa's fat pad is markedly hyperreflective throughout *(black asterisk)*.

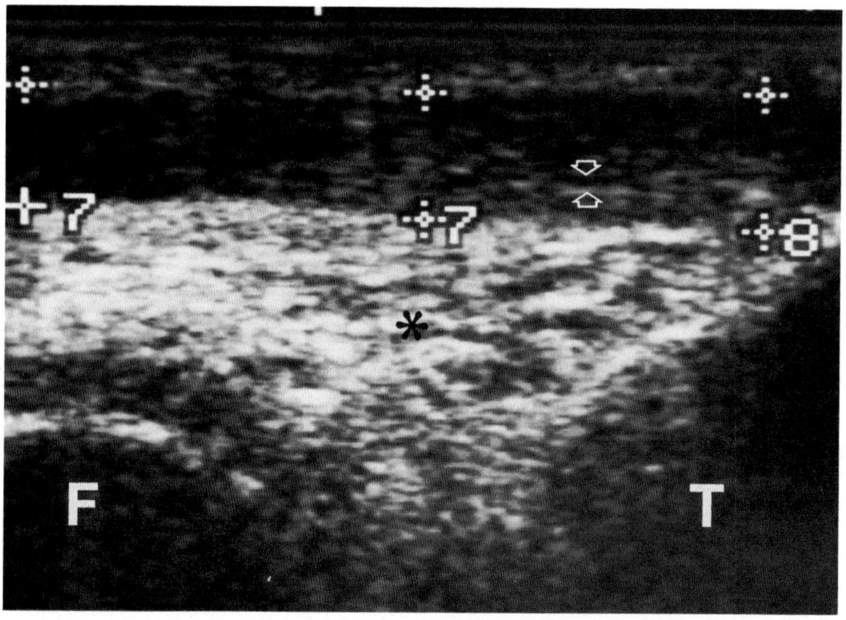


**FIG 4–26.**
Global patellar tendonitis: detail of a longitudinal sonogram. The *left* side of the image is the proximal portion of the tendon and the distal insertion is seen to the *right*. The sagittal diameters are given in millimeters. There is widening of the distance between tendon fibers *(small arrows)* and the tendon is hypoechoic compared with fat *(black asterisk)*. Abbreviations: femoral condyle *(F)*, tibia *(T)*.

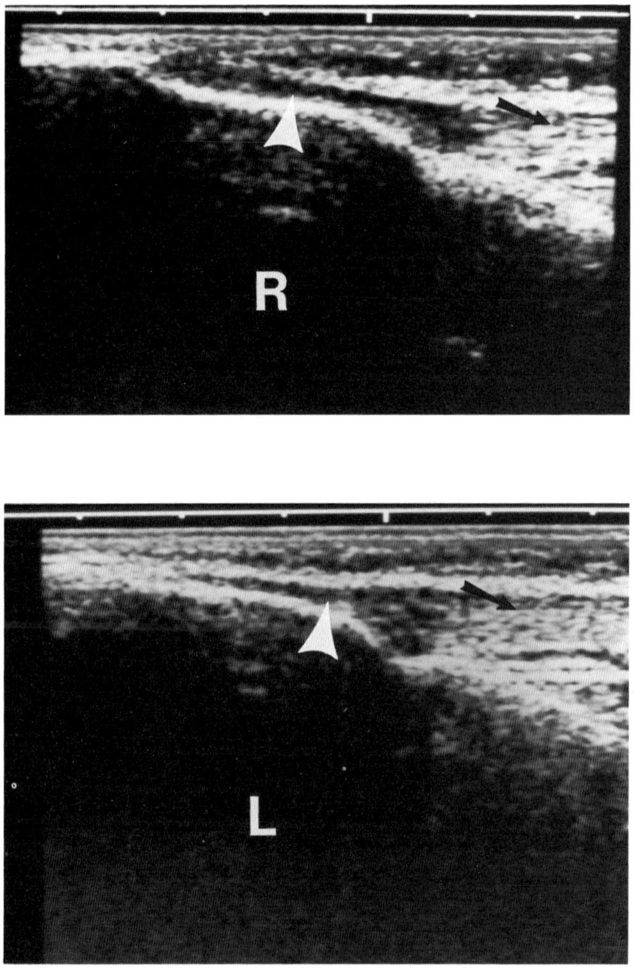


**FIG 4–27.**
Insertion tendonopathy of the distal patellar tendon: longitudinal sonograms. A 23-year-old soccer player with pain over the tibial tuberosity. The transducer is positioned in the longitudinal axis of the patellar tendon. Distal lies on observer's left. Thickening and decreased echogenicity are seen in the tendon insertion of the right knee *(R)*. The tendon insertion on the left is normal. At surgery this was found to be edema of cartilage and inflammation of the perichondrium. No necrosis was present. This did not require surgical treatment. Tendon insertion *(arrowhead)* and tendon fibers *(arrow)*.

patellar tendonitis sonographically. An oval anechoic fluid-filled structure is identified adjacent to the patellar tendon. These lesions differ in both prognosis and therapeutic management. Therefore, obtaining a definitive diagnosis is mandatory. Bursitis will be discussed in more detail in Chapter 5.

Insertion tendonopathy (Merkel et al., 1982) is another entity included in the differential diagnosis of patellar tendonitis. This disease occurs almost exclusively in athletes. The disease is limited to the fibrocartilaginous insertion of the tendon. The diagnosis of insertion tendonopathy is made if enlargement of the cartilaginous insertion is identified (Fig 4–27). Differentiation of this disease process from tendonitis is very important because there is no indication for surgery in insertion tendonopathy.

Achilles tendonitis is found somewhat less frequently than patellar tendonitis. The population at greatest risk is once again athletes (Palvalgyi and Balint, 1981; Clement et al., 1984). In our institution, the mean age of those afflicted with Achilles tendonitis is 33.5 years, 13.5 years older than the population with patellar tendonitis. However, closer examination of the data reveals a bimodal age distribution of individuals with Achilles tendonitis.

The older group, with a mean age of 35 years, demonstrates tendon thickening in the proximal portion of the tendon. This comprises 85% of all cases of Achilles tendonitis (Figs 4–28 through 4–30). Individuals with lesions in the distal 5 cm of the tendon fall into the second age peak, which has a mean age of 25 years. These patients represent 14.8% of all cases. An associated retrocalcaneal bursitis (Canoso, 1984) is almost always present in these cases (Figs 4–31 and 4–32). The prognosis for patients with distal lesions is significantly worse than that seen in patients with proximal lesions. Proximal lesions in Achilles tendonitis are amenable to treatment by surgical release. In our experience, surgical therapy has not been successful in distal lesions.

The remaining 0.2% of cases of Achilles tendonitis affect only the epitendineum and surrounding loose areolar tissues (Clement et al., 1984). This diagnosis is made when a fluid film is seen surrounding the tendon. The tendon itself appears normal (Fig 4–33). Surgical pathology demonstrates edema and inflammation affecting the epitendineum and adjacent mesenchymal tissues (Fig 4–34).

The differential diagnosis of Achilles tendonitis includes Haglund's disease, retrocalcaneal bursitis, and insertion tendonopathy. The pump-bump of Haglund's disease is due to thickening of the dermis and hypoderm (Fig 4–35). This is often accompanied by a bony excrescence on the posterosuperior aspect of the calcaneus. During exacerbations, a retrocalcaneal or subcutaneous bursitis develops. This is easily differentiated sonographically and is discussed further in chapter 5 (see Fig 5–11).

Isolated disease of the retrocalcaneal bursa will emulate Achilles tendonitis clinically. This condition can be the result of trauma, psoriasis, Reiter's disease, or rheumatoid arthritis (see Figs 5–2, 5–6,

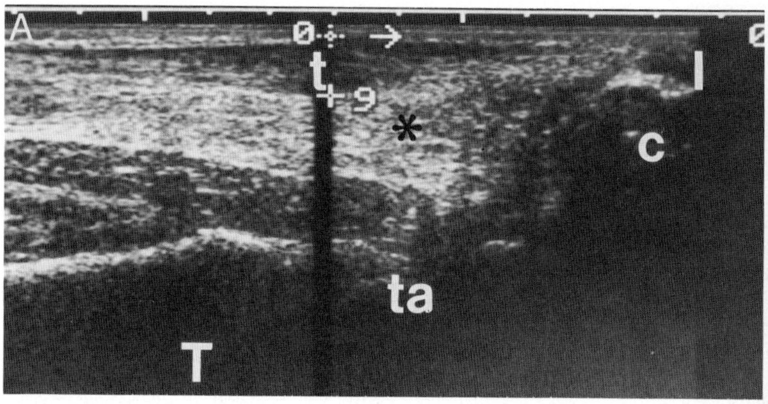


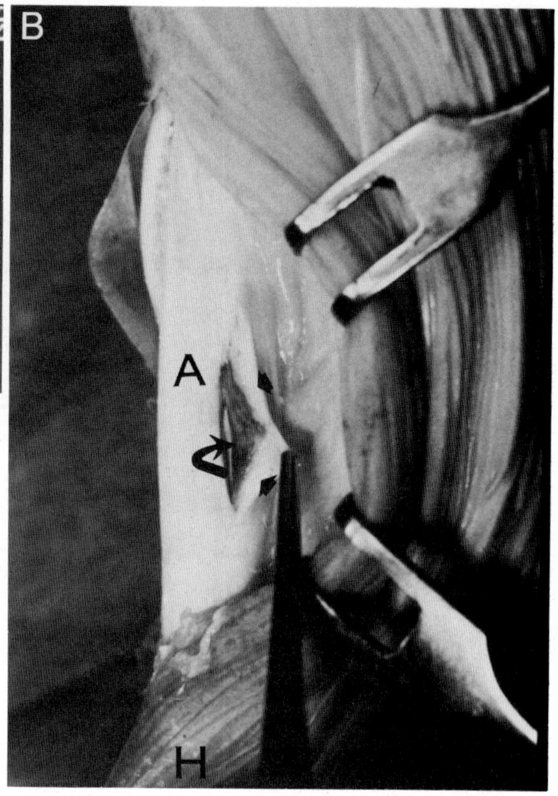


**FIG 4–28.**
**A,** achilles tendonitis, proximal type: longitudinal sonogram. A 24-year-old long distance runner with pain in the Achilles tendon at the beginning of exercise, which persisted. Swelling of the tendon was evident after exercise. The transducer is positioned posteriorly over the Achilles tendon. An image is constructed of the entire Achilles tendon using image memory to juxtapose two images. Maximal thickening of the tendon *(t)* is seen 6 cm above the tendon insertion *(I)*. The tendon has a hypoechoic appearance with a maximal thickness of 9 mm. Kager's fat pad appears hyperechoic *(black asterisk)*. Abbreviations: tibia *(T)*, talus *(ta)*, calcaneus *(c)*. **B,** intraoperative photograph of Achilles tendonitis. Same patient as in **A**. The peritenon has been incised *(short arrows)* and is being retracted with a pair of toothed forceps. Within the tendon, fibromyxoid and necrotic tissue is seen *(curved arrow)*. Abbreviations: heel *(H)*, Achilles tendon *(A)*.

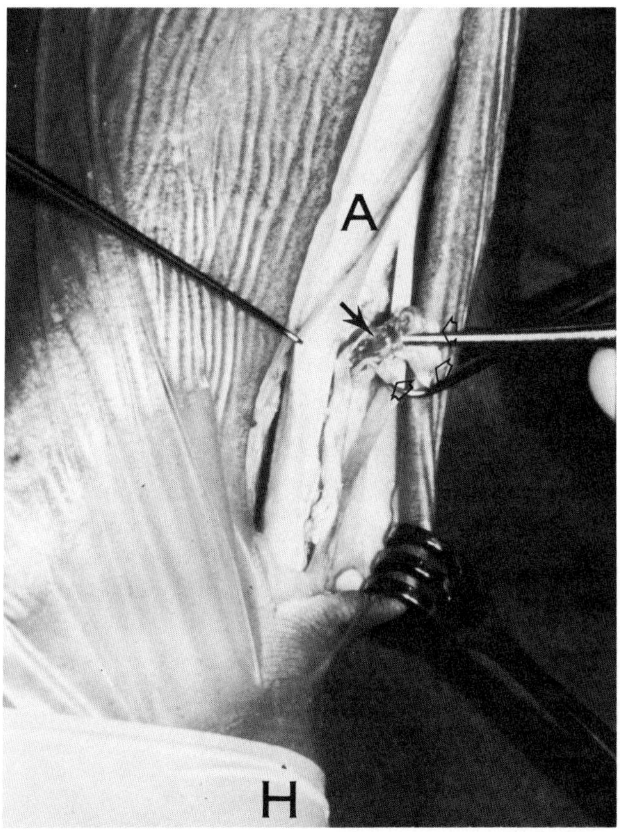


**FIG 4–29.**
Intraoperative photograph of another patient with Achilles tendonitis. Again, the peritenon has been incised and retracted. Granulation tissue is seen within the tendon *(black arrow)*. Several ruptured fiber bundles are reflected posteriorly *(open arrows)*.

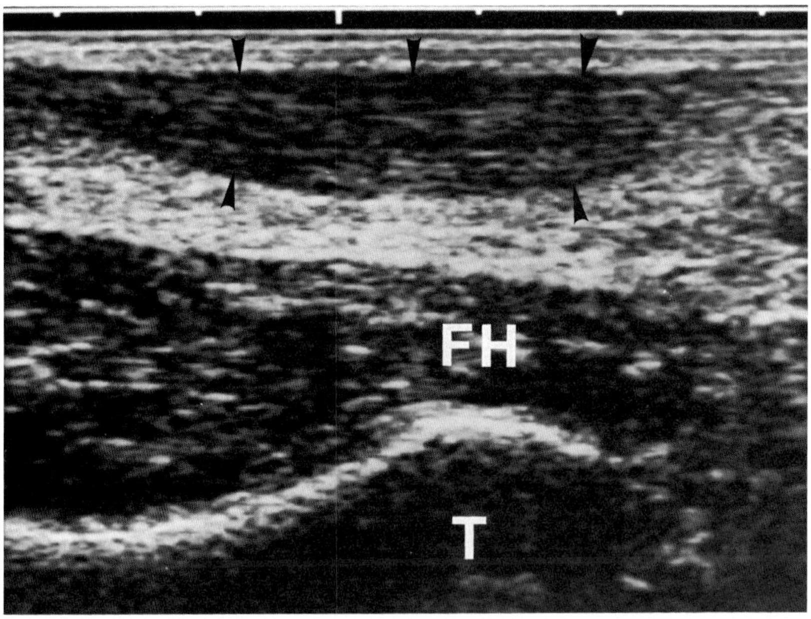


**FIG 4–30.**
Achilles tendonitis, proximal type: longitudinal sonogram. A jogger with continuous pain in the Achilles tendon. The image is a detail of a longitudinal sonogram through the Achilles tendon at the level of the distal tibia *(T)* and approximately 6 cm above the insertion. There is fusiform swelling of the tendon *(arrowheads)*. The tendon has a hypoechoic appearance, and the distance between longitudinal tendon fascicles is widened. The surrounding fat pad is hyperechoic and compressed between the tendon and the flexor hallucis longus muscle *(FH)*. A hyperechoic region is seen in the flexor hallucis longus, deep to the region of maximal tendon thickening.

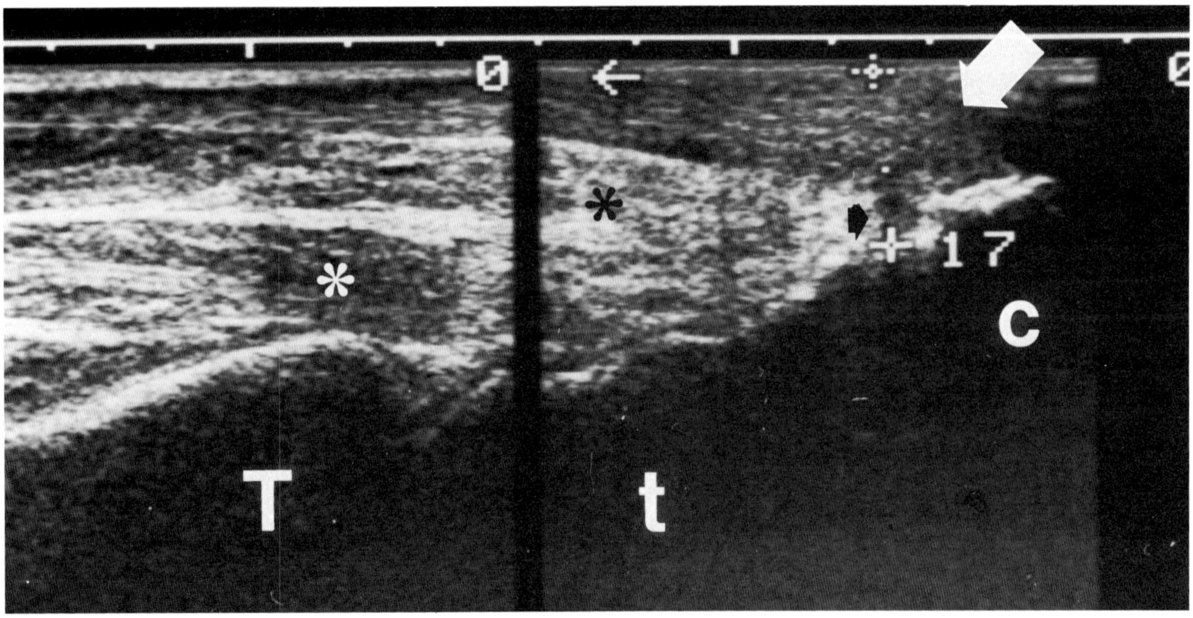


**FIG 4–31.**
Achilles tendonitis, distal type: longitudinal sonogram. After minor injury to the posterior aspect of the heel, this 20-year-old long distance runner developed pain located at the distal insertion of the Achilles tendon. The pain was worse at the beginning and end of exercise. The transducer is positioned dorsally over the Achilles tendon. An image of the entire tendon is produced using the image memory to juxtapose two images. Maximal thickening of the Achilles tendon is seen just superior to the calcaneus *(white arrow)*. In addition, there is inflammation of the retrocalcaneal bursa *(black arrow)*. The normal retrocalcaneal bursa is flat. Here it has an ovoid appearance. Abbreviations and symbols: pretendonous fat pad *(black asterisk)*, tibia *(T)*, talus *(t)*, calcaneus *(c)*, flexor hallucis longus muscle *(white asterisk)*.

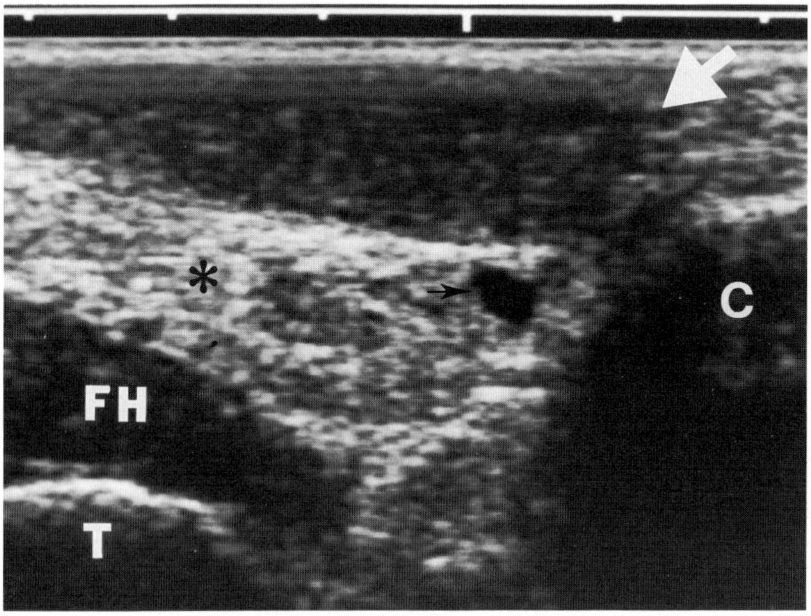


**FIG 4–32.**
Achilles tendonitis, distal type: longitudinal sonogram. Pain, erythema, and swelling were present over the Achilles tendon insertion in this 20-year-old basketball player. The image is a detail of a longitudinal sonogram through the Achilles tendon *(white arrow)* at the level of the calcaneus *(C)*. Fusiform swelling of the distal 5 cm of the tendon is present. The retrocalcaneal bursa is enlarged *(black arrow)*. Note the increased echogenicity of Kager's fat pad *(black asterisk)*. Abbreviations: tibia *(T)*, flexor hallucis longus muscle *(FH)*.

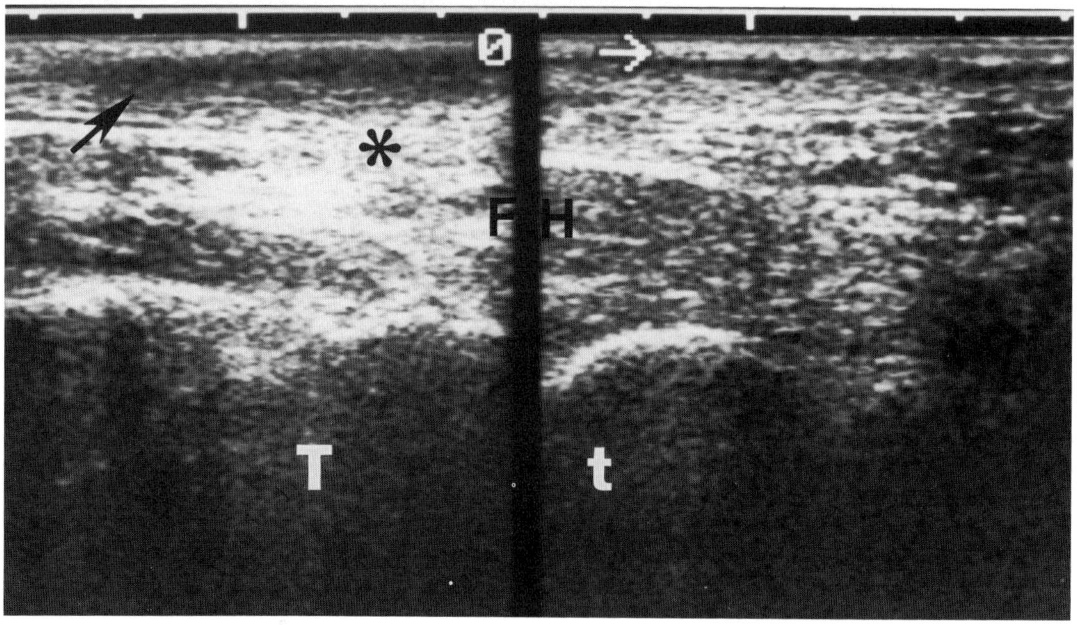


**FIG 4–33.**
Peritendonitis of the Achilles tendon: longitudinal sonogram. The patient is a long distance runner with pain in the Achilles tendon. The intensity of pain is related to the distance he runs. He has been forced to decrease his mileage due to the pain. The transducer is positioned in the long axis of the Achilles tendon over the dorsal aspect of the heel. Posterior to the tendon an abnormal hypoechoic region *(black arrow)* is present with increased through transmission. This was seen to extend laterally, surrounding the posterior aspect of the tendon. The posterior margin of the tendon is displaced anteriorly and is indistinct. Abbreviations and symbols: fat pad *(asterisk)*, flexor hallucis longus *(FH)*, tibia *(T)*, talus *(t)*.

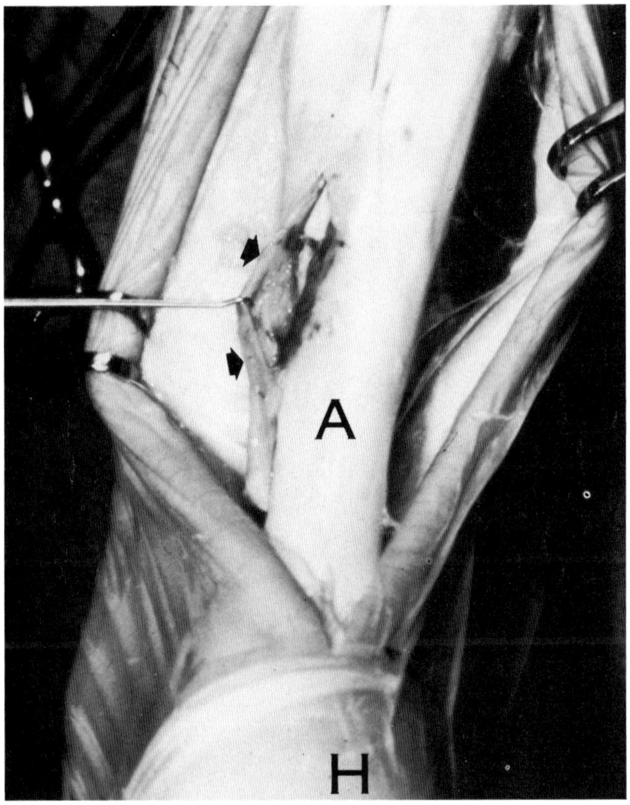


**FIG 4–34.**
Intraoperative photograph of peritendonitis of the Achilles tendon. Same patient as in Fig 4–33. The hypoechoic layer surrounding the posterior portion of the Achilles tendon represents the incised band of fibrous tissue being retracted laterally *(short arrows)*. This is secondary to a chronic inflammatory reaction in the peritenon.

5–13, and 5–27). In cases of inflammatory bursitis, irregular thickening of the wall of the bursa can often be detected sonographically. Traumatic bursitis carries a better prognosis than either tendonitis or the inflammatory bursitis.

An insertion tendonopathy is another item in the differential diagnosis of Achilles tendonitis. It is usually the result of traumatic overload of the tendon. The lesion is restricted to the fibrocartilage at the insertion of the Achilles tendon on the calcaneus. This region appears as a hypoechoic triangle at the junction of the tendon with the calcaneus. In cases of insertion tendinopathy, sonography shows thickening of the cartilaginous insertion of the Achilles tendon; in some cases calcific deposits are detected in an inflamed cartilaginous insertion. Therefore, we believe that the insertion tendinopathy of the Achilles tendon is closely related with calcific tendonitis. This type of lesion is not an indication for surgery (Fiamengo et al., 1982).

Tendonitis in other locations can be demonstrated sonographically. In the upper extremity the tendons of the brachial flexors and extensors may be examined. The quadriceps (suprapatellar) tendon, tendon of the hamstrings, adductor tendons, and pes anserinus can be examined in the lower extremity. Careful side to side comparison of identical segments is essential in these cases. In our experience, the sensitivity of sonographic examination in these regions is less than that for other more accessible areas. Hypoechoic swelling is seen in the majority of the cases. Echogenic foci with or without acoustic shadowing are present in cases of calcific tendonitis.

### *Tendon Rupture*

Rotator cuff tears are a frequent finding in the elderly and may be entirely asymptomatic. Autopsy studies of individuals more than 55 years old have discovered rotator cuff tears in 32% (Petersson et al., 1983). However, a true traumatic tear is a rare finding. These lesions are found in athletes who repeatedly

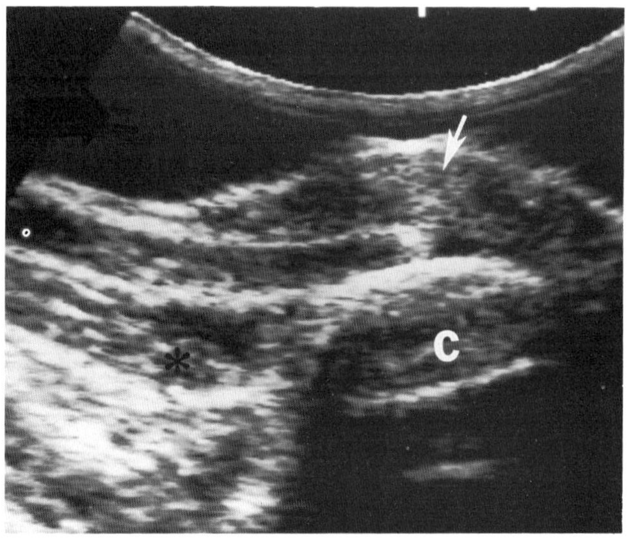


**FIG 4–35.**
Haglund's disease: longitudinal sonogram. An 18-year-old girl with pain and chronic swelling over the distal Achilles tendon. For this examination, a 5-MHz radial array transducer with standoff pad was utilized *(black arrow)*. The field of view is larger than that of a 5-MHz linear array, but images of the Achilles tendon and the tendon insertion suffer from decreased spatial resolution and anisotropy. The characteristic pump-bump of Haglund's disease is seen as thickening of the skin and subcutaneous tissues *(white arrow)*. Abbreviations: prominent calcaneus *(C)*, fat pad *(black asterisk)*.

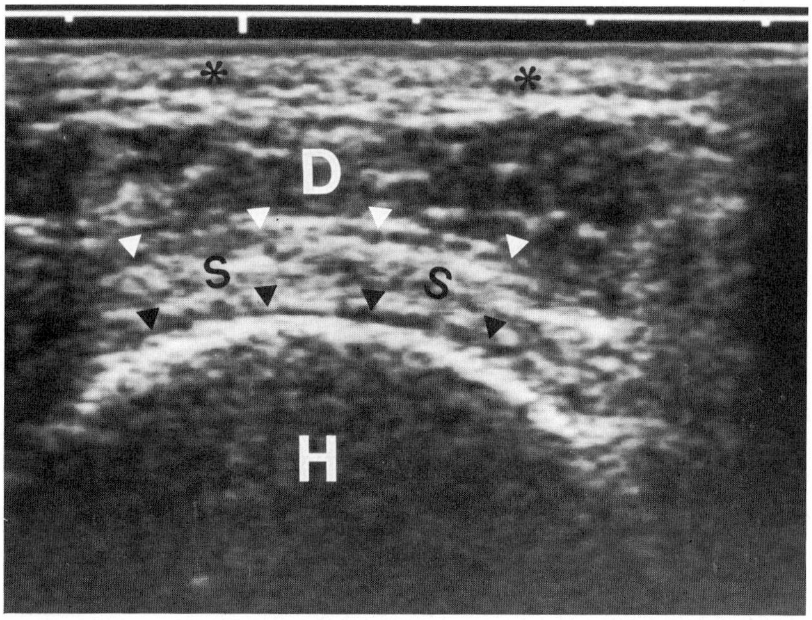


**FIG 4–36.**
Normal shoulder: axial sonogram. Asymptomatic young individual. The transducer is positioned horizontally over the shoulder joint, immediately inferior to the lateral end of the acromion. Thus, the image is comparable with a computed tomography (CT) image. The anatomy of this region has the sonographic appearance of a multilayer structure. The deltoid muscle *(D)* lies superficially, and the supraspinatus tendon *(S)* is slightly more reflective lying against the humerus *(H)*. Three other tissues are seen as boundaries around these muscles. The hypoderm *(asterisk)* is seen between the transducer and the deltoid muscle. The subdeltoid bursa *(white arrowhead)* is seen as a small hypoechoic layer between the deltoid muscle and supraspinatus tendon. The articular cartilage of the humeral head is a hypoechoic layer closely applied to the bone.

perform a throwing motion (e.g., baseball pitchers, javelin throwers, and football quarterbacks).

The principal sonographic findings (Fig 4–36) (Bretzke et al., 1985; Linnen et al., 1985) are fluid-filled clefts in the supraspinatus in axial images (Middleton et al., 1985a) and increased fluid in the subdeltoid bursa (Fig 4–37). In the elderly, tears may be visualized sonographically as focal areas of atrophy or as a hyperechoic streak in the rotator cuff. In some chronic cases of rotator cuff tear, the supraspinatus muscle layer may be completely absent (Mack et al., 1985). The diameter of the tear measured sonographically is always smaller than that observed with arthroscopy, plain film arthrography, or CT arthrography (Beltran et al., 1986) (Fig 4–38). In all of these invasive diagnostic modalities the joint is distended with fluid, air, or both. This leads to an exaggeration of the dimensions of the tear and degree of muscle retraction (Ahovuo et al., 1984). Tendon tears can be very small. In some cases, they will not extend to the joint capsule or the subacromial-subdeltoid bursa. These intrasubstance tears are interesting for diagnostic purposes, but they cannot be repaired surgically (Fig 4–39). Sonography, being entirely noninvasive, demonstrates the true extent of the lesion in vivo. This allows the surgeon to more accurately plan the procedure preoperatively.

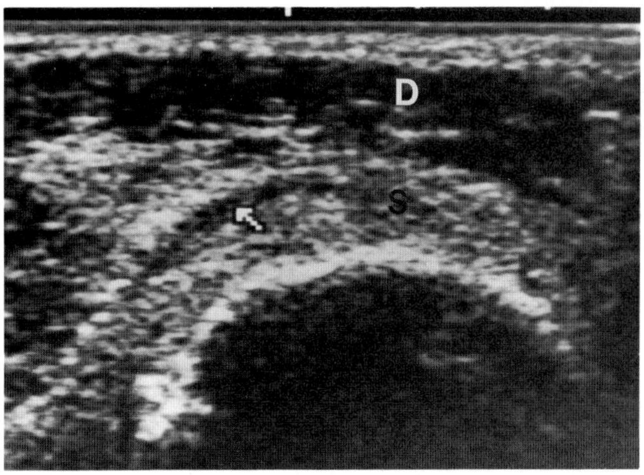


**FIG 4–37.**
Fluid-filled subdeltoid bursa: axial sonogram of the shoulder. A javelin thrower with pain in his right shoulder. Axial ultrasound examination with the transducer position identical to that in Figure 4–36. The thickness of the hypoechoic layer between the deltoid *(D)* and supraspinatus *(S)* is increased *(white arrow)*. This abnormal fluid collection is seen in inflammation of the rotator cuff resulting from impingement. In some patients, increased fluid within the subdeltoid bursa results from a rotator cuff tear communicating with that bursa. In this case, note that no fluid is present within the joint.

In cases of small tendon tears, the repair is often made by simple approximation of the two ends with sutures. The prognosis in these lesions is quite good. Large tears, greater than 5 cm, require more extensive procedures. Certain situations may require graft material. In these cases the prognosis is poor.

Three other disease processes are included in the differential diagnosis of soft tissue lesions of the shoulder. These are biceps tendonitis (Fig 4–40), biceps tendon dislocation (see Fig 4–9,C), and edema of the rotator cuff (Jobe and Jobe, 1983). Biceps ten-

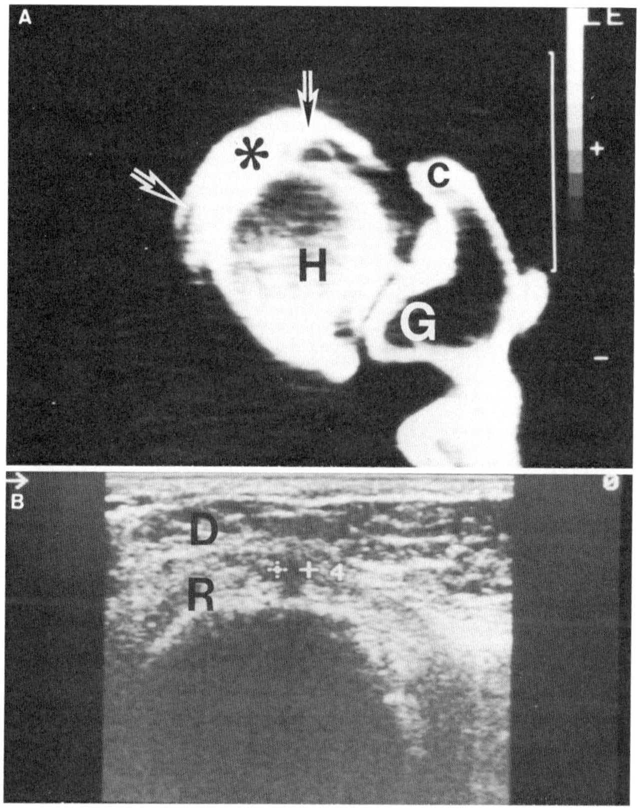


**FIG 4–38.**
**A,** rotator cuff tear: CT arthrogram. Contrast material fills the subdeltoid bursa through a large gap in the tendons of the rotator cuff *(asterisk)*. The torn ends *(arrows)* are separated by approximately 26 mm. Abbreviations: coracoid *(c)*, glenoid *(G)*, humeral head *(H)*. **B,** rotator cuff tear: axial sonogram. Same patient as in **A.** A hypoechoic cleft is seen in the rotator cuff *(crosses)* measuring 4 mm. The patient underwent surgical repair of this lesion and the size of the tear was approximately 4 mm. Arthrography grossly overestimates the extent of the lesion due to distention from contrast injection. Ultrasound, being an entirely noninvasive examination, accurately defines the extent of the lesion. Abbreviations: deltoid *(D)*, rotator cuff *(R)*.

donitis and biceps tendon dislocation have already been discussed in detail. The diagnosis of rotator cuff edema can be made sonographically when there is swelling of the rotator cuff sometimes in association with distention of the subacromial-subdeltoid bursa (Fig 4–41). Edema of the supraspinatus tendon is seen as enlargement of the affected tendon with a decrease in its echogenicity. This is the initial phase of the impingement syndrome (see chapter 13). Patients with chronic impingement eventually develop rotator cuff tears, some after 10 to 15 years of shoulder disease. It is generally believed that more than 95% of the rotator cuff tears are due to impingement

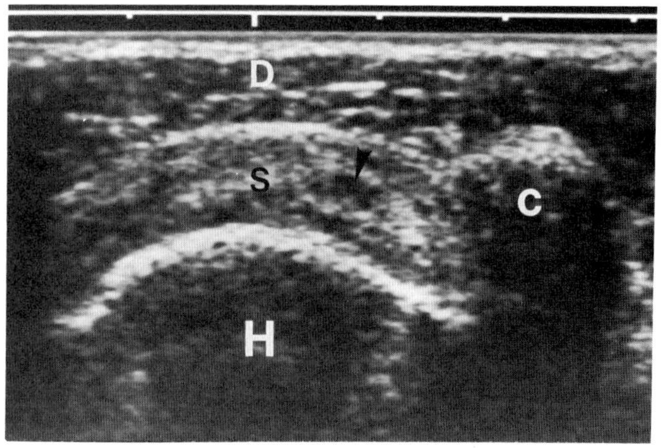


**FIG 4–39.**
Intratendonous supraspinatus tear: transverse sonogram of the right shoulder. A tennis player, 50 years old, with pain in the right shoulder worsened by abduction. Clinical examination demonstrated a painful arc. The transducer is positioned horizontally lateral to the tip of the acromion. The bony landmarks are the surface of the humeral head *(H)* and the coracoid *(c)*. The shoulder anatomy in this area is seen as a bilayer structure. The most superficial is the deltoid *(D)*. Directly overlying the bone is the supraspinatus tendon *(S)*, which is slightly more reflective than the deltoid. In the anterior portion of the tendon there is a hypoechoic region representing an incomplete tear *(black arrowhead)*. A normal subdeltoid bursa is seen between the deltoid muscle and the supraspinatus.

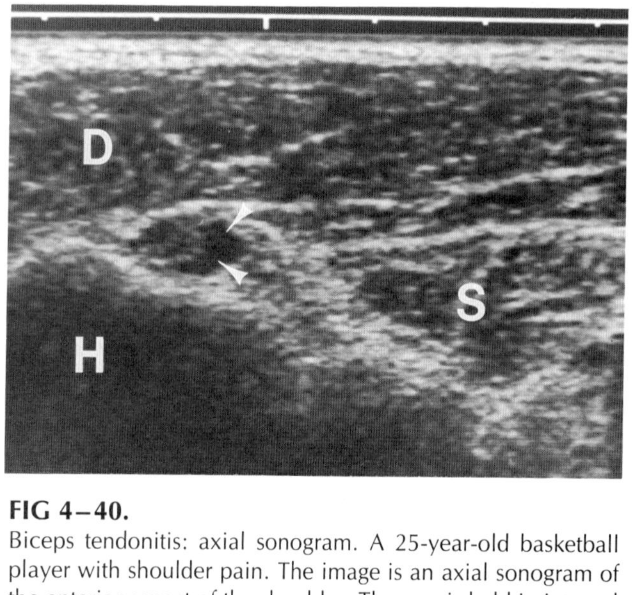


**FIG 4–40.**
Biceps tendonitis: axial sonogram. A 25-year-old basketball player with shoulder pain. The image is an axial sonogram of the anterior aspect of the shoulder. The arm is held in internal rotation, and the transducer is positioned over the bicipital groove of the humerus *(H)*. In the bicipital groove, the biceps tendon is seen surrounded by fluid *(white arrowheads)*. Bicipital tendonitis is a frequent finding in impingement syndrome. Abbreviations: deltoid muscle *(D)*, subscapularis muscle *(S)*.

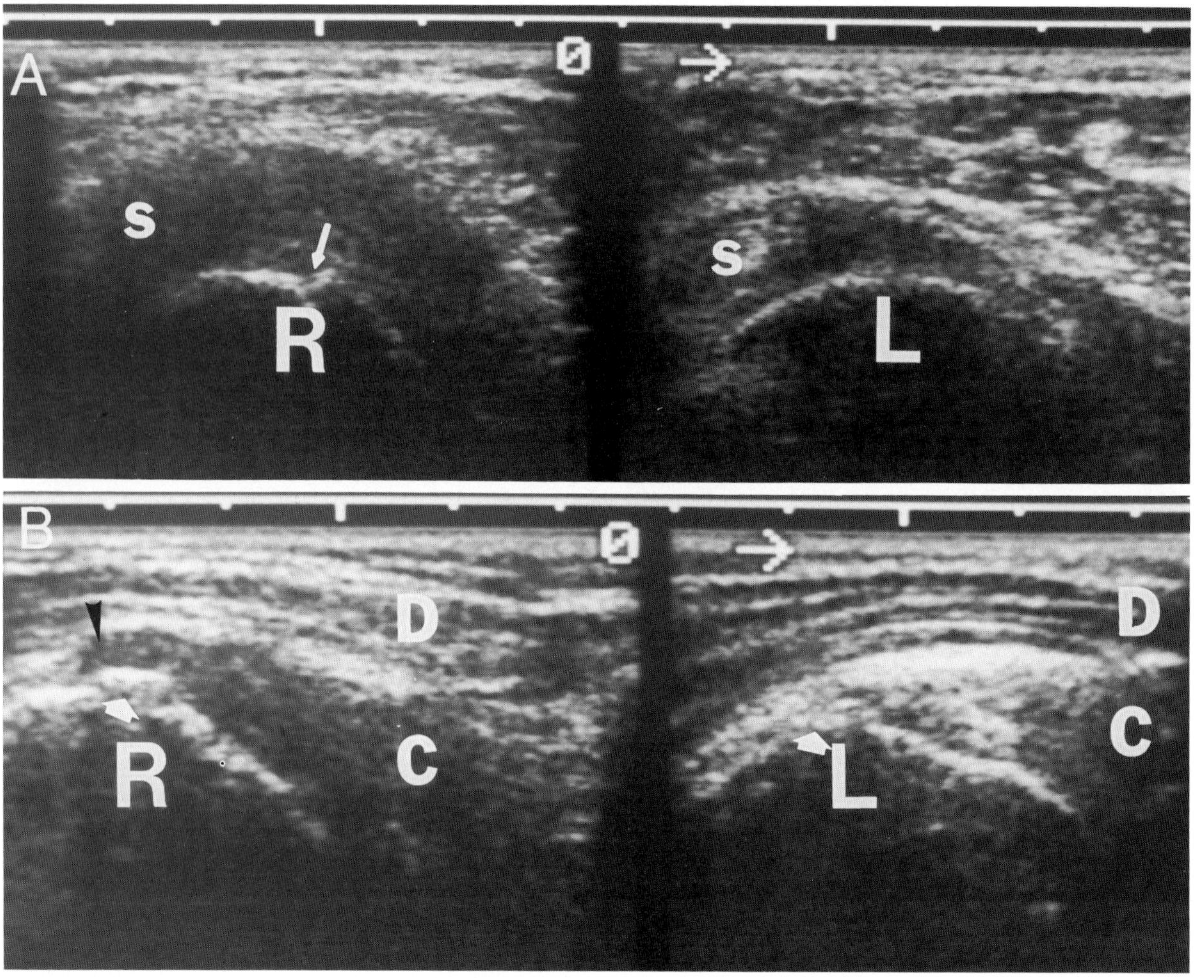


**FIG 4–41.**
**A,** impingement syndrome with edema of the rotator cuff: transverse sonograms. An axial sonogram of the right *(R)* and left *(L)* shoulders. The patient is a house painter with pain in his right shoulder. Clinically, he has a painful arc and is tender over the rotator cuff. The transducer is positioned lateral to the acromion. The supraspinatus tendon *(s)* of the symptomatic right shoulder is swollen, greater than 1 cm in diameter. The edematous supraspinatus tendon has a hypoechoic appearance relative to the deltoid and contralateral supraspinatus tendon. The right humeral head has an irregular outline *(arrow)*. **B,** impingement syndrome with edema of the rotator cuff: coronal sonogram. Same patient as in **A.** The transducer is positioned perpendicular to that used in **A** to obtain coronal images of the shoulder. The normal left rotator cuff has a triangular shape. It is slightly more reflective than the overlying deltoid muscle *(D)*, and the cuff inserts on the greater tuberosity *(arrow)*. The edematous right rotator cuff *(C)* is hypoechoic and swollen. The insertion bulges over the insertion area on the greater tuberosity *(black arrowhead)* and has a more irregular appearance. The insertion is more rounded, losing its triangular shape. The greater tuberosity *(arrow)* on the *right* is more irregular in shape. Abbreviations: right *(R)*, left *(L)*.

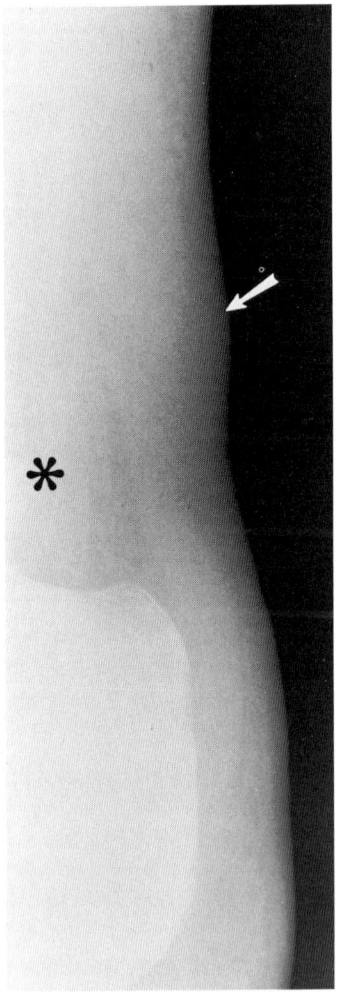


**FIG 4–42.**
Tendon rupture: Gevaert MRL radiograph. A patient with known tendonitis has increasing pain in his right Achilles tendon. He cannot stand on the tips of his toes. The Thompson sign was inconclusive, and a gap could not be palpated in the tendon. A lateral radiograph demonstrates thickening of the Achilles tendon *(arrow)*, and Kager's fat pad *(black asterisk)* appears fuzzy. With radiography it is not possible to differentiate between tendonitis and tendon rupture.

of the cuff between the humeral head and the acromioclavicular joint (Neer, 1983; Jobe and Jobe, 1983).

In contrast to rotator cuff tears, rupture of the Achilles tendon is almost exclusively traumatic in origin. These lesions are found in athletes engaged in high-level competition. The diagnosis is usually made clinically but may be overlooked if the rupture is incomplete. The plain film findings are nonspecific (Fig 4–42). In a series of 14 cases of Achilles tendon ruptures, 3 were referred to us with the clinical diagnosis of tendonitis. Early diagnosis is imperative as delay can lead to permanent functional impairment.

Sonographic finding in acute cases of Achilles tendon rupture are interruption of tendon fibers and fluid-filled clefts within the tendon (Fig 4–43). In most cases, the cleft will be located approximately 6 to 10 cm proximal to the tendon insertion (Figs 4–44 and 4–45). Surrounding hematoma will be minimal due to the poor vascularity of the tendon. The plantaris tendon is often seen to be intact, even in cases of complete Achilles tendon rupture. This is identified as a small strand at the medial aspect of the Achilles tendon (Fig 4–46). Occasionally, fusiform contracture of the distal portion of the tendon can be seen (Figs 4–47 and 4–48). This is often misinterpreted as tendonitis. On making this finding, the examiner should always check for an interruption in the tendon more proximally. Surrounding fat is often injured at the time of tendon rupture. Observation of a cleft within the fat helps to support the diagnosis (Figs 4–48 and 4–49). Changes in echotexture of the fat can be seen during and after healing of the tendon, sometimes months following trauma (Figs 4–50 and 4–51).

Herniation of fat through the site of rupture may occur in chronic cases. This will be seen as a small hyperechoic nodule protruding through the cleft in the tendon (Figs 4–52 through 4–55). A high-quality conventional radiograph of the soft tissues will demonstrate this as radiolucent deposits within the

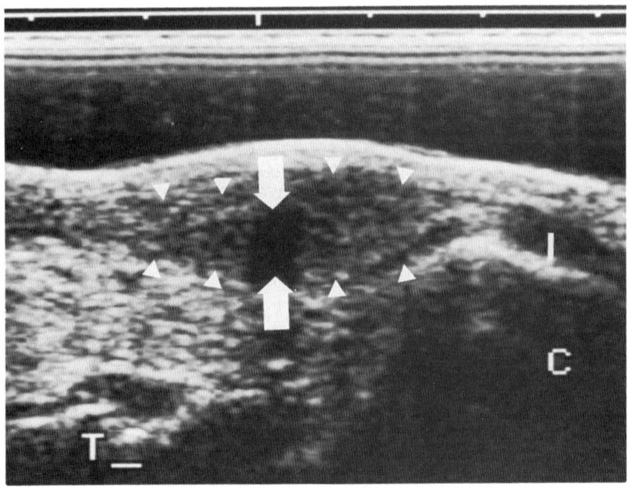


**FIG 4–43.**
Complete Achilles tendon rupture: longitudinal sonogram. Same patient as in Figure 4–42. The 7.5-MHz linear array transducer is positioned dorsally over the Achilles tendon. The distal portion of the tendon, the insertion *(I)*, is at the observer's right. Tendon rupture is seen as a hypoechoic gap in the tendon *(white arrows)*. There is fusiform contraction of the tendon segments, greater in the distal portion *(white arrowheads)*. Abbreviations: tibia *(T)*, calcaneus *(c)*.

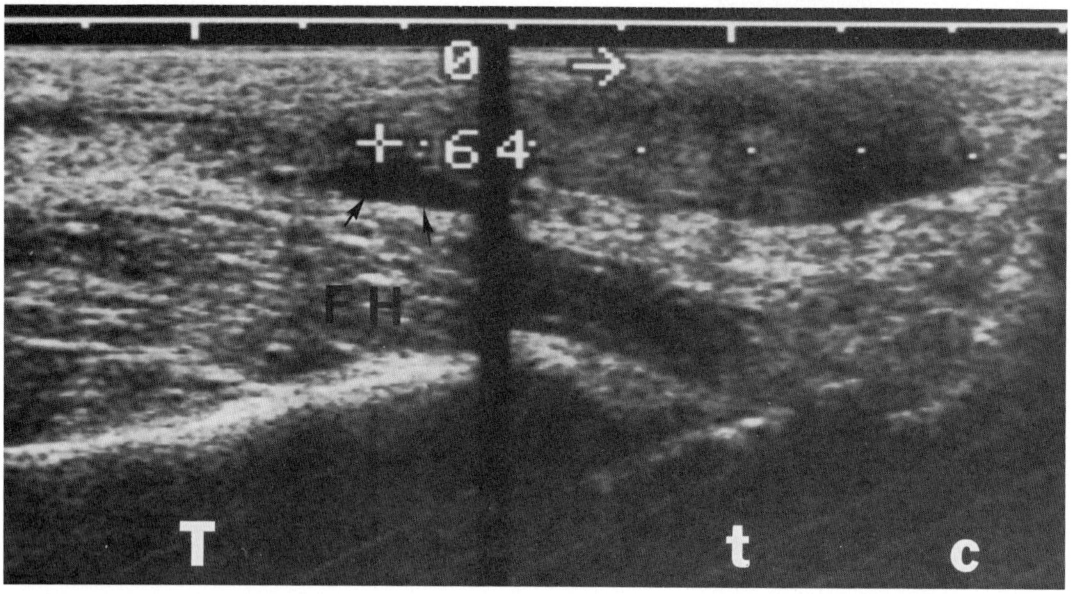


**FIG 4–44.**
Complete Achilles tendon rupture: longitudinal sonogram. A 50-year-old woman with asthma taking corticosteroids for chronic therapy. She felt a sudden sharp pain in the Achilles tendon while walking up stairs. Physical examination revealed absence of plantar flexion and a positive Thompson sign. The clinician palpated a gap in the Achilles tendon. The transducer is positioned dorsally over the Achilles tendon. Using the image memory, an image was constructed of the entire Achilles tendon. Ultrasound determines the exact location of the tendon rupture *(white cross)*, and the distance from the rupture to the tendon insertion can be accurately measured. The distance is 64 mm in this case. Fusiform contraction of the distal tendon fragment is also seen *(white arrowheads)*. A small amount of blood surrounds the area of rupture *(small arrows)*. Oblique white lines in the lower portion of the figure are artifact due to electrical interference. Abbreviations: tibia *(T)*, talus *(t)*, calcaneus *(c)*, flexor hallucis longus muscle *(FH)*.

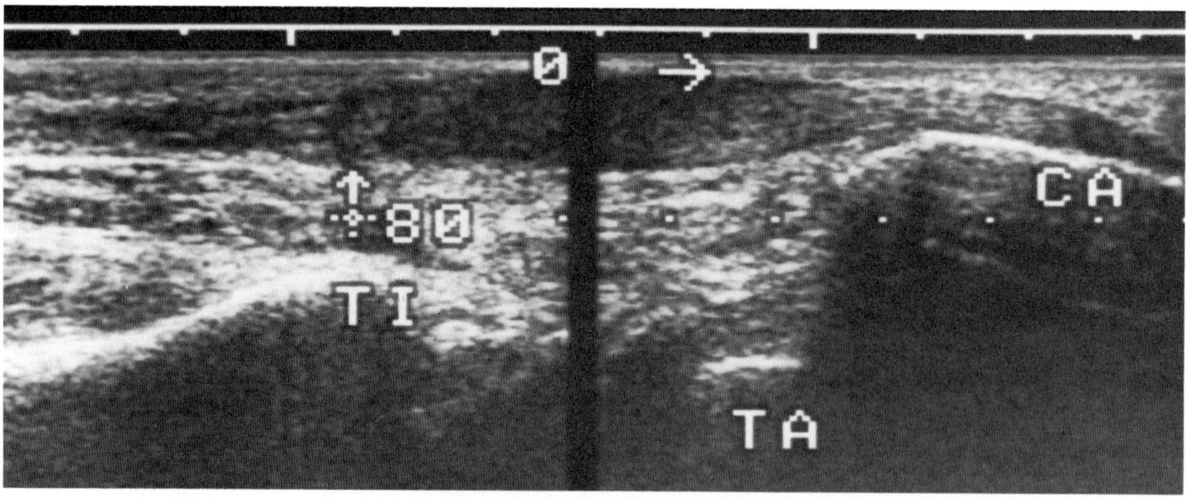


**FIG 4–45.**
Complete Achilles tendon rupture: longitudinal sonogram. This 40-year-old tennis player felt a crack in his heel during a vigorous vertical jump. Clinical examination revealed some swelling of the Achilles tendon, but the clinician could not palpate a gap in the tendon. The Thompson sign was positive. Examination technique was identical to that used in Figure 4–44. An interruption in the tendon fibers is seen 80 mm proximal to the tendon insertion. There was minimal bleeding, and therefore the tendon rupture is seen as a small hypoechoic band *(white arrow)*. As you can see, most of the tendon ruptures occur at the level of the tibial epiphysis *(TI)*. Abbreviations: talus *(TA)*, calcaneus *(CA)*.

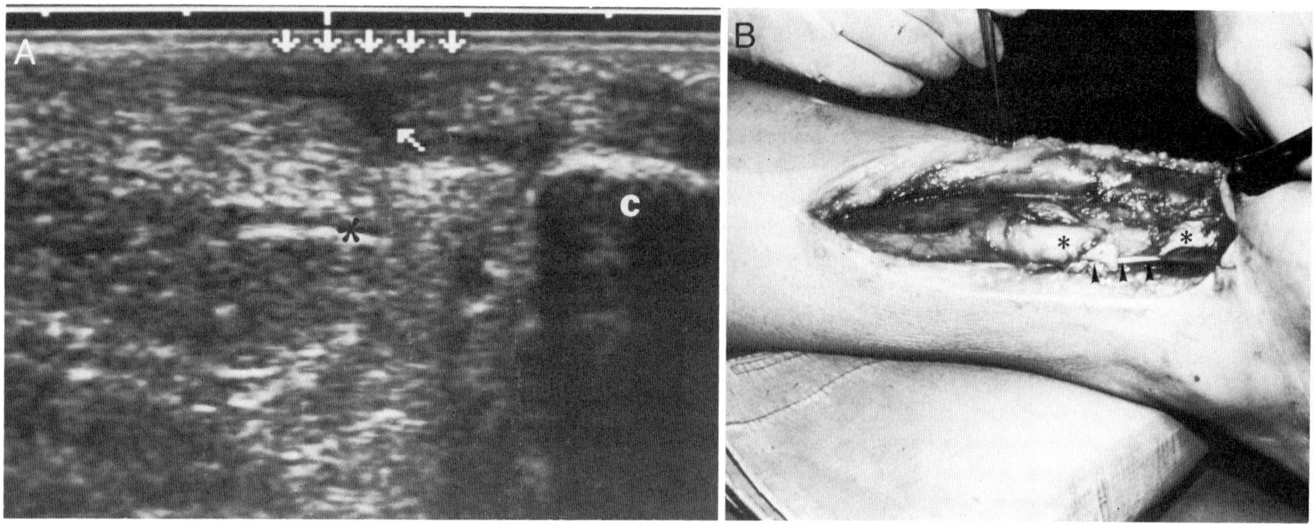


**FIG 4–46.**
**A,** complete Achilles tendon rupture: longitudinal sonogram. This figure is a magnification sonogram of the site of a complete tendon rupture. The rupture of the tendon *(oblique arrow)* was called complete by the surgeon, although the plantaris tendon was intact *(small arrows)*. Symbols and abbreviations: fat pad *(asterisk)*, calcaneus *(c)*. **B,** complete Achilles tendon rupture: intraoperative photograph. Same patient as in **A.** The operative site is viewed from the medial with the skin being retracted laterally. Contracted proximal and distal segments of the ruptured Achilles tendon are seen centrally in the surgical field *(asterisk)*. The intact plantaris tendon just medial and deep to the ruptured Achilles tendon *(arrowheads)*.

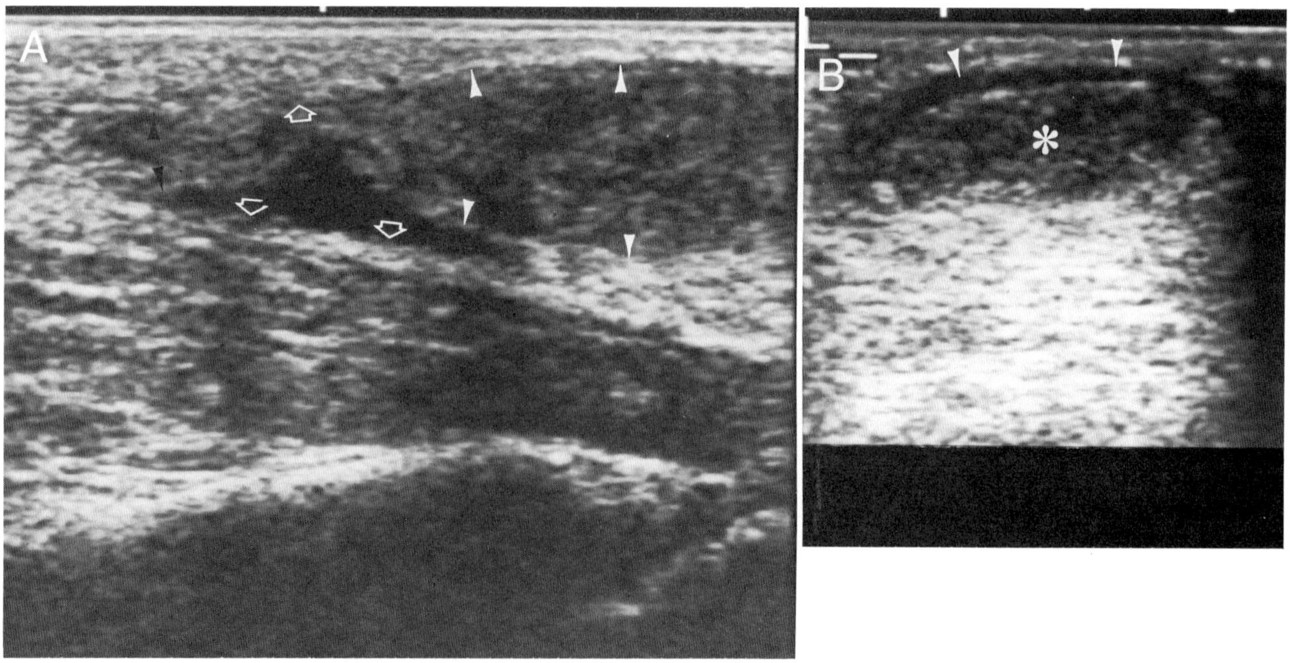


**FIG 4–47.**
**A,** complete rupture of the Achilles tendon: longitudinal sonogram. This is a detailed sonogram of the site of rupture. The retracted distal portion *(solid white arrowheads)* is generally thicker than the retracted proximal portion *(black arrowheads)*. Hematoma surrounds the two ends of the ruptured tendon. In this case, the hematoma is confined within the bounds of the peritenon *(open white arrows)*. Flexor hallucis longus muscle *(FH)*, tibia *(T)*. **B,** complete rupture of the Achilles tendon: transverse sonogram. Same patient as in **A.** The transducer is oriented perpendicular to the long axis of the tendon. This is a detailed view of the proximal portion of the tendon above the site of rupture. The retracted thickened proximal tendon *(white asterisk)* is seen surrounded by hematoma within the peritenon *(arrows)*.

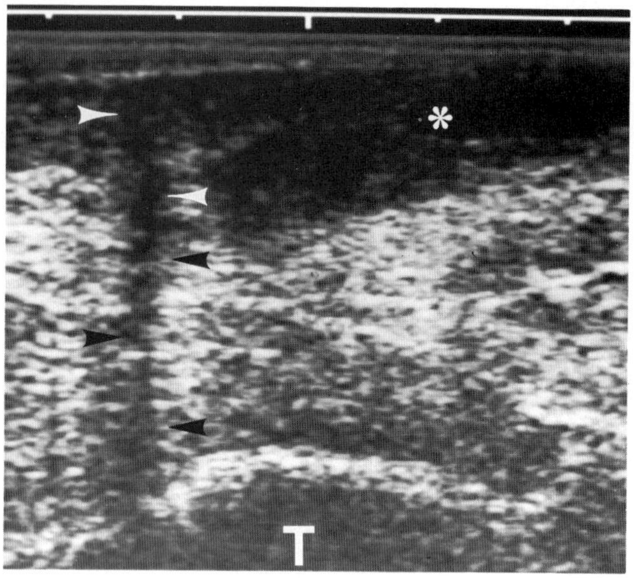


**FIG 4–48.**
Complete Achilles tendon rupture: longitudinal sonogram. A detail of the site of rupture demonstrates the hypoechoic gap in the tendon *(white arrowheads),* which extends into the peritendonous fat as a hypoechoic band *(black arrowheads).* This is a frequent finding in complete tendon rupture. Whether this represents a sonographic artifact or a true gap in the fat pad is unclear. Tibia *(T),* contracted distal portion of the tendon *(white asterisk).*

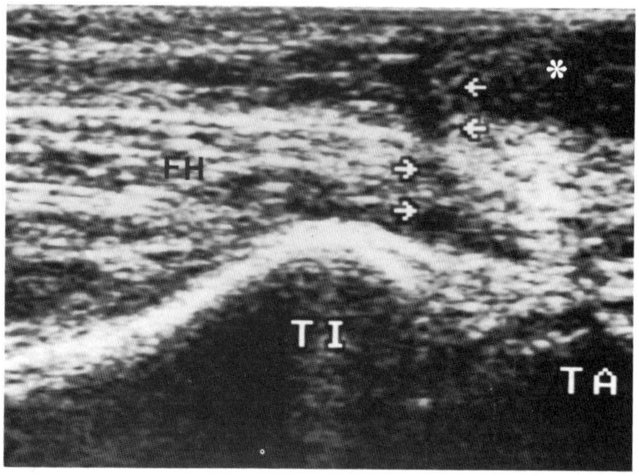


**FIG 4–49.**
Complete Achilles tendon rupture: longitudinal sonogram. Another patient with complete Achilles tendon rupture. This image provides a larger field of view and demonstrates the same sonographic findings as seen in Figure 4–48. Extension of the hypoechoic gap in the tendon into the fat pad is indicated by the *white arrows.* Symbols and abbreviations: contracted distal portion of the tendon *(white asterisk)* tibia *(TI),* flexor hallucis longus *(FH),* talus *(TA).*

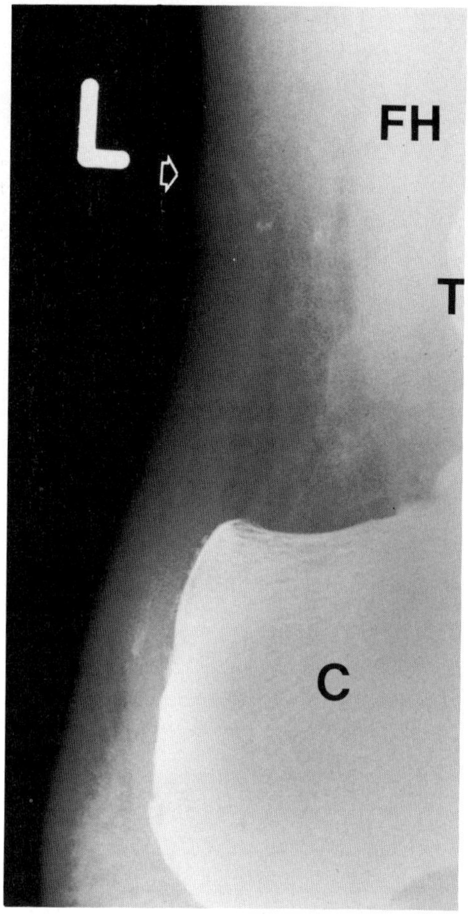


**FIG 4–50.**
Achilles tendon healing: Gevaert MRL radiograph (lateral view). The patient sustained a tendon rupture 6 months ago. He was treated conservatively with cast immobilization. The area of tendon rupture is still swollen *(open arrow).* Calcifications are present at the site of rupture and also at the tendon insertion. Abbreviations: flexor hallucis longus *(FH),* tibia *(T),* calcaneus *(c).*

Achilles tendon. Healing of these lesions occurs only after removal of the interposed fat.

Quadriceps tendon rupture is a rare finding, even in athletes (Roels and Martens, 1978). A complete rupture of the tendon is often the end stage of patellar tendonitis, especially if further overload and weakening of the tendon are facilitated by corticosteroid injections. Prior to the use of ultrasound in the evaluation of musculoskeletal pathology, this diagnosis could be proved only by arthrography (Aprin and Broukhim, 1985) or even more invasive exploratory surgery. Now, sonography can demonstrate the discontinuity with an examination that is fast, noninvasive, cost effective, and reliable.

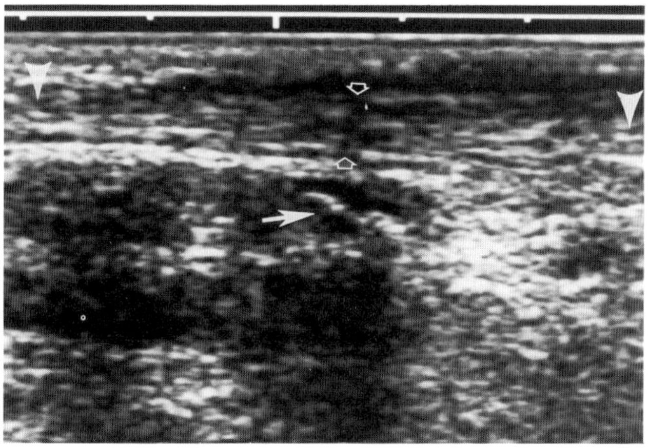

**FIG 4–51.**
Achilles tendon healing: longitudinal sonogram. Same patient as in Figure 4–50. The image is a detail of a longitudinal sonogram of the Achilles tendon 6 months following injury. The hypoechoic cleft *(open arrow)* is still seen, but now some tendon fibers cross the cleft. Structure of the tendon in this area is hypoechoic compared with the more normal proximal and distal tendon segments *(white arrowheads)*. The calcifications *(white arrow)* seen in Fig 4–50 are seen in this image to lie within the fat pad. Tendon rupture appears to result in damage to the fat pad as well.

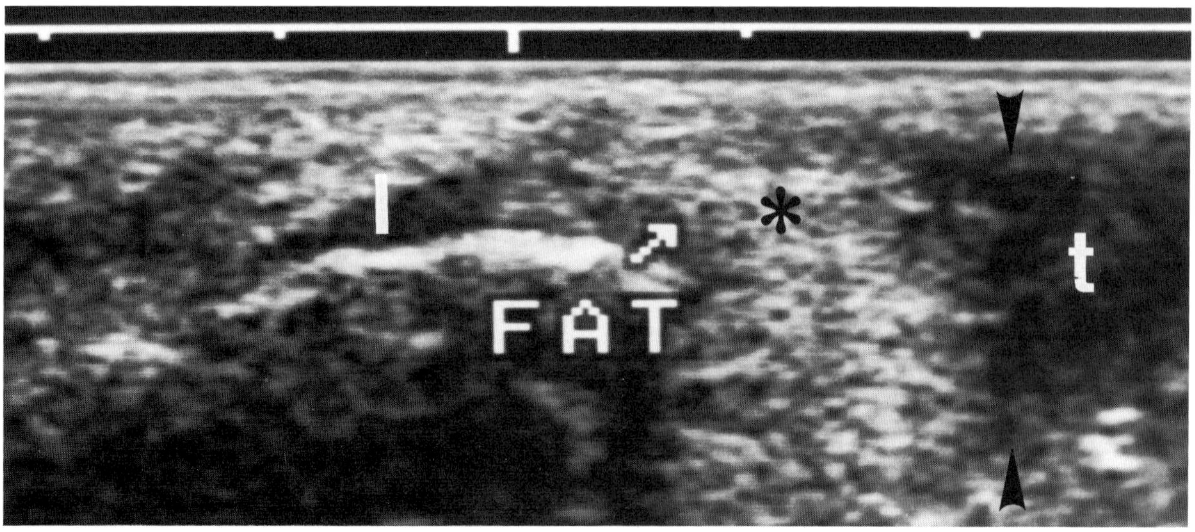


**FIG 4–52.**
Chronic Achilles tendon rupture: longitudinal sonogram. This patient ruptured his left Achilles tendon 1 year ago. The rupture was not recognized immediately, and treatment was started 1 month after the accident. Now, 1 year later, the patient is unable to plantarflex his foot. A longitudinal sonogram was obtained by positioning the transducer over the dorsal aspect of the tendon. The insertion *(I)* is seen on the *left*, and the proximal part of the tendon is at the *right*. The ruptured tendon is widened and appears hypoechoic *(black arrowheads)*. Compared to the ruptured tendon, the surrounding fat is hyperechoic *(black asterisk)*. This fat has herniated through the gap in the tendon at the site of rupture *(arrow)*, thus impairing healing. Retracted proximal portion of the tendon *(t)*.

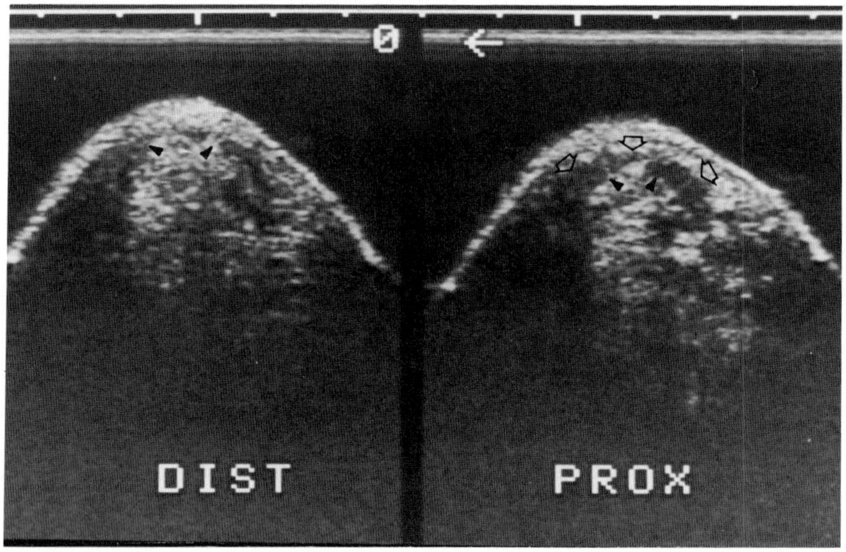


**FIG 4–53.**
Chronic Achilles tendon rupture: transverse sonogram. Same patient as in Figure 4–52. The transducer is now positioned perpendicular to the long axis of the tendon. The image on the *left* depicts the site of tendon rupture in cross section *(DIST)*. The image on the *right* is made proximal to the site of rupture *(PROX)*. The free distal end of the tendon is displaced medially by hyperechoic fat *(black arrowheads)*. Centrally, in the proximal portion of the ruptured tendon *(open arrows),* hyperechoic fat is found *(black arrowheads)*. The tendon is again deviated medially.

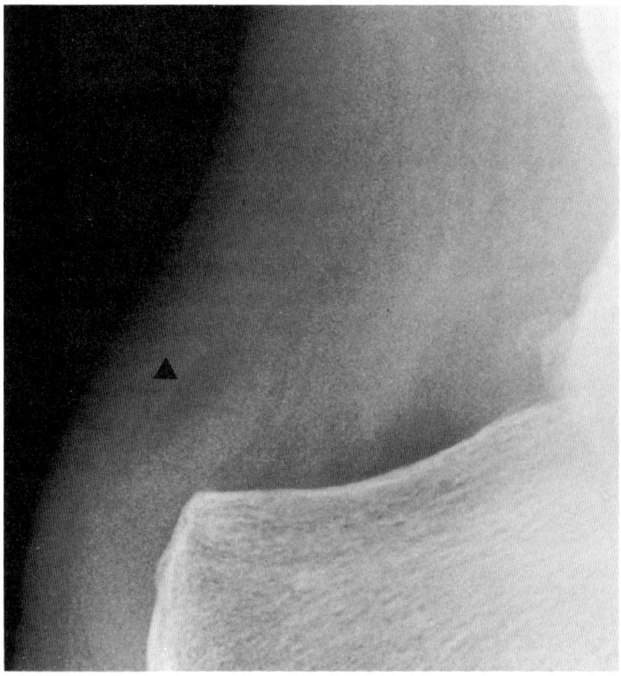

**FIG 4–54.**
Chronic Achilles tendon rupture: Gevaert MRL radiograph (lateral view). Same patient as in Figures 4–52 and 4–53. Fat is seen as a radiolucent area *(arrowhead)* in the substance of the distal tendon.

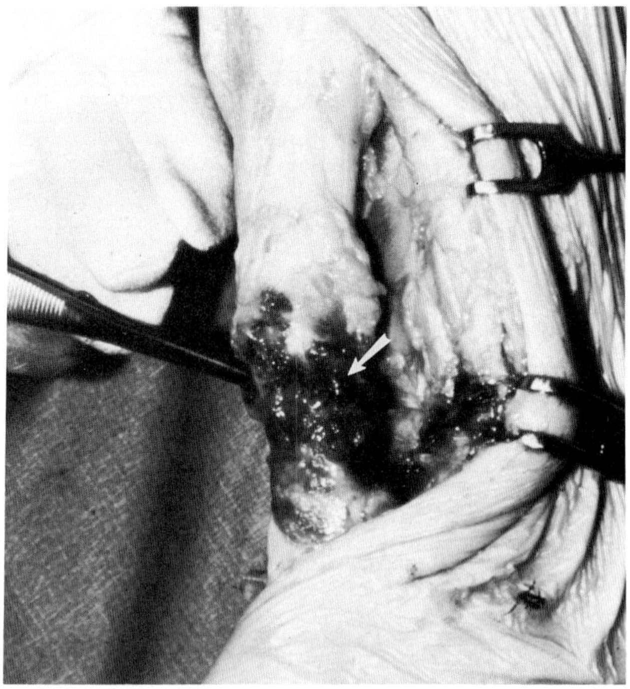

**FIG 4–55.**
Chronic Achilles tendon rupture: intraoperative photograph. Same patient as in Figures 4–50, 4–53, and 4–54. Fat was found separating the ends of the ruptured tendon at the time of surgery *(arrow)*. Necrotic debris was also present at the rupture site.

## REFERENCES


Ahovuo J, Paavolainen P, Slatis P: The diagnostic value of arthrography in rotator cuff tears. *Acta Orthop Scand* 1984; 55:220–223.

Aisen AM, McCune WJ, MacGuire A, et al: Sonographic evaluation of the cartilage of the knee. *Radiology* 1984; 153:781.

Aprin H, Broukhim N: Early diagnosis of acute rupture of the quadriceps tendon by arthrography. *Clin Orthop* 1985; 195:185–190.

Beltran J, Gray LA, Bools JC, et al: Rotator cuff lesions of the shoulder: Evaluation by direct sagittal CT arthrography. *Radiology* 1986; 160:161–165.

Blei CL, Nirschl RP, Grant EG: Achilles tendon: Ultrasound diagnosis of pathologic conditions. *Radiology* 1986; 159:765–767.

Bock E, Colavita N, Cotroneo AR, et al: Xeroradiography of tenomuscular traumatic pathologic conditions of the limbs. *Diagn Imag Clin Med* 1981; 50:235–248.

Bretzke LA, Crass JR, Craig EV, et al: Ultrasonography of the rotator cuff: normal and pathologic anatomy. *Invest Radiol* 1985; 20:311–315.

Bruce RK, Hale TL, Gilbert SK: Ultrasonographic evaluation for ruptured Achilles tendon. *J Am Pediatr Med Assoc* 1982; 72:15–17.

Bywaters EGL: Lesions of the bursae, tendons and tendon sheaths. *Clin Rheum Dis* 1979; 5:885.

Canoso JJ: Aspiration of the retrocalcaneal bursa. *Ann Rheum Dis* 1984; 43:308–312.

Chui MT: Shoulder ultrasound finds more than just cuff tears. *Radiol Today* 1988; 5:20.

Church CC: Radiographic diagnosis of acute peroneal tendon dislocation. *AJR* 1977; 129:1065–1068.

Clement DB, Taunton JE, Smart GW: Achilles tendonitis and peritendonitis: Etiology and treatment. *Am J Sports Med* 1984; 12:179–187.

Cooper RR, Misol S: Tendon and ligament insertion. A light and electron microscopic study. *J Bone Joint Surg (Am)* 1970; 52:1–20.

Crass JR, Craig EV, Thompson RC, et al: Ultrasonography of the rotator cuff: Surgical correlation. *J Clin Ultrasound* 1984; 12:487–491.

Crass JR, Craig EV, Feinberg SB: Sonography of the postoperative rotator cuff. *AJR* 1986; 146:561–564.

Das De S, Balasubramaniam P: A repair operation for recurrent dislocation of peroneal tendons. *J Bone Joint Surg (Br)* 1985; 67:585–587.

Demarais Y, Houles JP, Parier JH, et al: Echoscannographie dans les tendinites achilléennes et rotuliennes specialement chez le sportif. *Cinesiologie* 1984; 23:249–256.

Dillehay GL, Deschler T, Rogers LF, et al: The ultrasonographic characterization of tendons. *Invest Radiol* 1984; 19:338–341.

Eichelberger RP, Lichtenstein P, Brogdon BG: Peroneal tenography. *JAMA* 1982; 247:2587–2591.

Feretti A, Puddu G, Mariani PP, et al: The natural history of jumper's knee. Patellar or quadriceps tendonitis. *Int Orthop* 1985; 8:239–242.

Fiamengo SA, Warren RF, Marshall JL, et al: Posterior heel pain associated with a calcaneal step and Achilles tendon calcification. *Clin Orthop* 1982; 167:203–211.

Fornage B, Rifkin MD, Touche DH, et al: Sonography of the patellar tendon: Preliminary observations. *AJR* 1984; 143:179–182.

Fornage B: Achilles tendon: US examination. *Radiology* 1986; 159:759–764.

Hardaker WT Jr, et al: Foot and ankle injuries in theatrical dancers. *Foot Ankle* 1985; 6:59–69.

Jobe FW, Jobe CM: Painful athletic injuries of the shoulder. *Clin Orthop* 1983; 173:117–124.

Linnen C, et al: US evaluation. of the rotator cuff. *Radiology* 1985; 157:205–209.

Mack LA, Matsen FA, Kolcoyne HF, et al: US evaluation of the rotator cuff. *Radiology* 1985; 157:205–209.

Martens M, Wouters P, Burssens A, et al: Patellar tendinitis: Pathology and results of treatment. *Acta Orthop Scand* 1982; 53:445–450.

McConkey JP, Favero KJ: Subluxation of the peroneal tendons within the peroneal tendon sheath. *Am J Sports Med* 1987; 15:511–513.

Merkel KHH, Hess H, Kunz M: Insertion tendonopathy in athletes. *Pathol Res Pract* 1982; 173:303–309.

Middleton WD, Edelstein G, Reinus WR, et al: Sonographic detection of rotator cuff tears. *AJR* 1985a; 144:349–353.

Middleton WD, Reinus WR, Melson GL, et al: Pitfalls of rotator cuff sonography. *AJR* 1986a; 146:555–560.

Middleton WD, Reinus WR, Totty WG, et al: US of the biceps tendon apparatus. *Radiology* 1985b; 157:211–215.

Middleton WD, Reinus WR, Totty WG, et al: Ultrasonic evaluation of the rotator cuff and biceps tendon. *J Bone Joint Surg (Am)* 1986b; 68:440–450.

Neer CS: Impingement lesions. *Clin Orthop Rel Res* 1983; 173:70-77.

Niepel GA, Sit AJ : Enthesopathy. *Clin Rheum Dis* 1979; 5:857–872.

O'Donoghue DH: Subluxing biceps tendon in the athlete. *Clin Orthop* 1982; 164:26–29.

Oden RR: Tendon injuries about the ankle resulting from skiing. *Clin Orthop* 1987; 216:63–69.

Palvalgyi R, Balint BJ: Radiographic examination of the large tendons of the knee and ankle joints. *Arch Orthop Trauma Surg* 1981; 98:19–24.

Petersson CJ, Gentz CF: Ruptures of the supraspinatus tendon. *Clin Orthop* 1983; 174:000.

Petersson CJ: Spontaneous medial dislocation of the tendon of the long biceps brachii. *Clin Orthop* 1986; 211:224–227.

Roels J, Martens M: Patellar tendonitis (jumper's knee). *Am J Sports Med* 1978; 6:362–368.

van Holsbeeck M: Medische Beeldvorming en Diagnose van Sportletsels, in *Praktische Behandeling van Sportletsels.* CED, Samson, 1987, chapter 3.

Warwick R, Williams PL: *Gray's Anatomy* ed 36. Philadelphia, WB Saunders Co, 1980.

Zuinen C, et al: Les lesion chirurgicales du tendon d'Achille. *Sport Med* 1983; 25:7–11.

# 5

# Sonography of Bursae

The word bursa is derived from the Greek, meaning a wine skin. In modern usage, bursa refers to a variety of structures that have several features in common. They are all saclike structures with a lining similar to that found in diarthrodial joints. In addition, they are situated to facilitate movement of musculoskeletal structures (Canoso, 1981). Bursae are found in areas where a considerable degree of motion is required, yet a cartilaginous joint is not necessary.

The analogy of a wine skin is quite appropriate; both have their greatest dimensions in length and width. This provides a large surface area occupying little volume under normal circumstances (Codman, 1931). When positioned between two structures, this configuration allows mobility and gliding of one structure on the other. Bursae can develop de novo in response to friction and pressure. Examples of these are over the bony prominence of a hallux valgus, exostosis, or an irregular spinous process (Baastrup's and Pott's disease).

Some anatomists (Gray, Piersol) believe that bursae are "sacs filled with fluid." This is a misconception resulting from studies performed by injecting bursae with various materials (Fig 5–1). Physicians who perform bursography and bursoscopy often share this misconception because they distend the bursae with contrast material or irrigation fluid. In reality, bursae contain only a thin film of viscous fluid, which serves as a lubricant. The walls are separated by a fluid film approximately 1 mm thick. Therefore, bursae are really potential spaces, only becoming fluid filled sacs under pathological conditions (van Holsbeeck and Introcaso, 1989).

Bursae are divided into two groups, communicating and noncommunicating, depending on their relationship to a joint space. Noncommunicating bursae are more common in humans. Further categorization of bursae may be made based on their location: subcutaneous or deep (Table 5–1) (Canoso, 1981). Subcutaneous bursae are located between a bone and the overlying skin, such as the prepatellar and olecranon bursae. Deep bursae are located in various places deep to the investing fascia. They separate the joint capsule, tendons, ligaments, and fascial planes (iliotibial tract).

## EXAMINATION TECHNIQUE

Selection of a transducer is based on the location of the bursa being examined. Subcutaneous bursae are best examined using 7.5- or 10-MHz transducers. In addition, these superficial bursae may require a standoff for optimal evaluation. In obese patients, a 5-MHz transducer may be required. Deep bursae usually require a 5-MHz transducer. On rare occasions, a 3-MHz transducer will be required due to the patient's body habitus, limiting the examination. In all cases a linear array transducer is utilized. Peritendinous bursae are most easily examined by scanning in the long axis of the tendon, an example being the retrocalcaneal bursa (Fig 5–2). Evaluation of the quantity of fluid within the bursa is best assessed relative to the asymptomatic side.

## NORMAL SONOGRAPHIC BURSA STRUCTURE

All bursae share the fundamental attributes described earlier. A thorough knowledge of the anatomical location of the major bursae is essential. Other-

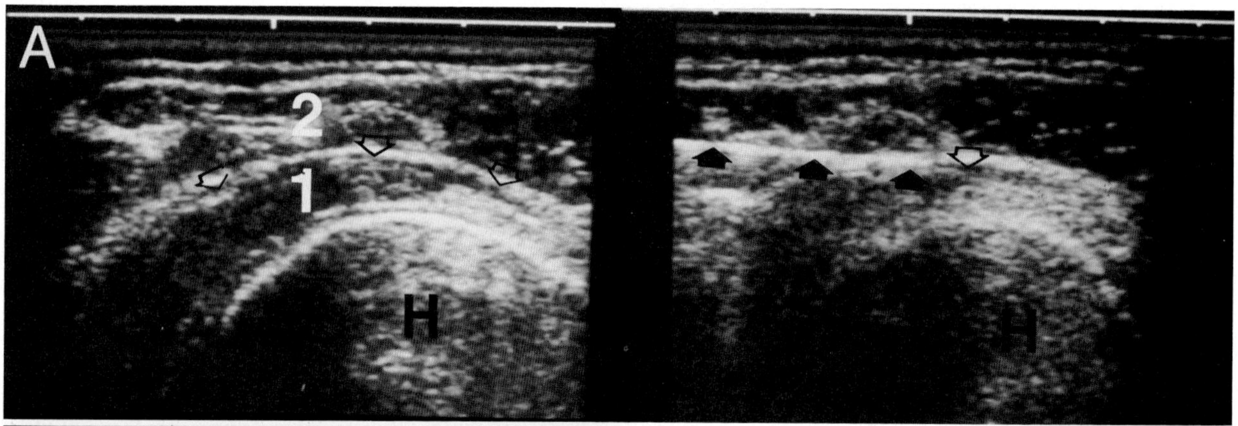


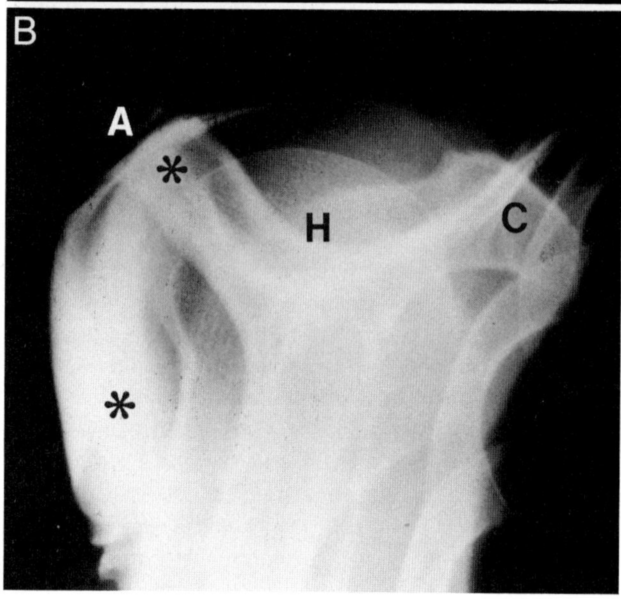


**FIG 5–1.**
**A,** normal bursa: transverse sonogram. One of the largest bursae in humans is the subacromial-subdeltoid bursa. The surfaces of this bursa separate a large area between the rotator cuff, the acromion, and the deltoid muscle. The cross-sectional diameter of the bursa is only 1 to 2 mm and is seen as a hypoechoic line *(open arrows)* between the supraspinatus tendon *(1)* and the deltoid muscle *(2)*. The cross-sectional image at the *right* shows a spinal needle *(solid arrows)* advanced with its tip in the bursa *(open arrow)*. **B,** normal bursa: bursogram. A bursogram was obtained in the same patient as in **A** after injection of 10 mL of water-soluble iodinated contrast medium. The bursa *(asterisk)* appears as a sac due to its distention with contrast medium. Sonography demonstrates the bursa in its natural state, a potential space **(A).** Abbreviations: humerus *(H)*, coracoid *(C)*, acromion *(A)*. (From van Holsbeeck M, Introcaso J: *AJR* 1988; 152:202. Used by permission.)

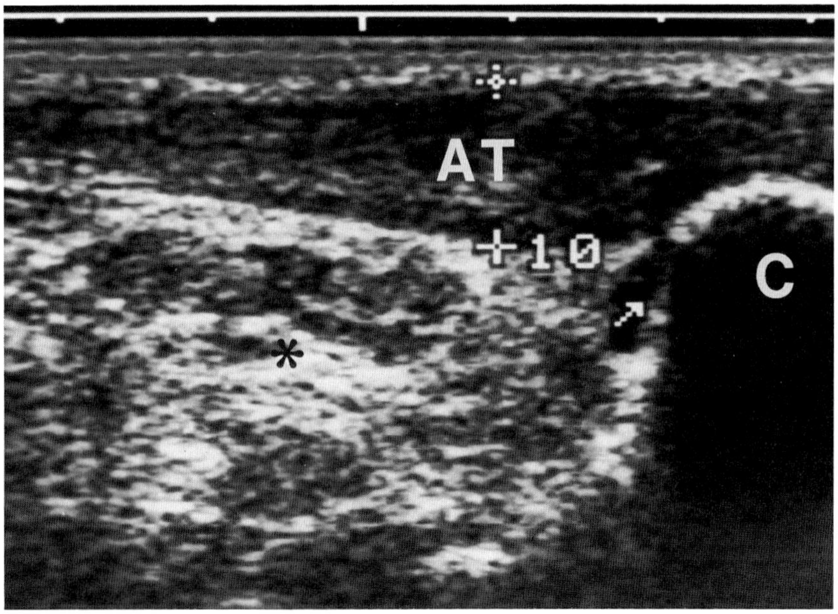


**FIG 5–2.**
Retrocalcaneal tendinobursitis: longitudinal sonogram. The patient is a 20-year-old male tennis player who suffered 6 months of continuous ankle pain, which caused him to stop all athletic activity. The transducer is oriented along the length of the Achilles tendon with the caudal extent touching the bursal projection of the calcaneus. Swelling and decreased echogenicity are noted in the distal 4 cm of the tendon. The maximal anteroposterior dimension of the tendon is 10 mm, twice normal. The retrocalcaneal bursa *(arrow)* is seen as a comma-shaped hypoechoic fluid-filled cleft between the Achilles tendon *(AT)* and the calcaneus *(C)*. Symbols: Kager's fat pad *(asterisk)*.

**TABLE 5–1.**
Sites of Bursitis Detected With Sonography

| | |
|---|---|
| Subcutaneous bursae | |
| Elbow: | Olecranon bursa |
| Spine: | Baastrup's bursa |
| Hip: | Subcutaneous trochanteric bursa |
| Knee: | Prepatellar bursa, subcutaneous infrapatellar bursa, subcutaneous tibial tuberosity bursa |
| Ankle: | Subcutaneous Achilles tendon bursa |
| Foot: | First metatarsal bursa |
| Deep bursae | |
| Shoulder: | Subacromial bursa,* subdeltoid bursa,* subscapular bursa† |
| Hip: | Obterator internus bursa, iliopsoas bursa,* deep trochanteric bursa |
| Knee: | Iliotibial tract bursa, fibular collateral ligament bursa, tibial collateral ligament bursa, subfascial prepatellar bursa, anserine bursa, popliteus bursa,* deep infrapatellar bursa, gastrocnemius-semimembranosus bursa,† suprapatellar bursa† |
| Ankle: | Retrocalcaneal bursa |

*Communication with the joint found in less than 50% of autopsies.
†Communication with the joint found in greater than 50% of autopsies.

wise, when pathologically enlarged, they may be mistaken for soft tissue tumors (Baudrillard et al., 1986). In our discussion of normal bursa structure, we will use the prepatellar bursa as a prototype. It is a superficial noncommunicating bursa. Figure 5–3 is a sagittal section through an anatomical specimen of the knee. The prepatellar bursa lies anterior to the patella with its base firmly fixed to the cortex of the base of the patella. The opposing wall is tightly bound to the deep surface of the skin. These thin walls of synovial tissue separated by a minute amount of viscous fluid allow the skin to smoothly glide over the bony prominence of the patella.

Sonographic images depict a bursa as a hypoechoic cleft in the soft tissues, often bounded by a hyperechoic line. This is demonstrated in Figure 5–4A, which is a sonogram of the specimen in Figure 5–3. The hyperechoic boundaries of the bursa do not directly correspond to the synovial lining. Synovial tissue forming the walls of a bursa is microscopically thin, beyond the resolution of clinical ultrasound units. Therefore, these hyperechoic lines represent the tissue-fluid interfaces of the bursa (van Holsbeeck and Introcaso, 1989). The viscous fluid that separates the walls of the bursa forms a lubricating film approximately 1 mm thick and is seen as a hypoechoic

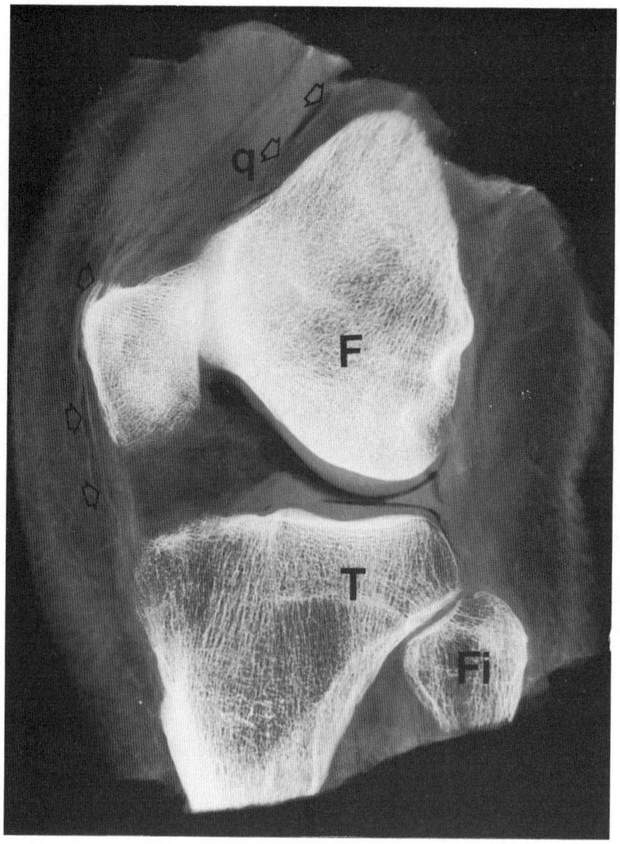


**FIG 5–3.**
Anatomical section through the lateral third of the knee: sagittal section. Two bursae are visualized on this section *(open arrows):* the suprapatellar and prepatellar bursae. The suprapatellar bursa is located deep to the quadriceps tendon *(q).* This is a communicating bursa that is continuous with the knee joint. The prepatellar bursa is a noncommunicating bursa separating the skin from the patella and patellar tendon. Abbreviations: femur *(F),* tibia *(T),* fibula *(Fi).*

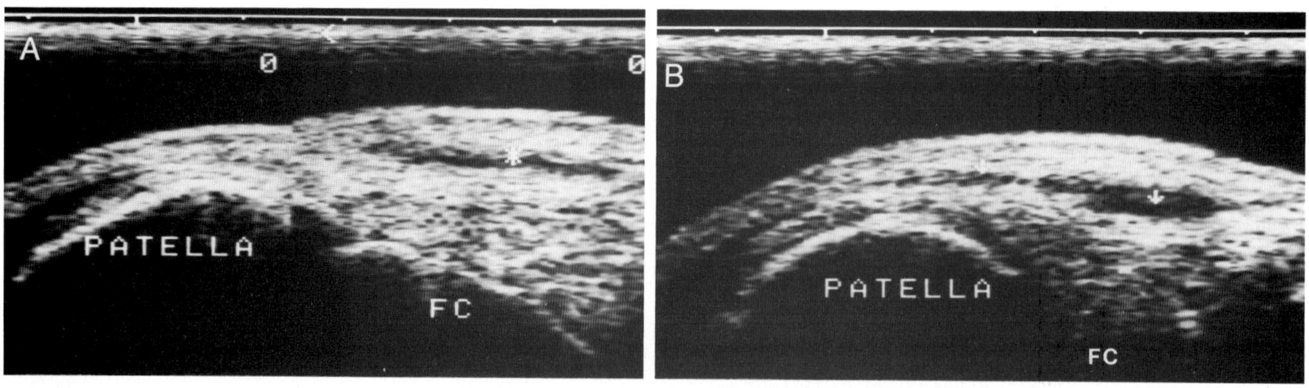


**FIG 5–4.**
**A,** normal prepatellar bursa: longitudinal sonogram. It is not always possible to image the normal prepatellar bursa using ultrasound. In this longitudinal image, the transducer is positioned over the patella 5 mm lateral to the patellar tendon. The fluid film within the bursa *(asterisk)* measured 1.5 mm thick. Abbreviations: lateral femoral condyle *(FC).* **B,** inflamed prepatellar bursa: longitudinal sonogram. The patient is a 19-year-old long jumper with pain in the right knee and point tenderness over the patellar tendon. He was given the clinical diagnosis of patellar tendonitis. Transducer position is identical to that in **A.** The amount of fluid within the bursa is abnormally increased consistent with frictional bursitis *(arrows).* The patellar tendon is normal.

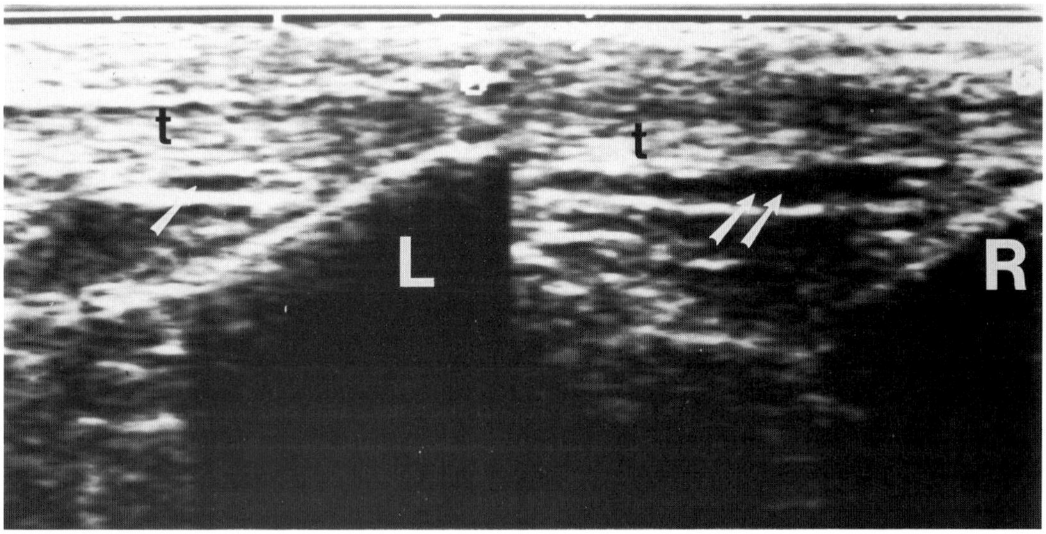


**FIG 5–5.**
Bursitis of the deep infrapatellar bursa: longitudinal sonogram. A 23-year-old soldier had right knee pain following a long march. The transducer is positioned in the long axis of the patellar tendon with the caudal aspect touching the tibial tuberosity at the observer's right. The normal deep infrapatellar bursa *(single white arrow)* is on the left located dorsal to the patellar tendon *(t)*. It demonstrates the normal fluid film within the bursa 1 to 2 mm thick. On the *right,* an inflamed bursa *(two white arrows)* appears larger, containing an abnormally increased amount of fluid. Note that the fluid is anechoic, and the bursal walls are not thickened in this case of frictional bursitis. The borders of the bursa are well defined by a hyperechoic line. Abbreviations: left tibial tuberosity *(L),* right tibial tuberosity *(R).*

line. Side to side comparison (Fig 5–5) is the best way to evaluate subtle variations in the quantity of bursal fluid. However, in rare cases, the examiner may be misled by bilateral disease. The hypoechoic fluid film within a normal bursa should not exceed 2 mm.

## PATHOLOGY OF BURSAE

### Noncommunicating Bursae

#### *Acute Traumatic Bursitis*

A bursa may become acutely inflamed and symptomatic from either direct pressure or excessive repetitive activity of adjacent structures (Fig 5–6). This condition is called acute traumatic bursitis, or frictional bursitis. It is frequently found in athletes whose sports require repetitive motion such as runners, tennis players, and oarsmen. Bursae adjacent to joints with irregular bony contours and hypertrophic tendon insertions are predisposed to the development of frictional bursitis. In the upper extremity, the subacromial-subdeltoid and olecranon bursae (Fig 5–7) are most frequently involved. The trochanteric bursa (Fig 5–8) is the most common site of frictional bursitis of the hip. Bursae of the knee most commonly involved are the prepatellar and deep infrapatellar bursae (Fig 5–9). In the foot, the superficial Achilles tendon bursa, retrocalcaneal and calcaneal bursae are the most frequent sites of frictional bursitis (Fig 5–10 and 5–11).

The pathophysiology is that of a typical acute inflammatory reaction. Initially there is a brief period of vasoconstriction followed by hyperemia. The hyperemia results from dilatation of arterioles, capillaries, and postcapillary venules. This hyperemic state leads to transudation and exudation. The bursa becomes painful and distended with a watery or mucoid fluid unlike the viscous fluid normally found within the bursa (Nicholas and Hershman, 1986).

Ultrasound provides the clinician with a noninvasive means of examining the bursae. In acute traumatic bursitis, the primary value of sonography is to confirm that disease is limited to the bursa. Associated tendons, ligaments, and joint space are easily examined to exclude bursitis secondary to pathology originating in these structures (Fig 5–12).

Acute traumatic bursitis is easily recognized by an increase in the volume of the bursa, without other changes in bursa structure. As stated previously, comparison with the asymptomatic side is extremely valuable. The character of the fluid within the bursa differs markedly from that found in a normal bursa. In acute traumatic bursitis, the fluid is most com-

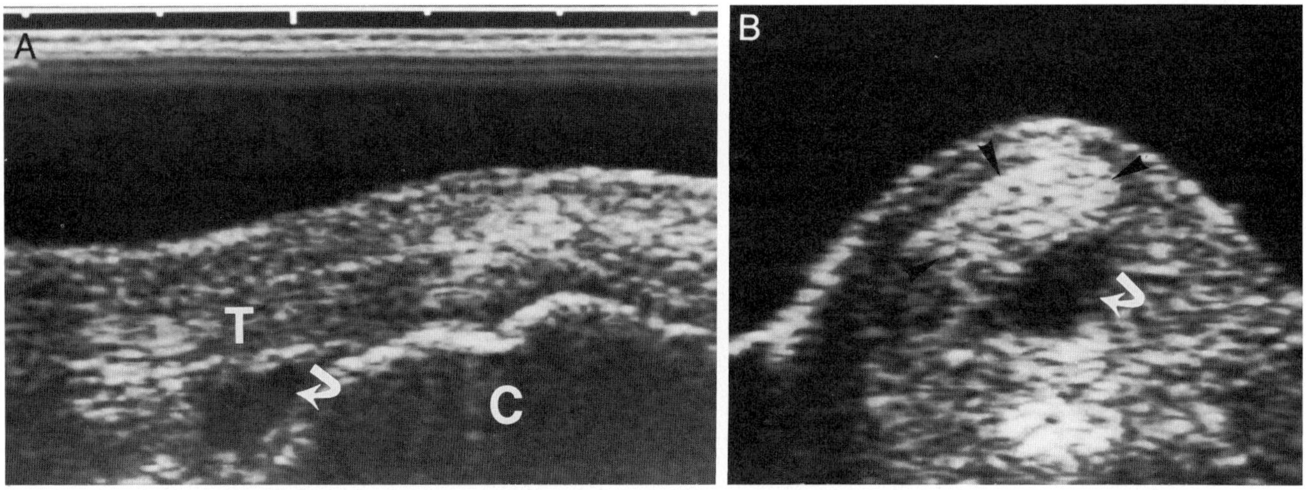


**FIG 5–6.**
**A,** frictional bursitis of the retrocalcaneal bursa: longitudinal sonogram. The patient is a 22-year-old long distance runner with pain and point tenderness over the distal left Achilles tendon. He experienced temporary relief of pain following steroid injection. The transducer aligned in the long axis of the Achilles tendon, and a standoff pad was utilized. The tendon insertion is at the observer's right. A distended fluid filled retrocalcaneal bursa *(curved arrow)* is noted between the Achilles tendon *(T)* and the calcaneus *(C)*. **B,** frictional bursitis of the retrocalcaneal bursa: transverse sonogram. Same patient as in **A.** The transducer is oriented transverse to the Achilles tendon, slightly superior to the calcaneus. At the *left* is the medial aspect of the ankle. The Achilles tendon *(black arrowheads)* has an ovoid appearance in this plane and is normally rotated slightly to the medial. The distended retrocalcaneal bursa *(curved arrow)* is seen deep to the tendon. Note the enhanced through-transmission deep to the bursa.

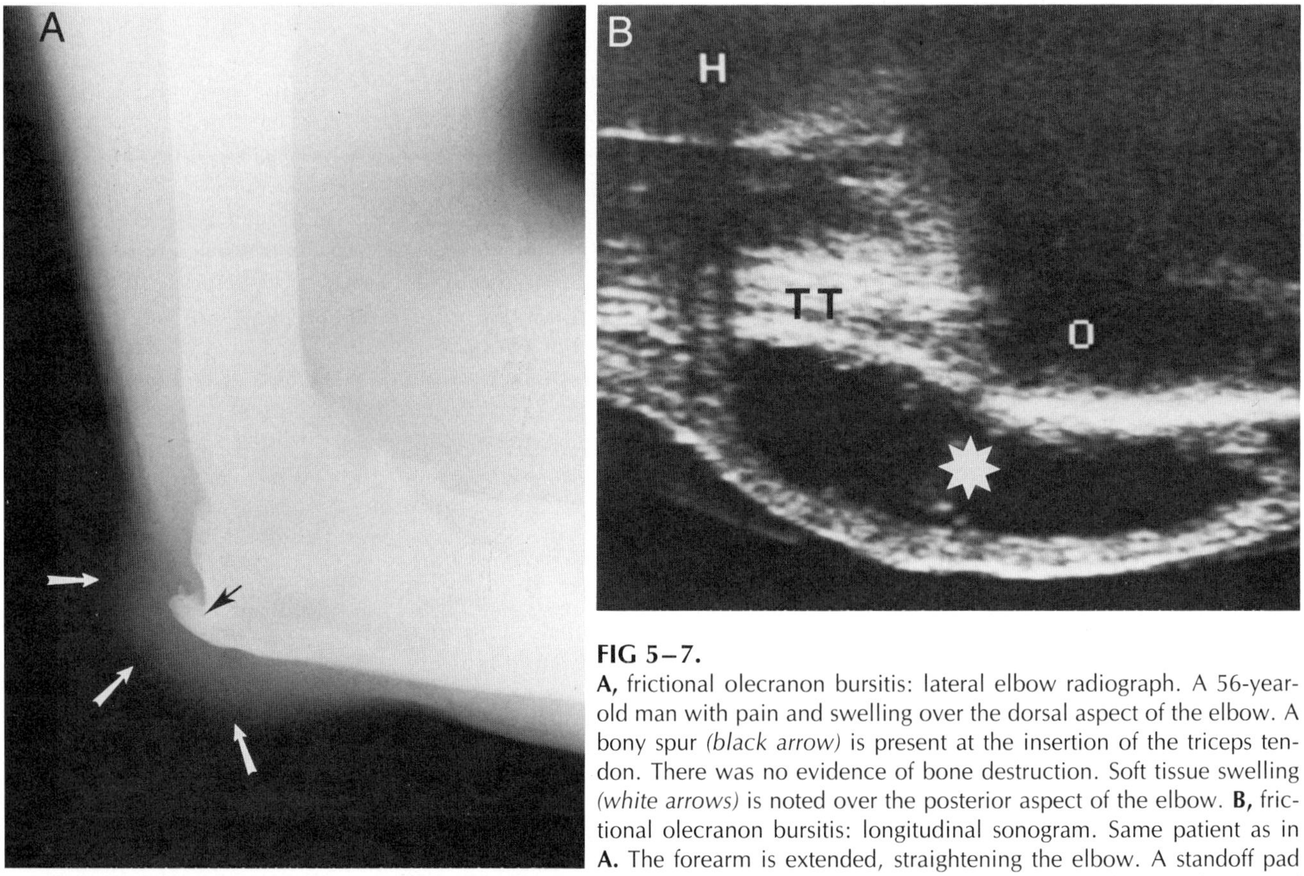


**FIG 5–7.**
**A,** frictional olecranon bursitis: lateral elbow radiograph. A 56-year-old man with pain and swelling over the dorsal aspect of the elbow. A bony spur *(black arrow)* is present at the insertion of the triceps tendon. There was no evidence of bone destruction. Soft tissue swelling *(white arrows)* is noted over the posterior aspect of the elbow. **B,** frictional olecranon bursitis: longitudinal sonogram. Same patient as in **A.** The forearm is extended, straightening the elbow. A standoff pad was employed with the transducer oriented along the triceps tendon. Proximal lies to the observer's left, and the dorsal aspect of the elbow is at the bottom of the image. The triceps tendon *(TT)* inserts on the olecranon at the center of the image. A concave spheroidal olecranon bursa *(asterisk)* overlies the olecranon and distal triceps tendon. The fluid is essentially anechoic. Abbreviations: humerus *(H)*, olecranon *(O)*.

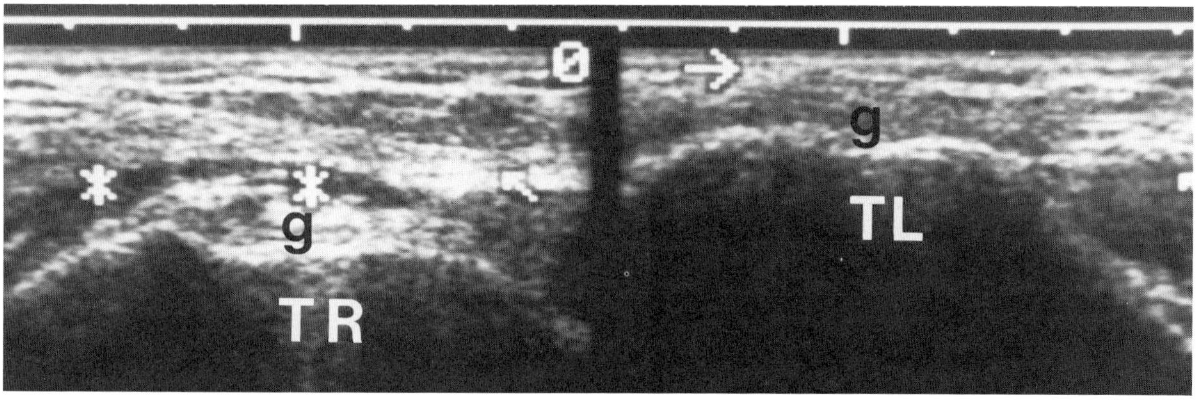


**FIG 5–8.**
Frictional bursitis of the trochanteric bursa: coronal sonogram. A 22-year-old medical student training for a marathon. He experienced persistent right hip pain and was instructed to discontinue training. The exam was performed with the patient in the lateral decubitus position. Scanning was performed in the coronal plane over the greater trochanter. The symptomatic right side is displayed on the left half of the image; the asymptomatic left side is to the right. An abnormal fluid-filled superficial trochanteric bursa *(asterisk)* is noted on the patient's right. This is a bursa located between the tendon of the gluteus muscles *(g)* and the fascia *(arrow)*. Abbreviations: right greater trochanter *(TR)*, left greater trochanter *(TL)*.

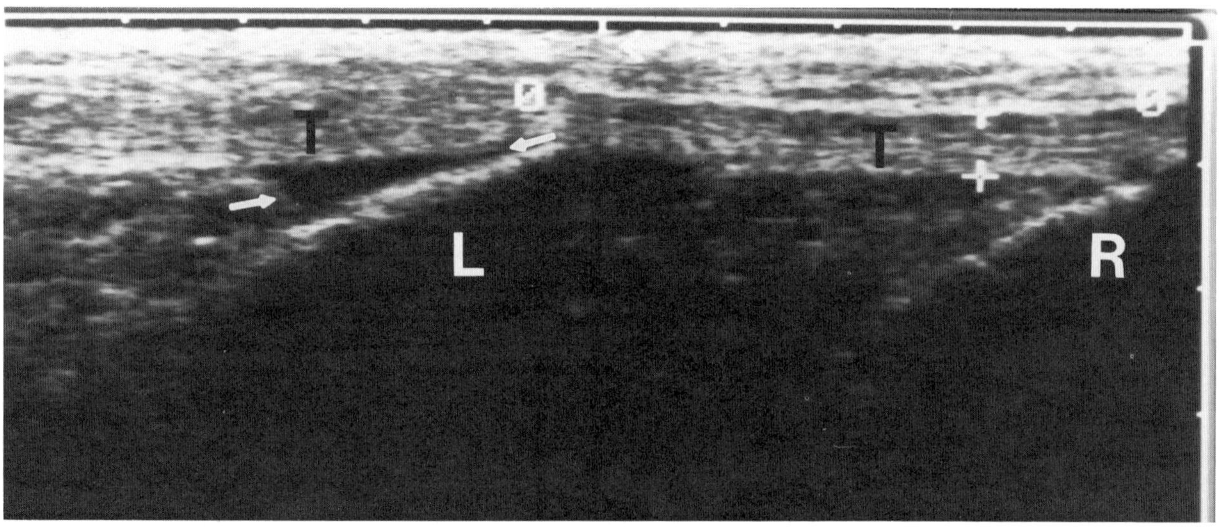


**FIG 5–9.**
Frictional bursitis of the deep infrapatellar bursa: longitudinal sonogram. The patient is a soccer player with the clinical diagnosis of jumper's knee on the left. The transducer is positioned in the long axis of the patellar tendon. This case demonstrates the necessity of comparison with the asymptomatic side. A triangular distended fluid-filled bursa *(white arrows)* is noted on the left side overlying the tibia *(L)*. The fluid is anechoic. The patellar tendons *(T)* are normal bilaterally, excluding the diagnosis of jumper's knee. This examination changed the course of therapy. In the absence of tendon injury, steroids can be safely administered. Abbreviations: right tibia *(R)*.

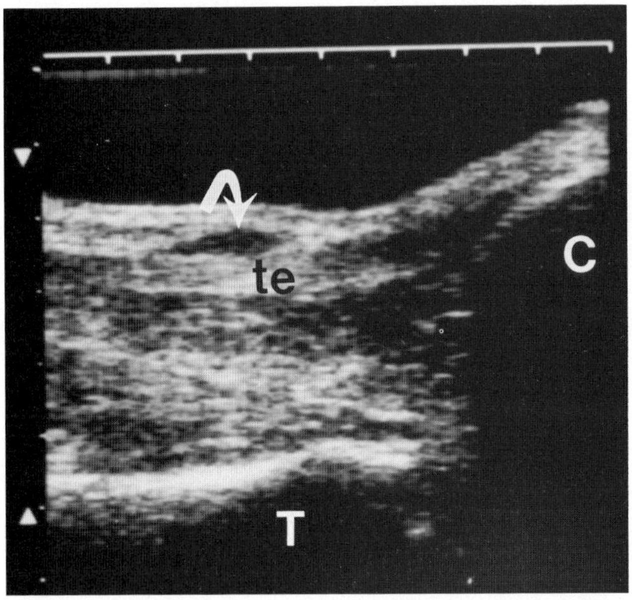


**FIG 5–10.**
Frictional bursitis of the superficial Achilles tendon bursa: longitudinal sonogram. A 40-year-old woman, previously the indoor long jump champion, who presented with pain and swelling of the ankle. The transducer was positioned along the length of the Achilles tendon *(te)* with the patient in the prone position. The superficial Achilles tendon bursa *(curved arrow)* is likely an adventitious bursa because its position relative to the calcaneus *(C)* varies considerably. Abbreviations: distal tibial metaphysis *(T)*.

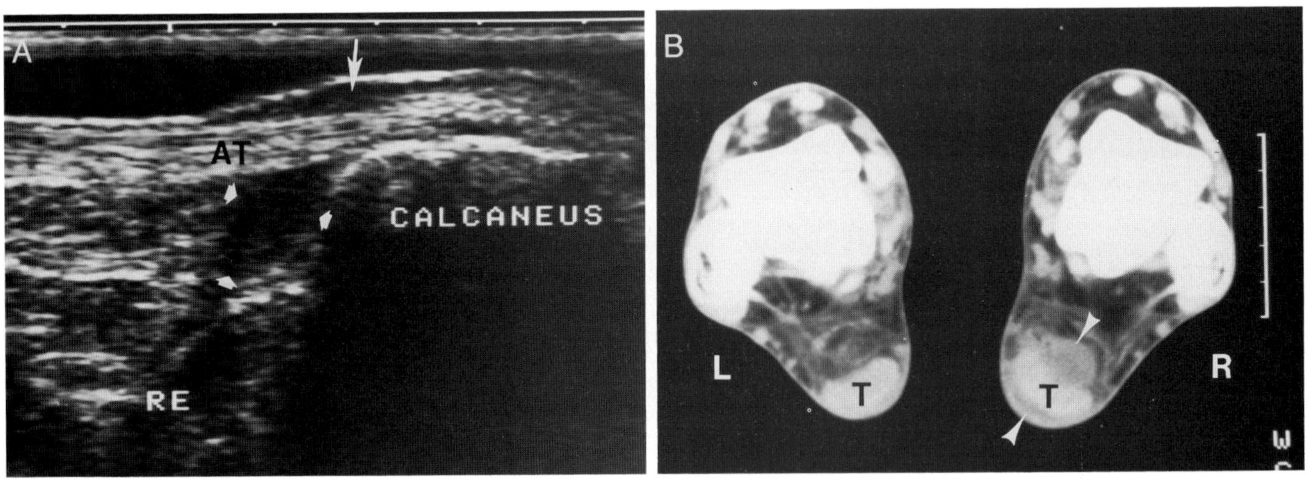


**FIG 5–11.**
**A,** chronic ankle bursitis: longitudinal sonogram. A 30-year-old woman who frequently wore tight-fitting high heel shoes. She presented with chronic right ankle pain and a "pump-bump" deformity. Transducer position is identical to that in Figure 5–10. Two inflamed bursae surround the Achilles tendon *(AT)*. The retrocalcaneal bursa *(short arrows)* lies deep to the Achilles tendon, and the superficial Achilles tendon bursa *(large arrow)* lies superficially. Multiple inflamed bursae at the same site are findings characteristic of chronic bursitis. We site this example at this point because it so nicely demonstrates the bursae of the ankle most frequently involved in traumatic bursitis. **B,** chronic ankle bursitis: computed tomography findings. Same patient as in **A.** This is a transverse axial image of the ankles at the level of the distal Achilles tendon. The Achilles tendon *(T)* is identified between the distended fluid-filled retrocalcaneal and superficial Achilles tendon bursae *(arrowheads)*. Although smaller than the retrocalcaneal bursa, the superficial Achilles tendon bursa can cause significant ankle swelling. Abbreviations: patient's left *(L)*, right *(R)*.

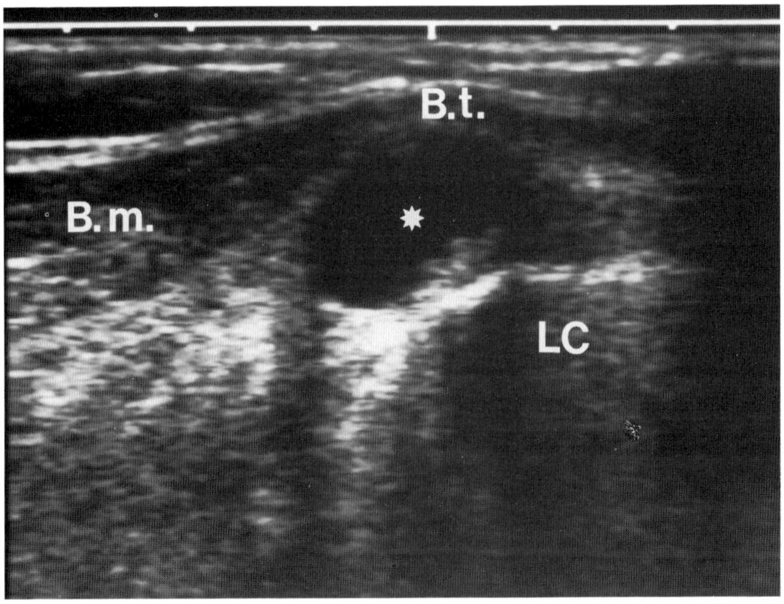


**FIG 5–12.**
Bursitis at the lateral aspect of the knee: coronal sonogram. An 18-year-old soccer player with tenderness over the lateral femoral condyle following a game. The clinical differential diagnosis was biceps femoris tendonitis and tensor fascia lata syndrome. A coronal image was obtained with the transducer over the posterior medial aspect of the knee. An oval anechoic mass elevates the biceps tendon *(Bt)* consistent with frictional bursitis of the bursa *(asterisk)* separating the fibular collateral ligament and the biceps tendon. Abbreviations: biceps femoris muscle *(Bm),* lateral femoral condyle *(LC).*

monly anechoic and is markedly hypoechoic compared with normal bursal fluid. Acoustic enhancement deep to the inflamed bursa is commonly seen. Walls of the bursa are unchanged in their appearance (Fig 5–13). This is an important factor in differentiating acute traumatic bursitis from the chronic form. Figures 5–6 through 5–13 illustrate the sonographic features of acute traumatic bursitis discussed earlier.

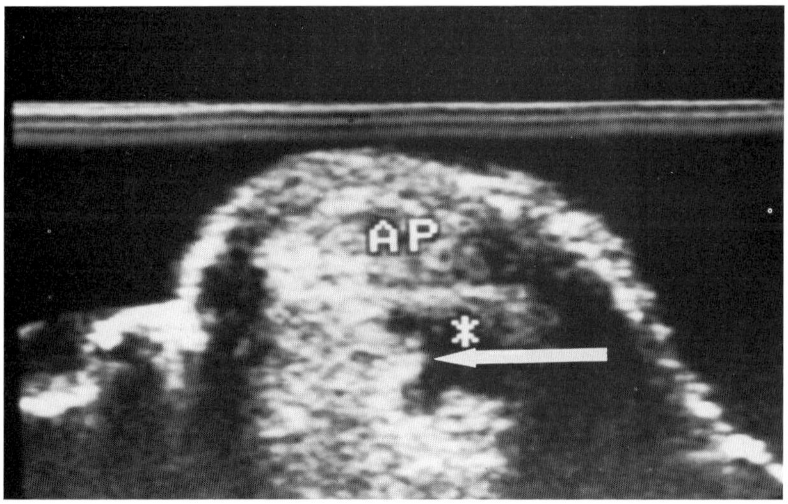


**FIG 5–13.**
Frictional bursitis of the retrocalcaneal bursa: transverse sonogram. The transducer position is identical to that in Figure 5–6,B. A slightly distended retrocalcaneal bursa *(asterisk)* is identified deep to the Achilles tendon *(AP).* This demonstrates fat indenting *(large arrow)* the medial wall of the bursa, a common configuration seen when the bursa is minimally distended.

### *Chronic Traumatic Bursitis*

If the inciting factor persists, an acute traumatic bursitis may become chronic. In the process of evolution, the synovial walls of the bursa become thickened, and it becomes filled with a fibrinous exudate. Calcifications are sometimes seen within or surrounding the chronically inflamed bursa. A subtype of chronic traumatic bursitis is the formation of an adventitious bursa in an area subjected to chronic frictional irritation. In this condition, inflammation of the adjacent areolar connective tissue forms a focus of fibrinoid necrosis (Gardner, 1965). A cystic structure filled with cellular debris, extracellular fluid, altered ground substance, and inflammatory exudate results (Nicholas and Hershman, 1986). This process is the formation of a bursa de novo (Fig 5–14). A common adventitious bursa is found in women over the first metatarsal head. It is a frequent complication of wearing narrow pointed shoes (Fig 5–15) (Canoso, 1981).

The sonographic features of chronic traumatic bursitis are demonstrated in Figures 5–16 through 5–19. As in acute traumatic bursitis, the bursa is distended with fluid. However, the fibrinous exudate found in chronic traumatic bursitis produces definite echos within the bursa. In addition, the synovial walls of the bursa are markedly thickened and easily visible as a millimeter-thick hyperechoic layer. The outline of the bursa is often very irregular (see Fig 5–17). If present, calcifications are seen as hyperechoic foci with or without shadowing. The presence of shadowing is dependent on the size of the calcifications and the frequency of the transducer utilized. Adjacent bursae are often involved in chronic bursitis, as well as the adjacent tendons (Figs 5–20 and 5–21).

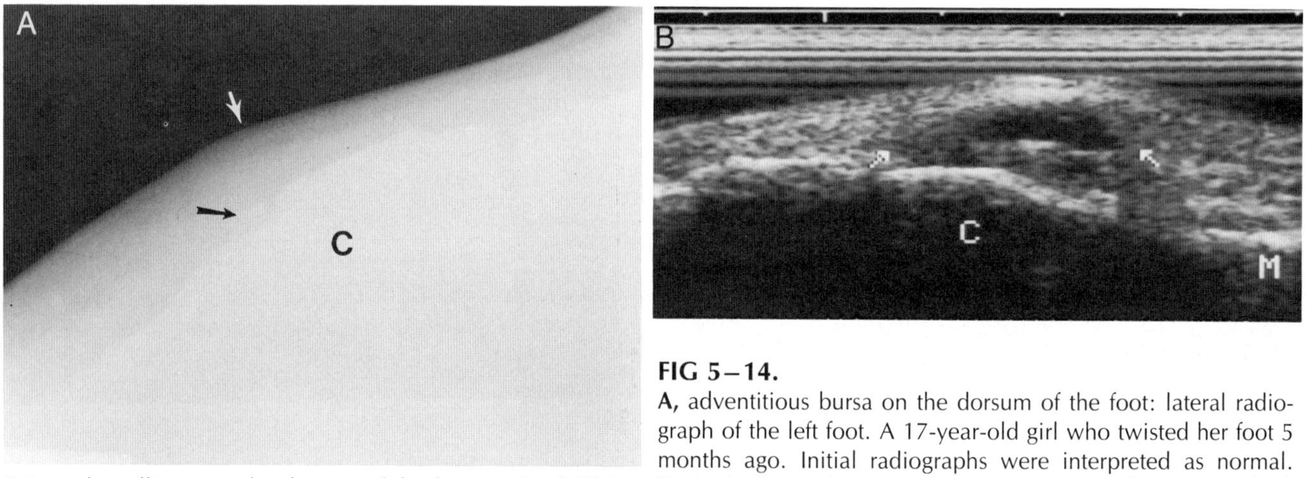


**FIG 5–14.**
**A,** adventitious bursa on the dorsum of the foot: lateral radiograph of the left foot. A 17-year-old girl who twisted her foot 5 months ago. Initial radiographs were interpreted as normal. Pain and swelling over the dorsum of the foot persisted. This radiograph obtained using soft tissue technique demonstrates soft tissue swelling *(white arrow)* and an avulsed fragment *(black arrow)* over the anterior medial aspect of the medial cuneiform bone *(C).* **B,** adventitious bursa: longitudinal sonogram. Same patient as in **A.** The transducer is positioned over the base of the first metatarsal oriented along its length. The linear hyperechoic structure superficial to the medial cuneiform is the avulsed fragment demonstrated in **A.** Note that only faint acoustic shadowing is present due to volume averaging artifact. An ovoid bursa *(arrow)* is present separating the skin and the bony fragment. Surgery demonstrated a small avulsed bone fragment with an adjacent adventitious bursa. Abbreviations: medial cuneiform *(C),* first metatarsal *(M).*

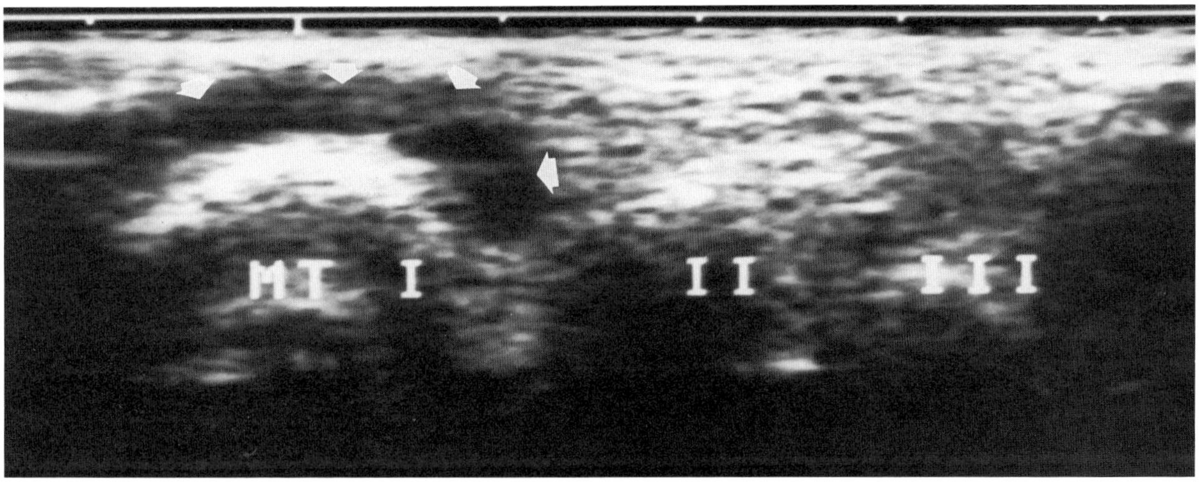


**FIG 5–15.**
Adventitious bursa over the first metatarsal head: transverse sonogram. The patient is a 35-year-old woman with a hallux valgus deformity and recurrent swelling over the head of the first metatarsal. The transducer is positioned transversely over the heads of the first through third metatarsals *(MTI–III)*. An adventitious bursa *(white arrows)* is identified as a hypoechoic curvilinear band over the first metatarsal head.

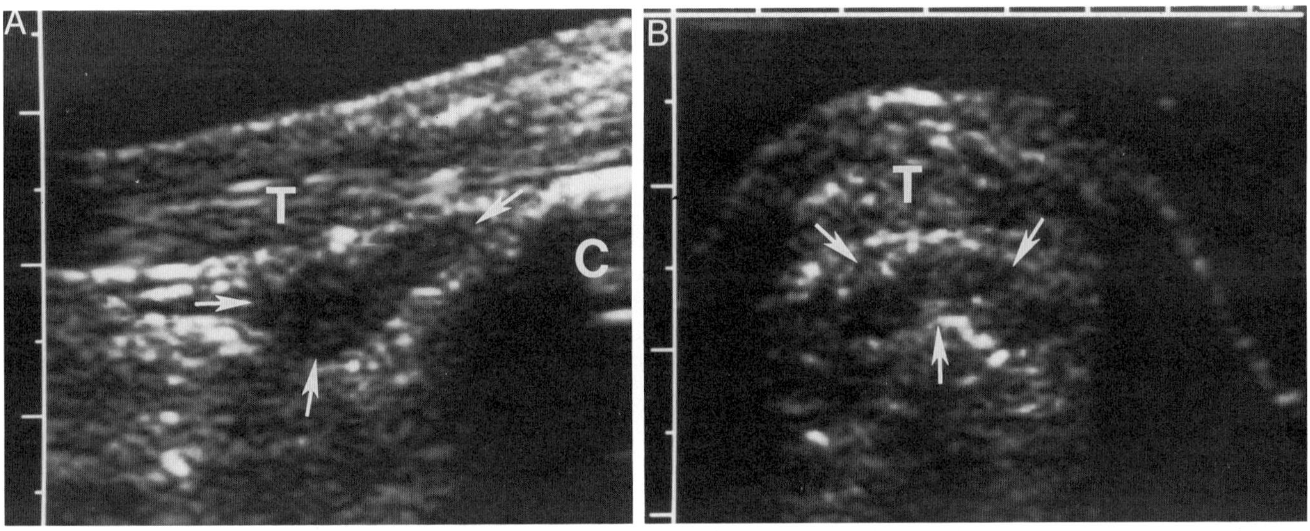


**FIG 5–16.**
**A,** chronic bursitis of the retrocalcaneal bursa: longitudinal sonogram. An 18-year-old girl who frequently wore tight fitting shoes. Over the past year she experienced chronic burning pain over the heel, worst at the end of the day. Physical examination revealed point tenderness over the bursal projection of the calcaneus. The insertion of the Achilles tendon *(T)* on the calcaneus *(C)* is at the observer's right. A distended retrocalcaneal bursa *(white arrows)* with definite internal echos is located deep to the Achilles tendon and superior to the calcaneus. **B,** chronic bursitis of the retrocalcaneal bursa: transverse sonogram. Same patient as in **A.** The illustration is a transverse image obtained with the transducer over the distal Achilles tendon. Deep to the Achilles tendon *(T)* is the distended retrocalcaneal bursa *(white arrows)*. The echogenicity of the bursal contents makes the borders indistinct. Definite echos within the bursa distinguish this as a case of chronic bursitis as opposed to acute bursitis.

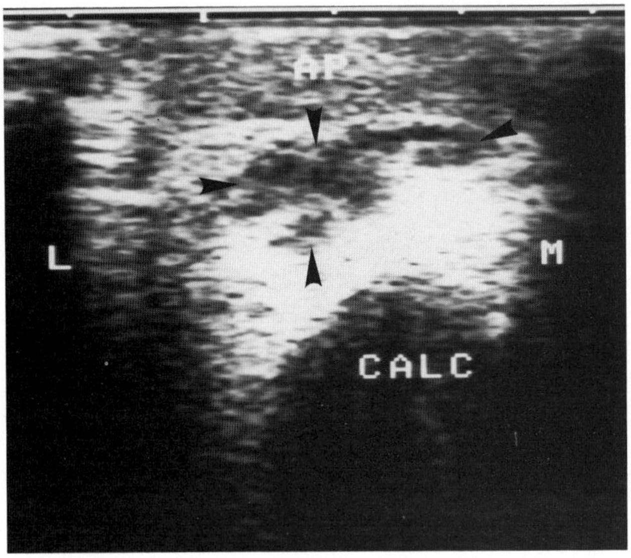


**FIG 5–17.**
Chronic retrocalcaneal bursitis: transverse sonogram. The patient is a 25-year-old woman who sustained a blow to the posterior ankle 18 months ago. Pain over the posterior ankle became progressively worse, and she now must wear an open shoe. Transducer position is identical to that in Figure 5–16,B. Located between the calcaneus *(CALC)* and the Achilles tendon *(AP)* is an enlarged retrocalcaneal bursa *(arrowheads)*. The bursa has a very irregular outline, and it contains fluid with definite internal echos. Abbreviations: medial *(M),* lateral *(L).*

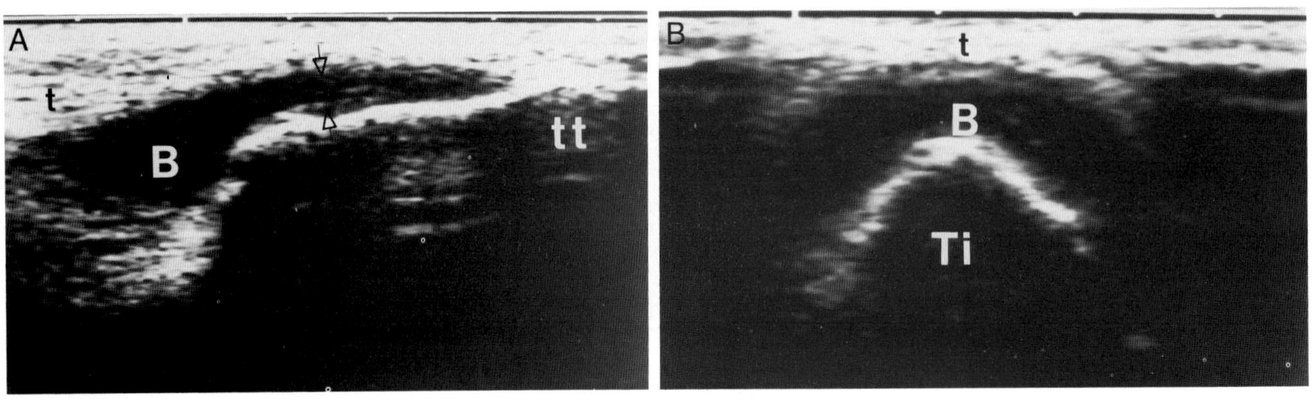


**FIG 5–18.**
**A,** chronic bursitis of the deep infrapatellar bursa: longitudinal sonogram. A 17-year-old male soccer player with pain, tenderness, and swelling over the tibial tuberosity. He previously had the diagnosis of Osgood-Schlatter disease. The transducer is oriented in the longitudinal axis of the patellar tendon *(t).* Distal structures are at the observer's right. The inflamed bursa *(B)* has a bilobed appearance, measuring approximately 5 cm long. Relative to the surrounding fat and tendon, the bursa appears essentially anechoic. Synovial wall thickening is also observed *(arrows).* Abbreviations: tibial tuberosity *(tt).* **B,** chronic bursitis of the deep infrapatellar bursa: transverse sonogram. Same patient as in **A.** The transducer is rotated 90 degrees to give a transverse image. In this image, the bursa has a hypoechoic bean-shaped configuration positioned between the tibia *(Ti)* and the patellar tendon *(t).* A rim of synovial thickening is again seen around the bursa.

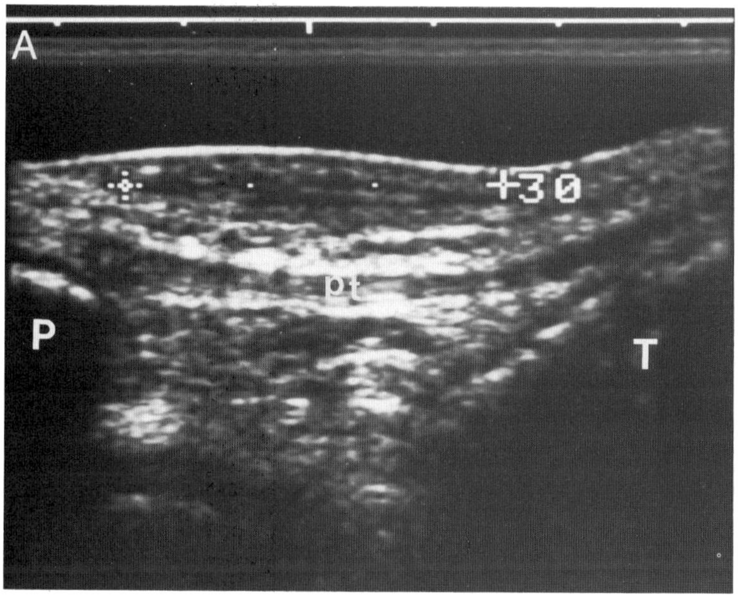


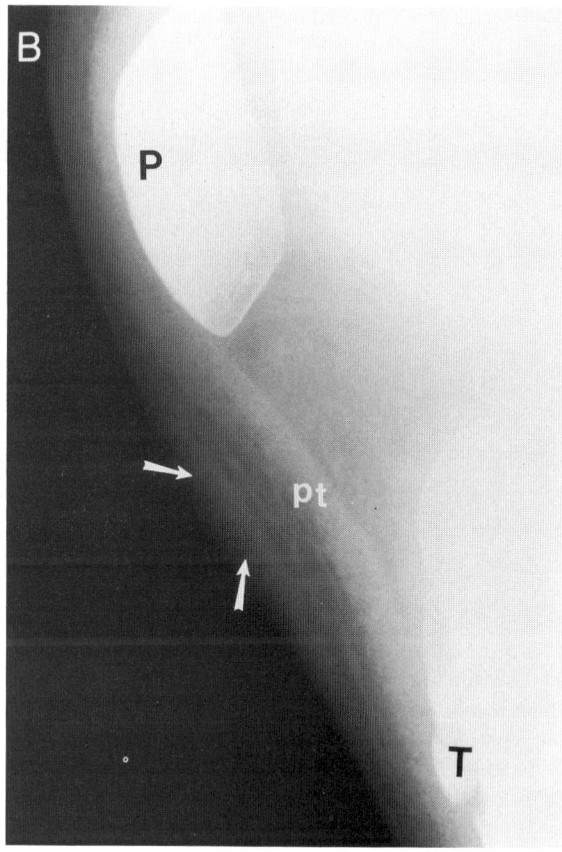


**FIG 5–19.**
**A,** chronic bursitis of the subcutaneous infrapatellar bursa: longitudinal sonogram. Pain and swelling over the anterior knee was clinically evident in this 14-year-old girl. The transducer is positioned over the patellar tendon along its length just inferior to the patellar apex *(p)*. The knee is extended with the quadriceps relaxed, which results in the curved appearance of the patellar tendon *(pt)*. A spindle-shaped bursa is located in the subcutaneous tissues measuring approximately 30 mm long. Fluid is noted centrally within the bursa surrounded by the markedly thickened bursal walls. **B,** chronic bursitis of the subcutaneous infrapatellar bursa: low-kilovoltage radiograph. Same patient as in **A.** Soft tissue swelling *(arrows)* is seen between the patellar tendon *(pt)* and the skin. Abbreviations: patella *(P)*, tibial tuberosity *(T)*.

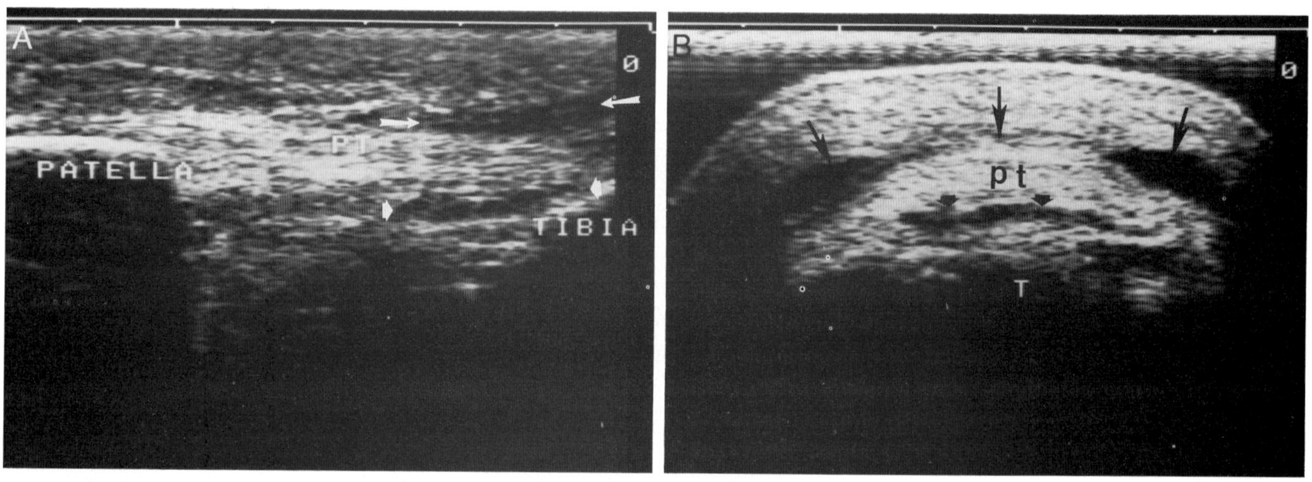


**FIG 5–20.**
**A,** chronic bursitis around the distal patellar tendon: longitudinal sonogram. A 62-year-old man had chronic knee pain following a below knee amputation. The transducer is positioned over the patellar tendon in the longitudinal axis of the knee. A normal patellar tendon *(PT)* is seen. Two distended fluid-filled bursae are noted around the distal patellar tendon. These are the deep infrapatellar bursa *(short arrows)* and the subcutaneous tibial tuberosity bursa *(long arrows)*. **B,** chronic bursitis around the distal patellar tendon: transverse sonogram. Same patient as in **A.** Transducer position is now transverse over the distal patellar tendon *(pt)*. The inflamed subcutaneous tibial tuberosity *(long arrows)* and deep infrapatellar bursae *(short arrows)* surround the patellar tendon. Essentially all of the fluid in the subcutaneous tibial tuberosity bursa is seen in the lateral recesses of the bursa.

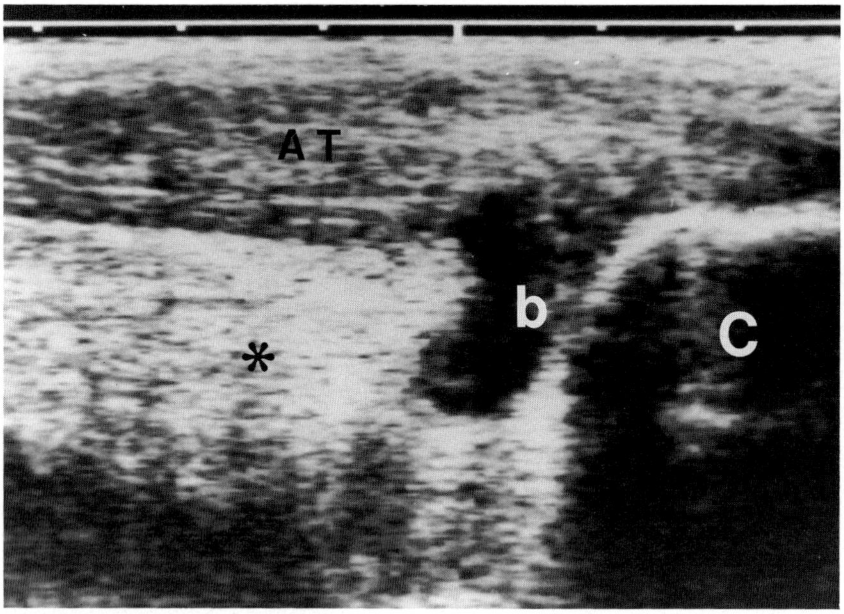


**FIG 5–21.**
Chronic tendonobursitis of the distal Achilles tendon insertion: longitudinal sonogram. A 19-year-old football player presented 6 months following trauma to the heel. During this period he continued to exercise, despite experiencing increasing pain. The transducer is oriented along the length of the Achilles tendon with the distal tendon at the observer's right. A chronic bursitis of the retrocalcaneal bursa *(b)* is often associated with tendonitis of the distal Achilles tendon *(AT)*. Fusiform swelling of the distal Achilles tendon is noted. The chronic nature of the bursitis is indicated by the wall thickening of the bursa. Abbreviations and symbols: calcaneus *(C)*, Kager's fat pad *(asterisk)*.

### *Hemorrhagic Bursitis*

A powerful blow to a bursa, fracture of an adjacent bone, or rupture of a tendon can cause hemorrhage into the bursa. If a significant amount of blood fills the bursa, it may form a clot (Peterson and Renström, 1986). Chemical irritation of the bursa occurring during resolution of the hemorrhage will result in formation of adhesions. Later, loose bodies and calcifications may be found within the bursa (Peterson and Renström, 1986).

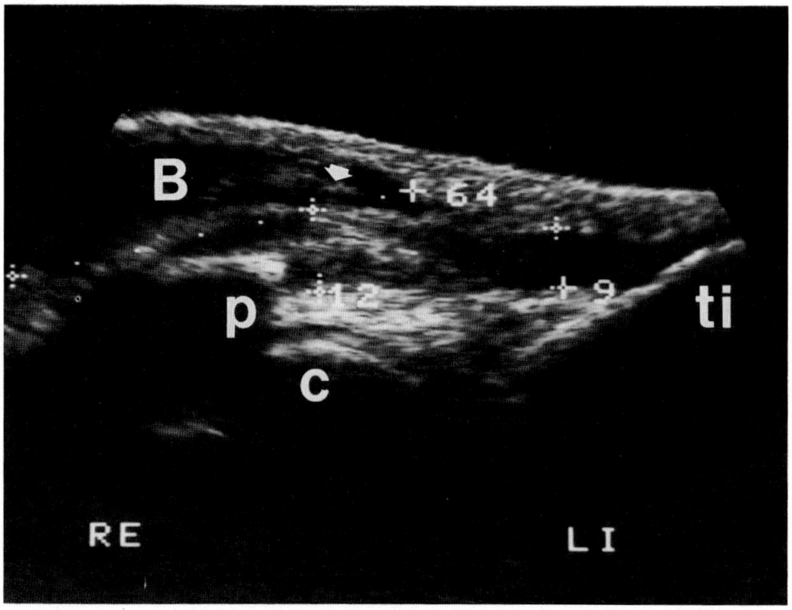


**FIG 5–22.**
Chronic hemorrhagic bursitis of the prepatellar bursa: longitudinal sonogram. This patient is a 16-year-old girl with chronic psychosis. She was admitted with a red swollen knee, and the clinical concern was infection. The transducer is positioned in the longitudinal axis of the knee with proximal at the observer's left. A standoff pad was utilized for optimal visualization of this superficial bursa. The swollen prepatellar bursa *(B)* is seen as an oval-shaped hypoechoic mass overlying the patella. Free-floating hyperechoic debris *(small arrow)* was identified within the bursa during real-time imaging. Aspiration of the bursa yielded blood, and observation of the patient confirmed the diagnosis of automutilation. Thickening of the patellar tendon was also noted, with the tendon measuring 12 mm at its thickest point, twice the normal thickness. The tendon was hypoechoic relative to the surrounding fat pad consistent with the diagnosis of tendonitis. Abbreviations: femoral condyle *(c)*, patella *(p)*, tibia *(ti)*.

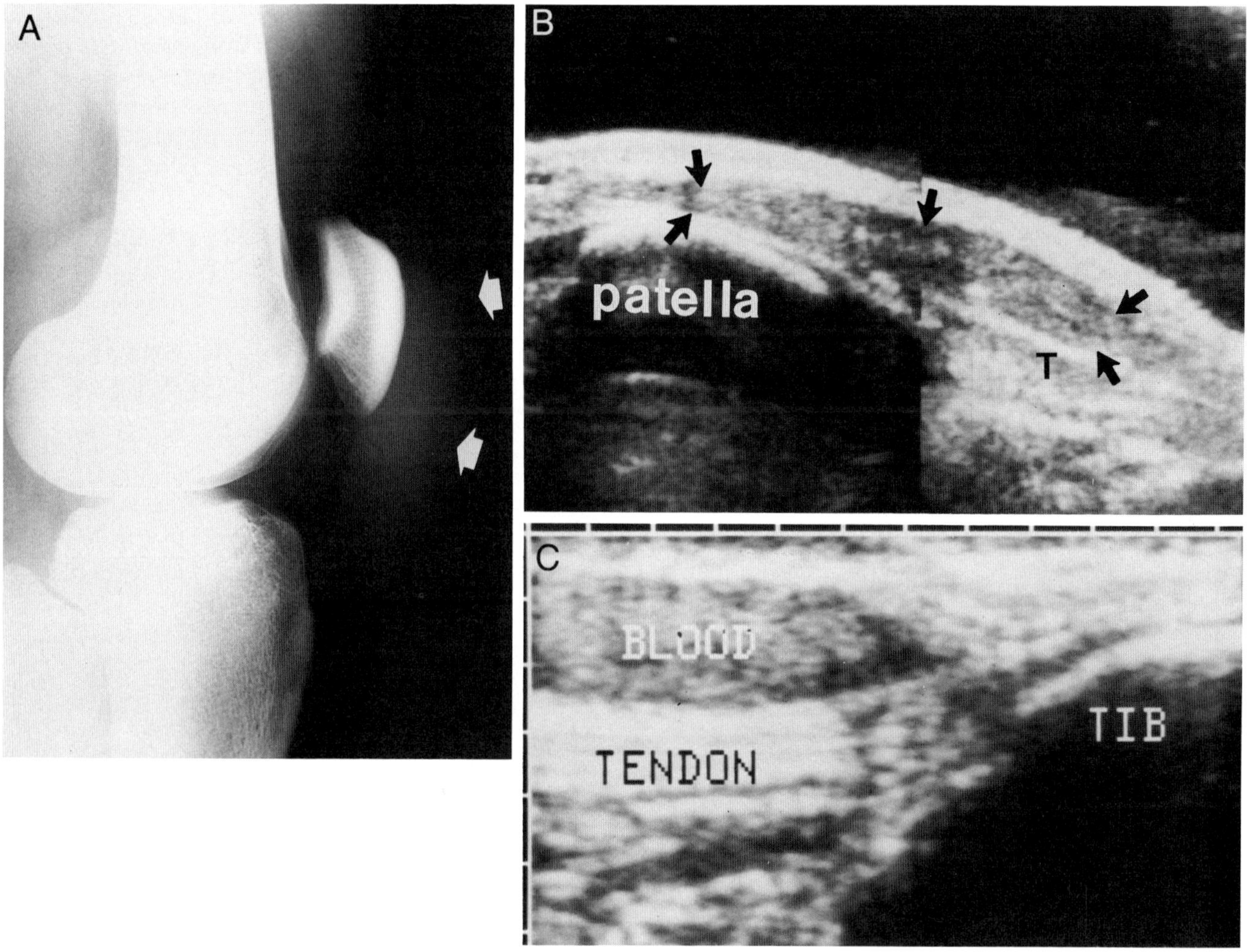


**FIG 5–23.**
**A,** hemorrhagic bursitis of the prepatellar bursa: conventional radiograph. The patient is a 21-year-old football player who was hit on the knee by a helmet and presented with pain and swelling over the patella. No fractures are present, but marked soft tissue swelling is noted over the patella *(white arrows)*. **B,** hemorrhagic bursitis of the prepatellar bursa: longitudinal sonogram. Same patient as in **A.** Ultrasound examination performed 3 hours following injury revealed swelling of the prepatellar bursa *(arrows)*, which is filled with homogeneous hypoechoic material. Surrounding subcutaneous tissue is diffusely decreased in echogenicity compared with the normal subcutaneous tissues. Abbreviations: patellar tendon *(T)*. **C,** This image of the same patient is a detail demonstrating a normal patellar tendon inserting on the tibial tuberosity *(TIB)* with the inferior aspect of the prepatellar bursa filled with blood.

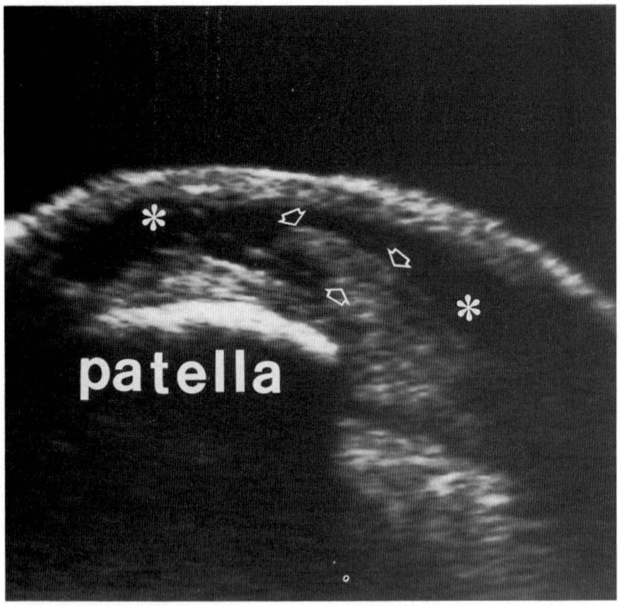


**FIG 5–24.**
Hemorrhagic bursitis of the prepatellar bursa: longitudinal sonogram. Same patient as in Figure 5–22. After blood clot retraction, the lumen is filled with anechoic fluid *(asterisk)* and fibrin clots *(open arrows)*.

Hemorrhagic bursitis affects bursae of the hands, knees, hips, and shoulders. This condition is frequently found in athletes who participate in sports that involve repeated contact with a hard surface, such as handball, volleyball, and football. In football players, the incidence of hemorrhagic bursitis has increased with the use of artificial surfaces. Automutilation should be suspected when hemorrhagic bursitis is diagnosed several times within a period of weeks or months. If other signs of acute or chronic trauma are found on further examination, suspicion should be increased (Fig 5–22). However, the patient may be a rugby player.

Sonography demonstrates a distended bursa forming a spheroid (prepatellar bursa), concave spheroid (olecranon bursa), beanlike (retrocalcaneal bursa), or bilobed (deep infrapatellar bursa) structure. The bursa is usually more distended than is seen in frictional bursitis. Fluid within the bursa has definite internal echos corresponding to the cellular components of blood. Immediately following injury, the bursa is filled with diffusely echogenic material (Fig 5–23). After clot retraction, blood and fibrin clots can be seen within the bursa as irregularly shaped hyperechoic masses floating within the anechoic serum (Fig 5–24). It is important to note that these masses are mobile and not attached to the synovial walls (Fig 5–25). No thickening of the synovial wall is present. In later stages of resolution adhesions to the synovial walls can develop. This usually occurs several months after the initial injury.

### *Chemical Bursitis*

A chemical bursitis results from the accumulation within the bursa of substances from metabolic disorders, inflammatory or degenerative processes of the surrounding tendons (Peterson and Renström, 1986). A resolving hemorrhagic bursitis may also be categorized as a chemical bursitis. The deposition of urate

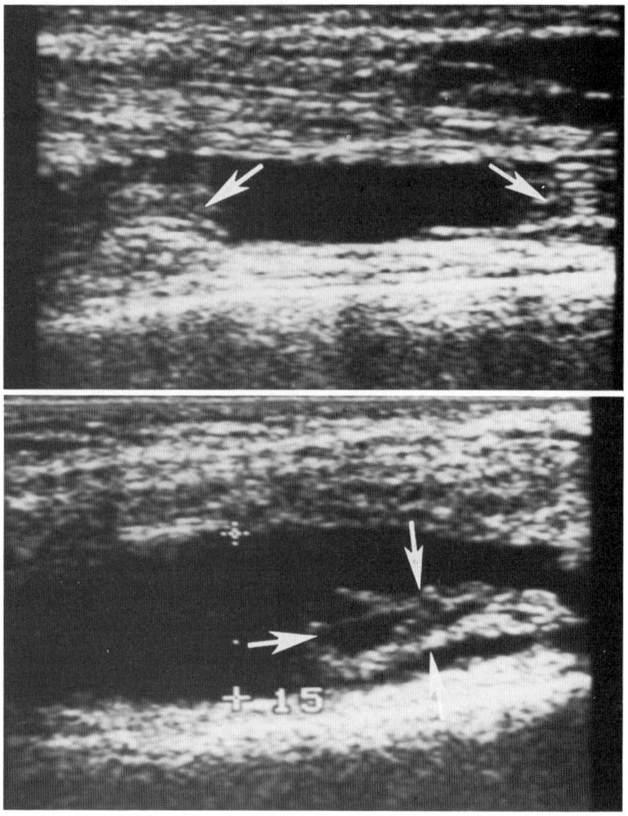


**FIG 5–25.**
Hemorrhagic bursitis of the deep trochanteric bursa: coronal sonogram. A 23-year-old man with pain and swelling over the right greater trochanter following direct trauma to this region. The patient was examined in the left lateral decubitus position with the transducer placed over the right greater trochanter in the coronal plane. Anechoic fluid is seen within the bursa in the *top image*. Clumps of debris are noted within the bursa *(arrows)*. Reduced pressure applied with the transducer and compression of the lateral recesses of the bursa by the examiners free hand demonstrated free-floating clots within the bursa *(bottom image)*. The diameter of the bursa is increased to 15 mm.

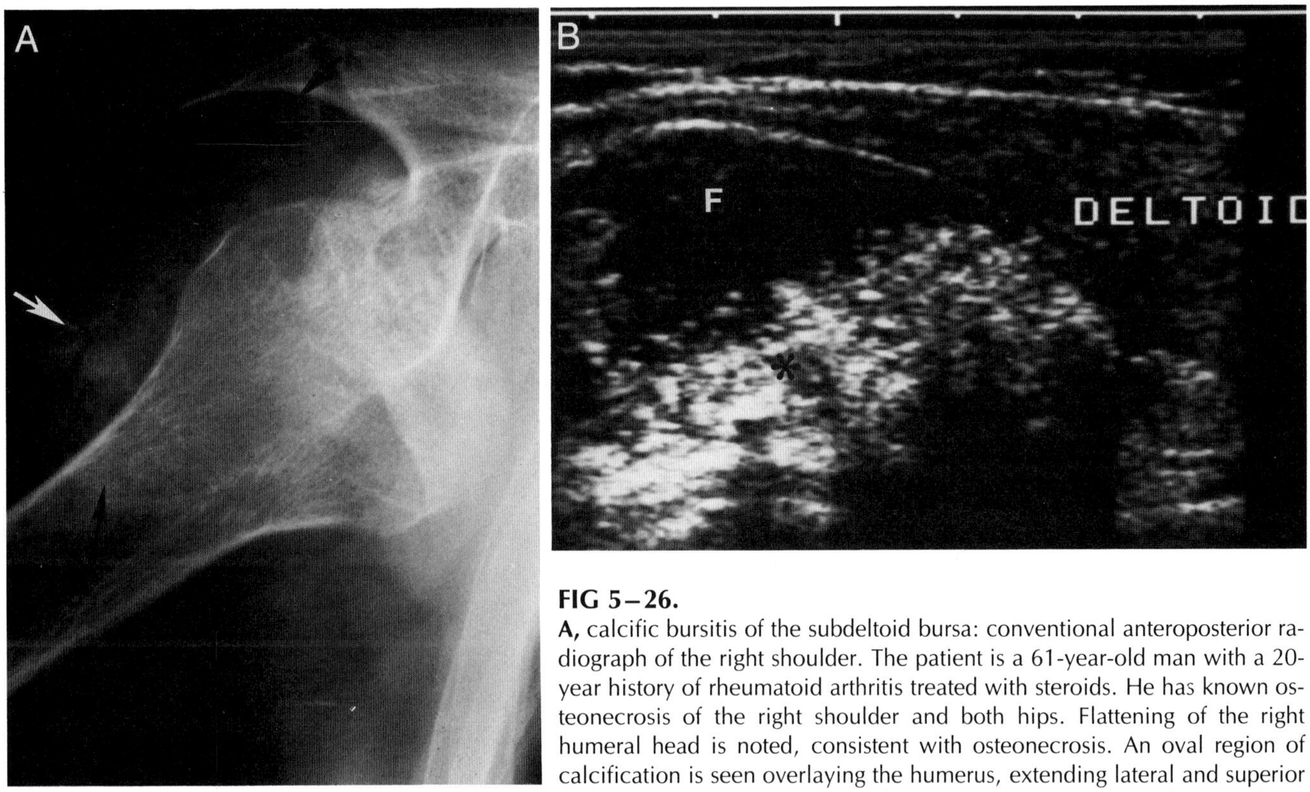


**FIG 5–26.**
**A,** calcific bursitis of the subdeltoid bursa: conventional anteroposterior radiograph of the right shoulder. The patient is a 61-year-old man with a 20-year history of rheumatoid arthritis treated with steroids. He has known osteonecrosis of the right shoulder and both hips. Flattening of the right humeral head is noted, consistent with osteonecrosis. An oval region of calcification is seen overlaying the humerus, extending lateral and superior to the humeral head *(arrows).* **B,** calcific bursitis of the subdeltoid bursa: coronal sonogram. Same patient as in **A.** The transducer is positioned along the length of the humerus with the acromion just out of the field to the left. An anechoic fluid collection *(F)* is present in the subdeltoid bursa. Numerous calcifications *(asterisk)* are noted dependently in the bursa, some demonstrating acoustic shadowing.

crystals within bursae, seen in gout, is the most common metabolic disorder causing chemical bursitis. Gout most frequently involves the subcutaneous bursae of the dorsal aspect of the elbows and the ventral aspect of the knees.

Inflammatory and degenerative causes of chemical bursitis have their greatest incidence in the fifth and sixth decades of life. The subdeltoid bursa is the most commonly involved. Crystals found most frequently are composed of calcium hydroxyapatite (Fig 5–26).

On ultrasound images, the bursa is distended with fluid that appears markedly hypoechoic or anechoic. Hyperechoic foci may be seen within the bursa or in the surrounding tendons corresponding to calcifications. Once again, shadowing may or may not be observed dependent on the size of the calcifications and the frequency of the transducer utilized.

### *Rheumatoid and Septic Bursitis*

The sonographic findings in rheumatoid (Fig 5–27) and septic bursitis (Fig 5–28) are identical to those seen in chronic traumatic bursitis. The contents of the bursa are turbid, definite internal echos are observed, and the synovial walls are often thickened. Therefore, the differential diagnosis of these three entities cannot always be narrowed using solely sonographic criteria. Cases of septic bursitis with gas-producing organisms will demonstrate the characteristic gas bubbles with dirty shadowing. Otherwise, clinical criteria and diagnostic aspiration of the bursal fluid must be utilized to establish a definitive diagnosis. In chronically ill patients, such as those with rheumatoid arthritis or chronic renal failure, septic bursitis can rapidly lead to systemic sepsis and demise of the patient. Early diagnosis of septic bursitis in individuals with compromised immune function is essential.

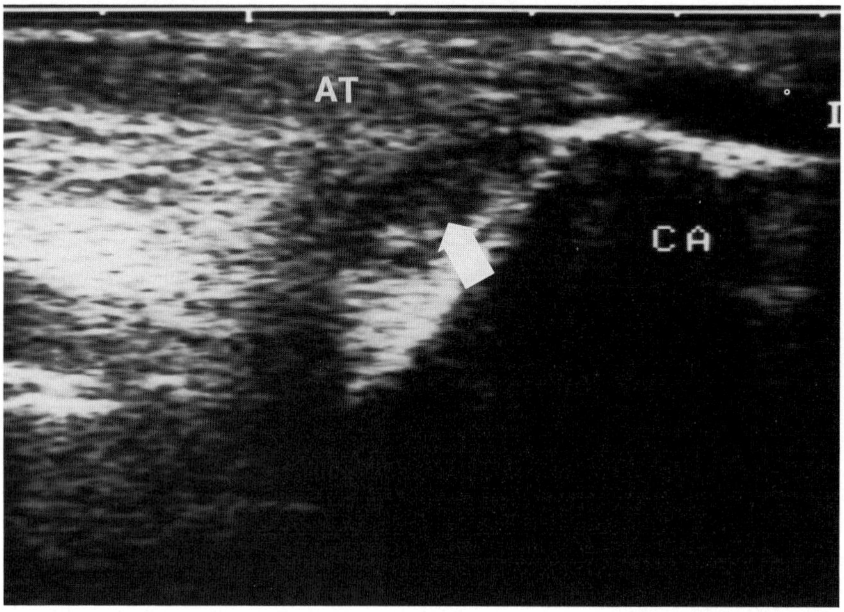


**FIG 5–27.**
Rheumatoid bursitis of the retrocalcaneal bursa: longitudinal sonogram. This 53-year-old woman presented with a 12-year history of symmetrical polyarthritis, and she is rheumatoid factor positive. Her current complaint is pain in the heels bilaterally of 2 months' duration. The transducer is oriented along the Achilles tendon *(AT)*. The distal Achilles tendon and its insertion *(I)* on the calcaneus *(CA)* are normal. The retrocalcaneal bursa *(arrow)* is enlarged and contains definite internal echos. The anterior synovial wall is thickened.

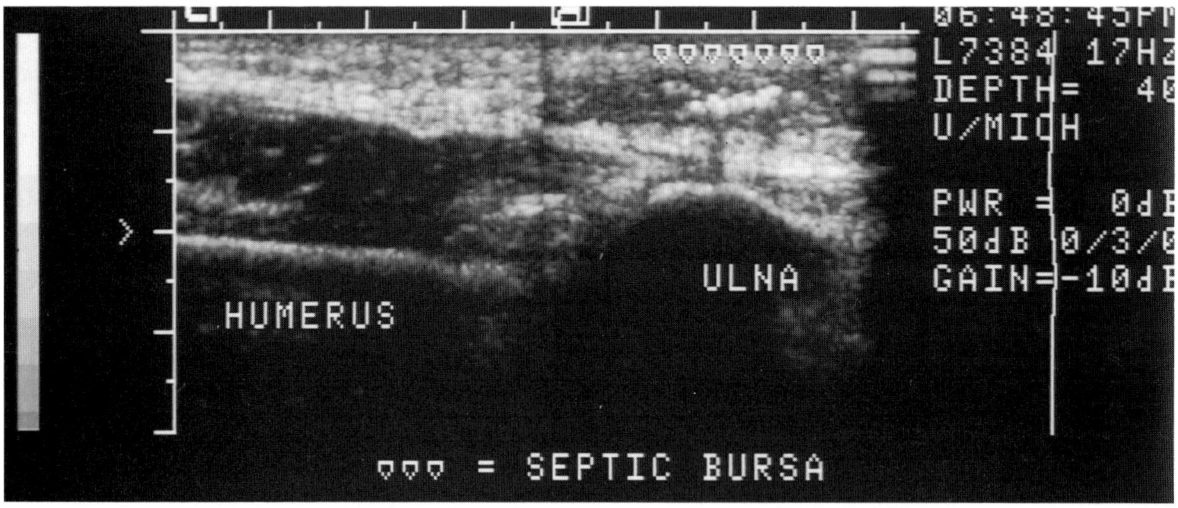


**FIG 5–28.**
Septic bursitis of the olecranon bursa: longitudinal sonogram. The patient is a 38-year-old construction worker who presented with pain and swelling of the right elbow. The transducer is positioned in the long axis of the arm over the olecranon process. A distended olecranon bursa *(arrowheads)* containing hypoechoic fluid is noted. Numerous hyperechoic foci with dirty shadowing are seen within the bursa, giving the definitive diagnosis of septic bursitis. The bursa was aspirated and *Staphylococcus epidermidis* cultured.

## Communicating Bursae

### *Bursal Fluid and Intra-articular Disease*

Communicating bursae are not present at birth but are acquired during growth and development from noncommunicating bursae. The passage between the gastrocnemius-semimembranosus bursa and the knee joint is never found in children less than 10 years of age and is found in 50% of knees by the fifth decade of life (Lindgren and Willen, 1977) (Fig 5–29). The hip joint communicates with the iliopsoas bursa (unilaterally or bilaterally) in approximately 20% of adults (Chandler, 1934) (Figs 5–30 through 5–32). Communication of the subacromial bursa with the glenohumeral joint is possible only in the presence of a rotator cuff tear (Fig 5–33 and 5–34). The incidence of rotator cuff tears in adults more than 50 years of age is approximately 13% (De Palma, 1950). This communication has significantly greater incidence in patients with rheumatoid arthritis. Approximately 30% of patients with long-standing rheumatoid arthritis have complete rotator cuff tears. Pannus (Fig 5–35) due to inflammation of the glenohumeral joint erodes the undersurface of the rotator cuff. This process progresses rapidly if there is inflammation of the subacromial-subdeltoid bursa or the patient is being treated with steroids (Resnick and Niwayama, 1988).

Large bursal effusions found in communicating bursae are usually the result of synovial fluid being pumped into bursae during motion. Communicating bursae serve as potential reservoirs for joint effusions and thus minimize the deleterious effects of the effusion. The result is a "cyst." Physiological studies indicate a valvular mechanism is responsible for cyst development. Although the exact mechanism has not been elucidated, anatomical studies have demonstrated the pathways of communication with the joint space to be narrow channels between the ligaments and tendons (Jayson, 1981). Movement of the joint is thought to open and close the valve. Pressure within the bursa is considerably higher than that within the joint space. Weight bearing on the joint will force fluid into the bursa against a pressure gradient while the valve is open (Jayson, 1981).

Sonography demonstrates a distended bursa containing hypoechoic or anechoic fluid. The definitive diagnosis can be made by identification of the slitlike channel connecting the bursa with the joint space. In an examination of the subdeltoid bursa, the communication with the joint space is a tear in the rotator cuff. Patients with rheumatoid arthritis will have

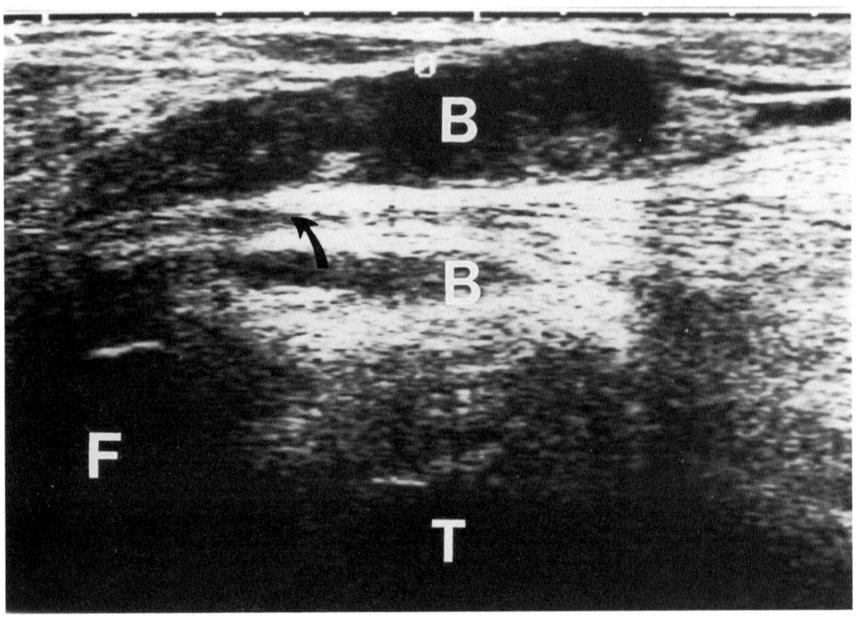


**FIG 5–29.**

Rheumatoid Baker's cyst: longitudinal sonogram. A 62-year-old woman with rheumatoid arthritis for 6 years presented with an acute exacerbation of synovitis of the right knee. The transducer is positioned in the long axis of the extremity over the medial gastrocnemius tendon *(curved arrow)*. A Baker cyst *(B)* surrounds the medial gastrocnemius tendon. Characteristics of a rheumatoid bursa demonstrated here are the irregular outline of the bursa and internal debris. More than 90% of patients with rheumatoid arthritis and involvement of the knee have Baker's cysts. Abbreviations: femur *(F)*, tibia *(T)*.

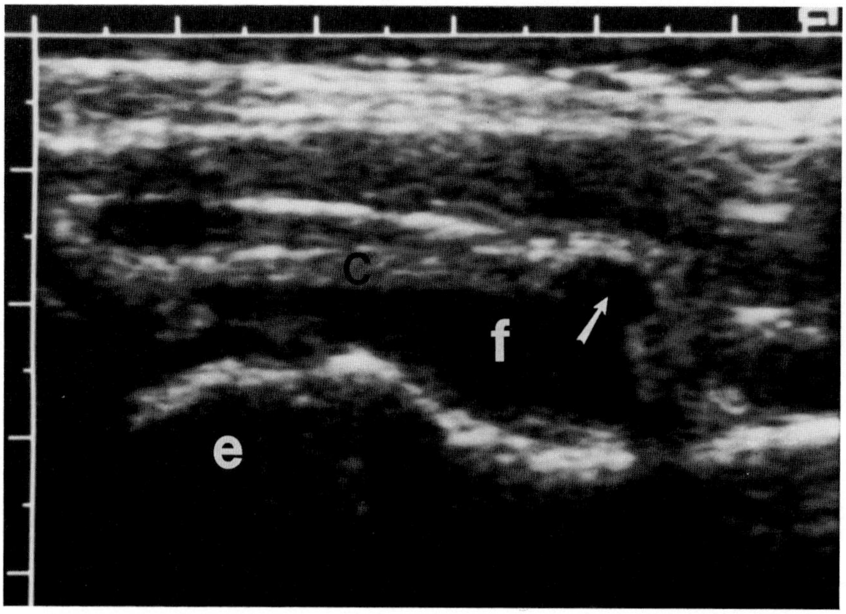


**FIG 5–30.**
Filling of the iliopsoas bursa in transient synovitis: longitudinal sonogram. A 7-year-old boy with hip pain. The transducer is positioned along the femoral neck. Fluid in the hip joint *(f)* communicates with the iliopsoas bursa *(arrow)* through an opening in the joint capsule *(C)*. Communicating iliopsoas bursae are present in about 2% of joints in children less than age 10 years. The incidence of communication increases with age, reaching 20% to 30% in patients more than 50 years old. Abbreviations: femoral epiphysis *(e)*.

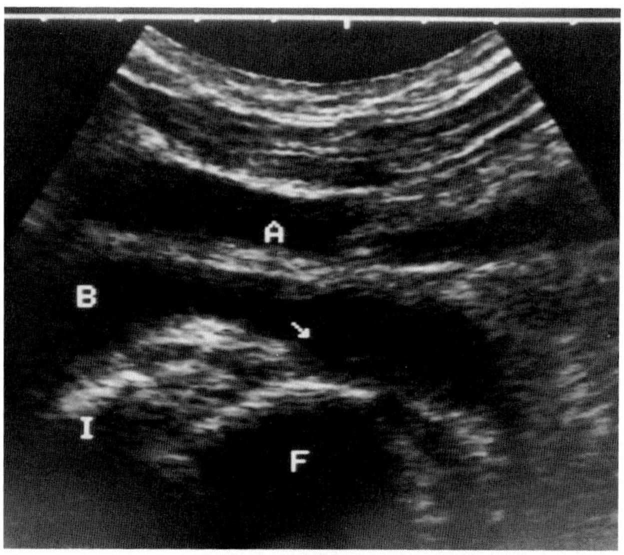


**FIG 5–31.**
Large iliopsoas bursa in an adult: longitudinal sonogram. A pulsatile mass could be palpated clinically in the left groin of this 60-year-old woman. The clinical diagnosis was femoral pseudoaneurysm. The transducer is positioned in the parasagittal plane just medial to the femoral neck. A large fluid-filled bursa *(B)* extends deep to the femoral artery *(A)* and communicates with the hip joint *(arrow)*. The femoral artery is clearly normal. Abbreviations: femur *(F)*, ilium *(I)*.

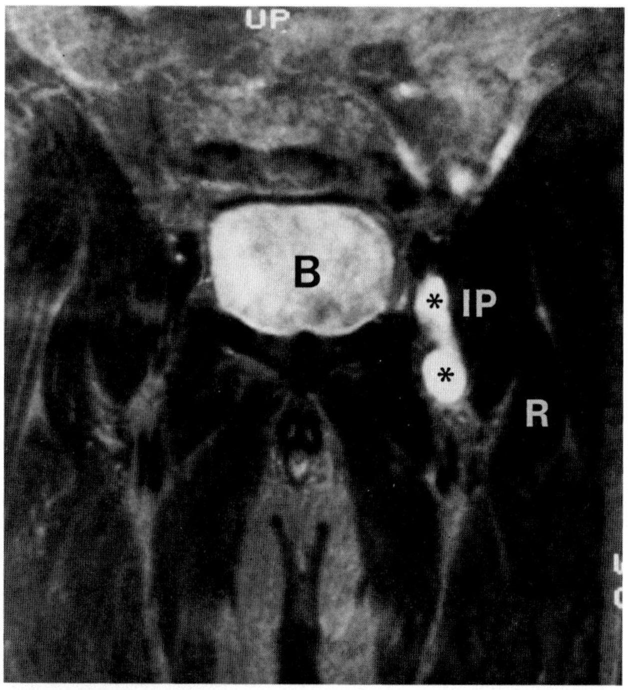


**FIG 5–32.**
Large iliopsoas bursa: MRI of the pelvis (coronal plane). Same patient as in Figure 5–31. The image shown is a T2-weighted sequence in the coronal plane through the iliopsoas and the proximal rectus, anterior to the hips. A bilobed iliopsoas bursa *(asterisk)* is indented by the ilioinguinal ligament. Fluid within the bursa demonstrates very high signal, more intense than the urinary bladder *(B)*.

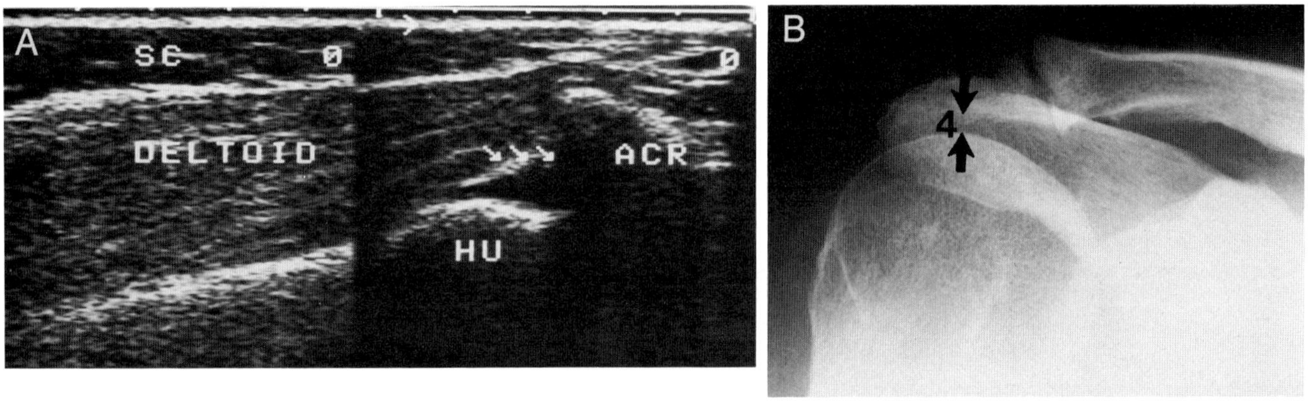


**FIG 5–33.**
**A,** supraspinatus tear with fluid in the subacromial-subdeltoid bursa: coronal sonogram. A 69-year-old man presented with limited abduction of the shoulder. The transducer is positioned along the humerus with the acromion *(ACR)* at the observer's right. The supraspinatus tendon is retracted and not seen. Fluid within the subacromial-subdeltoid bursa *(small arrows)* occupies the space vacated by the supraspinatus tendon. The deltoid muscle covers the shoulder and originates from the acromion. Abbreviations: subcutaneous fat *(sc),* humerus *(HU).* **B,** supraspinatus tear with fluid in the subacromial-subdeltoid bursa: conventional anteroposterior supine radiograph. Same patient as in **A.** The subacromial space is narrowed to 4 mm, and the acromioclavicular joint shows evidence of osteoarthritis. These findings are consistent with a complete supraspinatus tear.

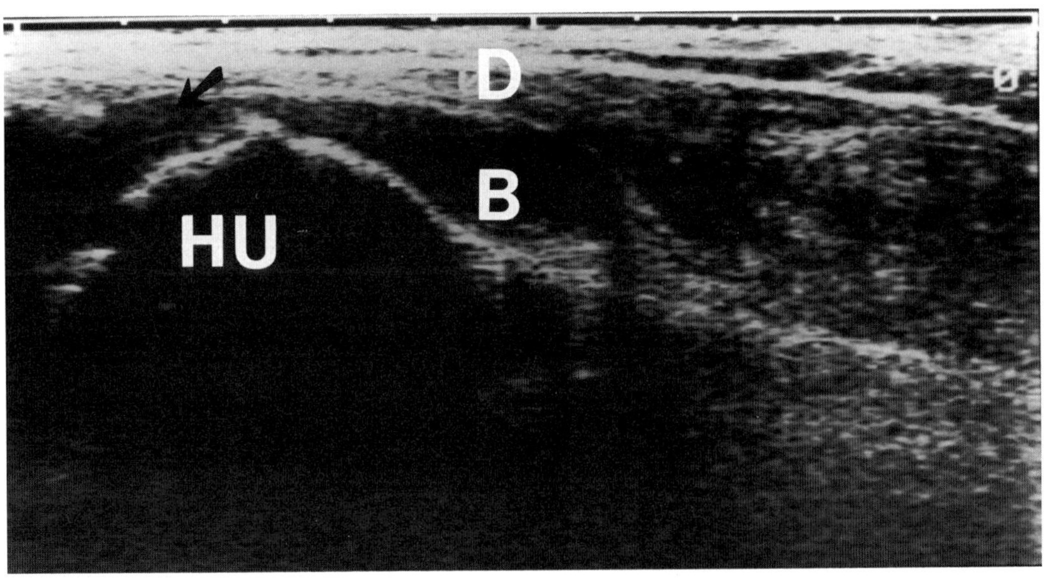


**FIG 5–34.**
Distention of the subacromial-subdeltoid bursa: coronal sonogram. The patient is a 45-year-old woman with a 2-year history of rheumatoid arthritis who presents with left shoulder pain. Transducer position is identical to that in Figure 5–33,A. Proximal lies to the observer's left. A significantly distended subacromial-subdeltoid bursa *(B)* is noted, which is characteristic of rheumatoid arthritis. Aspiration is indicated for differentiation between chronic inflammatory disease and infection when the clinical picture is unclear. The bursa communicates with the joint space under the acromion *(curved arrow).* Abbreviations: deltoid *(D),* humerus *(HU).*

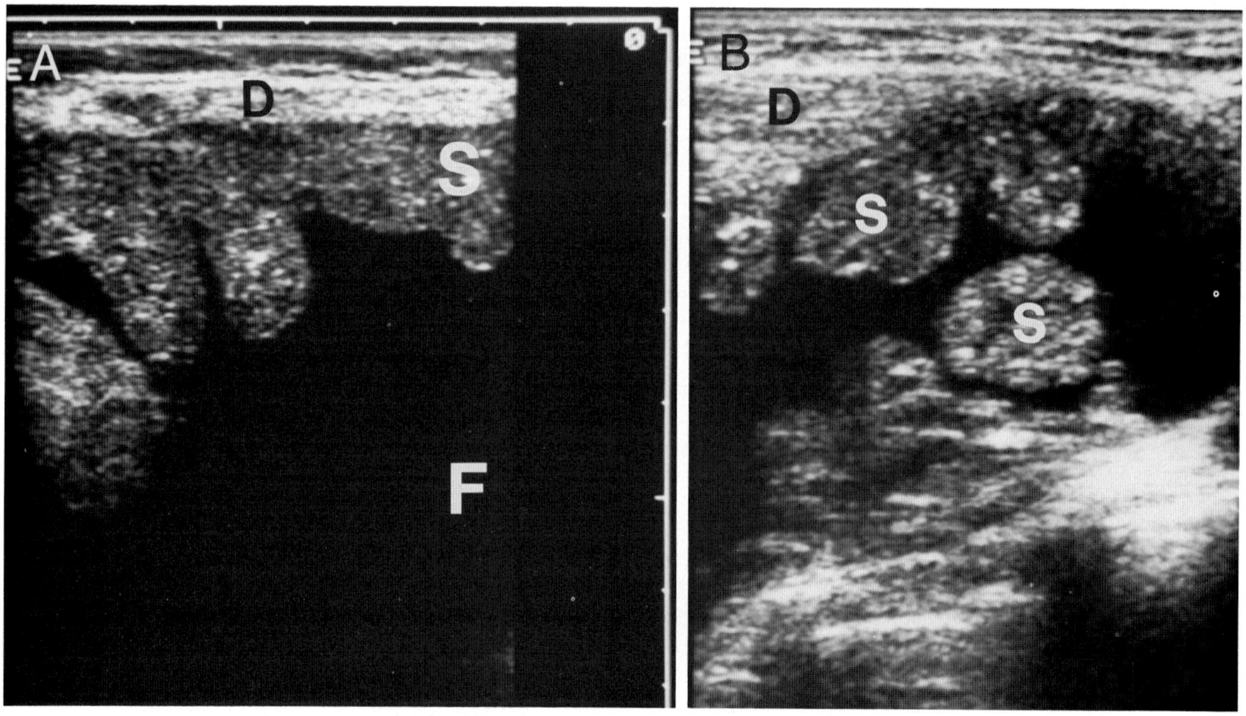


**FIG 5–35.**
**A,** pannus in the subacromial-subdeltoid bursa: coronal sonogram. A 51-year-old woman with a 24-year history of rheumatoid arthritis who presents with palpable swelling and tenderness of the right shoulder. Transducer position is identical to that in Figure 5–34. The deltoid muscle *(D)* is atrophic. Distention of the subacromial-subdeltoid bursa is due to both fluid *(F)* and synovial proliferation *(S)*. The hypertrophic synovial tissue has a polypoid appearance and is typically referred to as *pannus.* **B,** pannus in the subacromial-subdeltoid bursa: transverse sonogram. Same patient as in **A.** The transducer is rotated 90 degrees, yielding in a transverse image. Again noted is the polypoid appearance of the hypertrophic synovial tissue *(S)*. The pannus is hypoechoic relative to the surrounding tissue and is surrounded by anechoic fluid. Abbreviations: deltoid *(D)*.

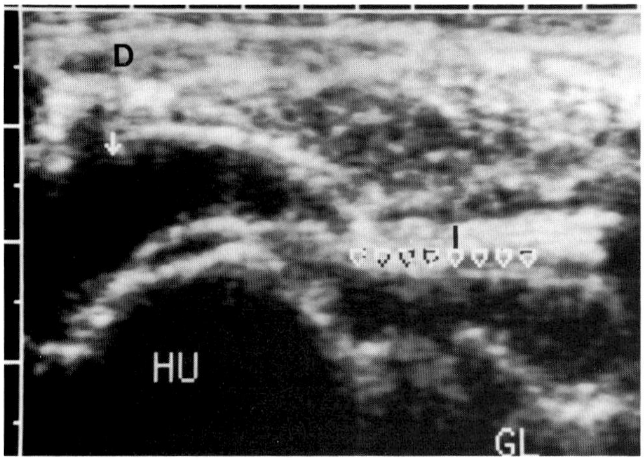


**FIG 5–36.**
Septic arthritis communicating with the subacromial-subdeltoid bursa: transverse sonogram. This 62-year-old man with a painful shoulder treated 3 weeks ago with steroids presents now with increased pain and soft tissue swelling. He was afebrile. The lateral aspect of the shoulder is at the observer's right. A distended subacromial-subdeltoid bursa *(white arrow)* filled with hypoechoic fluid is present between the deltoid muscle *(D)* and the humerus *(HU)*. The supraspinatus tendon is absent, indicative of a complete rotator cuff tear. Definite internal echos are present within the fluid, but the bursal walls are neither thickened nor irregular. Fluid is also seen in the glenohumeral joint between the infraspinatus tendon *(I)* and the glenoid *(GL)*. Aspiration and culture revealed *Staphylococcus aureus* infection.

irregularity and thickening of the synovial lining of the bursa (see Fig 5–35). Septic arthritis and other inflammatory processes will have definite internal echos within the fluid and inhomogeneous filling of the bursa (Fig 5–36).

### *Baker's Cyst*

Baker's cysts will be discussed as a separate entity because of the very high frequency of enlargement of the gastrocnemius-semimembranosus bursa. The popliteal cyst was first described by Adams in 1840 (Adams, 1840). He correctly identified the cyst as an enlarged bursa found medially in the popliteal fossa. In 1877, Baker described the lesion again and established the causal relationship between joint effusions and the popliteal cysts (Baker, 1877).

The location of the gastrocnemius-semimembranosus bursa is constant, found at the medial aspect of the popliteal fossa. Communication with the joint space of the knee develops after birth, likely during adolescence. Morphologically, the bursa is divided into three portions: the base, superficial extent, and the neck (Canoso, 1981). Located between the joint capsule and the gastrocnemius tendon is the base of the bursa. Beneath the fascia is the superficial portion of the bursa. The neck of the bursa lies between the gastrocnemius and semimembranosus tendons, thus the name of this bursa (Fig 5–37 through 5–39).

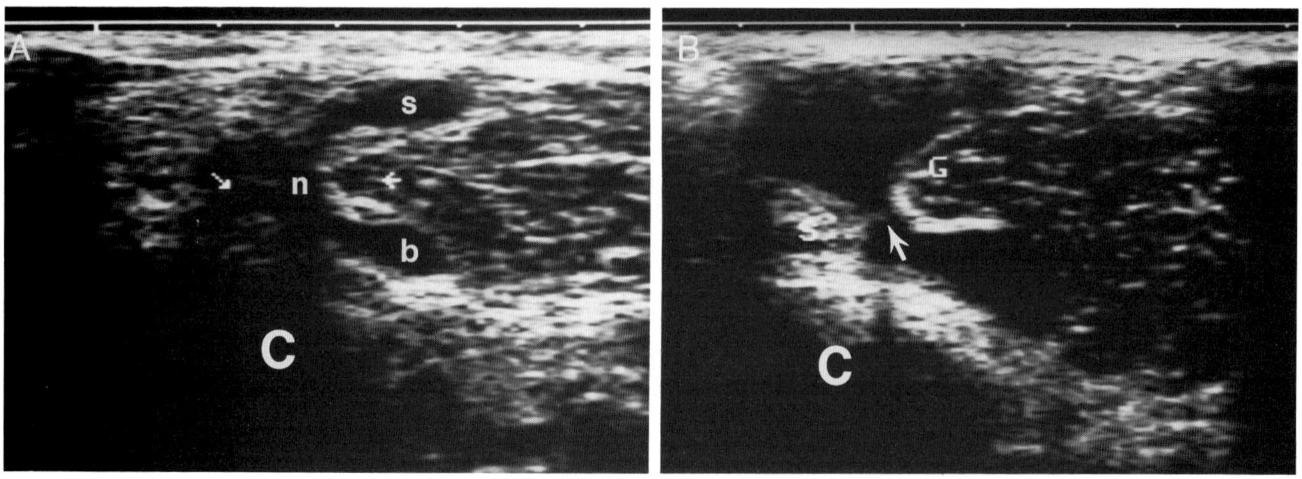


**FIG 5–37.**
**A,** small Baker's cyst: transverse sonogram with knee in slight flexion. The patient is a 39-year-old man with a medial meniscal tear. The transducer is positioned transversely over the medial aspect of the popliteal fossa just above the joint space. A Baker cyst is divided into three anatomical parts: the base, neck, and superficial portion. Separating the joint capsule and the medial femoral condyle *(C)* from the medial gastrocnemius tendon *(horizontal arrow)* is the base *(b)* of a small Baker cyst. The neck *(n)* of the Baker cyst is a slitlike channel between the medial gastrocnemius tendon and the semimembranosus tendon *(oblique arrow)*. Its superficial portion *(s)* extends posteriorly between the fascia and the medial gastrocnemius tendon. **B,** same patient and transducer position as **A** but with the knee in extension. With the knee in extension, the neck of the bursa is narrowed *(arrow)*, and a greater amount of fluid is seen in the superficial extent of the bursa. This supports the theory of a pump and valve mechanism in the formation of a Baker cyst. Abbreviations: semimembranosus tendon *(S)*, medial gastrocnemius tendon *(G)*, medial femoral condyle *(C)*.

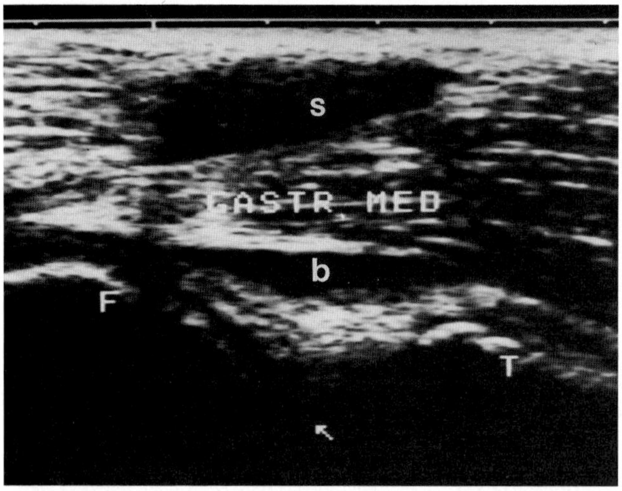


**FIG 5–38.**
Small Baker's cyst: longitudinal sonogram. Same patient as in Figure 5–37. In this image the transducer is oriented along the length of the medial gastrocnemius tendon. The medial gastrocnemius tendon is surrounded by the base *(b)* and superficial portion *(s)* of the Baker cyst. Deep to the base of the Baker cyst lies the posterior extent of the joint capsule and the medial meniscus *(arrow)*. Abbreviations: medial femoral condyle *(F)*, proximal tibia *(T)*.

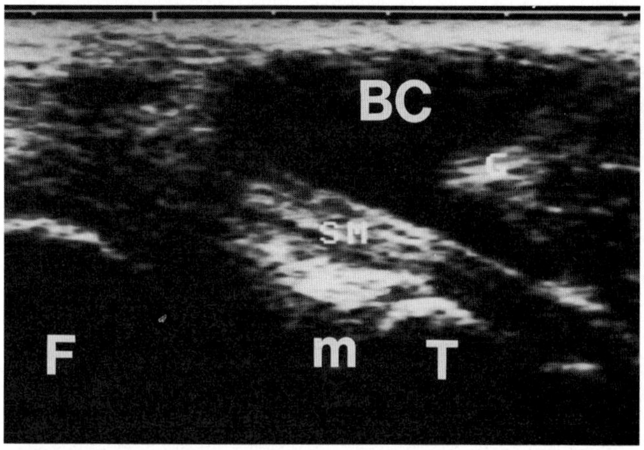


**FIG 5–39.**
Small Baker's cyst: longitudinal sonogram. Same patient as in Figure 5–37. The transducer is moved approximately 1 cm medial to that in Figure 5–38 to demonstrate the semimembranosus tendon *(SM)* in its long axis. Its insertion on the posterior aspect of the tibial epiphysis *(T)* is seen on the *right*. A small amount of joint fluid is present between the tendon and the posterior aspect of the medial meniscus *(m)*. The neck of the bursa is located between the gastrocnemius tendon *(G)* and the semimembranosus tendon. The largest portion of the Baker cyst is its superficial extent *(BC)*.

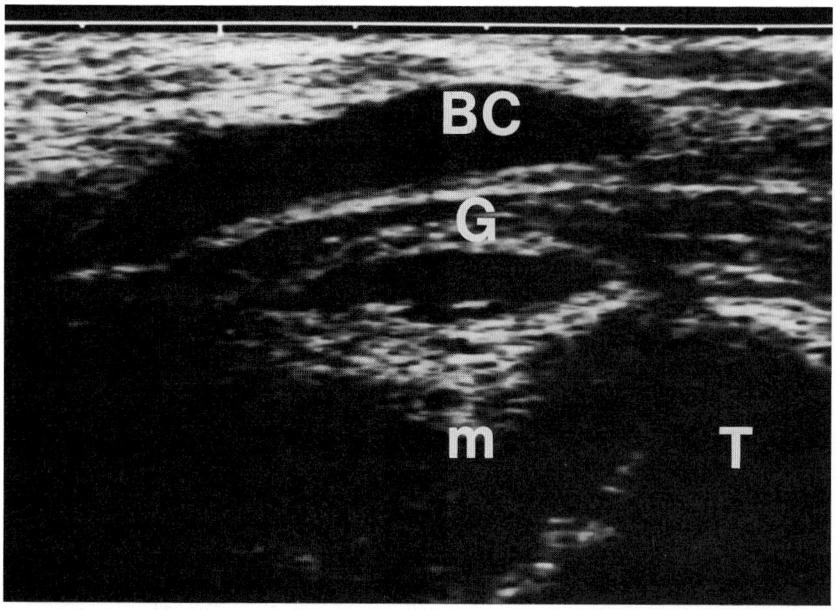


**FIG 5–40.**
Baker's cyst: longitudinal sonogram. Two years following a meniscal injury, this 29-year-old woman continued to experience recurrent painful swelling of the right knee in the popliteal fossa, especially after jogging. Again, the transducer is oriented along the length of the medial gastrocnemius tendon *(G)*. In Baker's cysts that arise as a result of noninflammatory intra-articular disease, the fluid is uniformly anechoic. Clear fluid is obtained on aspiration of these joints. The synovial lining of these cysts is very smooth and regular. Abbreviations: Baker's cyst *(BC)*, medial meniscus *(m)*, proximal tibia *(T)*.

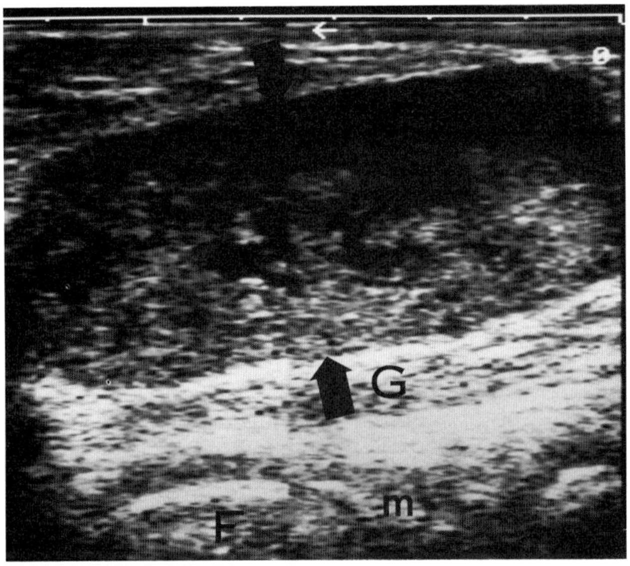


**FIG 5–41.**
Rheumatoid Baker's cyst: longitudinal sonogram. The patient is a 53-year-old man with a 10-year history of symmetrical arthritis involving the hands, feet, shoulders, and right knee. He was rheumatoid factor positive. At this time he presented with increasing pain in the right calf. Transducer position is identical to that in Figure 5–40. No fluid is present in the base of the Baker cyst between the medial gastrocnemius *(G)* and the meniscus *(m)*. However, the superficial portion of the cyst is markedly distended with fluid *(large arrows)*. Definite internal echos are identified moving within the fluid dependently. Synovial proliferation predominated along the anterior surface of the superficial extent of the bursa.

The slitlike channel of communication between the bursa and the joint space is closed during extension of the knee. This is accomplished by compression of the neck of the bursa between the medial femoral condyle and the gastrocnemius tendon. Flexion of the knee separates these structures and opens the channel (Canoso, 1981).

Two general categories of disease can result in bursal distention: pathology associated with increased intra-articular fluid and the inflammatory arthropathies. The group of diseases that result in increased intra-articular fluid share one common feature. They all result in irregularity of joint surfaces. Diseases included in this group are osteochondritis dissecans, osteochondral fractures, osteonecrosis, osteoarthritis, cartilage defects, meniscal lesions, and loose bodies. In these cases, the fluid within Baker's cyst is almost always anechoic (Fig 5–40).

Inflammatory arthropathies are characterized by increased fluid production and synovial proliferation within the joint and communicating bursae. Baker's cysts developing secondary to an inflammatory arthropathy tend to be much larger than those associated with other types of joint pathology (Fig 5–41). Rheumatoid Baker's cysts characteristically have marked irregularity of the synovial lining with numerous internal echos identified within the fluid (van Holsbeeck et al., 1988) (Fig 5–42). These large Baker cysts can dissect the soft tissues in the legs in two different ways. The most common presentation is extension of the Baker cyst dorsal to the gastrocnemius muscle in a subcutaneous location (Fig 5–43). The other presentation seen in rheumatoid Baker's cysts is dissection of the fascial planes between the soleus and gastrocnemius muscles, extending between the muscles of the posterior compartment (Fig 5–44). The pathogenesis of these differing presentations is easily understood. The gastrocnemius-semimembranosus bursa surrounds the medial gastrocnemius tendon (see Fig 5–38). Swelling of the bursa can then extend ventral or dorsal to this tendon. The inter-

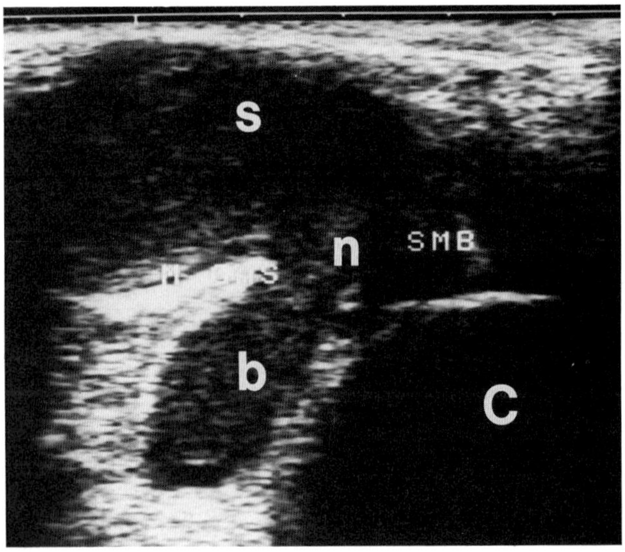


**FIG 5–42.**
Rheumatoid Baker's cyst: transverse sonogram. This 62-year-old woman presented with inflammatory type of pain in both knees. Swelling of the knees was greater on the left with fullness in the left popliteal fossa. She was rheumatoid factor positive. Synovial biopsy results were consistent with chronic synovitis, consistent with rheumatoid arthritis. All portions of the Baker cyst are markedly distended: base *(b)*, neck *(n)*, and superficial extent *(s)*. The neck of the cyst is much wider than that seen in cases due to noninflammatory intra-articular disease. Wide separation of the medial gastrocnemius *(MGAS)* and semimembranosus tendons is observed. The synovial proliferation is seen diffusely as thickening of the walls and debris is noted throughout. Abbreviations: medial femoral condyle *(C)*.

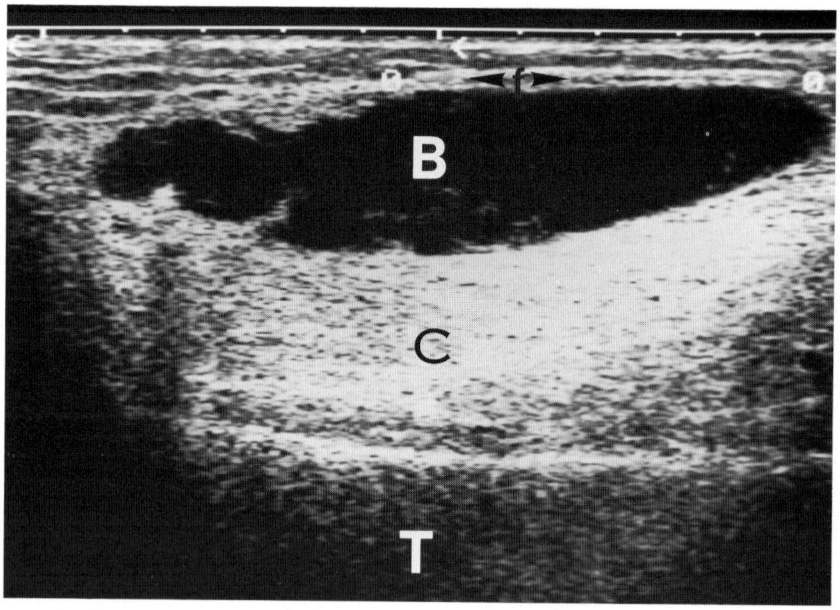


**FIG 5–43.**
Superficial extent of rheumatoid Baker's cyst: longitudinal sonogram. A 66-year-old woman with long-standing rheumatoid arthritis presented with pain and swelling of the right calf. This case demonstrates the location of dissection of the superficial portion of the rheumatoid Baker cyst. It is the most frequent presentation of an enlarging rheumatoid Baker cyst. The large superficial extent of this cyst *(B)* lies between the fascia *(f)* and calf muscles *(C)*, displacing the muscles anteriorly. Rheumatoid Baker's cysts tend to be the largest, this one measuring more than 10 cm long. Again, irregularity of the synovial wall is noted. Abbreviations: tibia *(T)*.

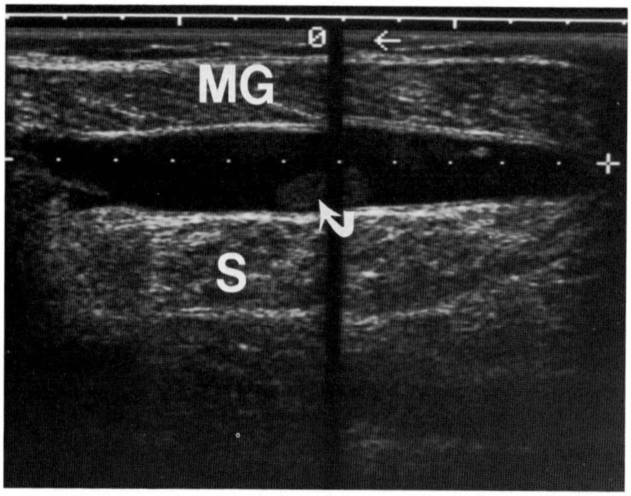


**FIG 5–44.**
Deep dissection of rheumatoid Baker's cyst: longitudinal sonogram. The patient is a 50-year-old man with polyarticular disease of 4 months' duration who presented with pain and swelling of the left leg. A 15-cm long Baker's cyst extends into the calf deep to the medial gastrocnemius *(MG)* and superficial to the soleus *(S)*. The bursa is filled with fluid and pannus *(curved arrow)*, which is characteristic of rheumatoid disease.

muscular extension is far less frequent, probably due to higher pressure within the muscle compartment.

Fibrin clots may also be seen as hyperreflective masses within a rheumatoid bursa (see Fig 5–43). The largest of these rheumatoid Baker cysts, filled with pannus, may rupture resulting in changes within the surrounding fat and muscle (Figs 5–45 and 5–46). These changes are due to inflammation secondary to enzymes found in the pathological bursal fluid. When this occurs, it is often referred to as *pseudothrombophlebitis* because the clinical appearance of the leg is indistinguishable from acute thrombophlebitis (Jayson, 1972). Ultrasound will establish the correct diagnosis and is preferred over invasive studies such as contrast venography or arthrography.

Studies comparing sonography with contrast arthrography in the diagnosis of Baker's cyst show diagnostic accuracy to be equivalent (Resnick and Niwayama, 1988). In these patients, structural integrity of the joint cartilage is not an issue. Given the discomfort inherent in an invasive procedure, especially when the synovia are inflamed, ultrasound is the examination of choice. Magnetic resonance imaging (MRI) is comparable in diagnostic accuracy and is noninvasive. However, the examination requires the patient to lie motionless for at least 1 hour, far longer

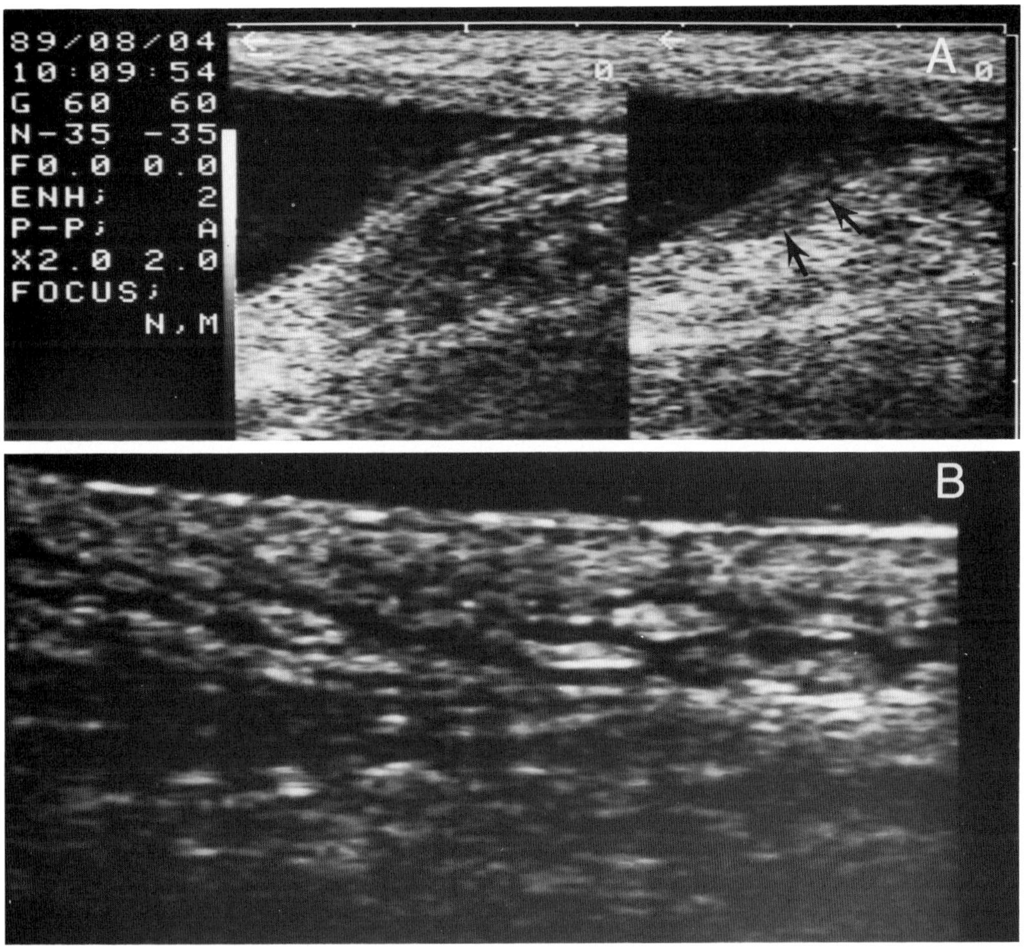


**FIG 5–45.**
**A,** ruptured rheumatoid Baker's cyst: longitudinal sonogram. The calf of the right leg of this 46-year-old man was erythematous and edematous. Clinical presentation was consistent with deep venous thrombosis. A Baker cyst that has ruptured is identified in the popliteal fossa. Note that the inferior border of the cyst is no longer rounded but instead is pointed. This is characteristic of a cyst that has ruptured. In the image on the *right,* pressure was applied with the transducer, and fluid was seen extravasating into the surrounding tissues *(arrows).* Both sonographic examination for venous thrombosis and contrast venography were normal. **B,** ruptured rheumatoid Baker's cyst: longitudinal sonogram of the subcutaneous fat. Same patient as in **A.** A standoff pad was utilized. Fluid is identified surrounding the subcutaneous fat lobules forming a reticular pattern. This is always seen in cases of ruptured Baker's cyst but may also be seen in deep venous thrombosis.

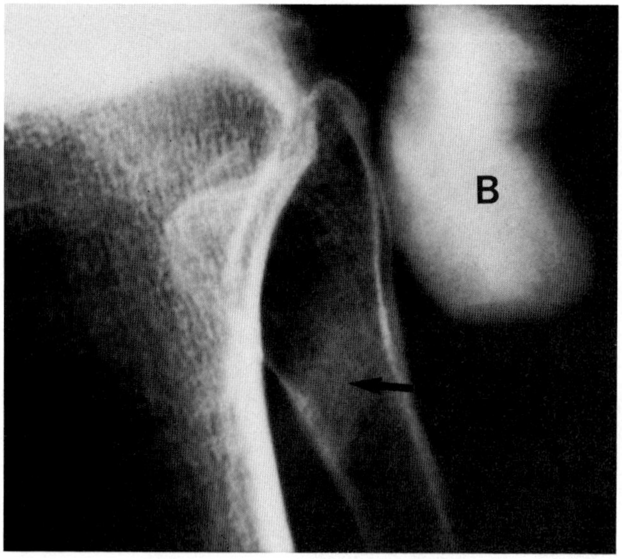


**FIG 5–46.**
Ruptured rheumatoid Baker's cyst: arthrography. Same patient as in Figure 5–45. Arthrography was performed to confirm the diagnosis. This image was obtained 45 minutes following injection of contrast material. Subtle leak of contrast material *(arrow)* is noted anterior and inferior to the Baker's cyst *(B)*.

than the 15 to 20 minutes required to perform a careful ultrasound examination. This factor makes ultrasound the preferred examination, particularly in patients with crippling rheumatoid arthritis. Cost factors also favor ultrasound.

The differential diagnosis of a mass in the popliteal fossa includes popliteal artery aneurysm, varices, lymphocele, abscess, schwannoma, or other soft tissue tumor with cystic characteristics, as well as Baker's cyst (Scott et al., 1977). Absence of pulsation on gray scale imaging helps to exclude aneurysm. Conventional Doppler and color flow imaging will further exclude vascular pathology. The definitive diagnosis of Baker's cyst is made by identification of the channel of communication with the joint space adjacent to the medial femoral condyle. This is pathognomonic for Baker's cyst. Differentiation between Baker's cyst secondary to monoarticular rheumatoid arthritis and pigmented villonodular synovitis is based on clinical criteria. These entities are similar in their sonographic appearance (Fig 5–47).

### *Traumatic Bursitis*

In exceptional cases, a communicating bursa can become pathologically distended with fluid in the absence of articular disease. An example of this is the development of gastrocnemius-semimembranosus cysts in children. These cysts result from overload of the adjacent tendons. There is no evidence of ligamentous or meniscal disease. Local traumatic inflammation of the synovial lining of the bursa is the most likely etiology of this type of bursitis (Fig 5–48).

Ultrasound examination in these cases reveals a distended bursa containing anechoic fluid. Communication with the joint is not detectable in children less than 10 years of age. The ligaments and menisci are normal.

### *Loose Body Entrapment*

Communicating bursae, such as the gastrocnemius-semimembranosus bursa, have valvular connections with the adjacent joint, as described earlier. Fluid pumped into the bursa contains glycoproteins. As solute is reabsorbed from the bursa and new fluid pumped in, the concentration of these glycoproteins progressively increases. The result is a collection of detritus within the bursa (Canoso, 1981). In addition, cartilage fragments may become entrapped within the bursa (Fig 5–49). Some of these loose bodies may calcify and become visible on conventional radiographs (Figs 5–50 and 5–51). However, sonography is more sensitive, identifying both calcified and noncalcified loose bodies within the bursa as hyperreflective foci with or without shadowing (Fig 5–52). Baker's cysts containing loose bodies tend to be intermittently painful and tender to palpation.

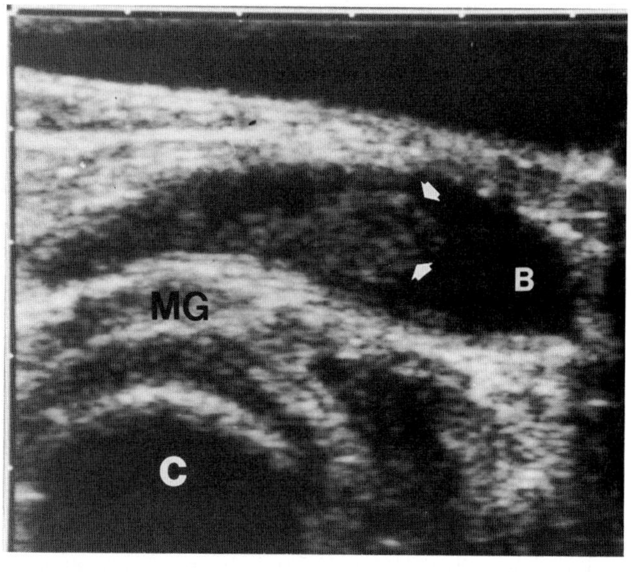


**FIG 5–47.**
Benign synovioma arising within Baker's cyst: longitudinal sonogram. A middle-aged man with a palpable mass in the popliteal fossa of the right knee. Again the transducer is aligned along the length of the medial gastrocnemius tendon. A polypoid soft tissue mass *(small arrows)* is seen protruding into the Baker cyst *(B)*. Histological examination of the synovectomy specimen revealed a benign synovioma. Abbreviations: femoral condyle *(C)*. *Courtesy of Dr. A. de Brabant, Milan, Italy.)*

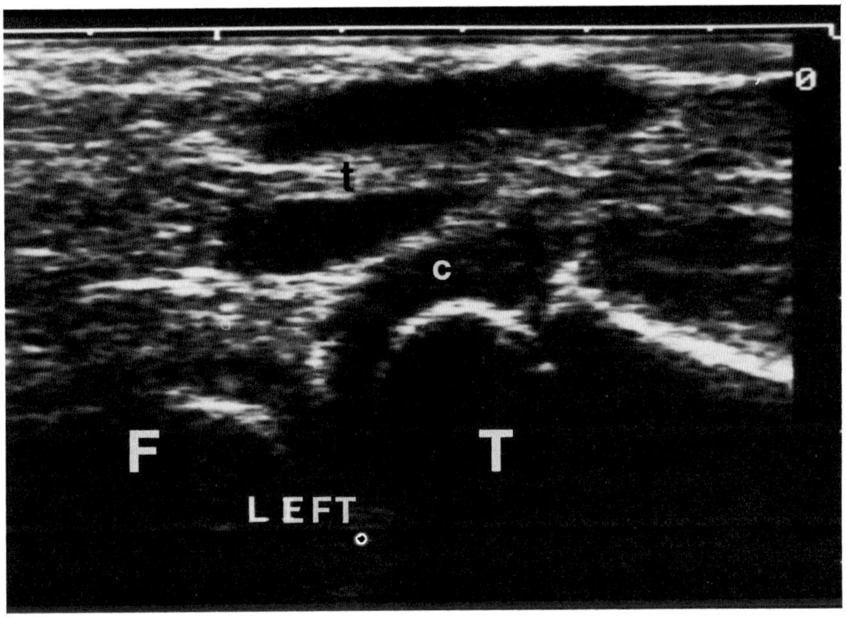


**FIG 5–48.**
Gastrocnemius-semimembranosus bursa in a child: longitudinal sonogram. This 7-year-old girl had swelling in the left popliteal fossa. Transducer position is identical to Figure 5–47. A distended gastrocnemius-semimembranosus bursa surrounds the medial gastrocnemius tendon *(t)*. This is the characteristic appearance of frictional bursitis. No communication with the joint can be demonstrated in children. An arthrogram performed on this patient was normal. There was no communication of the joint with the bursa. Abbreviations: femur *(F)*, tibia *(T)*, epiphyseal cartilage of the proximal tibia *(c)*.

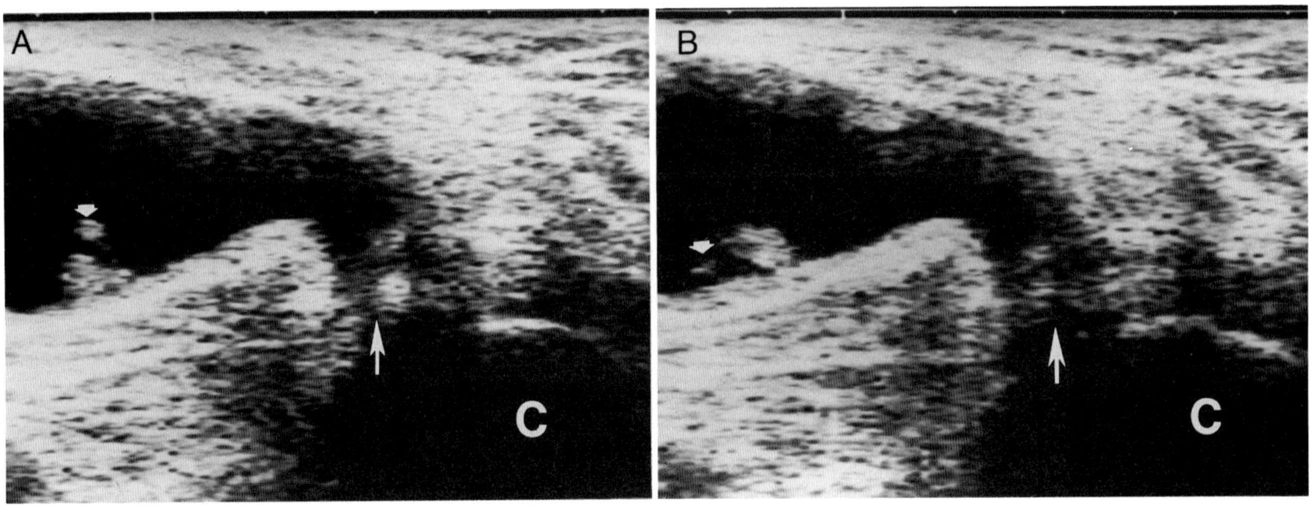


**FIG 5–49.**
Meniscal fragments within Baker's cyst: transverse sonogram. The patient is a 41-year-old man who sustained trauma to the left medial meniscus and presented with progressive swelling of the knee over the past 2 months. The transducer is positioned transversely over the medial aspect of the left popliteal fossa. Communication of the Baker cyst with the joint space is demonstrated *(large arrow)*. Several meniscal fragments *(small arrows)* are identified within the cyst. Pressure applied with the transducer demonstrated movement of the fragments within the cyst. Arthroscopy showed a medial meniscal tear and multiple meniscal fragments loose within the joint. Loose bodies are preferentially entrapped within Baker's cysts. Abbreviations: medial femoral condyle *(C)*.

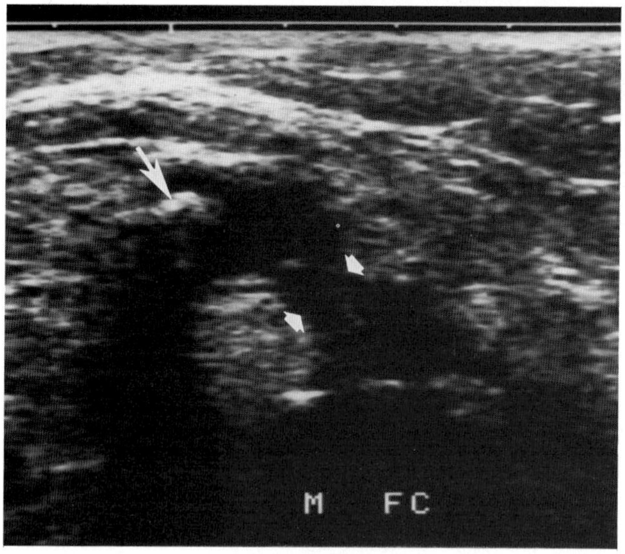


**FIG 5–50.**
Calcified loose body within Baker's cyst: transverse sonogram. A 50-year-old woman with pain and swelling in the right popliteal fossa. Transducer position is identical to that in Figure 5–49. A calcified loose body *(large arrow)* 8 mm in diameter is identified in the superficial portion of the bursa with prominent acoustic shadowing. The diameter of the loose body exceeds that of the neck of the Baker cyst *(arrowheads)*. This implies that it has increased in size since its entrapment within the cyst. Abbreviations: medial femoral condyle *(MFC)*.

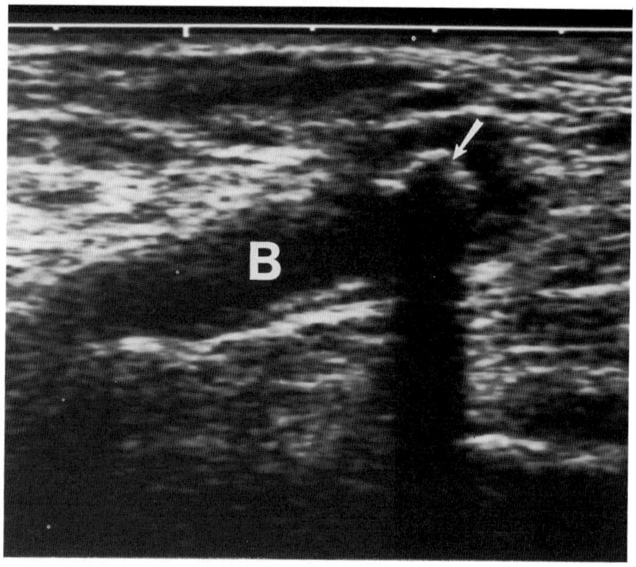


**FIG 5–51.**
Calcified loose body entrapped within Baker's cyst: longitudinal sonogram. This calcified loose body *(arrow)* caused pain and swelling in the popliteal fossa of a 75-year-old woman. Note the prominent acoustic shadow. Baker's cyst *(B)*.

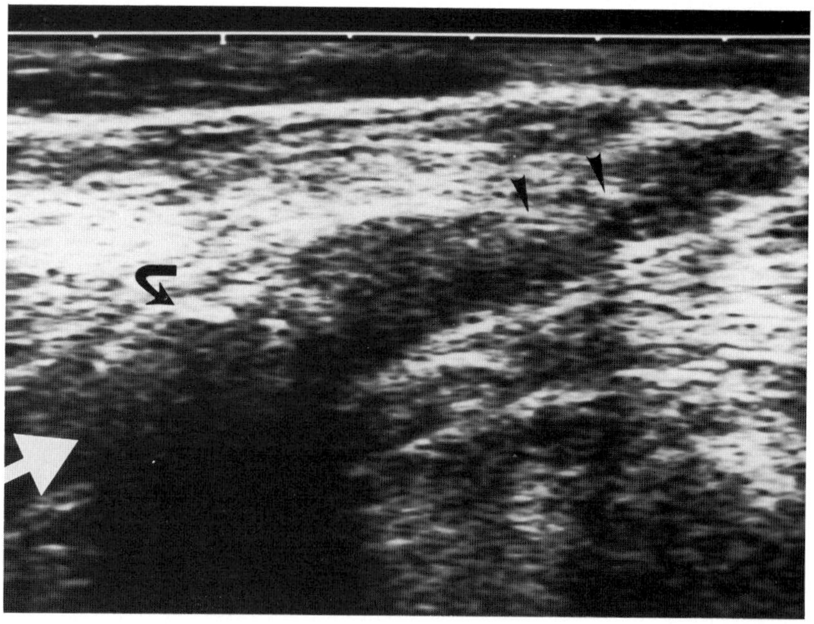

**FIG 5–52.**
Loose bodies entrapped within Baker's cyst: longitudinal sonogram. The patient is a 69-year-old man with pain in the left popliteal fossa. Numerous echogenic foci fill an elongated Baker cyst *(large white arrow)*. Some of these demonstrate marked acoustic shadowing *(curved arrow)*, whereas shadowing is more subtle with others *(arrowheads)*. Only the largest of these loose bodies were evident on the conventional radiograph. Shadowing does not necessarily indicate calcification.

## REFERENCES


Adams R: Chronic rheumatic arthritis of the knee joint. *Dublin J Med Sci* 1840; 17:520–522.

Baker WM: On the formation of synovial cysts in the leg in connection with disease of the knee joint. *St Bartholomew's Hosp Rep* 1877; 13:245–261.

Baudrillard JC, et al: Synovial cysts of unusual location. Two cases and review of the literature. *J Radiol* 1986; 67:201–207.

Canoso JJ: Bursae, tendons and ligaments. *Clin Rheum Dis* 1981; 7:189–221.

Chandler SB: The iliopsoas bursa in man. *Anat Rec* 1934; 58:235–240.

Codman EA: *The shoulder: Rupture of the Supraspinatus Tendon and Other Lesions in or About the Subacromial Bursa.* Boston, Thomas Todd Co, Printers, 1931.

De Palma AF: *Surgery of the Shoulder.* Philadelphia, JB Lippincott Co, 1950.

Gardner DL: *Pathology of the Connective Tissue Diseases.* London, Eduard Arnold, 1965.

Jayson MIV: Protective value of synovial cysts in rheumatoid knees. *Ann Rheum Dis* 1981; 31:179–182.

Lindgren PG, Willen R: Gastrocnemius-semimembranosus bursa and its relation to the knee joint. Anatomy and histology. *Acta Radiol* 1977; 18:497–512.

Nicholas JA, Hershman EB: The lower extremity and spine in sports medicine. St Louis, CV Mosby Co, 1986.

Peterson L, Renström P: *Sports Injuries.* Chicago, Year Book Medical Publishers, 1986.

Resnick D, Niwayama G: *Diagnosis of Bone and Joint Disorders.* Philadelphia, WB Saunders Co, 1988.

Scott WW, et al: B-scan ultrasound in the diagnosis of popliteal aneurysms. *Surgery* 1977; 81:436.

van Holsbeeck M, Introcaso J: Sonography of the postoperative shoulder. *AJR* 1989; 152:22.

van Holsbeeck M, van Holsbeeck K, Gevers G, et al: Staging and follow-up of rheumatoid arthritis of the knee. Comparison of sonography, thermography, and clinical assessment. *J Ultrasound Med* 1988; 7:561–566.

6

# Sonography of Ligaments

Traumatic injury of ligaments in the knee and ankle are quite common, most often seen in athletes participating in contact sports. Clinical examination is valuable only in cases of acute complete rupture but usually cannot be performed immediately following injury. Chronic ligamentous injury and meniscal tears associated with either acute or chronic trauma are poorly evaluated with clinical examination. Arthroscopy provides direct visualization of the cruciate ligaments and menisci, but extracapsular ligaments are not evaluated. Cruciate ligaments are often difficult to evaluate with arthrography, and like arthroscopy, arthrography examines only intra-articular structures. Computed tomography lacks sufficient contrast resolution to define ligamentous structures. In addition, its limitation to transverse axial imaging makes it inappropriate for evaluation of most ligaments. Ultrasound and magnetic resonance imaging (MRI) are the only diagnostic modalities well suited for examination of ligaments. Both of these imaging techniques provide multiplanar capability and demonstrate ligamentous anatomy well. The advantages of ultrasound over MRI are short examination time, capability for dynamic examination, cost, and availability. These factors are most significant in the setting of acute injury. Ultrasound examinations are universally available to emergency room physicians, but considerable delay is encountered in scheduling MRI examinations. Often the prognosis depends on prompt initiation of therapy.

## EXAMINATION TECHNIQUE

In the examination of ligaments, high-frequency transducers are of the utmost importance. Linear arrays of 10, 7.5 and 5 MHz are acceptable. The highest frequency available is preferred because ligaments are generally thin superficially located structures. Low-frequency transducers, less than 5 MHz, are not acceptable for the examination of ligaments. Since ligaments are usually located quite superficially, a standoff pad is employed to image extra-articular ligaments. Intra-articular ligaments, such as the cruciates of the knee, are well demonstrated without a standoff. Careful attention must be paid to gain and time gain compensation (TGC) settings. Oversaturation of images will result in failure to identify ligaments. Ligaments are normally hyperreflective and will blend in with the strong echos produced by cortical bone when gain or TGC are set too high. Identification and examination of ligaments is made considerably easier if the transducer is always aligned along the length of the ligament to be examined. Transverse images are of little diagnostic value. As in other musculoskeletal ultrasound examinations, comparison with the contralateral extremity is often valuable.

## NORMAL SONOGRAPHIC LIGAMENT STRUCTURE

Many ligaments of the hand, wrist, shoulder, hip, knee, and ankle can be examined sonographically. These include the vincula tendineum, collateral ligaments of the phalanges, triangular fibrocartilage of the wrist, collateral and annular ligaments of the elbow, coracoacromial ligament of the shoulder, iliofemoral ligament of the hip, collateral ligaments of the knee, and fibulotalar and deltoid ligaments of the ankle. They are composed of dense regular connec-

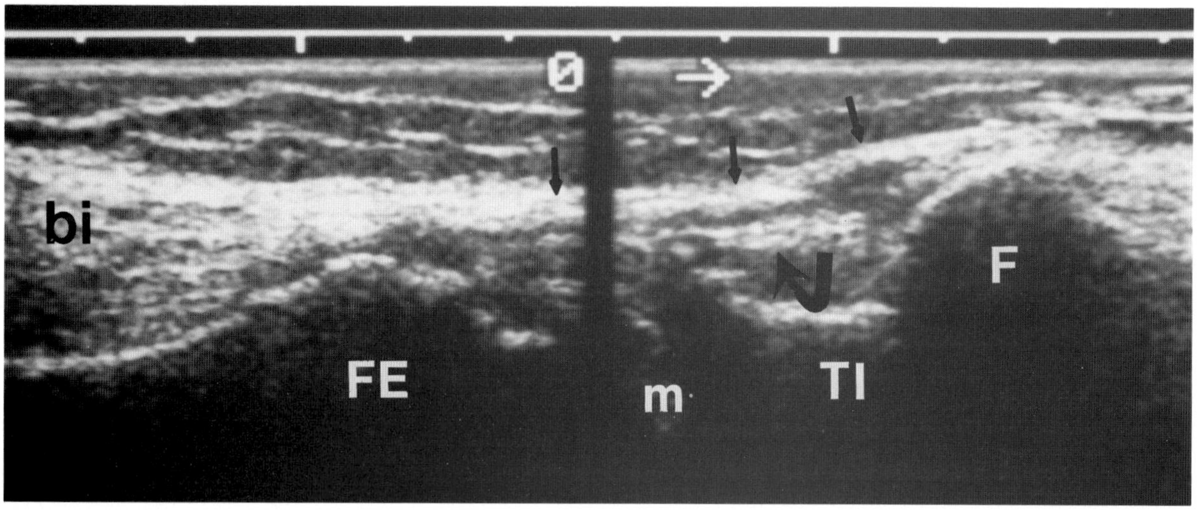


**FIG 6–1.**
Normal lateral collateral ligament: longitudinal sonogram. A normal lateral collateral ligament is identified in this 50-year-old woman. The transducer is positioned over the posterior lateral aspect of the knee. Inferior lies to the observer's right, with the caudal portion of the transducer over the fibular head *(F)*. The biceps femoris tendon *(arrows)* encircles the lateral collateral ligament *(curved arrow)* from the lateral side, appearing like a horseshoe on transverse images. This sometimes makes it difficult to distinguish these two as separate hyperechoic structures on longitudinal images. Abbreviations: biceps femoris muscle *(bi)*, distal femur *(FE)*, lateral meniscus *(m)*, proximal tibia *(TI)*.

tive tissue similar to tendons. Their structure differs from tendons in that more interweaving of collagen fibers is observed in ligaments, giving them a less regular histological and sonographic appearance (Bloom and Fawcett, 1980). All of these ligaments (Fig 6–1), with the exception of the medial collateral ligament of the knee, appear as homogeneous hyperechoic bands approximately 2 to 3 mm thick closely approximating the bony contours.

The medial collateral ligament of the knee (MCL) is a specialized structure and therefore requires a more detailed description. This ligament, also called the *tibial collateral ligament*, is a broad flat structure approximately 9 cm long that extends from the medial femoral condyle to the medial aspect of the proximal tibia (Fig 6–2). It is divided into superficial and deep components separated by a zone of loose areolar connective tissue (Fig 6–3). The superficial component is a band of dense regular connective tissue anchoring the medial femoral condyle to the proximal

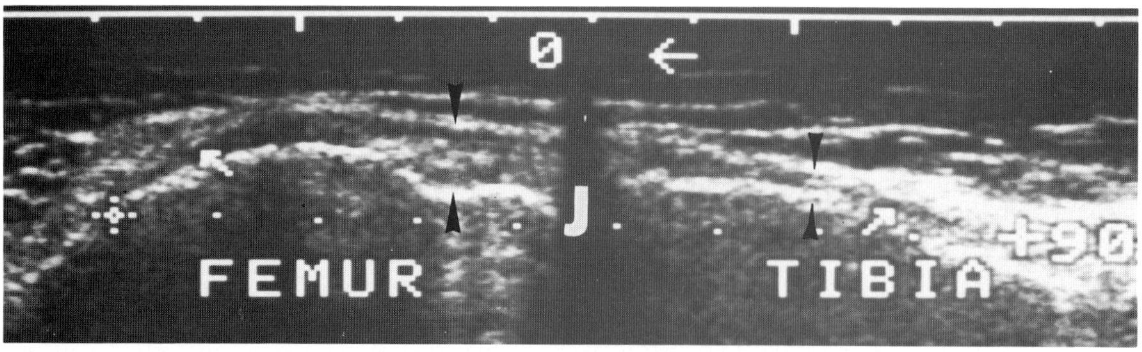


**FIG 6–2.**
Femoral and tibial insertions of the MCL: compound coronal sonogram. Images of the MCL are obtained with the patient in lateral decubitus position and the leg extended. A normal MCL is seen in its entire length (9.0 cm) in this longitudinal image of a normal young male. In this compound image, the joint space *(J)* is noted centrally. The locations for measuring the thickness of the proximal and distal segments of the ligament are noted *(arrowheads)*. Measurement of the proximal segment is obtained at the depth of the concavity of the medial femoral condyle. The distal measurement is made at a point half way between the tibial plateau and the distal most extent of the ligament. Note the normal trilaminar structure of the MCL, best seen at the site of proximal measurement.

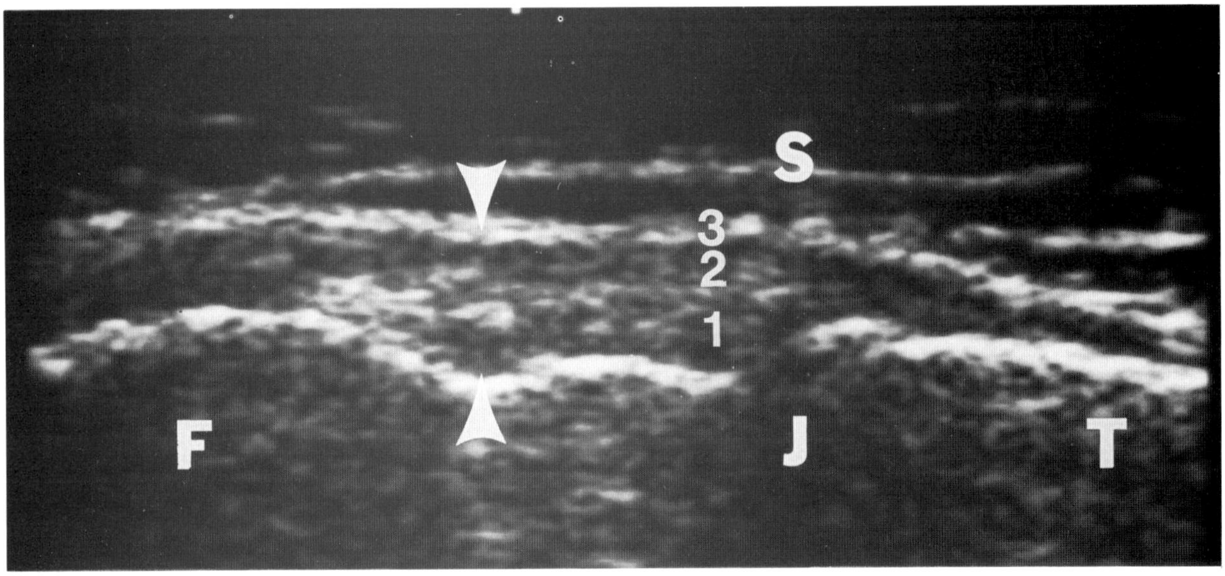


**FIG 6–3.**
Normal trilaminar structure of the MCL: coronal sonogram. This detailed image of the MCL better demonstrates the normal trilaminar structure of this ligament. The deep hyperechoic layer *(1)* is composed of the meniscofemoral and meniscotibial ligaments. Loose areolar connective tissue *(2)* separates the superficial and deep hyperechoic layers. Abbreviations: femur *(F)*, joint space *(J)*, proximal measurement *(arrowheads)*, superficial MCL *(3)*, tibia *(T)*.

tibia. Many physicians consider the deep component to be composed of two smaller ligaments connecting the medial meniscus to the femur and tibia. These are referred to as the *meniscofemoral* and *meniscotibial ligaments.*

Ultrasound demonstrates the MCL as a trilaminar structure: two hyperechoic layers separated by a hypoechoic zone. The hyperreflective bands correspond directly to the superficial and deep dense connective tissue ligaments. Loose areolar connective tissue forms the hypoechoic band that separates the superficial and deep components. In some patients a bursa is found within this hypoechoic zone of loose connective tissue (Anderson, 1983; Clemente, 1985). Figure 6–4 is a longitudinal sonogram of a cadaver knee specimen. Note the trilaminar appearance de-

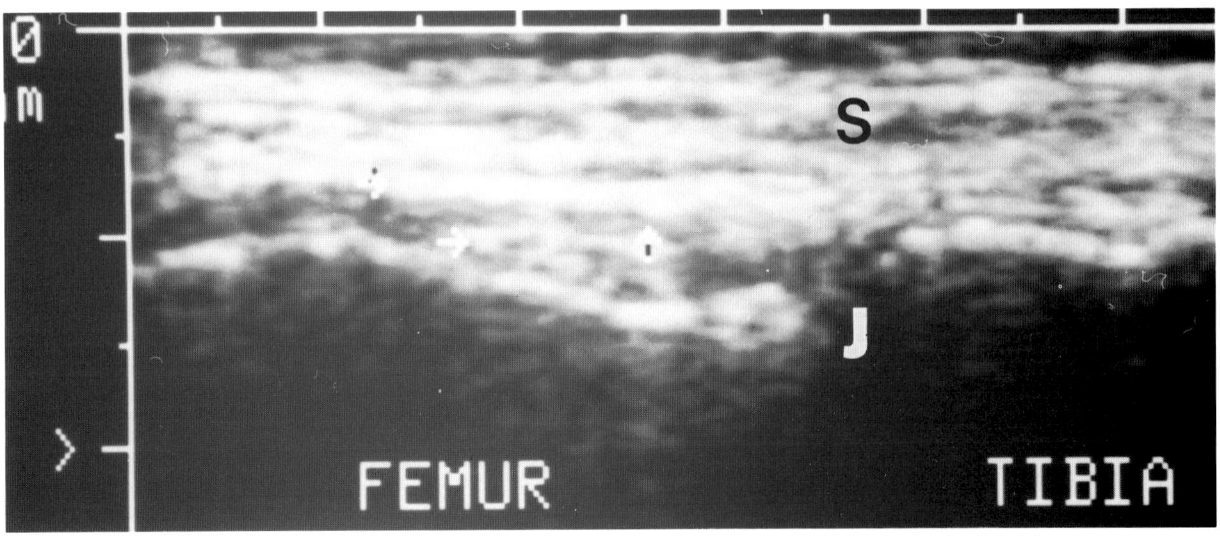


**FIG 6–4.**
Medial collateral ligament of an autopsy specimen: coronal sonogram. This longitudinal sonogram once again demonstrates the normal anatomy of the MCL. The two *horizontal arrows* indicate the hyperechoic dense fibrous connective tissue, and the *vertical arrow* indicates the intervening loose areolar connective tissue. Abbreviations: joint space *(J)*, subcutaneous fat *(S)*.

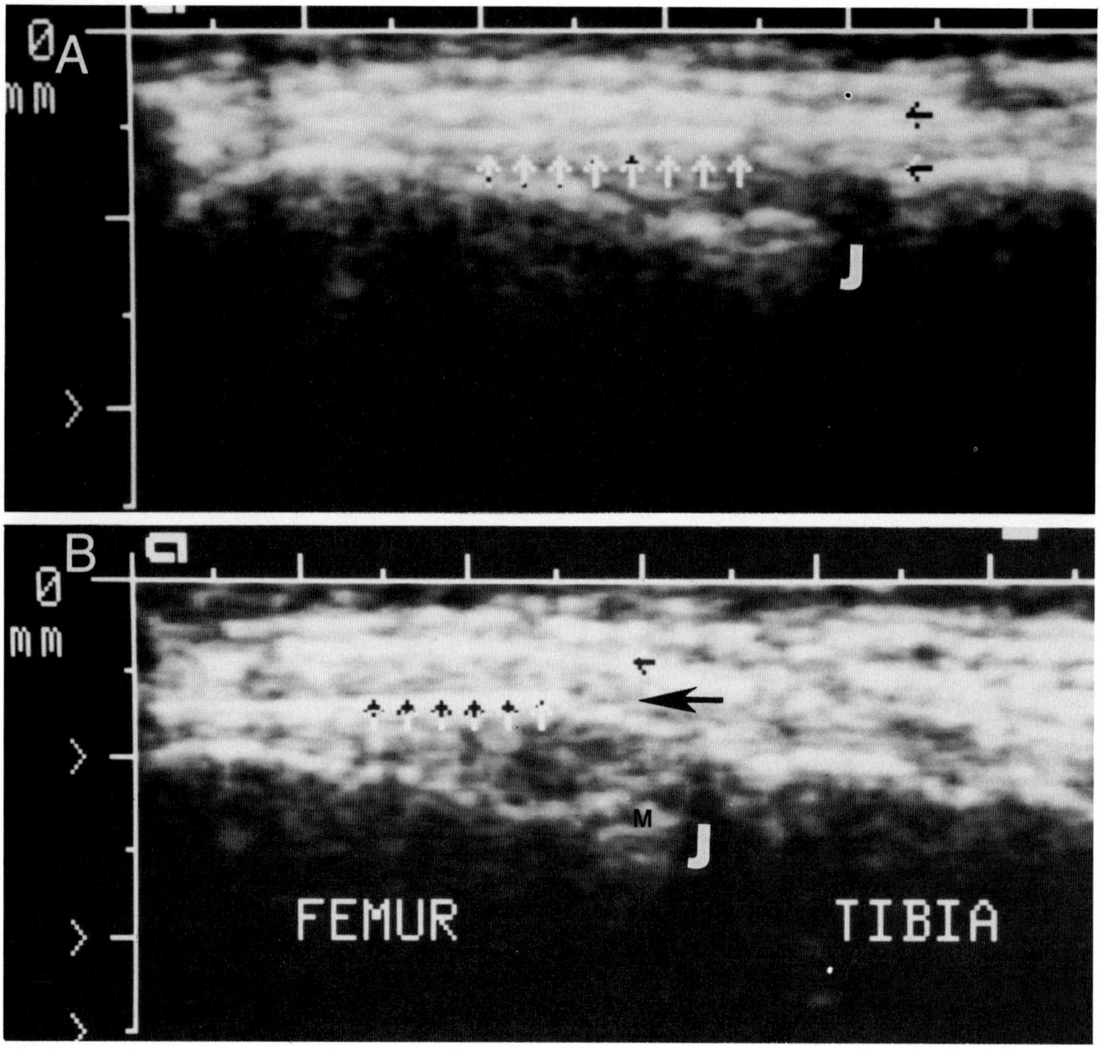


**FIG 6–5.**
**A,** MCL of an autopsy specimen: coronal sonogram. A 22-gauge spinal needle *(vertical arrows)* was inserted into the middle hypoechoic layer, bounded by the superficial and deep hyperechoic layers *(horizontal arrows)*. Barium sulfate was then injected into the hypoechoic layer. Abbreviations: joint space *(J)*. **B,** following the injection of barium sulfate *(large arrow)*, the superficial hyperechoic layer *(horizontal arrows)* is separated from the deep meniscofemoral and meniscotibial ligaments. Abbreviations: joint space *(J)*, meniscus *(M)*.

scribed earlier. Barium sulfate was injected into the middle hypoechoic layer within the ligament (Fig 6–5). A specimen radiograph was then obtained, demonstrating that the barium was within the substance of the MCL (Fig 6–6 and 6–7). Histological sections were then made of the MCL, which showed barium confined to the loose areolar connective tissue that separates the deep and superficial layers of the ligament.

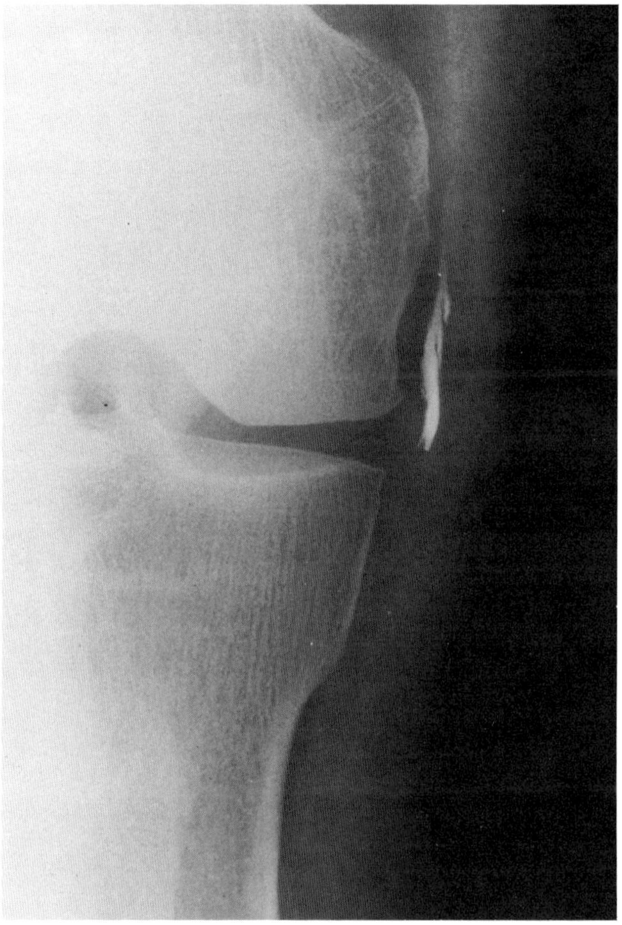

**FIG 6–6.**
Medial collateral ligament of an autopsy specimen: anteroposterior knee radiograph. It clearly demonstrates contrast material layering within an anatomically confined space.

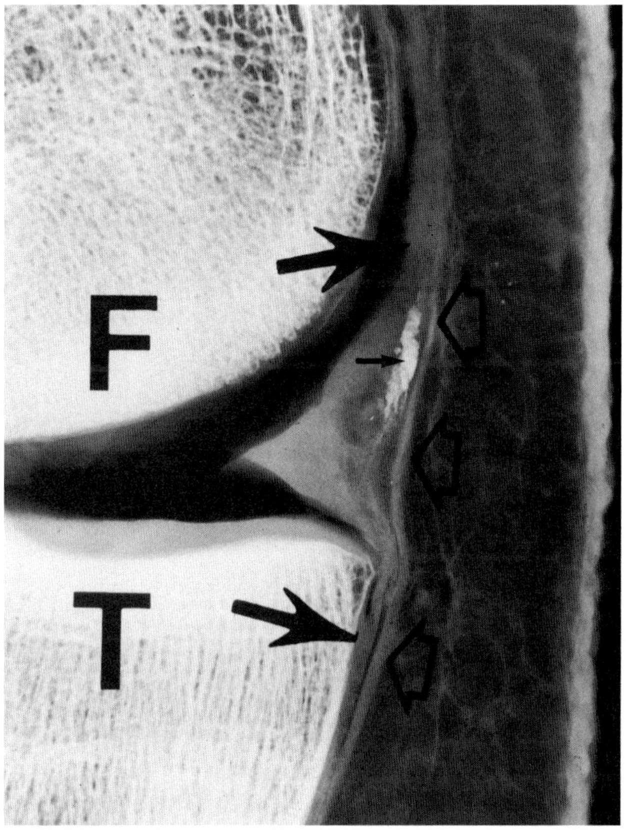


**FIG 6–7.**
Medial collateral ligament of an autopsy specimen: specimen radiograph. This knee was then sectioned to yield 1-cm thick coronal tissue specimens. A mammographic technique was used to obtain this specimen radiograph. Barium *(small arrow)* is identified layering within the MCL between the deep and superficial layers. The *large black arrows* indicate the meniscofemoral and meniscotibial ligaments, which stabilize the medial meniscus. Abbreviations: femur *(F)*, superficial component of the MCL *(open arrows)*, tibia *(T)*.

## LIGAMENT PATHOLOGY

### Intra-articular Ligaments

Ligaments are categorized by location: either intra-articular or extra-articular. This distinction is valuable in determining the best diagnostic approach. We have been successful in demonstrating tears of the intra-articular ligaments of the knee and wrist. Detection of tears of the cruciate ligaments or triangular fibrocartilage (Fig 6–8) requires careful technique and some patience. In the knee, care must be taken to distinguish the ligament from intercondylar fat (Figs 6–9 through 6–13). Approximately 60 degrees of flexion (Fig 6–14) is required to obtain an adequate sonographic window. Clearly, in the acutely traumatized patient, this is rarely possible. Other investigators have experienced similar difficulties in imaging intra-articular ligaments using pulse-echo ultrasound (Laine et al., 1987; Richardson et al., 1988). Transmission ultrasound has shown promise in imaging intra-articular ligaments, but equipment is not yet commercially available (Hentz et al., 1987). Therefore, we prefer to evaluate pathology of intra-articular ligaments with MRI at the present time.

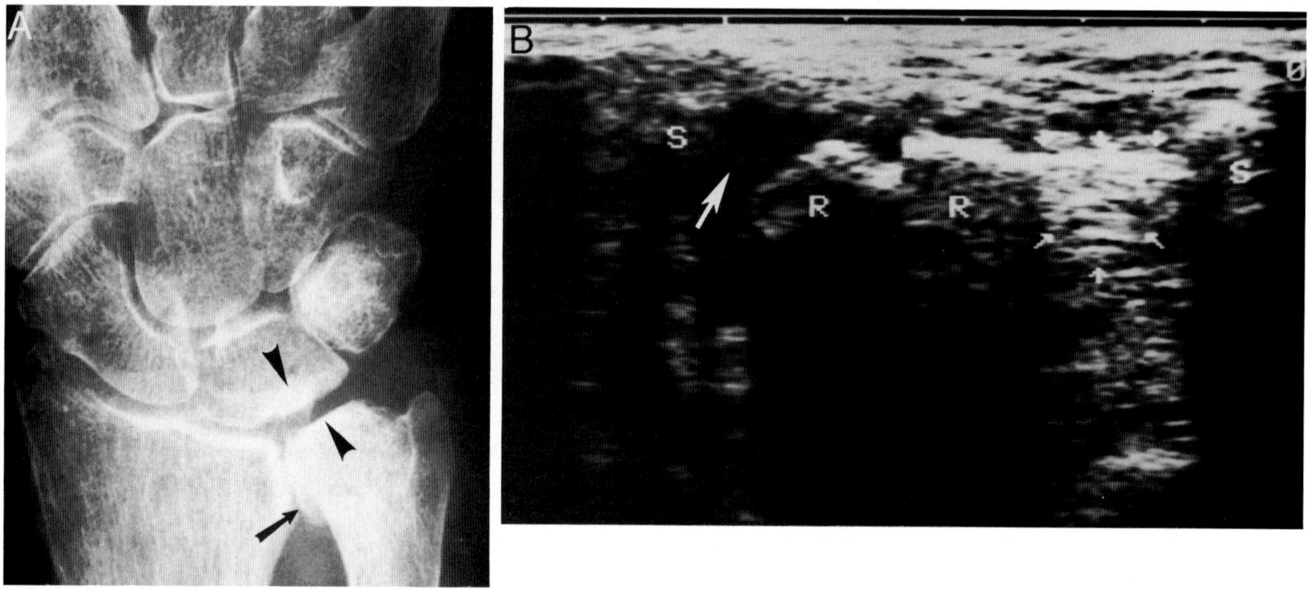


**FIG 6–8.**
**A,** triangular fibrocartilage tear in the wrist: radiograph. The patient is a 62-year-old woman with pain in the ulnar side of the wrist during pronation and supination. Sclerosis *(arrowheads)* is demonstrated in the proximal lunate and distal ulna, indicative of abnormal motion. Osteophytic change and narrowing of the radioulnar joint *(arrow)* is also present. **B,** triangular fibrocartilage tear in the wrist: transverse sonogram. Same patient as in **A.** The transducer is positioned over the volar aspect of the wrists, just distal to the ulnar head at the level of the ulnar styloid *(S)*. Side by side comparison is provided in this split screen image with the asymptomatic left wrist at the observer's right. A normal triangular fibrocartilage *(arrows)* is identified between the radius *(R)* and ulnar styloid *(S)* of the left wrist. A complete absence of fibrocartilage is seen on the symptomatic right wrist *(left image)*. Fluid *(white arrow)* fills the space previously occupied by the triangular fibrocartilage.

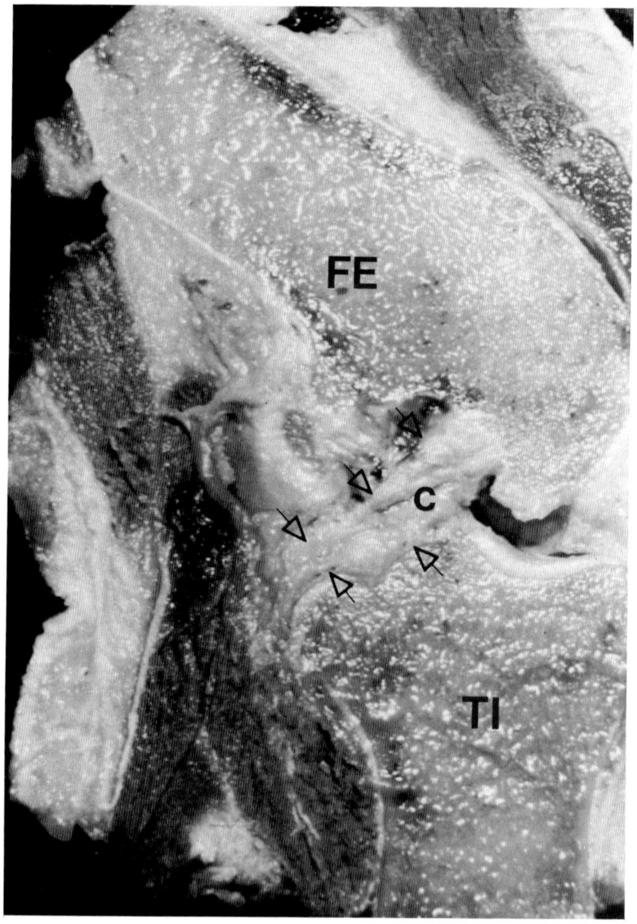


**FIG 6–9.**
Cadaver knee: longitudinal anatomical section through the posterior cruciate. An oblique section was obtained through this cadaver specimen to bisect the posterior cruciate ligament along its length. The posterior cruciate ligament *(arrows)* travels obliquely and anteriorly from the most posterior aspect of the tibia to the medial femoral condyle *(FE)*. Fat surrounds the ligament within the intercondylar fossa. Abbreviations: posterior cruciate ligament *(C)*, proximal tibia *(TI)*.

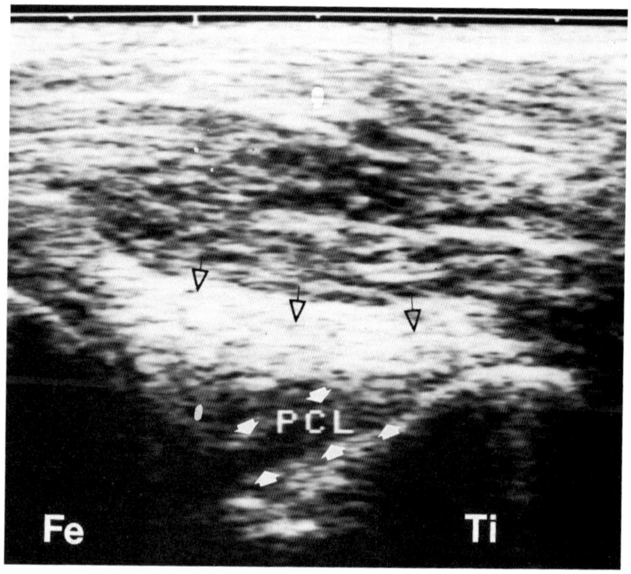


**FIG 6–10.**
Normal posterior cruciate ligament: longitudinal sonogram. A normal posterior cruciate ligament is seen in this 30-year-old man. The transducer is positioned over the popliteal fossa with the cranial end obliqued slightly toward the medial. The posterior cruciate ligament *(white arrows)* has a hypoechoic appearance relative to the surrounding fat. This is due in part to anisotropy. *Open arrows* indicate the normal posterior joint capsule, which appears markedly hyperechoic and curved anteriorly. Abbreviations: distal femur *(Fe)*, proximal tibia *(Ti)*.

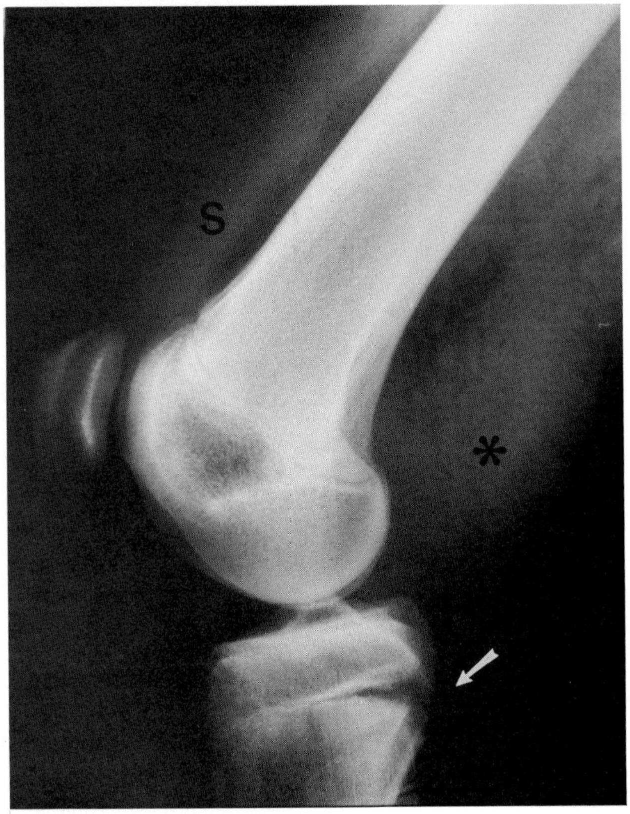


**FIG 6–11.**
Epiphysiolysis proximal tibia with associated posterior cruciate tear: lateral film of the knee. This 17-year-old boy felt a snap in his knee while hanging upside down with his foot caught in an elevator door. The tibial growth plate is markedly widened posteriorly, and an avulsed fragment of the tibial metaphysis is noted *(arrow).* Subcutaneous hemorrhage *(asterisk)* results in swelling of the soft tissues of the popliteal fossa. A distended suprapatellar bursa *(S)* is also well demonstrated.

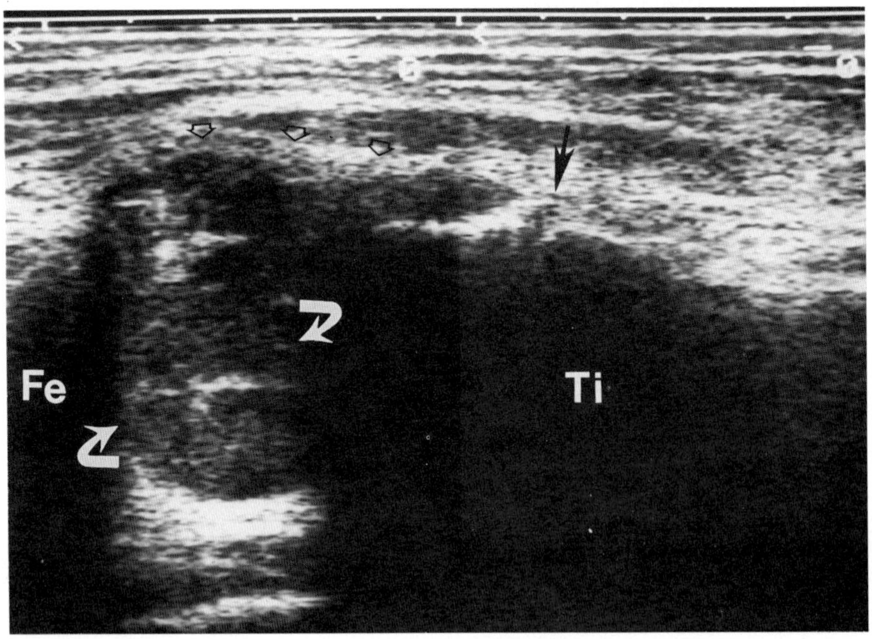


**FIG 6–12.**
Epiphysiolysis proximal tibia with associated posterior cruciate tear: longitudinal sonogram of the popliteal fossa. Same patient as in Figure 6–11. An ultrasound examination was requested to exclude vascular injury. The popliteal vessels were intact. Intercondylar hemorrhage is seen in the expected location of the posterior cruciate ligament *(curved arrows).* The posterior joint capsule *(open arrows)* bows posteriorly, displaced by hemorrhage. At the level of the tibial growth plate, a small osseous fragment *(black arrow)* is identified. Abbreviations: distal femur *(Fe),* proximal tibia *(Ti).*

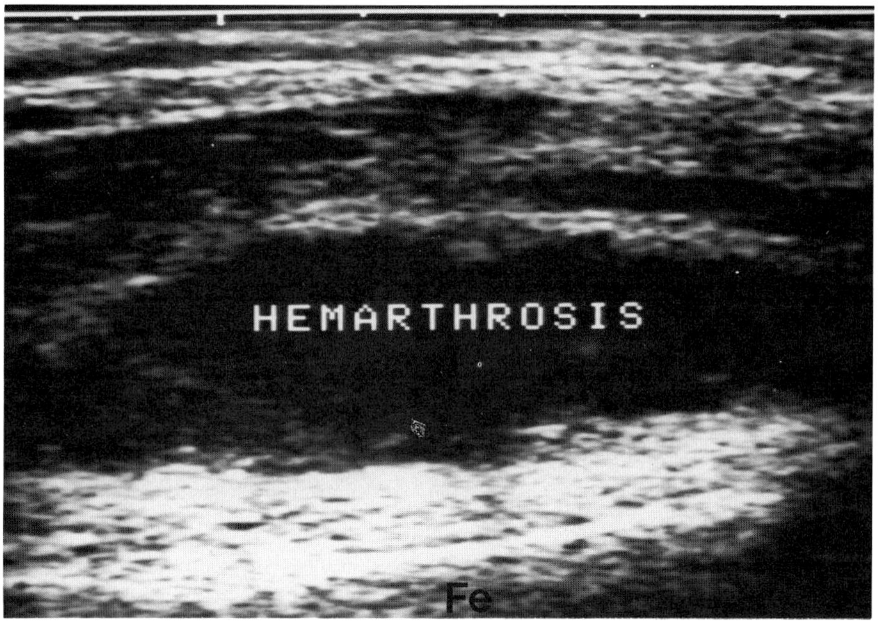


**FIG 6–13.**
Epiphysiolysis proximal tibia with associated posterior cruciate tear: longitudinal sonogram suprapatellar bursa. Same patient as in Figures 6–11 and 6–12. Significant intra-articular hemorrhage is seen within the suprapatellar bursa. This is a common finding in cruciate ligament tears.

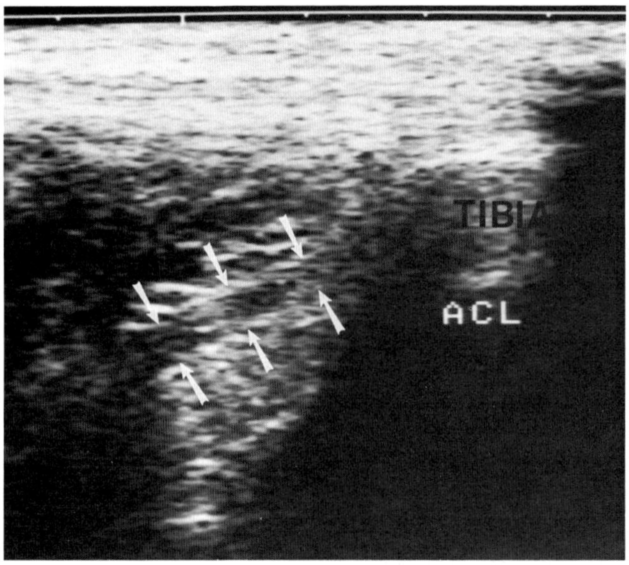


**FIG 6–14.**
Normal anterior cruciate ligament: longitudinal sonogram. A normal anterior cruciate ligament is seen in this asymptomatic 30-year-old man. The examination is performed with the knee in full flexion and the transducer positioned along the length of the anterior cruciate, the cranial end of the transducer obliqued to the lateral. Like the posterior cruciate, the anterior cruciate ligament *(arrows)* appears hypoechoic relative to the surrounding intercondylar fat.

## Extra-articular Ligaments

### *Acute Ligament Injury*

Clinical examination in detection of complete ligament tears is quite reliable, but often the diagnosis is considerably delayed. Stress tests for the detection of joint laxity are used to identify ligament injury. Pain and swelling present at the time of injury make these tests inappropriate in the acute setting. Several weeks after injury, when pain and swelling have subsided, these tests may be administered and the diagnosis made. This delay is undesirable because surgical repair, if indicated, should be performed as soon as possible following injury to ensure the best possible result.

Ultrasound provides a reliable means of detecting acute ligament injury, both complete and partial tears. The simple noninvasive nature of the examination and the short time required for examination make it ideal in the emergent setting. In cases of partial rupture, the involved area of the ligament appears markedly thickened and decreased in echogenicity. Complete ruptures are demonstrated as a discontinuity of the ligament with the free ends separated by hematoma. The fluid occupying the gap will appear hypoechoic or anechoic, depending on how long after injury imaging is performed. In general, the diagnosis is most easily made when imaged as soon after injury as possible. Some retraction of

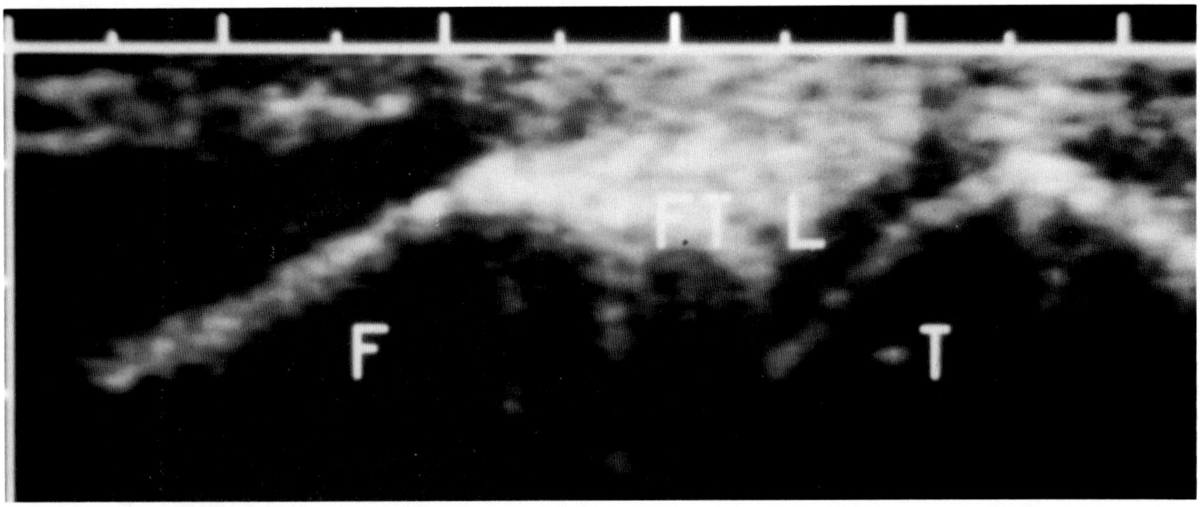


**FIG 6–15.**
Normal anterior fibulotalar ligament: oblique sonogram. Same patient as in Fig 6–14. To examine the anterior fibulotalar ligament, the transducer is positioned over the lateral maleolus, angled 45 degrees anteriorly relative to the long axis of the fibula. The normal anterior fibulotalar ligament *(FT L)* is a hyperechoic band that connects the distal fibula *(F)* with the talus *(T)*.

the ends of the ruptured ligament occurs, giving them a slightly rounded appearance.

In the ankle, the fibulocalcaneal and fibulotalar ligaments are most frequently involved. Two fibulotalar ligaments are present: anterior and posterior. The anterior fibulotalar ligament courses anteriorly and slightly inferiorly from the anterior aspect of the distal fibular epiphysis to insert on the lateral aspect of the talus. The posterior fibulotalar ligament is relatively short, running horizontally from the posterior aspect of the distal fibular epiphysis to the posterior talus. In 70% of ankle ligament ruptures, the anterior fibulotalar ligament (Figs 6–15 and 6–16) is the only ligament involved. Combined rupture of the anterior fibulotalar and fibulocalcaneal ligaments is seen in 20% of cases (Peterson and Renström, 1986). It is very rare that the posterior fibulotalar ligament is ruptured. The mechanism of injury to the ankle ligaments is nearly always supination and inward rotation of the foot. During the same examination in

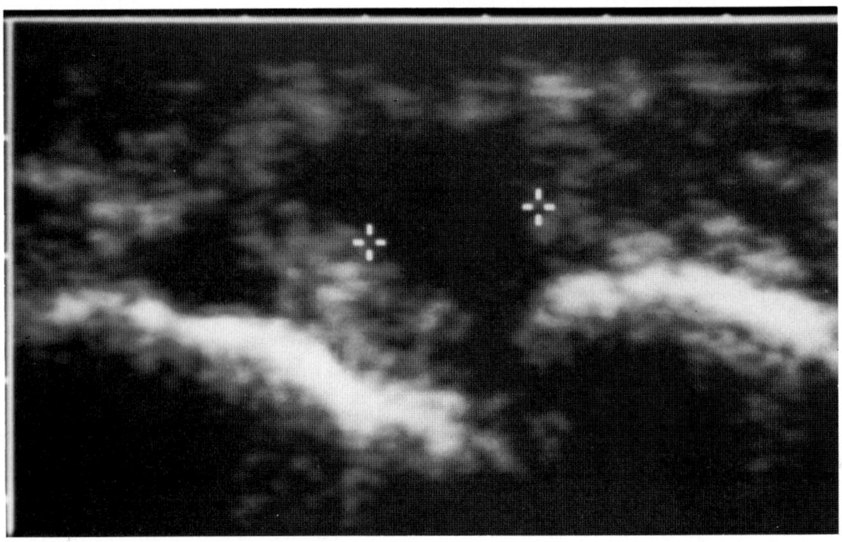

**FIG 6–16.**
Torn anterior fibulotalar ligament: oblique sonogram. This image was obtained approximately 8 hours following inversion injury of the right ankle. The soft tissues are diffusely swollen with hematoma separating the torn ends of the anterior fibulotalar ligament (+).

which integrity of the ligaments is evaluated, the peroneus tendons, Achilles tendon, and retrocalcaneal bursa should be examined for concurrent injury.

In the remaining 9% of acute ankle ligament ruptures, the deltoid ligament at the medial aspect of the ankle is involved. Pain and swelling over the medial aspect of the ankle may also be due to posterior tibial tendon rupture (see Fig 4–10,B). This tendon is one of the principle stabilizers of the hind foot, and lesions of the posterior tibial tendon may mimic deltoid ligament injury. Surgical therapy is sometimes indicated in these cases, but the approach for these two types of lesions differs considerably. Ultrasound is extremely valuable in differentiating deltoid ligament injury from tendon injury, facilitating surgical planning.

Sonography is also valuable in follow-up of healing ligament tears. Ligaments tend to heal quite slowly. The first evidence of restoration of ligament integrity is seen approximately 5 weeks following injury. Definite echos filling the gap in the ligament will be observed. After several months, the normal homogeneous hyperechoic appearance of the ligament will be restored. Serial ultrasound examinations are valuable in helping to determine when normal activity may be resumed.

Injury to the collateral ligaments of the knee is quite common in athletes. Of these injuries, the MCL is most frequently involved. Complete rupture of the MCL is often accompanied by rupture of the anterior cruciate ligament and tear of the medial meniscus (Hughston et al., 1976). The mechanism of injury is usually a blow to the lateral aspect of the knee, frequent in football and soccer (Peterson and Renström, 1986). Isolated MCL tear may also result from overaggressive arthroscopic surgery if the capsule is perforated.

The normal sonographic appearance of the MCL was described in detail earlier. Acute rupture (Figs 6–17 through 6–20) of the MCL is identified as an interruption in the hyperechoic bands that form the ligament. Hypoechoic or anechoic fluid fills the gap, as seen in other ligament ruptures. Pressure applied with the transducer will displace fluid from the rupture site. The site of rupture most frequently observed is at the junction of the meniscofemoral and meniscotibial ligaments with the medial meniscus. This deep portion of the MCL is short and tightly binds these structures together. Therefore, traction applied to the ligament ruptures the deep portion first (see Fig 6–20) (Peterson and Renström, 1986). Hematoma may be observed confined to the hypoechoic layer if only the deep portion of the ligament is ruptured. If present within the middle hypoechoic layer, a hemorrhagic bursitis is possible. Clearly, functional disability is most severe with rupture of both the superficial and deep components of the ligament. As stated earlier, pain and swelling limit clinical examination immediately following injury. Sonography will quickly make the diagnosis and accurately assess the extent of injury, allowing prompt institution of appropriate therapy.

Partial ruptures of the MCL will result in thickening of the ligament (see Fig 6–19). Two standardized measurements of MCL thickness will help to detect subtle injury (Fig 6–21). Proximal and distal measurements of MCL thickness are obtained. The greatest depth of concavity of the medial surface of the femoral condyle is used as the proximal landmark for

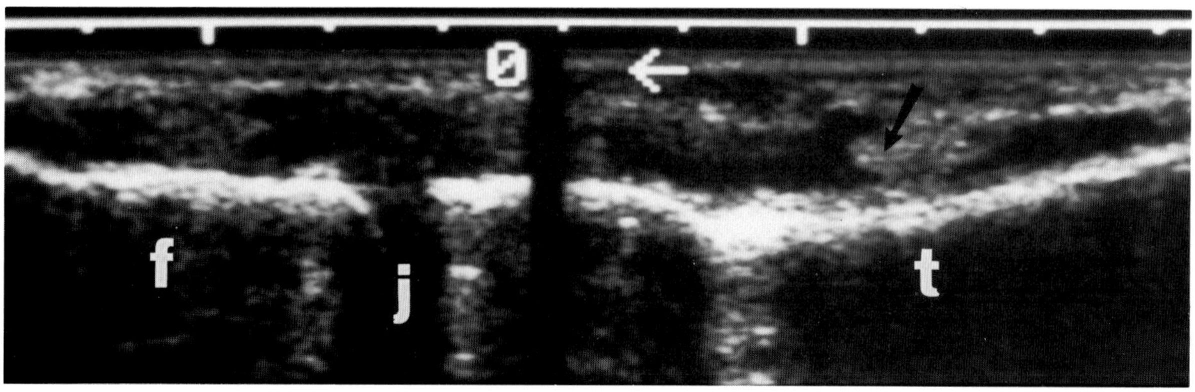


**FIG 6–17.**
Acute rupture of the MCL: coronal sonogram. This 15-year-old hockey player was hit on the lateral aspect of the knee. He presented with pain and swelling over the medial aspect of the knee. A good physical examination could not be performed due to the degree of pain and swelling; therefore, ultrasound examination was performed. Hematoma replaces a large portion of the structure of the MCL. Only a small fragment of retracted ligament *(black arrow)* is seen distally. Abbreviations: distal femur *(f)*, joint space *(j)*, proximal tibia *(t)*.

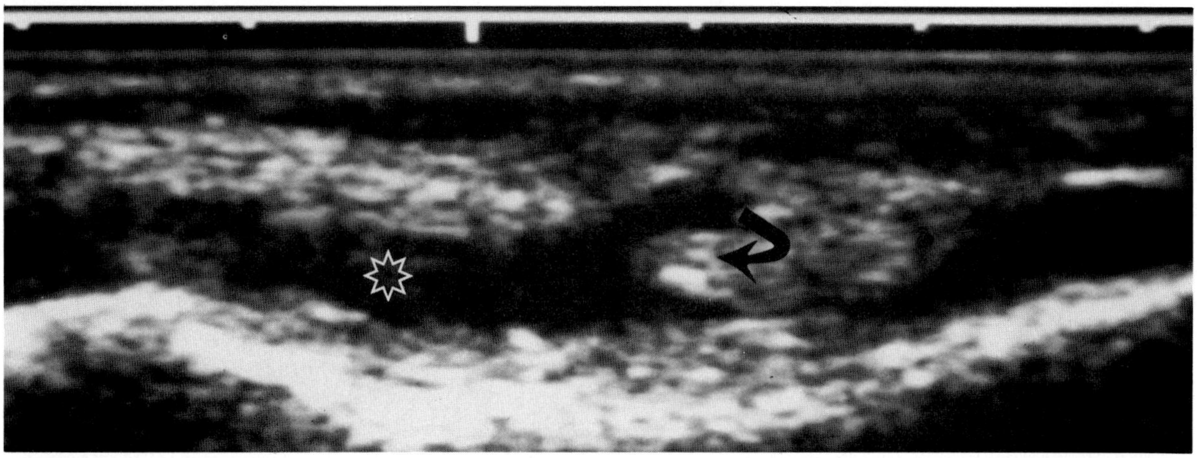

**FIG 6–18.**
Acute rupture of the MCL: detail of the tibial MCL insertion. Same patient as in Fig 6–17. No normal ligament structure is identified. Hematoma *(asterisk)* surrounds a small retracted fragment of the distal ligament *(curved arrow)*. Two weeks after this examination, physical examination could be performed and revealed a grade III MCL tear. Results of the anterior drawer test were also positive, indicative of anterior cruciate ligament rupture. Results of a MacMurray test were indicative of medial meniscal tear. This is a classic triad seen in acute MCL rupture.

measurement. Normal thickness of the MCI at this point is 3.6 mm (S.D. ± 0.5 mm) (van Holsbeeck et al., 1988). This is often the most useful parameter, because proximal swelling of the ligament is most frequently observed in MCL injury (van Holsbeeck et al., 1988).

The point at which the distal measurement of MCL thickness is obtained is determined by identifying a point equidistant from the tibial plateau and the distal most extent of the MCL. Thickness of the ligament at this point is normally 2.3 mm (S.D. ± 0.3 mm). Athletic conditioning without injury does not change the thickness or appearance of the ligament (van Holsbeeck et al., 1990). Acute changes in liga-

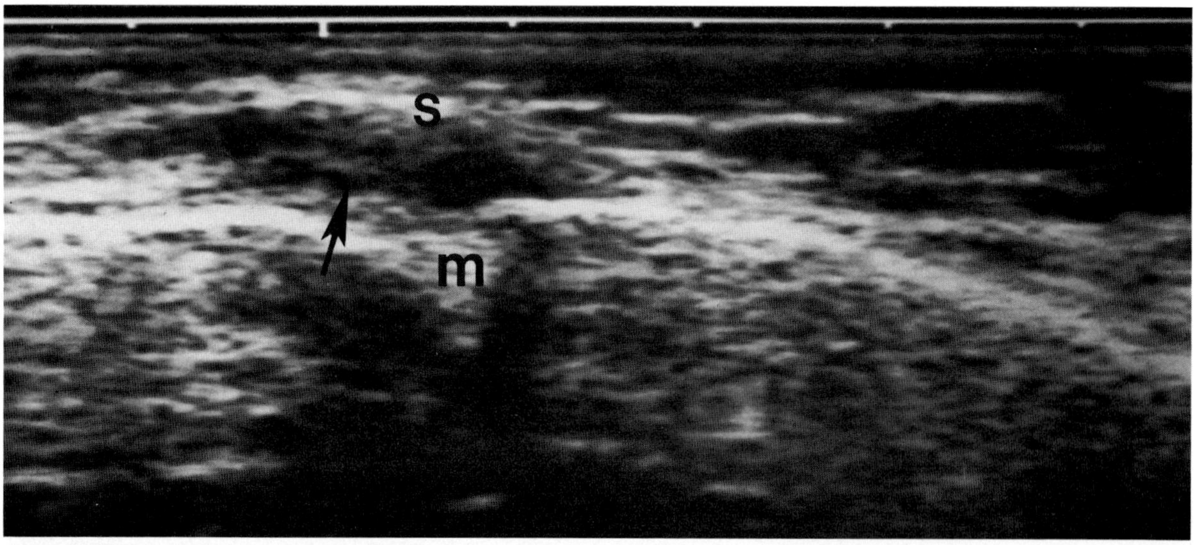


**FIG 6–19.**
Partial MCL rupture: coronal sonogram. Three days ago this 21-year-old soccer player was blocked while kicking. He presented here with pain and swelling over the medial femoral condyle. This coronal sonogram demonstrates an intact superficial portion *(s)* of the MCL, which is displaced medially by a swollen hypoechoic deep portion *(arrow)*. This appearance is typical of partial rupture of the MCL. These lesions usually involve the proximal portion of the ligament to a greater extent than the distal and begin in the deep portion, involving only the superficial portion in more severe injuries. Abbreviations: normal medial meniscus *(m)*.

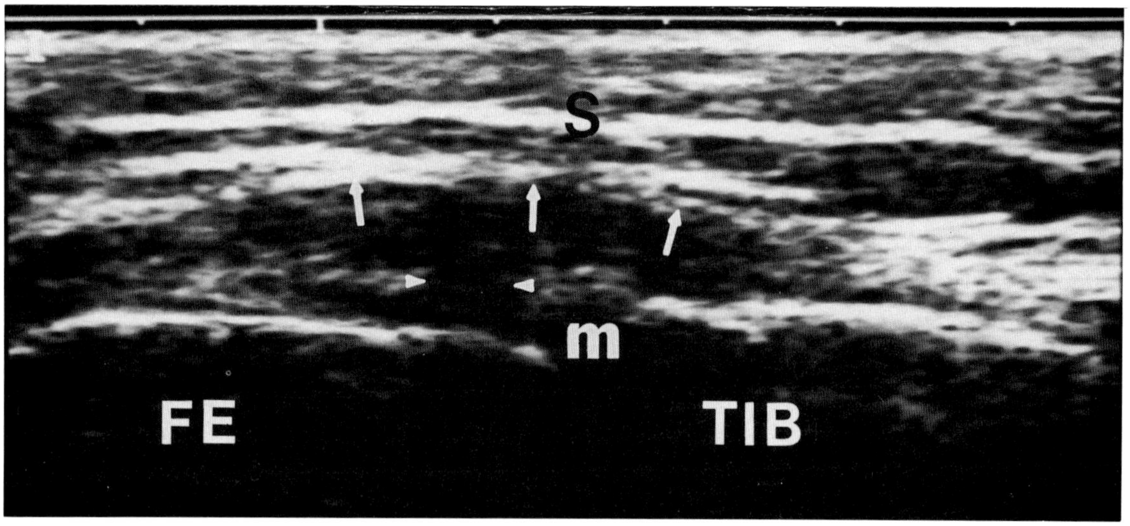


**FIG 6–20.**
Partial MCL rupture: coronal sonogram. This pedestrian was hit on the lateral side of the knee while walking between two cars stopped in traffic. A gap is observed in the meniscofemoral ligament *(arrowheads),* and swelling of the tissues beneath the superficial portion of the ligament displace it medially *(arrows).* Abbreviations: medial femoral condyle *(FE),* medial meniscus *(m),* proximal tibia *(TIB),* subcutaneous tissues *(S).*

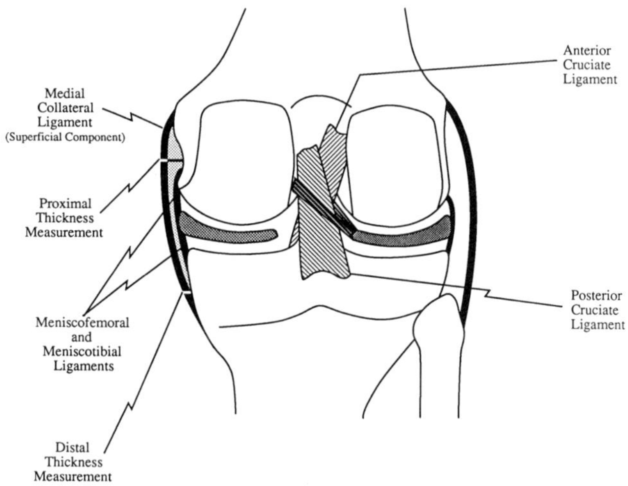


**FIG 6–21.**
Schematic diagram of ligaments of the knee. This posterior view of the knee demonstrates the trilaminar structure of the MCL. The hyperechoic superficial layer is separated from the deep meniscofemoral and meniscotibial ligaments by a hypoechoic zone of loose areolar connective tissue. Sites of standard measurement of the proximal and distal portions of the ligament are indicated. The proximal measurement is obtained at the point of greatest depth of the concavity in the medial femoral condyle. Measurement of the distal portion of the ligament is made at a point equidistant from the tibial plateau and the distal most extent of the ligament.

ment thickness must be interpreted in light of which layer or layers are responsible. Thickening and decreased echogenicity of either the deep or superficial hyperechoic layers without an identifiable gap must be considered a partial rupture of the involved ligament. If the central hypoechoic portion of the MCL is thickened, the diagnosis of meniscal cyst must be considered. In these cases, integrity of the deep hyperechoic portion of the MCL and the meniscus should be carefully examined. When the meniscofemoral ligament, meniscotibial ligament, and meniscus are demonstrated to be intact, the presence of an enlarged bursa within the central portion of the ligament should be suspected (Fig 6–22). Follow-up examination after resolution of acute swelling may prove beneficial in these cases.

Treatment of ligament rupture varies depending on the site of rupture and the extent of injury. Rapid diagnosis and assessment of the extent of injury is most critical in cases when the need for surgical repair is likely. The sooner the correct diagnosis is established and surgical repair accomplished, the more favorable the outcome (Nicholas and Hershman, 1986). In cases of less extensive injury, splint or cast immobilization for 5 to 6 weeks is the treatment most frequently employed (Peterson and Renström, 1986; Hastings, 1980). The direct and indirect consequences of immobilizing an athlete for 1.5 months are considerable. Therefore, establishing the definitive diagnosis with ultrasound is extremely important.

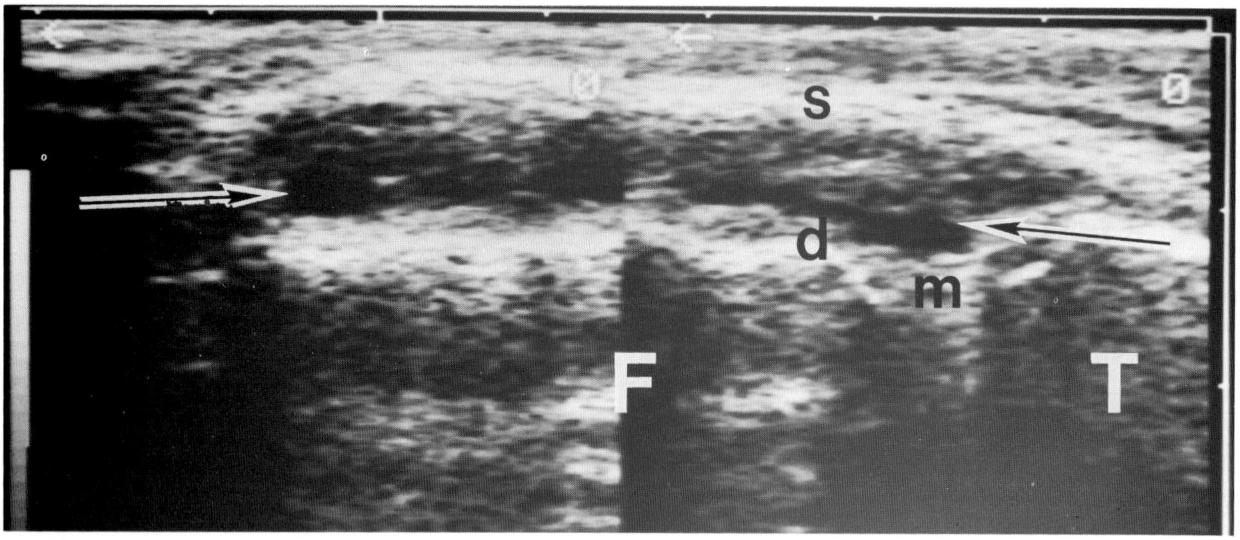


**FIG 6–22.**
Bursitis in the MCL: coronal sonogram. The patient is a 26-year-old man who was hit on the medial side of the knee during a judo match. Over the past 3 months, he continued to compete, with increasing pain and swelling over the medial aspect of the knee. A markedly hypoechoic fluid collection *(arrows)* separates the superficial *(s)* and deep *(d)* layers of the medial MCL. The medial meniscus *(m)* and dense connective tissue components of the ligament are intact. In a certain percentage of cases, a bursa will separate the deep and superficial components of the ligament. Traumatic bursitis may then be seen in these patients. In this case, aspiration of the lesion for palliative reasons yielded clear bursal fluid. Abbreviations: distal femur *(F)*, proximal tibia *(T)*.

### *Chronic Ligament Injury*

The normal repair mechanisms of ligaments usually result in restitutio ad integrum when adequate therapy is instituted (Frank et al., 1983; Kannus, 1988; Laws and Walton, 1988). However, when clinical examination alone is used for diagnosis, some lesions pass undetected and go untreated (Hastings, 1986). Healing will then be considerably impaired in a manner similar to the formation of a pseudarthrosis following fracture. For this reason, we have chosen to

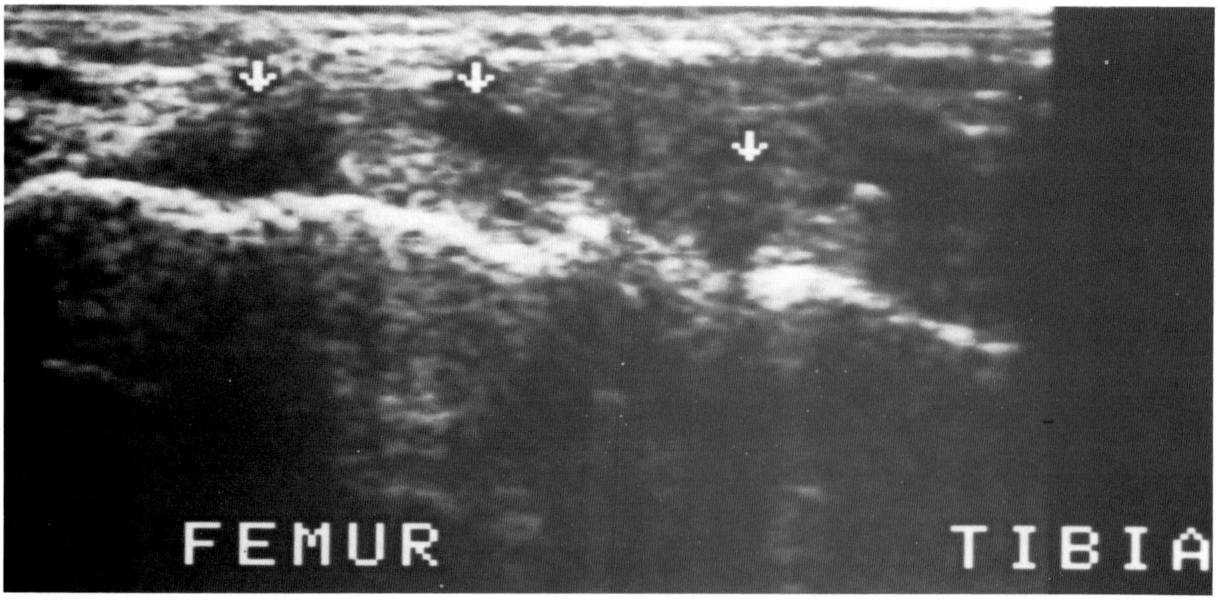


**FIG 6–23.**
Chronic reinjury of the MCL: coronal sonogram. Six months ago this 26-year-old man was hit by a car while cycling. At that time he sustained a grade III tear of the MCL. Since that time he reinjured the ligament twice. He presented here with pain and swelling over the medial surface of the knee. The normal trilaminar structure of the ligament is no longer recognizable. It is markedly swollen with three points of rupture *(arrows)* noted along the length of the ligament.

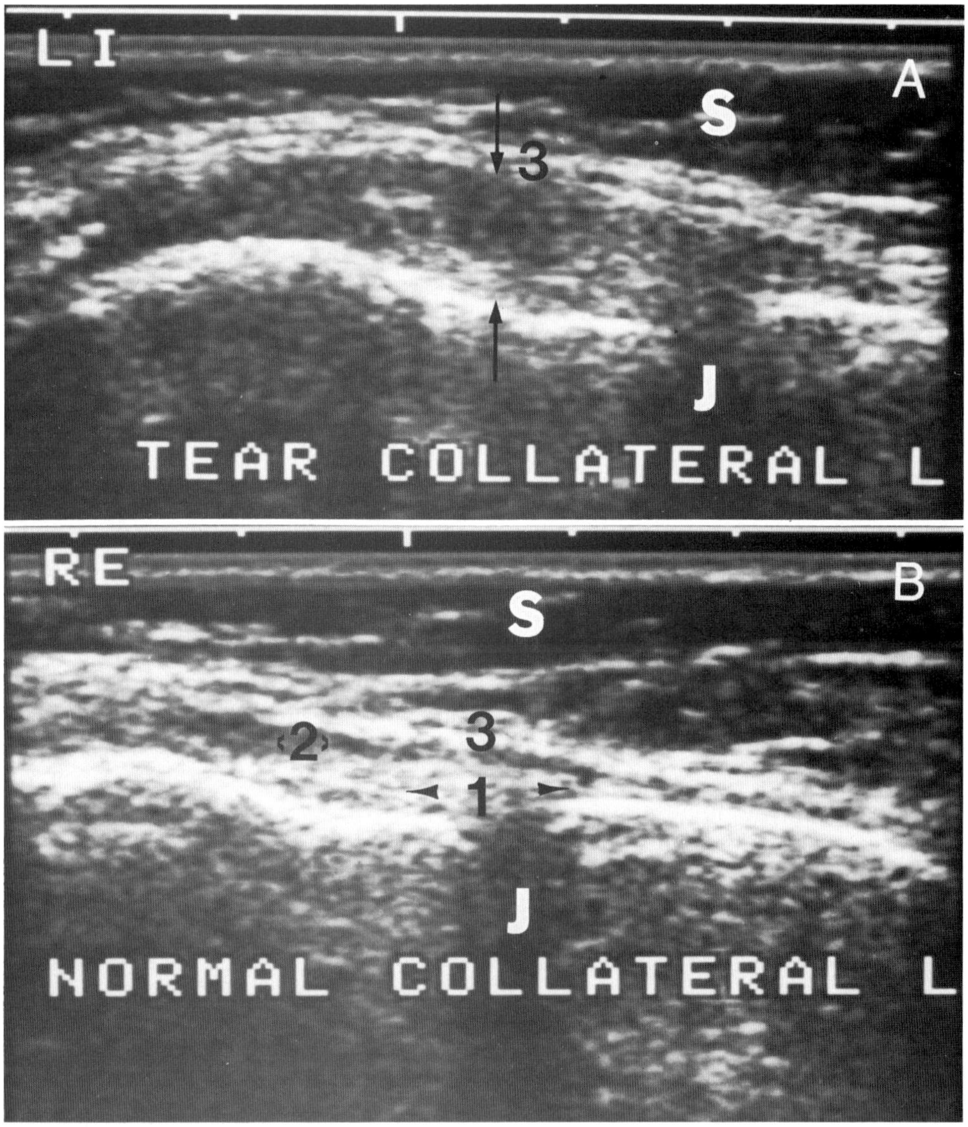


**FIG 6–24.**
Comparison of symptomatic with asymptomatic MCL: coronal sonogram. **A,** the patient is a 30-year-old soccer player who presented 5 months following a grade II rupture of the MCL. He continued to have pain, especially while kicking. The deep layer of the MCL is decreased in echogenicity and markedly swollen *(arrows)*, displacing the superficial layer *(3)* medially. Joint space *(J)*, subcutaneous tissues *(S)*. **B,** the same patient as in **A,** asymptomatic knee. A normal trilaminar MCL is observed. The deep layer *(1)* comprised of the meniscofemoral and meniscotibial lig ments is intact. A normal hypoechoic layer of loose areolar connective tissue *(2)* separates the superficial layer *(3)* from the deep layer. Note that the proximal portion of the ligament is thicker than the distal. The mean thickness of a normal MCL at the proximal site is 3.6 mm (S.D. ± 0.5) and 2.3 mm (S.D. ± 0.3) at the distal site. Abbreviations: joint space *(J)*, subcutaneous tissue *(S)*.

refer to this lesion as nonunion of a ligament. We have identified two types of ligamentous nonunion, which have been confirmed at the time of surgery. The first is characterized by ligament interruption. Little or no reparative activity is evident. Sonography demonstrates the persistence of gaps or focal thinning of the ligament. This form of nonunion is observed in slightly less than 10% of cases (Fig 6–23).

In the majority of cases of ligamentous nonunion, a hypertrophic response (Fig 6–24) with excessive formation of granulation tissue and mucoid degeneration is observed. The hypertrophic form of nonunion is a frequent complication of untreated partial ligament rupture. These chronic lesions are associated with pain and tenderness over the lesion. We have observed hypertrophic nonunion in sinus tarsi syndrome, chronic peroneal retinaculum rupture, and fibulotalar and deltoid ligament ruptures. Similar lesions are seen in the coracoacromial ligament in impingement syndrome of the shoulder, a common injury found in tennis players. Foci of granulation tissue in the plantar fascia of long distance runners have also been identified sonographically. The pathogenesis of these lesions is most likely the same mechanism. The most common site of hypertrophic nonunion is the knee, particularly the medial collateral ligament. Although it can be seen in football, basketball, and hockey players, it is characteristically found in soccer players. A soccer-style kicking motion applies a tremendous amount of valgus stress to the knee. The typical history is pain that is increased immediately before contact with the ball when kicking. This pain differs from that experienced with meniscal injury, which is characterized by an increase in pain immediately following contact with the ball (Nicholas and Hershman, 1986).

Hypertrophic ligamentous nonunion is easily recognized sonographically. A hypoechoic mass (Fig 6–25) is identified within the highly echogenic substance of the ligament. This is quite a striking finding since the normal ligament appears as a thin hyperreflective band of tissue. Few internal reflections are observed within the hypoechoic soft tissue component of the lesion. As a result, they may initially be mistaken for meniscal cysts in the knee. However, when pressure is applied with the transducer, the hypoechoic material will respond like soft tissue. In a meniscal cyst, fluid beneath the transducer will be expressed when pressure is applied.

The sonographic picture of a hypertrophic nonunion of the MCL is characterized by thickening and disruption of the normal trilaminar structure of the ligament (Fig 6–26). In our experience, the thickening is more prominent in the proximal portion of the MCL (van Holsbeeck et al., 1988). This was observed in 91% of our patients. Calcium deposition in Pellegrini-Stieda disease is found in a similar distribution, suggesting that this is part of the same disease process (Peterson and Renström, 1986; Mink et al., 1987; Ricklin et al., 1964). Calcifications were observed sonographically in 14% of the patients in our series (Fig 6–27 through 6–29). In all of our patients with hypertrophic nonunion, the capsular portion of the ligament was involved, whereas only one third had involvement of both the superficial and deep components. This supports the hypothesis that tears of the MCL start in the deep component (Mink et al., 1987; Peterson and Renström, 1986).

Unfortunately, these lesions are frequently misdiagnosed without the assistance of ultrasound (Fig 6–30). Arthroscopy is not valuable in the evaluation of these lesions. The arthroscopist can detect only

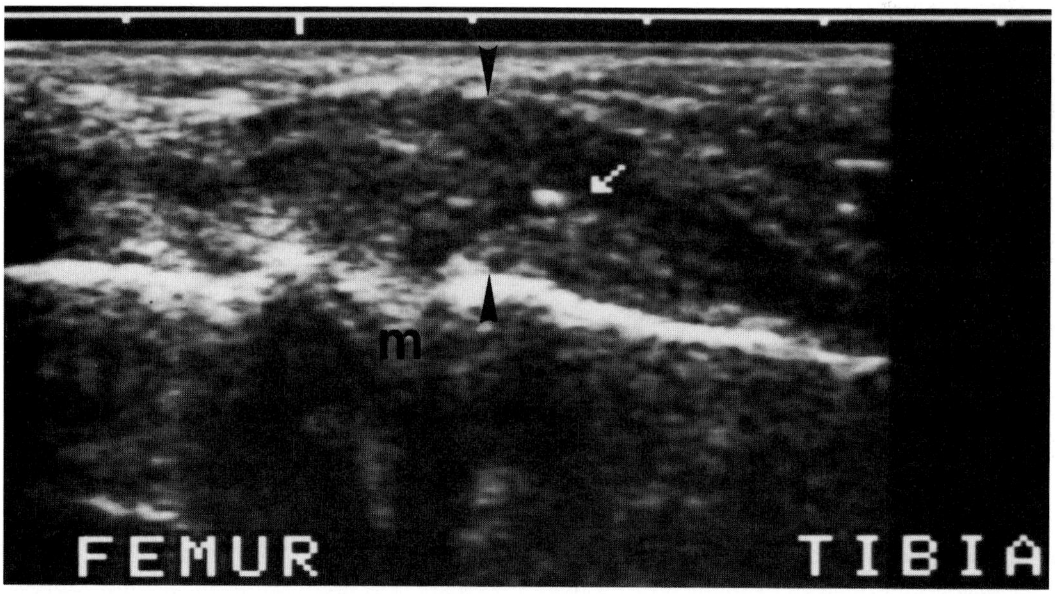


**FIG 6–25.**

Chronic calcified MCL lesion: coronal sonogram. A 19-year-old wrestler injured his right knee 5 months ago and continued to have pain. Global swelling of the MCL is noted, predominantly over the proximal tibia *(arrowheads)*. A small calcification *(arrow)* is seen in the swollen middle portion of the ligament. These are familiar to radiologists as Pellegrini-Stieda disease, but this understates the magnitude of the problem of chronic MCL injury. Calcifications are seen in only 14% of patients with chronic MCL injury. Abbreviations: medial meniscus *(m)*.

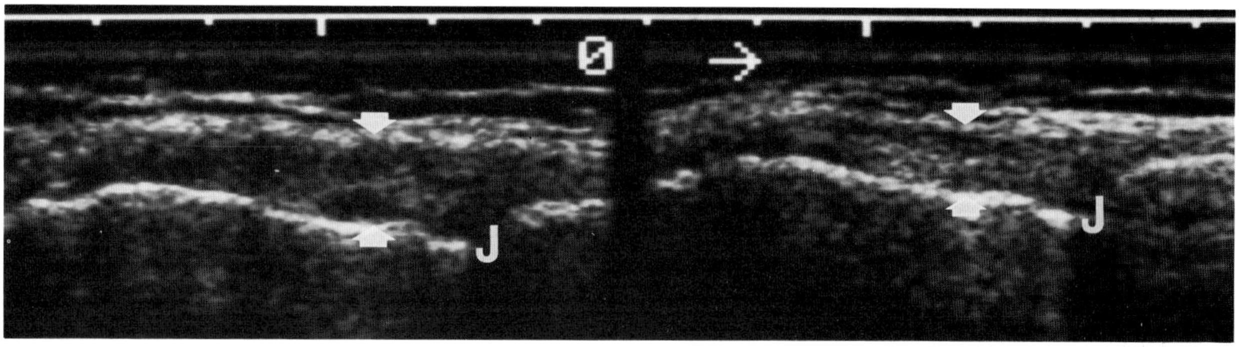


**FIG 6–26.**
Chronically painful MCL: coronal sonograms. One year ago, this farmer was hit on the lateral aspect of the left knee by an angry cow. He continued to have medial knee pain, especially when walking on irregular surfaces. Arthroscopy was normal. The symptomatic left MCL is on the left in this split screen image. The deep portion of the proximal ligament *(arrows)* is decreased in echogenicity and thickened when compared with the asymptomatic right side. Abbreviations: joint space *(J)*.

indirect signs of these MCL lesions, such as "floating" of the medial meniscus (Mink et al., 1987). When arthroscopy is performed in these cases, it is often complicated by extravasation of large volumes of irrigation fluid. If unrecognized, this will result in a compartment syndrome with compression of the neurovascular bundle (Mink et al., 1987; Casscells, 1984; Jackson and Dandy, 1976; Noyes et al., 1980). In professional soccer players, 60% of patients with chronic medial knee pain and a negative arthroscopy have hypertrophic nonunion of the MCL (van Holsbeeck et al., 1990). It is not unusual for these patients to present for sonographic examination having undergone medial shelf resection with no relief of symptoms. Surgical repair of the hypertrophic MCL lesion with curettage of the granulation tissue and side to side suturing of the ligament margins results in relief of symptoms (Fig 6–31). Normal healing ensues following proper immobilization. Histological examination of the surgical specimen reveals disruption of collagen fibers with foci of necrosis, vascular ingrowth, granulation tissue, and areas of mucoid degeneration (van Holsbeeck et al., 1988).

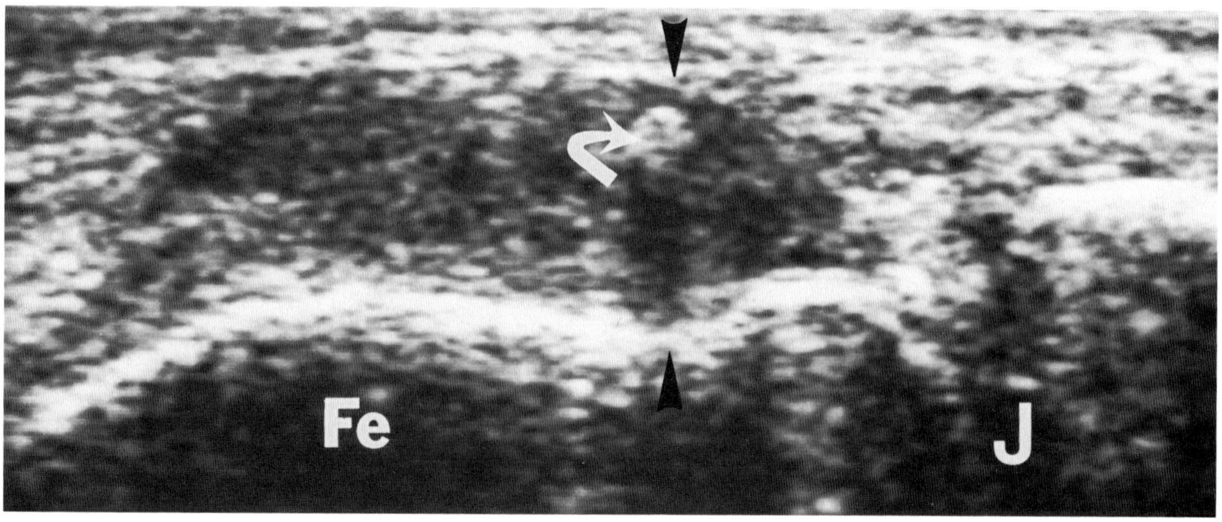


**FIG 6–27.**
Chronic MCL injury: coronal sonogram. At routine physical examination prior to conclusion of a trade to another team, this coronal image of the MCL in a 25-year-old football player demonstrates marked thickening *(arrowheads)* of the proximal ligament with loss of the normal trilaminar structure in this region. A small calcification *(curved arrow)* with incomplete acoustic shadowing is also noted. Abbreviations: medial femoral condyle *(Fe)*, joint space *(J)*.

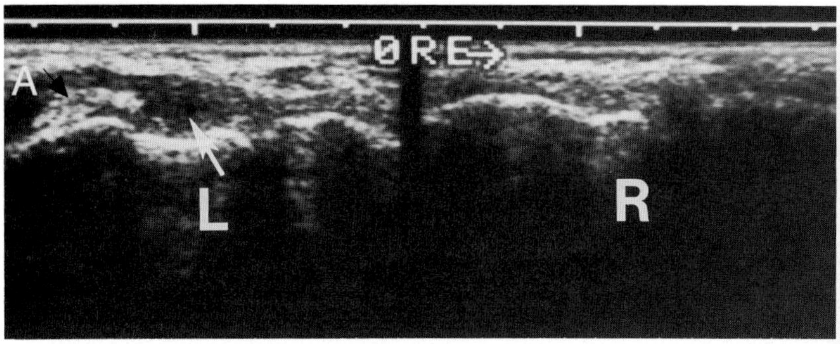


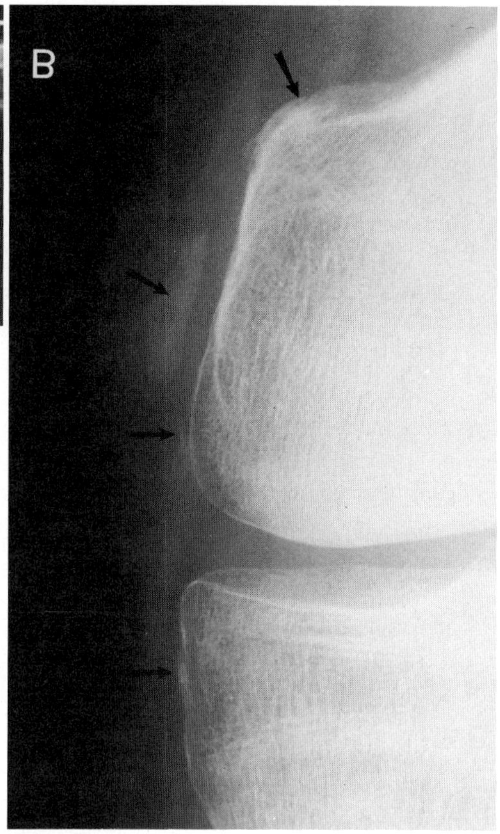


**FIG 6–28.**
**A,** calcification in a chronic MCL lesion: coronal sonograms. The patient is a 19-year-old basketball player who injured his left knee three times in the past 6 months. He presented with persistent swelling and pain over the medial aspect of the left knee. The left MCL is markedly swollen and globally decreased in echogenicity *(white arrow)* with loss of the normal trilaminar structure. Echogenic calcifications *(black arrow)* were noted in several locations throughout the ligament. Abbreviations: symptomatic left knee *(L)*, asymptomatic right knee *(R)*. **B,** calcification in a chronic MCL lesion: anteroposterior radiograph of the knee. Same patient as in **A.** Multiple foci of calcification and ossification *(arrows)* are noted within the substance and insertions of the ligament.

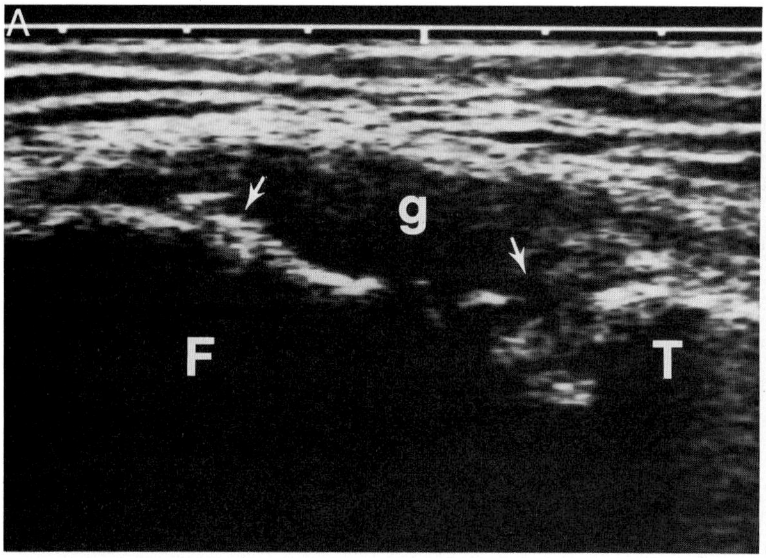


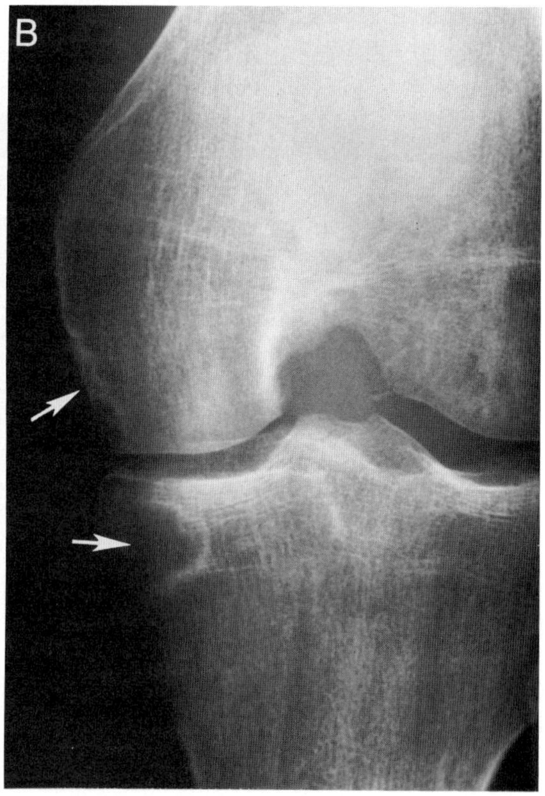


**FIG 6–29.**
**A,** bone changes in chronic MCL injury: coronal sonogram. This 27-year-old soccer player repeatedly injured his left MCL over the past 2 years. Bone underlying the swollen hypoechoic MCL is markedly irregular *(arrows)*. At the time of surgery, granulation tissue *(g)* was found replacing the deep portion of the ligament. Abbreviations: medial femoral condyle *(F)*, proximal tibia *(T)*. **B,** bone changes in chronic MCL injury: anteroposterior radiograph of the knee. Same patient as in **A,** Erosions of bone *(arrows)* underlying the diseased MCL are bordered by a region of sclerosis.

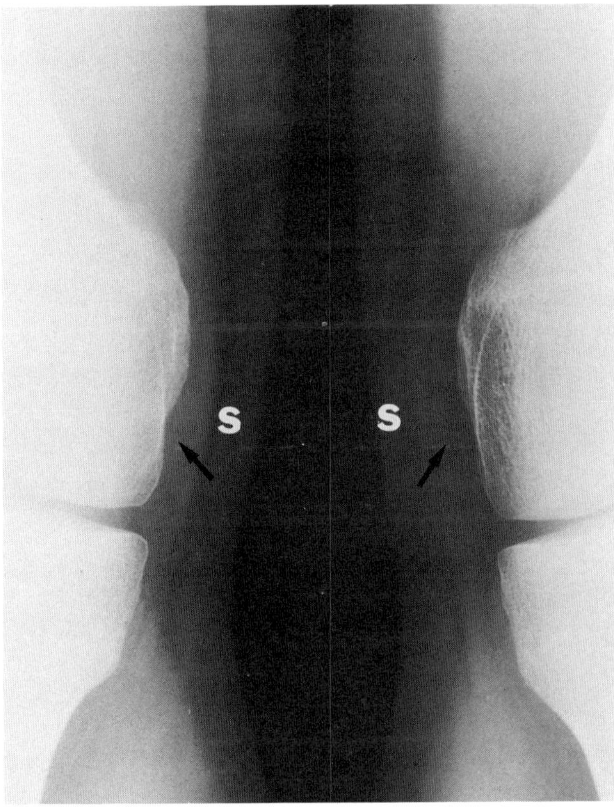


**FIG 6–30.**
Chronic MCL lesion: low-kilovoltage radiograph of the knees. Chronic MCL injury is poorly evaluated with radiographs. Calcifications are seen in only a small percentage of cases, and thickening of the ligament *(arrow)* is difficult to evaluate due to small variations in rotation. Abbreviations: subcutaneous tissue *(S)*.

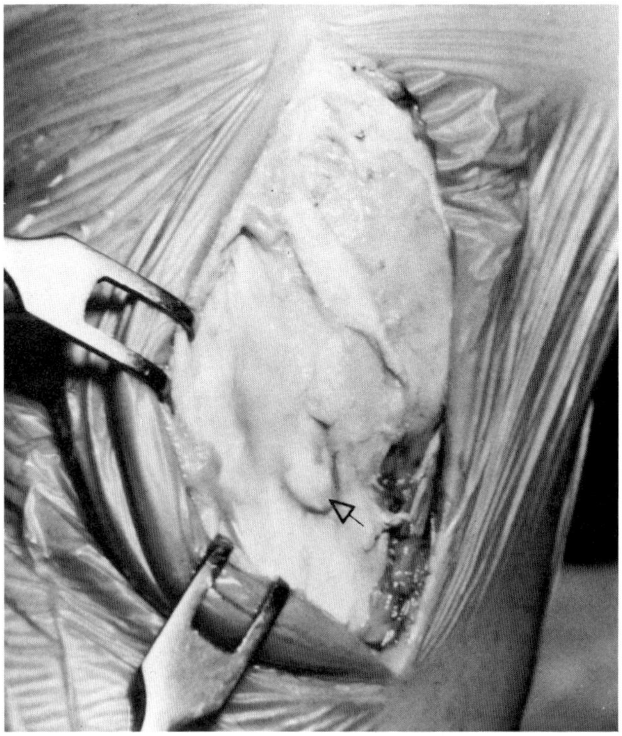

**FIG 6–31.**
Chronic MCL lesion: surgical findings. Foci of granulation tissue *(open arrow)* are often found in cases of chronic injury within the substance of the MCL at the time of surgery. Careful curettage and side to side suturing of the ligament margins usually result in favorable wound healing.

## REFERENCES


Anderson JE: *Grant's Atlas of Anatomy.* Baltimore, Williams & Wilkins Co, 1983.

Bloom W, Fawcett DW: *A Textbook of Histology,* ed 11. Philadelphia, WB Saunders Co, 1980.

Casscells SW: *Arthroscopy: Diagnostic and Surgical Practice.* Philadelphia, Lea & Febiger, 1984.

Clemente CD: *Gray's Anatomy.* Philadelphia, Lea & Febiger, 1985.

Frank C, et al: Healing of the medial collateral ligament of the knee. *Acta Orthop Scand* 1983; 54:917–923.

Hastings DE: The non-operative management of collateral ligament injuries of the knee joint. *Clin Orthop* 1980; 147:22–28.

Hastings DE: Knee ligament instability—a rational anatomical classification. *Clin Orthop* 1986; 208:104–107.

Hentz VR, Green PS, Arditi M: Imaging studies of the cadaver hand using transmission ultrasound. *Skeletal Radiol* 1987; 16:474–480.

Hughston JC, et al: Classification of knee ligament instabilities. I: The medial compartment and the cruciate ligaments. *J Bone Joint Surg [Am]* 1976; 58:159–172.

Jackson RW, Dandy DJ: *Arthroscopy of the Knee.* New York, Grune & Stratton, 1976.

Kannus P: Long term results of conservatively treated medial collateral ligament injuries of the knee joint. *Clin Orthop* 1988; 226:103–111.

Laine HR, et al: Ultrasound in the evaluation of the knee and patellar regions. *J Ultrasound* 1987; 6:33–36.

Laws G, Walton M: Fibroblastic healing of grade II ligament injuries. Histological and mechanical studies in sheep. *J Bone Joint Surg [Br]* 1988; 70: 390–396.

Mink JH, Reicher MA, Crues JV III: *Magnetic Resonance Imaging of the Knee.* New York, Raven Press, 1987.

Nicholas JA, Hershman EB: *The Lower Extremity and Spine in Sports Medicine.* St Louis, CV Mosby Co, 1986.

Noyes FR, et al: Arthroscopy in acute traumatic hemarthrosis of the knee. *J Bone Joint Surg [Am]* 1980; 62:685–695.

Peterson L, Renström P: *Sports Injuries.* Chicago, Year Book Medical Publishers, 1986.

Richardson ML, et al: Ultrasonography of the knee. *Radiol Clin North Am* 1988; 26:63–75.

Ricklin P, Ruttiman A, Del Buono MS: *Meniscal Lesions: Practical Problems of Clinical Diagnosis, Arthrography and Therapy.* New York, Grune & Stratton, 1964.

van Holsbeeck M, et al: Sonography of the medial collateral ligament [abstract]. *Radiology* 1988; 169P:178.

van Holsbeeck M, Introcaso J, Martens M: Sonographic detection of chronic medial collateral ligament injury of the knee. (Manuscript in review.)

# 7

# Sonography of Large Synovial Joints

Sonographic examination of large synovial joints is frequently requested because these joints are difficult to evaluate clinically. The shoulder, hip, and knee are deep-seated joints in which physical examination is limited. Ultrasound is dramatically changing our approach to the evaluation of joint disease. In Europe, ultrasound has already established its place as the primary means of evaluating periarticular disease of synovial joints. It is not limited by the size of the joint. Even the smallest interphalangeal joints can be examined sonographically. Currently in the United States, arthroscopy is the method most frequently employed for evaluation of the knee. Arthroscopy is limited to the intra-articular structures. It permits the surgeon to examine the cartilage and perform corrective surgery in the same session. However, arthroscopy is an invasive technique not without risk to the patient. The risk of anesthesia and possibility of damage to the periarticular soft tissues is not insignificant.

The principle advantage of ultrasound over arthroscopy is its ability to examine the periarticular soft tissues. Many pain syndromes do not originate in bone or articular cartilage. Until now, these were presumptive clinical diagnoses. Sonography can be used to diagnose disease of the periarticular tissues with high sensitivity and specificity. It can detect tendinous, ligamentous, and muscular lesions. It can differentiate scar, granulation tissue, complete and incomplete tears. Hemorrhagic and frictional bursitis can also be differentiated by their sonographic appearance. In addition, ultrasound is entirely noninvasive and requires no anesthesia or premedication.

Intra-articular disease can also be accurately evaluated using ultrasound. It can quickly estimate the amount of intra-articular fluid. Changes in intra-articular fluid are easily detected, but this is a nonspecific finding seen in chondromalacia, osteoarthritis, rheumatoid arthritis, osteonecrosis, infection, intra-articular tumors, and chronic traumatic disease. In addition, ultrasound can identify synovial edema, hemarthrosis, and loose bodies within the joint. Cartilage and synovial thickness can be measured accurately, providing an excellent method to evaluate the effectiveness of treatment and the progression of disease (Aisen et al., 1984). Peripheral meniscal tears can also be detected, but in general, meniscal tears are better evaluated with magnetic resonance imaging (MRI) or arthroscopy.

In summary, ultrasound is a valuable tool in the diagnosis of both periarticular and intra-articular disease. It is the method of choice for evaluation of periarticular disease and the first step in the evaluation of intra-articular disease. If the intra-articular examination is normal, no further imaging studies are indicated. Abnormal sonographic examination is an indication for evaluation with MRI, arthrography, or arthroscopy.

## EXAMINATION TECHNIQUE

In the examination of superficially located joints, 5-, 7.5-, and 10-MHz transducers are utilized. These will provide excellent spatial resolution in the evaluation of the joint capsule and synovia in elbows, hands, and feet. In the majority of patients, these transducers can also be used for examination of the shoulders, hips, and knees. However, in obese patients, it may be necessary to utilize a 3.5 MHz-transducer for these deep-seated joints. In these cases, sensitivity in detecting small lesions will be decreased, but valuable information can still be obtained.

When large synovial joints are scanned for intra-articular pathology and synovial disease, the examination is directed toward the largest synovial recesses of the joint not protected by bone. The field of view can be further increased by flexion, extension, and internal or external rotation of the joint. Anteversion of the hip limits examination to the anterior approach. The shoulder joint is most accessible through the posterior synovial recess due to retroversion of the humeral head. Other approaches are restricted by bony protuberances. Figure 7–1 summarizes the recommended approaches to scanning the large synovial joints. In cases of extensive synovial disease, such as rheumatoid arthritis, the portal selected for scanning is not critical. Active rheumatoid disease of the shoulder enlarges the axillary and subcoracoid recesses, as well as the posterior synovial recess. Thus, the same diagnostic information can be obtained in rheumatoid patients regardless of which approach to the joint is selected. This is true for all large synovial joints with active disease. The progression of disease in the knee differs slightly from that in other joints. Early disease is best detected by examination of the suprapatellar recess, as shown in Figure 7–1,A. However, chronic disease will enlarge the gastrocnemius-semimembranosus bursa resulting in a Baker's cyst. Examination for Baker's cyst is performed using a posterior approach, scanning the medial aspect of the popliteal fossa.

## NORMAL SONOGRAPHIC ANATOMY

In a normal joint the bony contours are seen as hyperreflective lines with shadowing posteriorly. The epiphyseal growth plate is readily identified in children as a relatively hypoechoic band separating the epiphysis from the metaphysis. Cartilage covering the articular surfaces is depicted as a relatively hypoechoic homogeneous stripe with smooth contours covering the epiphyses. The thickness of the articular cartilage is easily measured using the electronic calipers of the ultrasound unit. A minimal amount of fluid is normally found in the synovial recesses. The

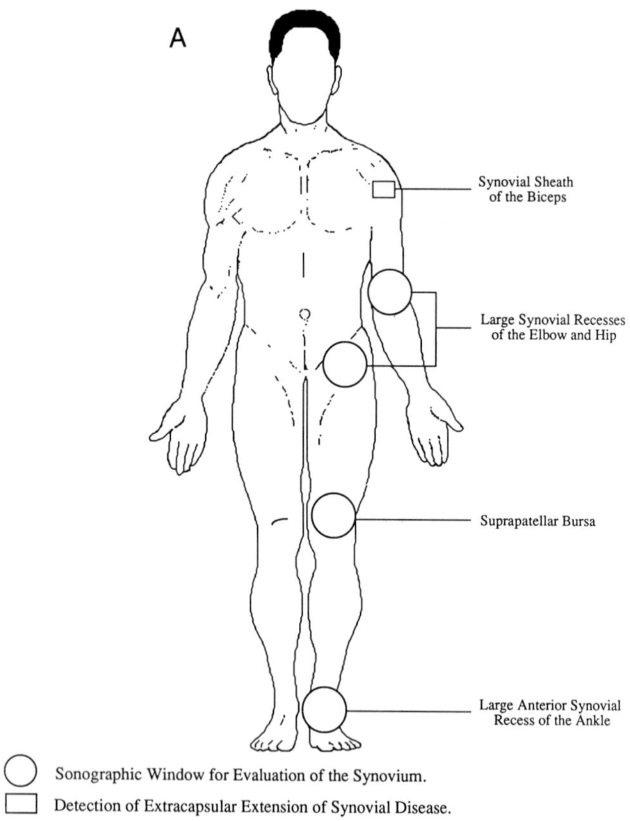


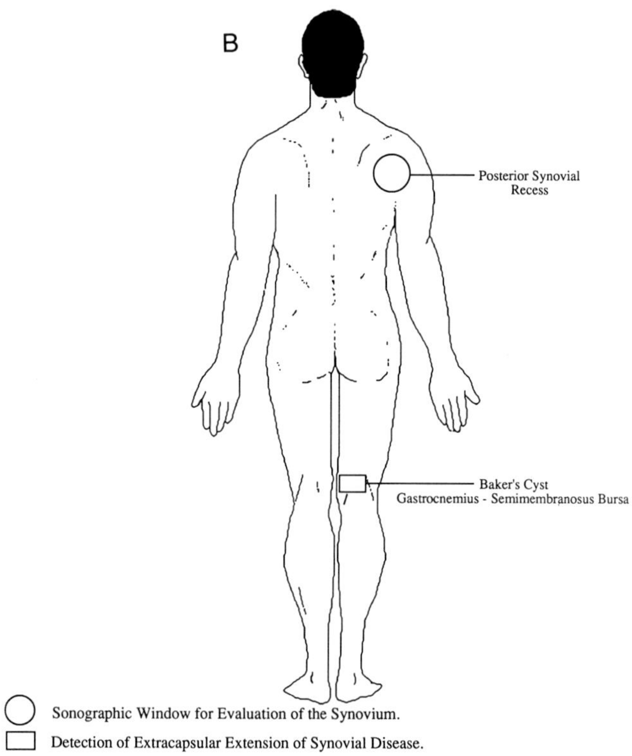


**FIG 7–1.**
Sonographic windows for examination of the large synovial joints: anterior approach **(A)** and posterior approach **(B)**.

synovial leaves are separated by a fluid film approximately 2 to 3 mm thick. Synovial tissue lining the joint is immeasurably thin with smooth regular contours. A clearly identifiable hyperechoic line defines the joint capsule. Supporting ligaments demonstrate a lamellar appearance with alternating narrow hyperechoic and hypoechoic bands. Figures 7–2 through 7–7 illustrate these findings in an adult with normal knees, a child's knee, and a child's hip.

## SONOGRAPHY OF ARTICULAR PATHOLOGY

Pathology of large synovial joints can be divided into those processes affecting the intra-articular structures and those involving the periarticular tissues. All structures lying inside of the synovial membrane, as well as the membrane itself, are considered intra-articular. Everything outside of the synovial membrane is considered periarticular.

### Intra-articular Disease

Increased intra-articular fluid is a nonspecific finding indicative of joint pathology. It is seen in inflammatory processes, osteonecrosis, osteoarthritis, and tumors (Resnick Niwayama, 1988; Gielen et al., 1990). A reactive synovitis, resulting in joint effusion, may be seen in cases of tumor adjacent to the joint. Despite the lack of specificity of this finding, joint effusion remains a valuable indicator of joint disease, and ultrasound is unquestionably the best means of detecting increased intra-articular fluid. Effusions as small as 1 mL are easily detected sonographically (Marchal and van Holsbeeck, 1987). Sonography has proved to be more sensitive than blind aspiration of the joint (Peck, 1986; Adam et al., 1986; Marchal et al., 1987).

#### *Transient Synovitis*

Acute joint effusion in children is a difficult diagnostic problem for the clinician. Occurring in the hip, greater than 95% of cases are the result of an idiopathic inflammatory process of the synovium. This process resolves spontaneously within 6 weeks without sequelae. Rarely, after repeated bouts of the disease, coxa magna may result. Transient synovitis is the preferred name associated with this disease because it clearly conveys the essential features of the disease. Toxic synovitis is synonymous but less meaningful.

In the past, the diagnosis of transient synovitis was one of exclusion. It was based on clinical history and the finding of increased joint space of the hip medially, detected on conventional radiographs of

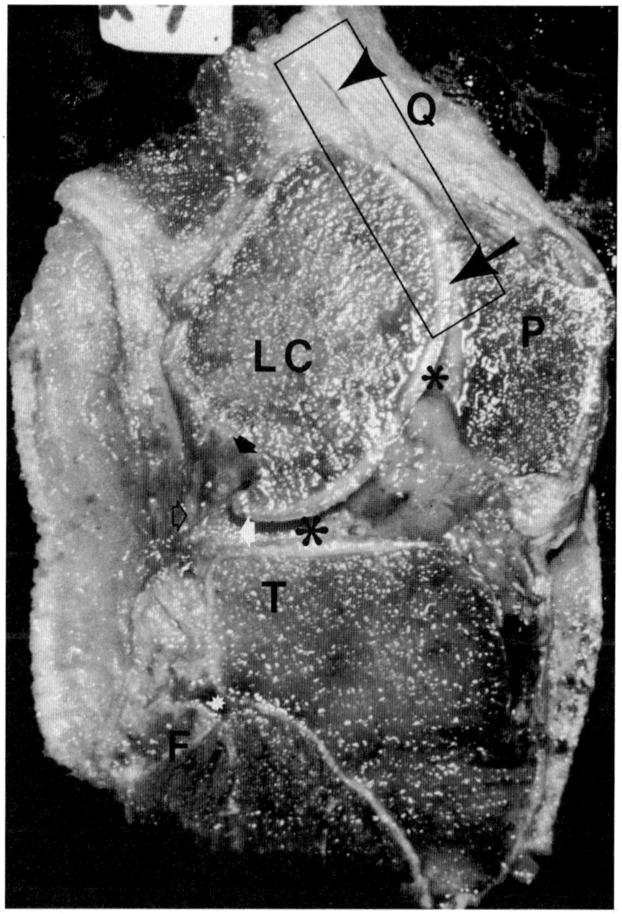


**FIG 7–2.**
Sagittal anatomical section through the lateral compartment of the knee. The rectangular area indicates the region of greatest interest for routine examination of the knee. Deep to the quadriceps tendon *(Q)* lies the suprapatellar bursa *(black arrowhead),* which is an excellent indicator of intra-articular pathology. Normally the walls of the suprapatellar bursa are extremely thin. Effusions, loose bodies, synovial edema, pannus, or tumor involvement change the volume and appearance of the bursa. Through the same sonographic window we can image the cartilage of the femoral condyle *(large black arrow).* In the popliteal fossa, ultrasound can demonstrate the posterior joint capsule *(open black arrow),* the posterior cruciate ligament *(short black arrow),* and the meniscus *(white arrow).* Areas that are poorly or not visualized include the cartilage of the patella, the tibial condyles, and the middle segment of the femoral condyles *(asterisk).* Abbreviations: lateral femoral condyle *(LC),* tibia *(T),* fibula *(F),* patella *(P).*

the pelvis. The finding of joint space widening is valid only if the radiograph is an orthogonal projection. This severely limits the clinical value of this sign, with fewer than one half of the cases (Marchal et al., 1987) of transient synovitis having joint space widening demonstrable on conventional radio-

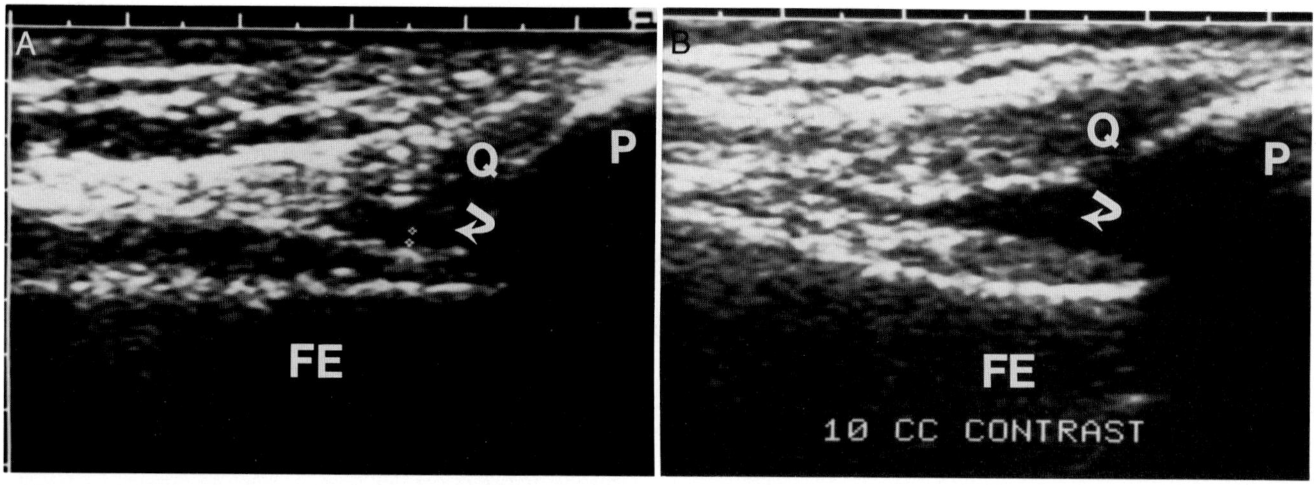


**FIG 7–3.**
**A,** normal suprapatellar bursa: longitudinal sonogram. The patient is a 20-year-old man with pain and tenderness inferior to the patella. Imaging was performed with the transducer oriented in the long axis of the quadriceps tendon using firm pressure. The inferior aspect of the transducer touches the middle of the basis patellae. A normal suprapatellar bursa *(curved arrow)* is approximately 1.5 to 3 cm long. It measures 2 to 3 mm in greatest thickness (+). Patella *(P)*, femur *(Fe)*, quadriceps *(Q)*. **B,** 10 mL of contrast material was introduced for arthrography, which was normal. Anteroposterior diameter of the bursa *(curved arrow)* is increased by a factor of 5. Abbreviations: patella *(P)*, femur *(Fe)*, quadriceps *(Q)*.

graphs. As a result, clinicians have adopted unguided joint aspiration to confirm the diagnosis. In transient synovitis, serous or serosanguineous fluid is obtained (Vandeputte et al., 1971). However, unguided joint aspiration is not reliable unless performed by a clinician with considerable experience. Even when unguided joint aspiration is performed by experienced orthopedic surgeons, sonography has proved to be more sensitive in detecting joint effusion (Adam et al., 1986; Peck, 1986). Careful sonographic comparison with the asymptomatic side is mandatory. An effusion is detected by observing anterior displace-

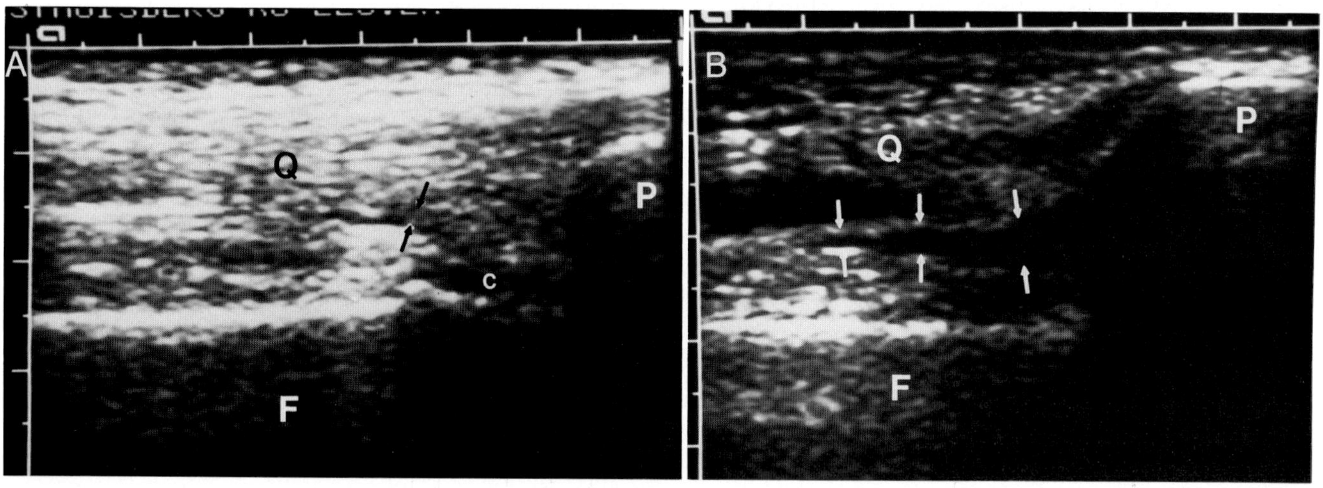


**FIG 7–4.**
**A,** normal suprapatellar bursa with the knee in extension: longitudinal sonogram. A normal 26-year-old man. Transducer position is identical to that in Figure 7–3. A slitlike suprapatellar bursa *(black arrows)* is identified within the fat deep to the quadriceps tendon *(Q)*. Abbreviations: patella *(P)*, femur *(F)*, femoral cartilage *(c)*. **B,** normal suprapatellar bursa with the knee flexed 20 degrees: longitudinal sonogram. Same patient and transducer position as in **A.** Note that the suprapatellar bursa *(white arrows)* is better demonstrated with the knee slightly flexed. The patella lies farther anterior relative to the femoral diaphysis when the knee is in flexion. Abbreviations: patella *(P)*, quadriceps *(Q)*, femur *(F)*.

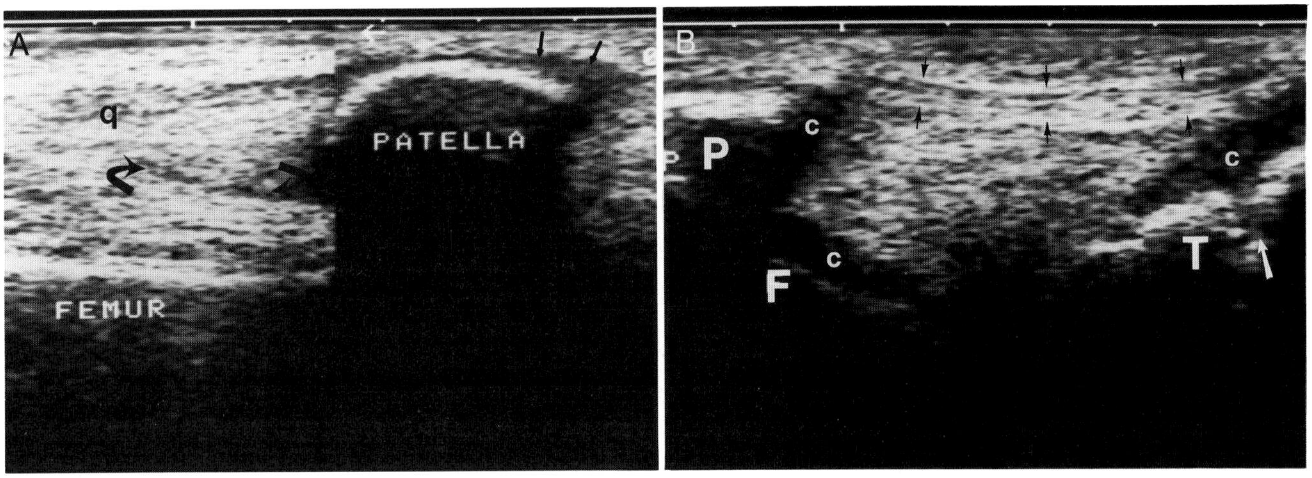


**FIG 7–5.**
**A,** normal adolescent knee: longitudinal sonogram. A healthy 12-year-old boy. The knee is smaller, allowing visualization of the distal femur, patella, and proximal tibia on the same image. Deep to the quadriceps tendon *(q)*, the suprapatellar bursa *(curved arrow)* extends 2 cm superior to the base of the patella. The anteroposterior dimension of the bursa is not more than 1 mm when firm pressure is applied with the transducer. Growth cartilage is still present around the apex of the patella *(arrows)*. **B,** the transducer is moved inferiorly, now overlying the patellar tendon. A thick layer of cartilage *(c)* is noted covering the patellar apex *(P)*, femoral condyle *(F)*, and the proximal tibia *(T)*. The patellar tendon *(black arrows)* joins the patellar apex and the tibial tuberosity. A hypoechoic band *(white arrow)* in the proximal tibial represents the proximal growth plate.

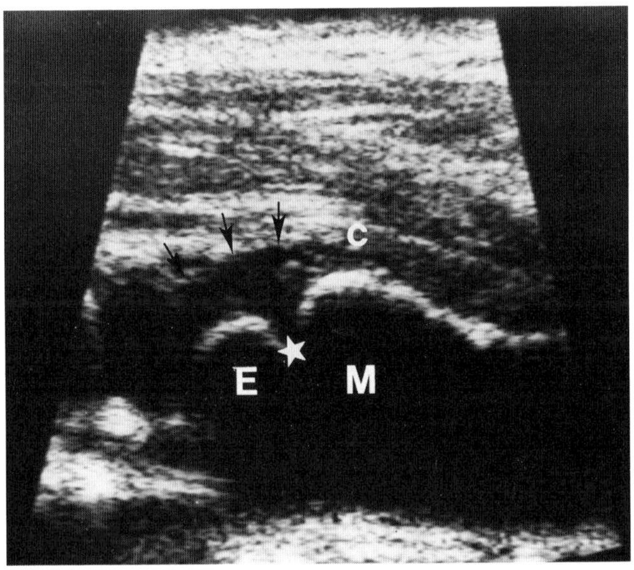


**FIG 7–6.**
Normal hip of a child: longitudinal sonogram. The patient is a normal 6-year-old boy. This image was obtained with the transducer aligned along the length of the femoral neck. Bony contours of the proximal femoral epiphysis *(E)* and metaphysis *(M)* are clearly demonstrated. The growth plate is identified as a hypoechoic band *(star)* through the femoral epiphysis. A hyperechoic band approximately 4 mm in thickness anterior to the femoral head represents the anterior joint capsule *(c)*, also called the *iliofemoral ligament*. The epiphyseal cartilage is seen as a hypoechoic zone covering the femoral head *(arrows)*.

ment of the iliofemoral ligament by a fluid collection. Asymmetry greater than 2 mm (Fig 7–8) or a capsule to bone distance greater than 3 mm is significant (Adam et al., 1986; Dörr, 1988; Peck, 1986; Marchal et al., 1987).

The differential diagnosis includes septic arthritis and Legg-Calvé-Perthes disease. Although these account for less than 4% of cases in our experience, they must be carefully considered in cases of acute hip pain with effusion in children. Ultrasound images in transient synovitis will demonstrate a joint effusion composed of anechoic or markedly hypoechoic fluid. Effusions in septic arthritis will be hypoechoic with definite internal echos. Marked synovial irregularity and thickening of the joint capsule will also be seen in septic arthritis (Dörr, 1988). In exceptional cases of septic arthritis, intra-articular gas bubbles may be identified as hyperechoic foci with "dirty" shadowing, indicative of infection with gas-producing organisms. Although the specificity of these findings is high, the decision to aspirate the joint to exclude infection must be based on clinical criteria. One case of early septic arthritis with anechoic fluid and minimal synovial changes has been reported (Marchal et al., 1987).

Sonographic characteristics of the joint effusions seen in transient synovitis and Legg-Calvé-Perthes disease are identical (Marchal et al., 1987; Wingstrand, 1986). Early diagnosis of Legg-Calvé-Perthes

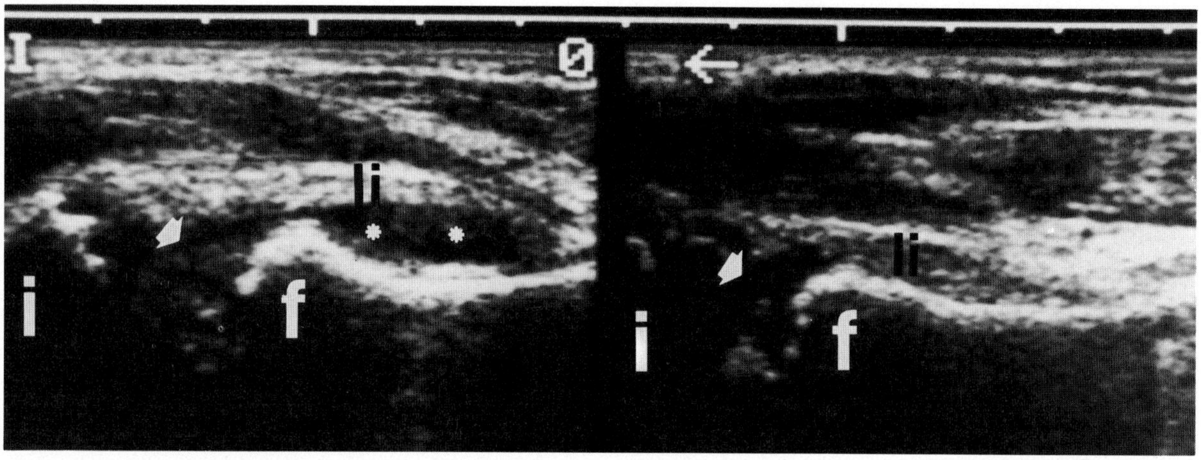


**FIG 7–7.**
Bilateral hips, unilateral effusion: longitudinal sonograms. The patient is an irritable 3-month-old boy who was not moving his left leg. Transducer position is identical to that in Figure 7–6. The anterior joint capsule is formed by the hyperechoic iliofemoral ligament *(li)* that connects the ilium and femur. Recognition of this structure is critical in making the diagnosis of hip effusion. A distance greater than 3 mm separating the iliofemoral ligament from the femoral cortex is diagnostic for an effusion. In this case, an effusion is identified in the left hip *(asterisk)* at the observer's left. The right hip, imaged for comparison, is normal. Epiphyseal cartilage *(white arrow)*. Abbreviations: ilium *(i)*, femur metaphysis *(f)*.

disease, before conventional radiographic findings are evident, is based on repeating the sonographic examination 6 weeks following the initial diagnosis of uncomplicated joint effusion. During this time, effusion resulting from transient synovitis will resolve. Persistence of a simple effusion is diagnostic of Legg-Calvé-Perthes disease. Cartilage edema, recognized as thickening of the cartilage, will also be seen in early stages of Legg-Calvé-Perthes disease. Later, flattening of the femoral head can be observed sono-

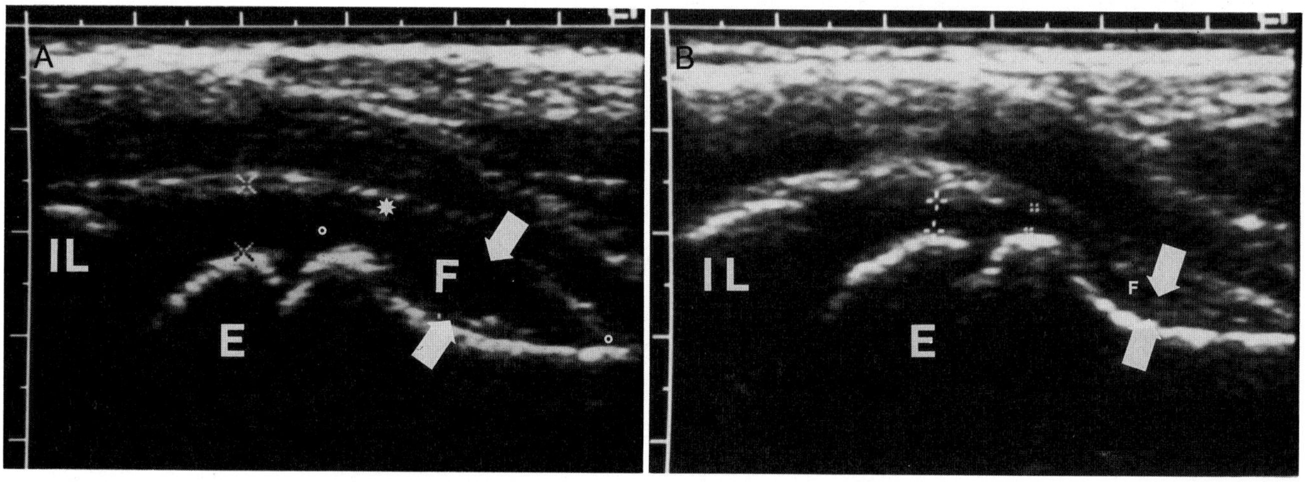


**FIG 7–8.**
**A,** transient synovitis in a child: longitudinal sonogram. An 8-year-old girl who refused to bear weight on the left leg. Transducer position is identical to Figure 7–7. The capsule to cortex distance *(arrows)* is measured from the internal aspect of the iliofemoral ligament *(asterisk)* to the cortical surface at a point equidistant between the growth plate *(open circle)* and the insertion of the iliofemoral ligament on the femoral neck *(open circle)*. In this case the capsule to cortex distance is greater than 3 mm, indicative of joint effusion *(F)*. A slightly thickened rim of synovium is seen covering the femoral neck. Abbreviations: femoral epiphysis *(E)*, ilium *(IL)*, epiphyseal cartilage *(x)*. **B,** transient synovitis: normal contralateral hip. Same patient as in **A.** The capsule to cortex distance is normal, 1 to 2 mm *(arrows)*. A tiny fluid film *(F)* is noted between the iliofemoral ligament and synovium. Abbreviations: ilium *(IL)*, femoral epiphysis *(E)*, epiphyseal cartilage *(x)*.

graphically and on conventional radiographs (Fig 7–9).

### *Follow-up of Joint Effusions*

In addition to cases of suspected transient synovitis, sonographic follow-up examinations are extremely valuable in cases of inflammatory arthritides: rheumatoid arthritis, psoriatic arthritis, and Reiter's disease. A decrease in the quantity of intra-articular fluid is the first sign of positive response to therapy. The response can almost always be detected within 48 hours of initiating treatment. Later, changes in thickness of the synovia can be seen (van Holsbeeck et al., 1988b). A more detailed discussion of the changes seen in rheumatoid arthritis is found in the chapter on rheumatoid disease.

### *Effusions in Trauma*

Ultrasound is very helpful in distinguishing simple effusions from other types (Figs 7–10 through 7–15). Hemarthrosis or lipohemarthrosis following intra-articular fracture, cruciate ligament, and meniscal tears can easily be identified. Free-floating blood clots or fat lobules are seen within the effusion. These lobules are mobile and easily deformed when pressure is applied by the examiner's hand or the transducer.

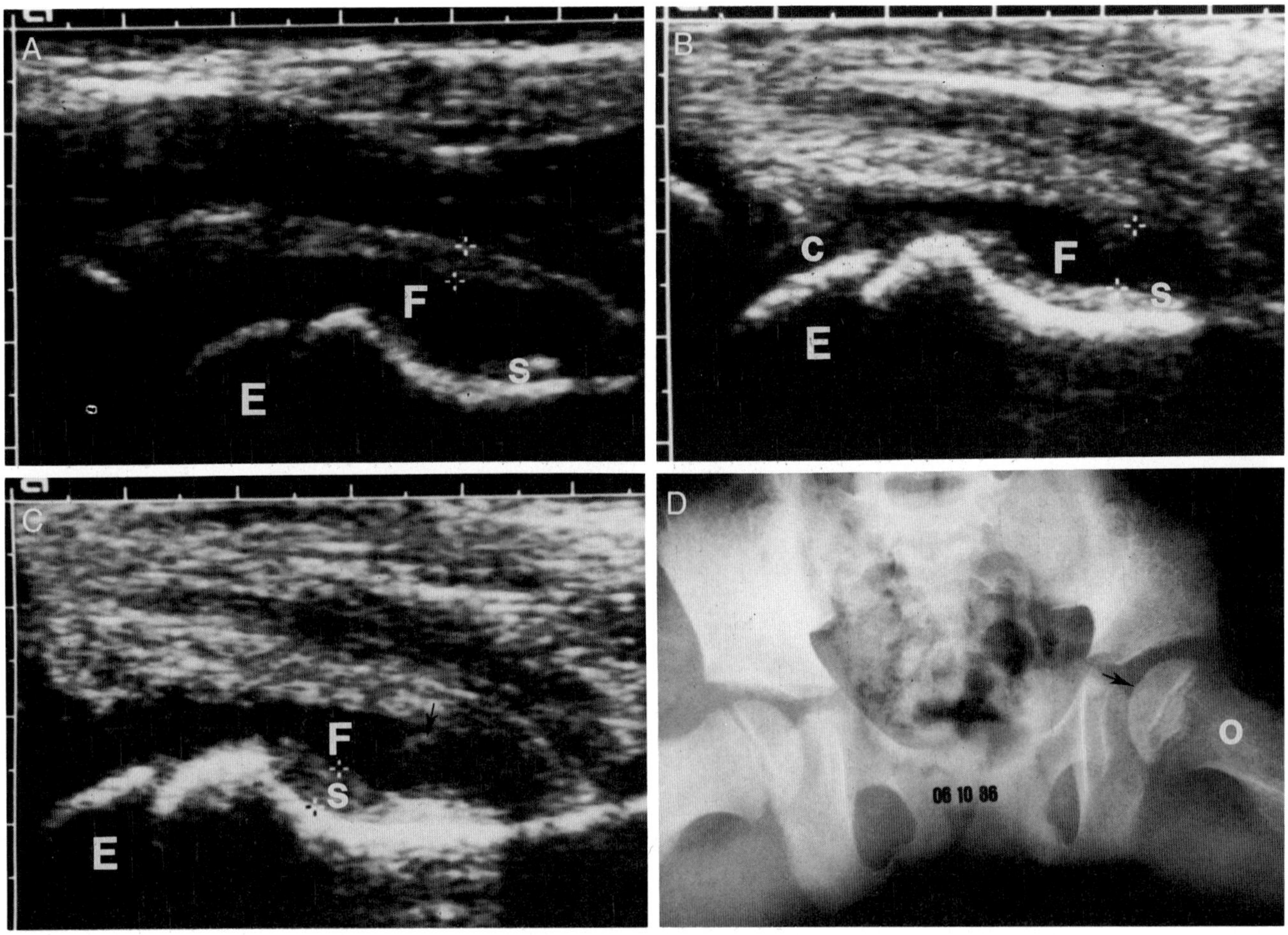


**FIG 7–9.**

**A,** hip effusion in Legg-Calvé-Perthes disease: longitudinal sonogram. The patient is an 11 year old boy with pain in the left leg over the past week. A large joint effusion *(F)* displaces the iliofemoral ligament (+) anteriorly, and a thin band of synovial tissue *(s)* is identified around the femoral neck. Abbreviations: femoral epiphysis *(E)*. **B,** 3 weeks later, the amount of fluid is decreased (+), but an effusion is still present. The synovial tissue *(s)* around the femoral neck has increased in thickness. Abbreviations: cartilage *(c)*, femoral epiphysis *(E)*, fluid *(F)*. **C,** 12 weeks later, the femoral epiphysis *(E)* appears flatter. No change is observed in the quantity of intra-articular fluid *(F)*. Further increase in synovial thickness (+) is noted, and the contour of the synovium has become quite irregular *(arrow)*. **D,** frog leg pelvic radiograph taken 12 weeks after the initial sonographic examination. The left femoral epiphysis is slightly flatter and more irregular than the right *(arrow)*. Density of the left femoral epiphysis is increased relative to the osteoporotic femoral metaphysis *(open circles)* indicative of avascular necrosis.

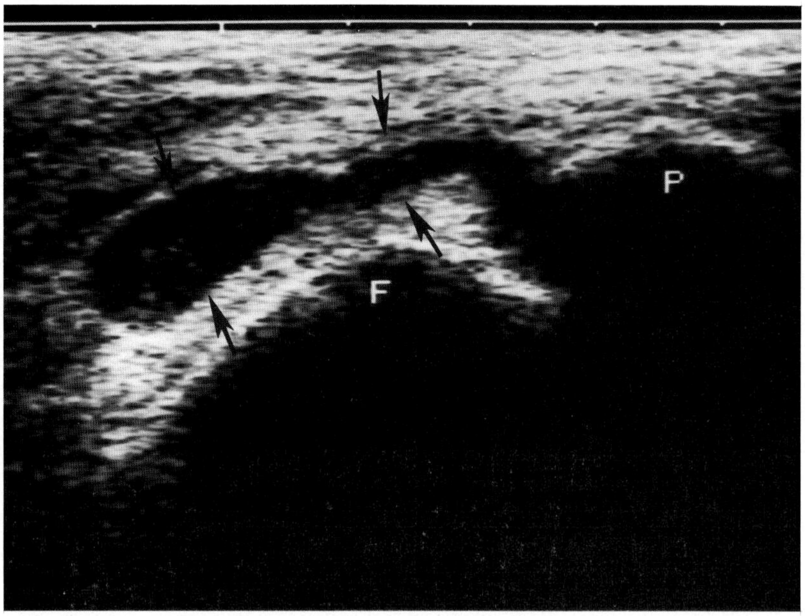


**FIG 7–10.**
Acute hemarthrosis of the knee: transverse sonogram. A 25-year-old man with trauma to the lateral aspect of the knee. Sonographic examination was performed about 10 hours following injury. The transducer is positioned transversely over the anterolateral aspect of the knee, approximately 2 cm above the joint space. The lateral joint recess *(arrows)* is filled with homogeneous hypoechoic material. Aspiration of the joint was negative, presumably due to obstruction of the needle by clot. Arthroscopy performed 2 months later demonstrated a large bucket handle tear of the medial meniscus. Abbreviations: lateral femoral condyle *(F)*, patella *(P)*.

In many of these cases, aspiration of the joint is negative, likely due to obstruction of the needle lumen by clot or debris (see Fig 7–15).

### *Synovial Edema and Pannus*

Thickened edematous synovia can be seen in inflammatory arthritides (Resnick and Niwayama, 1988), amyloidosis (Gielen et al., 1990) with joint involvement, and in cases of tumoral involvement of the synovia (Resnick and Niwayama, 1988). Pigmented villonodular synovitis and synovial chondromatosis are two relatively common synovial tumors associated with synovial edema and thickening. Hypertrophic synovia are often hypoechoic, recognized as a dark band between surrounding muscle or fat (Cooperberg et al., 1978). Compression applied with the transducer effects the edematous synovia greater than the adjacent muscle, tendons, ligaments, or fat. Therefore, compression is a helpful maneuver in identifying edematous synovial tissue (Fig 7–16). A reliable and reproducible measurement of synovial thickness can be made only at maximal compression (van Holsbeeck et al., 1988b). All free fluid is then expelled from the region, and the measurement most

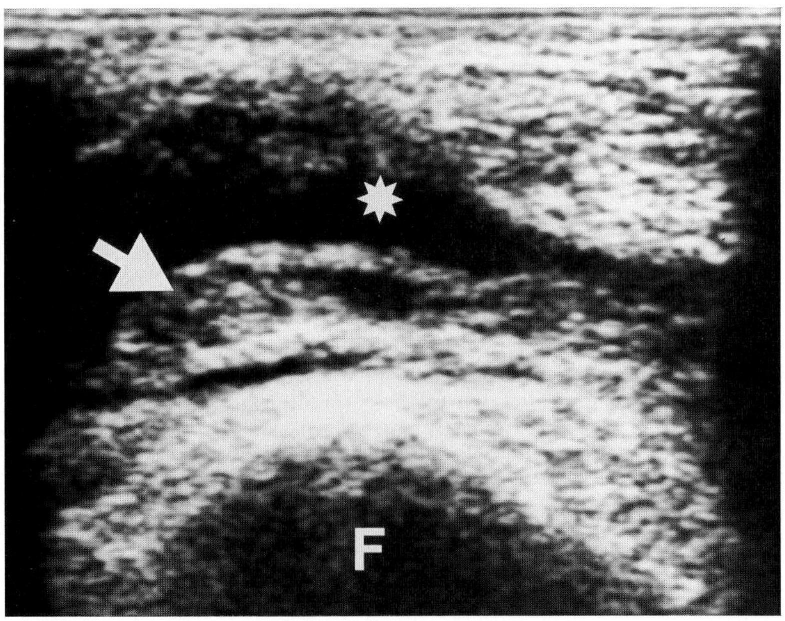


**FIG 7–11.**
Evolution of hemarthrosis in the knee: transverse sonogram. This 31-year-old man presented with an acute torsion injury and acute swelling of the knee. Ultrasound was performed 5 days following injury. Transducer position is identical to Figure 7–10. The lateral recess is distended with anechoic fluid *(asterisk)*. An echogenic mass *(arrow)* over the femoral condyle *(F)* was mobile and compressible with pressure applied with the transducer. This is characteristic of a blood clot within the joint. Aspiration of the joint yielded serosanguineous fluid. Clinical examination 3 weeks later demonstrated a positive anterior drawer test result, indicative of rupture of the anterior cruciate ligament.

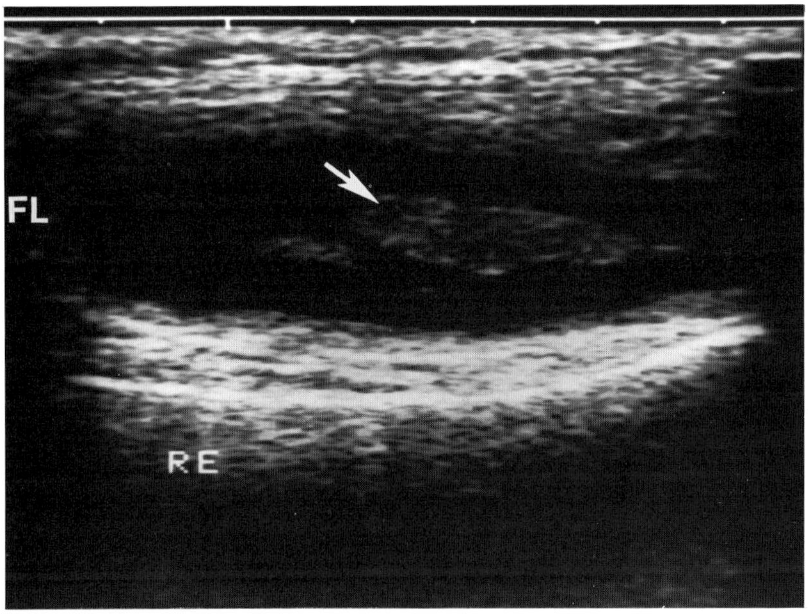


**FIG 7–12.**
Lipohemarthrosis of the knee: longitudinal sonogram. A 33-year-old man who was kicked on the medial aspect of the knee presented for sonographic examination 2 days later. Transducer is positioned over the quadriceps tendon, just superior to the patellar base. The suprapatellar bursa is markedly distended with anechoic fluid *(FL)*. A hypoechoic mass *(arrow)* is noted floating nondependently within the collection. Conventional radiographs demonstrated a lipohemarthrosis.

closely reflects true synovial thickness (Fig 7–17).

Pannus is most commonly seen in cases of inflammatory arthritides, the classic example being rheumatoid arthritis. However, it can also be seen with chronic infection: tuberculosis, brucellosis, Lyme arthritis, and fungal infection. Villous synovial proliferation is often seen when pannus is present. Synechiae and fibrous adhesions within the synovial recesses are identified in later stages of disease. Fine particulate debris floating in the articular fluid may also be seen in rheumatoid joints. This is probably due to aggregation of fibrin within the inflammatory effusion but may be seen following intra-articular corticosteroid administration (Worth et al., 1986) (Fig 7–18).

Evaluation of synovial changes in large synovial joints is most easily performed at the suprapatellar recess of the knee, anterior synovial recess of the hip, and the posterior synovial recess of the shoulder. All measurements of synovial thickness should be made with maximal compression applied by the transducer. This will yield the most accurate and reproducible measurements. Sonographic measurements of synovial thickness using this technique have been shown to be a reliable means of comparing the effectiveness of different therapies (van Holsbeeck et al., 1988b; Hammer et al., 1986).

### *Cartilage Pathology*

**Cartilage Structure.**—Articular cartilage is a gel of protein-polysaccharide complexes reinforced by a matrix of collagen fibers. The principle constituents of the polysaccharide portion are chondroitin and keratin sulfates, which are very hydrophilic (Sissons, 1987, in Taveras and Ferrucci, 1987). Chondrocytes are diffusely scattered throughout. The thickness of normal weight-bearing cartilage ranges from 1.2 to 1.9 mm (Richardson et al., 1988; Aisen et al., 1984). Cartilage of the growth plate differs principally in its arrangement of chondroblasts into regular pallisades of cells oriented in the long axis of the bone.

The sonographic appearance of articular cartilage is a thin hypoechoic layer juxtaposed with the subchondral cortical bone. This hypoechoic appearance is likely a reflection of the homogeneous hydrophilic structure of hyaline cartilage. Cartilage of the growth plate shares this homogeneous hypoechoic appearance. Normal cartilage has sharp margins at both the articular and deep surfaces, making measurement of thickness quite easy (Fig 7–19).

Fibrocartilage differs significantly from hyaline cartilage histologically and sonographically. In fibrocartilage the predominant component is collagen fibers, as the name implies. Densely packed bundles of collagen fibers are arranged in parallel. Cartilage cells that have differentiated from fibroblasts are scattered throughout this dense collagen matrix. In some cases they are found in rows between the collagen bundles. A minute amount of ground substance surrounds these cells but is absent elsewhere (Bloom and Fawcett, 1975). This structure of tightly packed collagen with interspersed cartilage cells and minimal ground substance explains the hyperechoic appearance of fibrocartilage in ultrasound images. The menisci of the knee are composed of fibrocartilage and demonstrate a homogeneous hyperechoic ap-

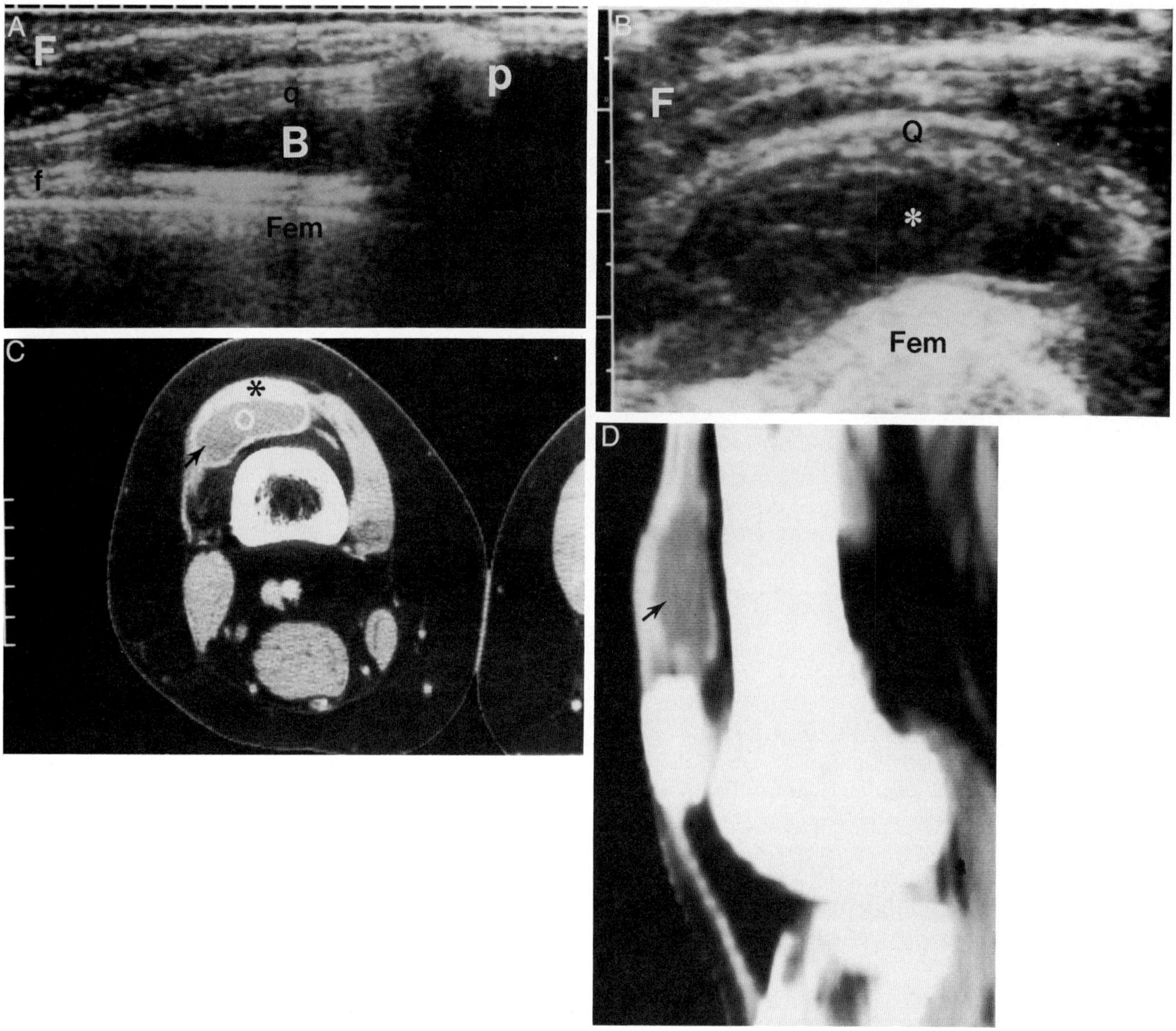


**FIG 7–13.**

**A,** acute hemorrhagic bursitis: longitudinal sonogram. The patient is a 19-year-old female basketball player who received a blow to the knee during practice 3 days ago. Aspiration of the joint performed in the orthopedic surgeon's office yielded no fluid. The transducer is positioned at the superior aspect of the patella along the quadriceps tendon. A distended suprapatellar bursa *(B)* filled with hypoechoic fluid is noted deep to the quadriceps tendon *(q)*. This homogeneous hypoechoic fluid is consistent with recent hemorrhage. The fat *(f)* superior and deep to the bursa is hyperechoic, whereas the subcutaneous fat *(F)* is less echogenic. Hemorrhage in the knee is usually due to intra-articular fracture or to meniscal or cruciate tears. Hemorrhage localized to the bursa secondary to direct trauma, as in this case, is quite unusual. Abbreviations: femoral diaphysis *(Fem)*, patella *(p)*. **B,** acute hemorrhagic bursitis: transverse sonogram. Same patient as in **A.** The transducer was rotated 90 degrees to yield a transverse image. The blood-filled suprapatellar bursa *(asterisk)* is again seen deep to the quadriceps tendon *(Q)*. Abbreviations: subcutaneous fat *(F)*, femur *(Fem)*. **C** and **D,** acute hemorrhagic bursitis: transverse and parasagittal computed tomograms (CT). Same patient as in **A** and **B.** The distended fluid-filled bursa *(arrows)* is demonstrated deep to the quadriceps tendon *(asterisk)*. Density measurement of the fluid *(circle)* was 37 Hounsfield units. Note on the parasagittal image that the posterior joint capsule *(small arrow)* is not displaced.

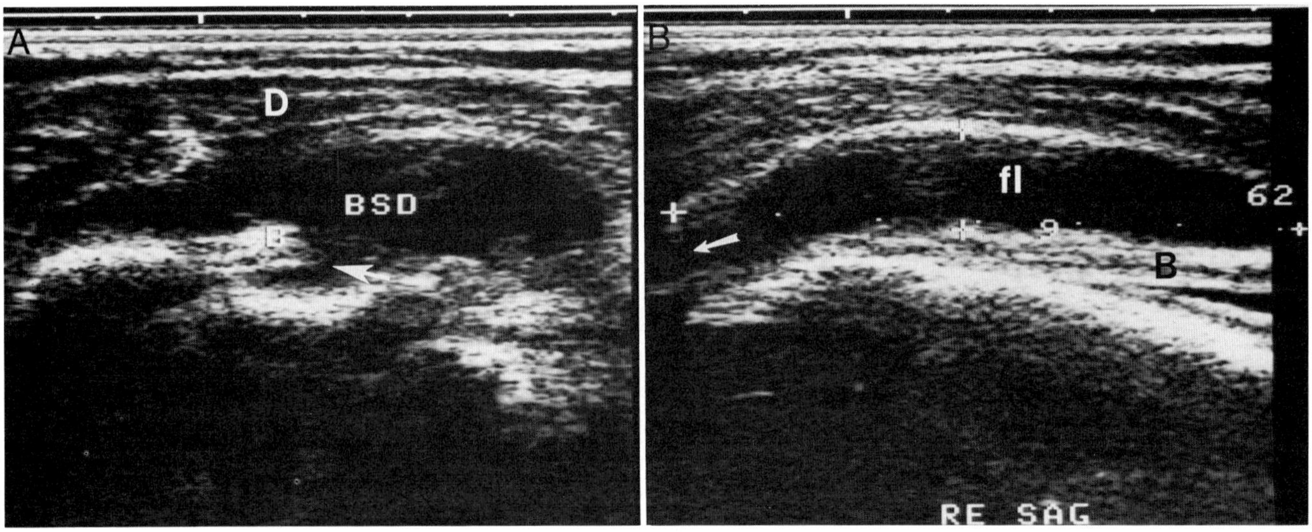


**FIG 7–14.**
**A,** shoulder hemarthrosis: transverse sonogram. A 28-year-old male javelin thrower developed acute pain and swelling of the right shoulder during competition. He presented 5 days later with persistent swelling. The transducer is positioned transversely over the bicipital groove with the shoulder in neutral position. A large fluid-filled subacromial-subdeltoid bursa *(BSD)* is demonstrated between the deltoid *(D)* and the long biceps tendon *(B).* A minimal amount of fluid is seen around the biceps tendon *(arrow)* within the tendon sheath. Further imaging demonstrated a complete rotator cuff tear. Aspiration of the joint yielded blood. Hemorrhage in the glenohumeral joint and subdeltoid bursa is a rare complication of acute rotator cuff tear. **B,** shoulder hemarthrosis: longitudinal sonogram. Same patient as in **A.** The transducer is aligned along the length of the long biceps tendon *(B).* The previously noted fluid collection *(fl)* within the subacromial-subdeltoid bursa is noted to be in direct communication with the joint space *(arrow).*

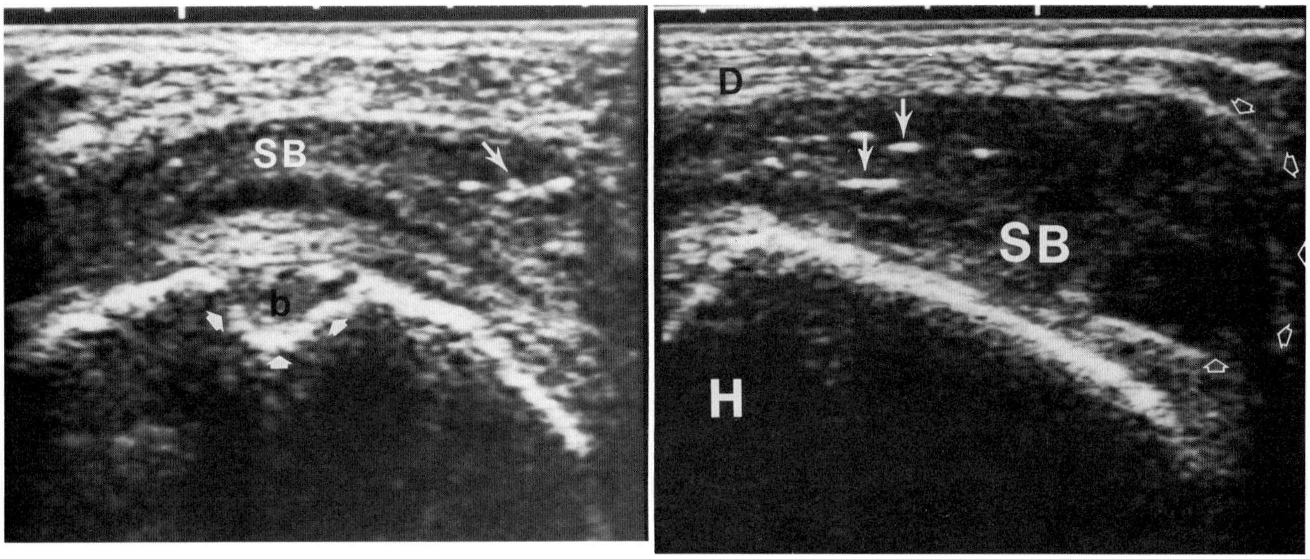


**FIG 7–15.**
**A,** septic arthritis of the shoulder: transverse sonogram. This 65-year-old man initially presented with a history of right shoulder pain of 1 week's duration. He was treated by his family physician with intra-articular steroid injection. This relieved his symptoms temporarily, but he returned with increased pain and swelling 1 week later. He was afebrile with minimal leukocytosis and an elevated erythrocyte sedimentation rate. Transducer position is identical to Figure 7–14,A. The subacromial-subdeltoid bursa *(SB)* is filled with fluid that is only slightly less echogenic than the overlying deltoid muscle. Highly echogenic foci *(large arrow)* within the fluid collection are consistent with gas bubbles. A normal biceps tendon *(b)* is seen in the bicipital groove *(small arrows).* Blind aspiration of the joint was negative. Surgical drainage of the joint was performed based on the sonographic findings, and *Staphylococcus epidermidis* was cultured. **B,** septic arthritis of the shoulder: coronal sonogram. Same patient as in **A.** A bulging subacromial-subdeltoid bursa *(SB)* is noted deep to the deltoid muscle *(D).* Note that the inferior extent of the bursa *(open arrows)* is larger than the superior aspect. A common error is to not identify the convex inferior margin of the distended bursa. Again noted are the echogenic gas bubbles *(large arrows).* Abbreviations: proximal humerus *(H).*

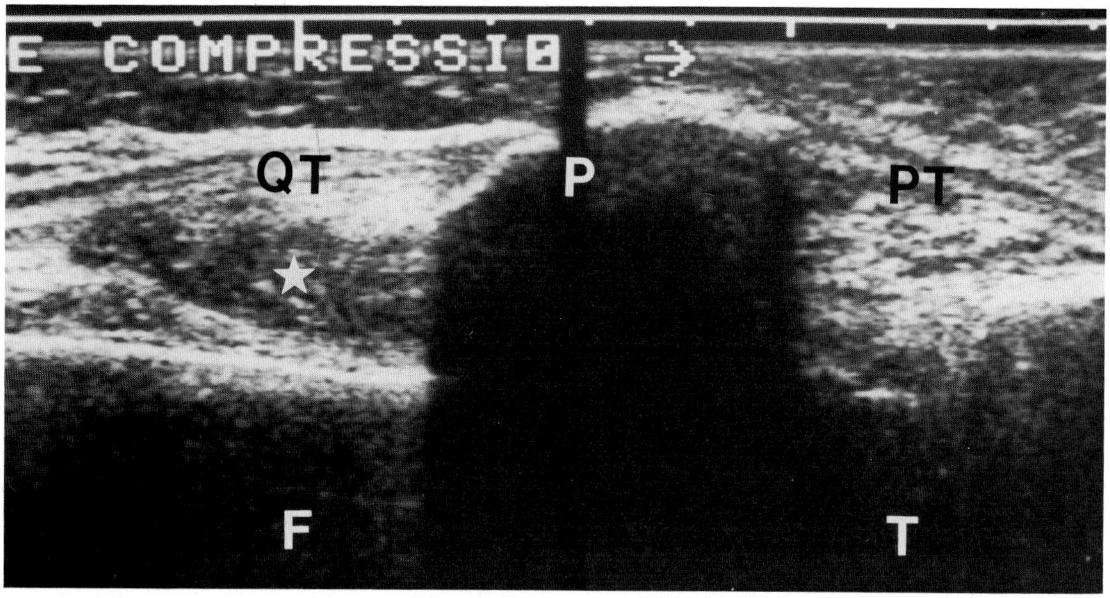


**FIG 7–16.**
Rheumatoid arthritis of the knee: longitudinal sonogram with compression. A 58-year-old woman with a 16-year history of rheumatoid arthritis presented with new onset of pain in the right knee. The image is a sagittal view of the knee centered on the patella. Firm compression was applied with the transducer to express all of the fluid from the suprapatellar bursa. This allows reproducible measurement of the total synovial thickness (TST), the sum of the anterior and posterior walls of the suprapatellar bursa. In this case, the TST measures 1.7 cm, approximately eight to nine times greater than normal. The bursa is seen as a hypoechoic mass *(star)* deep to the quadriceps tendon. Pannus remains hypoechoic relative to the surrounding soft tissues despite firm compression to express joint fluid. Abbreviations: femur *(F)*, quadriceps tendon *(QT)*, patella *(P)*, patellar tendon *(PT)*, tibia *(T)*.

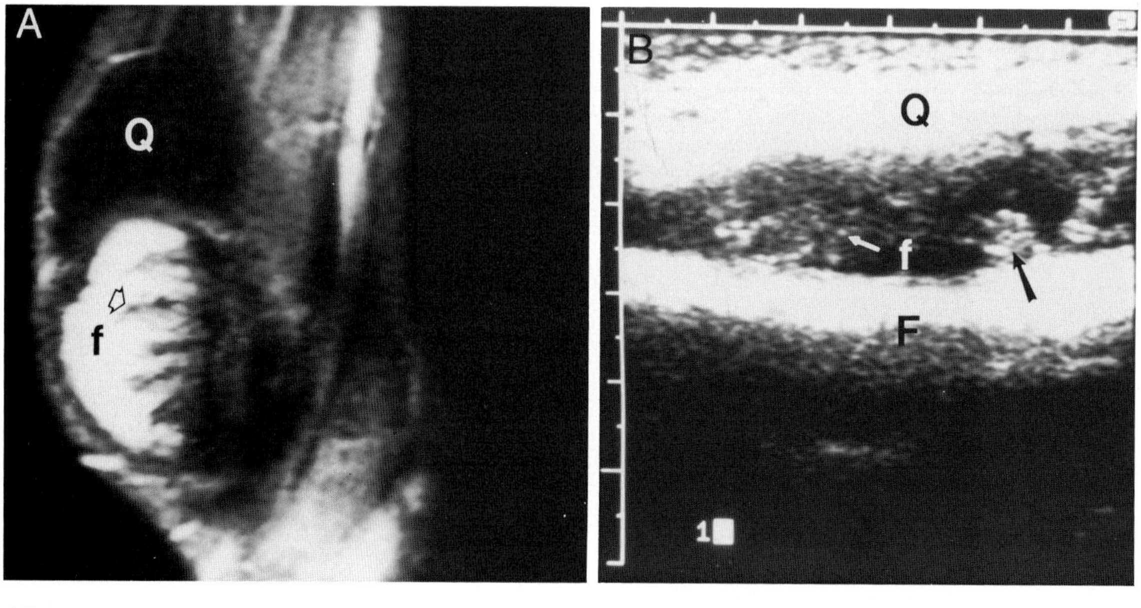


**FIG 7–17.**
**A,** chronic monoarticular knee swelling: parasagittal T2 MRI. This teenage boy presented with monoarticular swelling of the knee. A large joint effusion *(f)* and marked synovial proliferation *(open arrow)* is demonstrated on this parasagittal T2-weighted spin-echo image of the right knee through the lateral joint recess. Villous synovial proliferation is seen in the inflammatory arthritides, chronic infection, and synovial tumors. Biopsy of this lesion revealed pigmented villonodular synovitis *(PVNS)*. Abbreviations: quadriceps muscle *(Q)*. **B,** chronic inflammatory knee swelling: longitudinal sonogram. The patient is a 45-year-old woman with symmetrical polyarthritis due to rheumatoid disease. The transducer is positioned along the length of the quadriceps tendon *(Q)* just above the patella, out of view to the observer's right. Irregular synovial proliferation *(black arrow)*, fluid *(f)*, and floating debris *(white arrow)* are seen within the suprapatellar bursa. The cases shown here and in **A** demonstrate that pathology of the synovium is always accompanied by excess fluid production. Abbreviations: femur *(F)*.

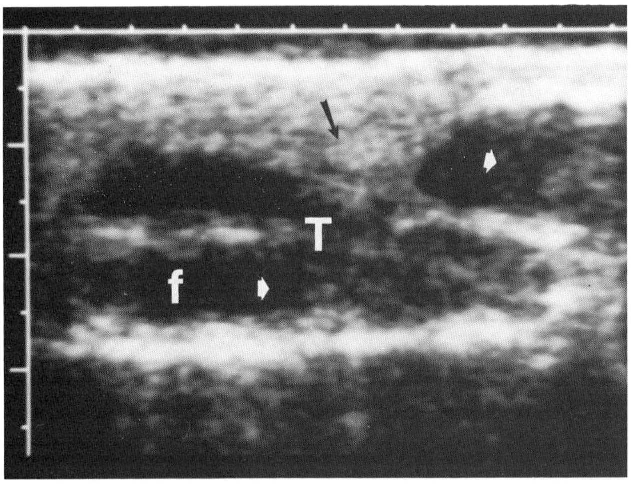


**FIG 7–18.**
Inflammatory distention of the biceps tendon sheath: longitudinal sonogram. A 62-year-old woman with rheumatoid arthritis for 12 years presented with pain, swelling, and fever after intra-articular steroid injection in the right shoulder. The transducer is oriented along the long biceps tendon *(T)*. Fluid within the synovial sheath *(f)* of the tendon displaces the tendon anteriorly. Compare this appearance with that of fluid within the subacromial-subdeltoid bursa shown in Figure 7–14,B. Debris *(small arrows)* is noted within the fluid-filled sheath. Adhesions *(black arrow)* between the visceral and parietal synovia are also demonstrated. Aspiration of the fluid yielded frank pus and *Staphylococcus aureus* was cultured.

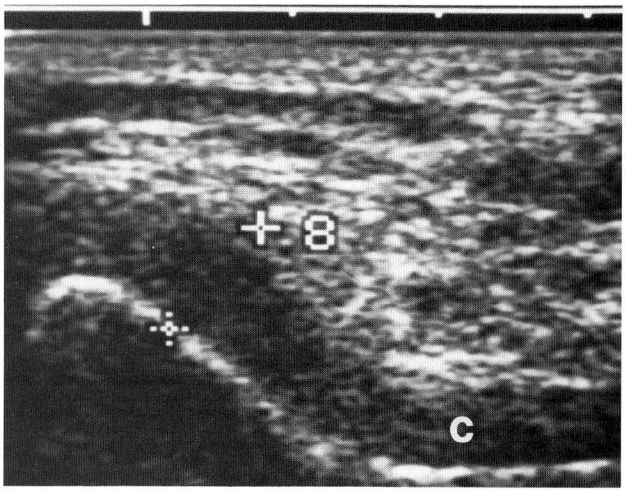


**FIG 7–19.**
Cartilage edema of the knee: transverse sonogram. This 66-year-old woman presented with a 6-year history of rheumatoid arthritis and 1 year of involvement of the right knee. The knee was flexed 90 degrees and the transducer placed transversely just above the patella. The hypoechoic articular cartilage *(c)* covering the femoral condyles measures approximately 8 mm thick, markedly thickened. Margins of the cartilage are indistinct, consistent with roughening of the articular surface.

pearance. This is also true of the triangular fibrocartilage of the wrist and the menisci of the temporomandibular joints.

**Cartilage Pathology.**—Edema of cartilage is the earliest sign of pathological change. The normal cartilage margins become ill defined, and the cartilage substance appears inhomogeneous (Aisen et al., 1984; Richardson et al., 1988). Thickness of the edematous cartilage increases slightly. Comparison with the asymptomatic side is usually very helpful. Later, the surface of the involved cartilage will appear rough, and there will be measurable loss in thickness. This progression of disease can be seen in both inflammatory arthritis and osteoarthritis. The rate of progression and response to therapeutic agents can be accurately monitored through serial sonographic examinations.

Chondral, osteochondral defects, and osteochondritis dissecans can also be diagnosed sonographically (Selby et al., 1986) (Fig 7–20). Most commonly these are the sequelae of trauma but may occur secondary to infarcts and epiphyseal osteonecrosis in older patients. Intra-articular fragments of cartilage are almost always present in these joints. These loose bodies may calcify and become visible on conventional radiographs. However, calcified and noncalcified loose bodies can be detected with ultrasound when there is increased intra-articular fluid. Acoustic shadowing is frequently seen with loose intra-articular cartilage fragments but may be absent if little degenerative change has occurred in the cartilage. Loose cartilage fragments that do not shadow are most commonly found following trauma in young, otherwise healthy patients. In the knee, loose bodies are preferentially entrapped in Baker's cysts. Therefore, examination of the popliteal fossa is mandatory in the evaluation of knees in which cartilage loss is observed (Figs 7–21 and 7–22).

Sonographic examination of the menisci of the knee has proved to be very valuable (Richardson et al., 1988; van Holsbeeck et al., 1988a). Posterior and peripheral meniscal tears are very well demonstrated using ultrasound. These are areas that are poorly evaluated with arthroscopy. Ultrasound examination is clearly indicated in patients with tenderness over the joint space, especially in light of a normal arthroscopic examination. Small tears along the internal meniscal margin are easily missed sonographically. Therefore, MRI remains the gold standard in the evaluation of meniscal tears. Cost, availability, and long examination time of MRI make ultrasound an

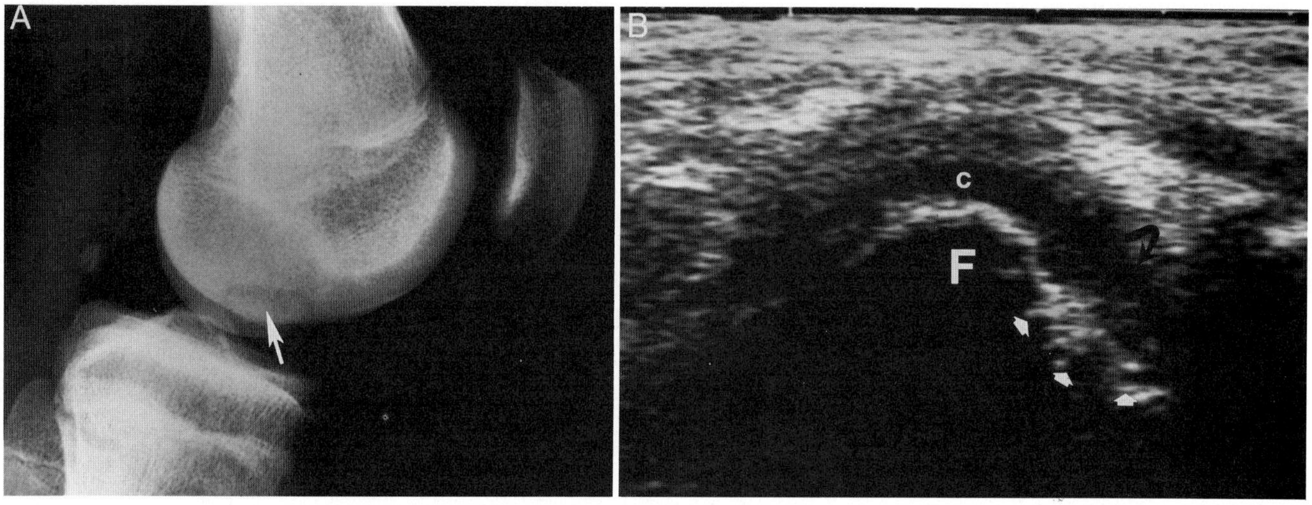


**FIG 7–20.**
**A,** osteochondritis dissecans medial femoral condyle: lateral knee radiograph. An ovoid osseous fragment *(arrow)* is seen within a defect of the medial femoral condyle in this 13-year-old soccer player. **B,** osteochondritis dissecans medial femoral condyle: longitudinal sonogram. Same patient as in **A.** The transducer is placed over the posteromedial aspect of the knee in full extension. Proximal lies to the observer's left. Cartilage *(c* and *curved arrow)* overlying the osseous fragment *(small white arrows)* is intact. Abbreviations: medial femoral condyle *(F).*

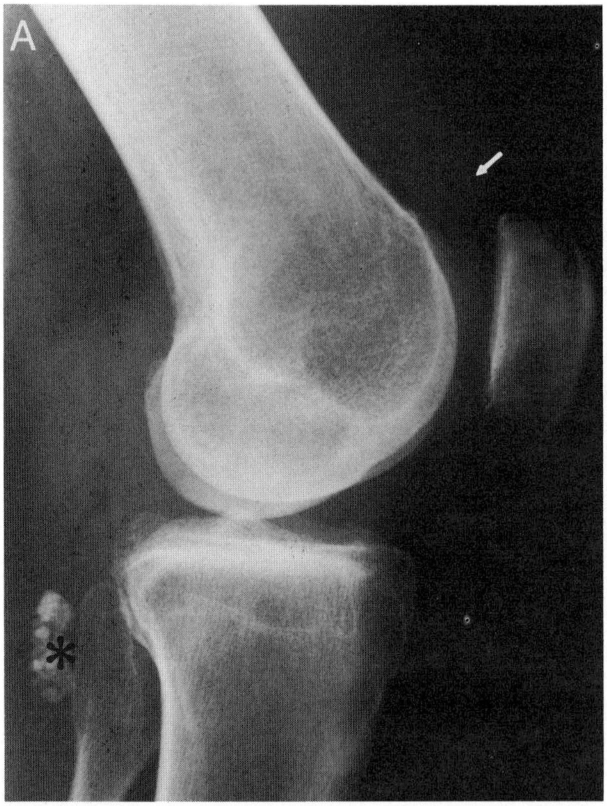


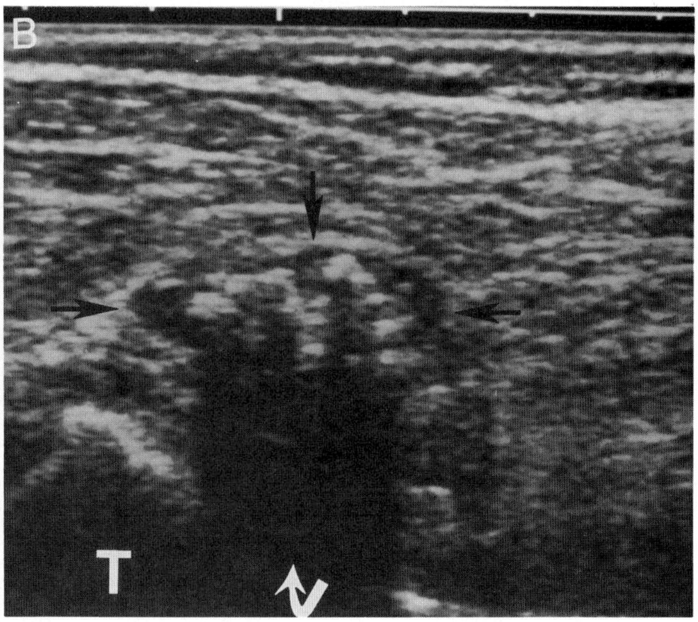


**FIG 7–21.**
**A,** detection of loose bodies: lateral knee radiograph. This 55-year-old man presented with right knee pain. The radiograph demonstrates degenerative changes of the patellofemoral joint and calcifications *(asterisk)* projected posterior to the fibular head. Enlarged suprapatellar bursa *(arrow).* **B,** sonographic localization of loose bodies: longitudinal sonogram. Same patient as in **A.** This longitudinal sonogram was obtained of the medial aspect of the popliteal fossa showing multiple calcification with shadowing *(curved arrow)* within a small Baker cyst *(black arrows).* These were demonstrated as separate calcifications on real-time examination. Abbreviations: tibia *(T).*

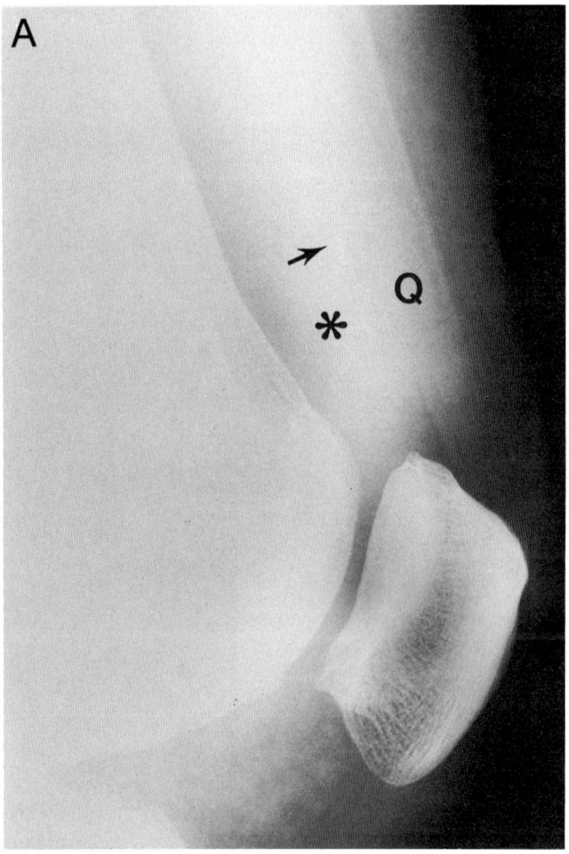


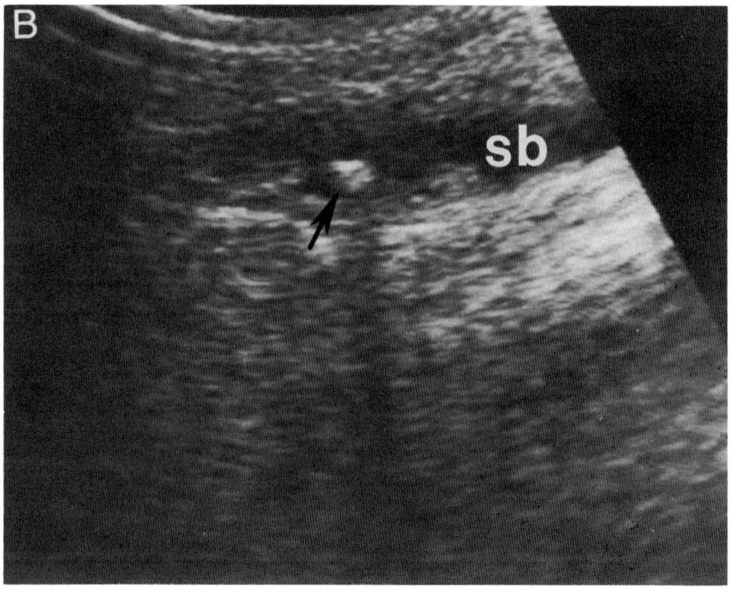


**FIG 7–22.**
**A,** detection of loose bodies: lateral knee radiograph. The patient is a 49-year-old woman with a known lesion of the medial meniscus who experienced recurrent swelling and locking of the left knee. Swelling of the suprapatellar bursa *(asterisk)* is noted dorsal to the quadriceps tendon *(Q)*. Tiny calcific opacities *(black arrow)* are noted in the region of the suprapatellar bursa. **B,** sonographic localization of loose bodies: longitudinal sonogram. Same patient as in **A.** The transducer is positioned longitudinally over the suprapatellar bursa *(sb)*. Fluid fills the bursa, and a loose body with acoustic shadowing *(arrow)* is identified in the superior aspect of the bursa. In this case, a 5-MHz radial array transducer was utilized, resulting in an image that is less desirable than that produced with a linear array.

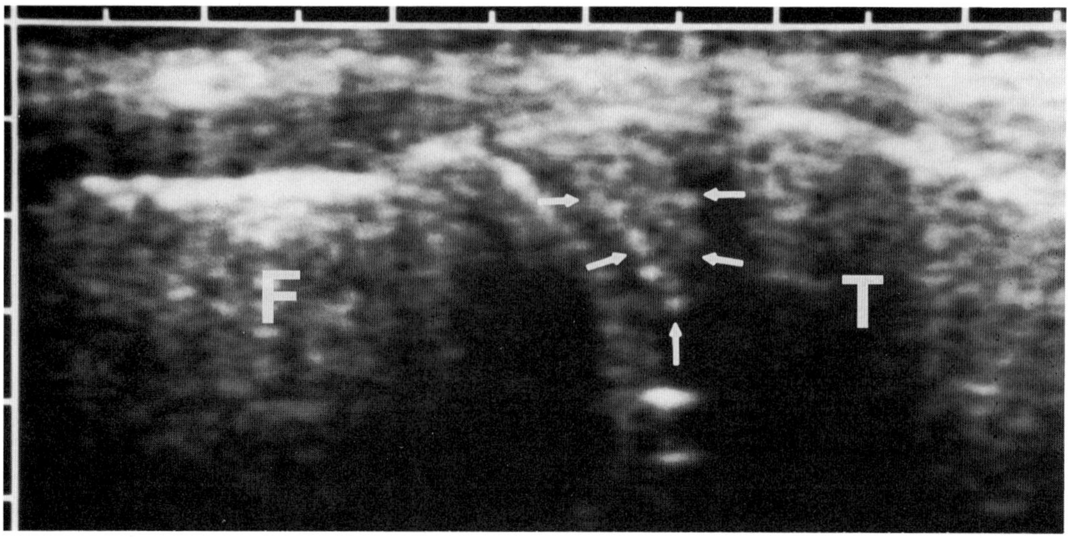


**FIG 7–23.**
Normal medial meniscus: coronal sonogram. This image of the medial meniscus was obtained in an asymptomatic 35-year-old man. It was obtained with the transducer in the coronal plane at the midpoint of the medial aspect of the knee. Slight valgus stress was applied with a cushion under the lateral aspect of the knee. The normal meniscus is seen as a hyperechoic triangle *(arrows)* between the femoral condyle *(F)* and the tibial plateau *(T)*.

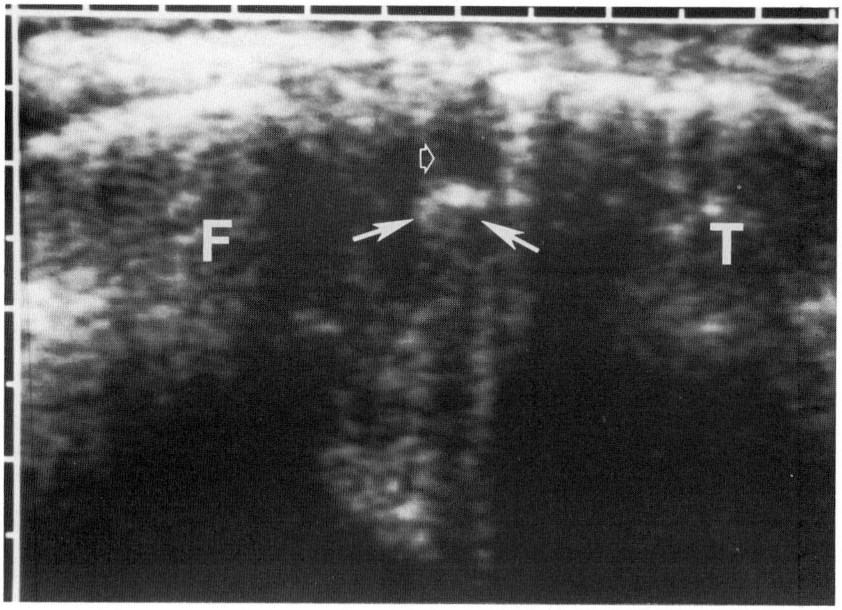


**FIG 7–24.**
Bucket handle tear of the medial meniscus: coronal sonogram. This patient experienced severe pain over the medial aspect of the knee and locking after a soccer game. Transducer position is identical to that in Figure 7–23. A triangular fragment *(large arrows)* of the meniscus is separated from its peripheral attachment by fluid *(open arrow)*. Arthroscopy confirmed a bucket handle tear of the medial meniscus.

essential first step in evaluation for meniscal tears (see Fig 15–5).

Sonographic examination of the menisci of the knee is performed with the patient in a lateral decubitus position with a cushion beneath the knee. The transducer is placed over the joint space in the long axis of the leg. Mild valgus stress is applied when the medial meniscus is examined. Likewise, varus stress is applied for examination of the lateral meniscus. Coronal images must be followed by transverse imaging. Once the meniscus is localized on coronal images, the transducer is rotated 90 degrees, parallel with the joint space. Examination of the menisci is most easily performed using an intra-operative fingertip linear array transducer. The meniscus is identified as a homogeneous hyperechoic triangle situated between the hypoechoic layers of articular hyaline cartilage (Fig 7–23). Small amounts of synovial fluid may be seen in the recesses of the joint around the meniscus. Once localized in longitudinal images, axial images of the meniscus are obtained. Meniscal tears are identified as hypoechoic clefts or gaps within the hyperechoic fibrocartilage of the meniscus (Fig 7–24 through 7–33). Richardson et al. (1988) have reported meniscal tears to be hyperchoic lesions. Their work was performed using cadaver specimens and the hyperechoic appearance of meniscal tears was likely the result of gas within the le-

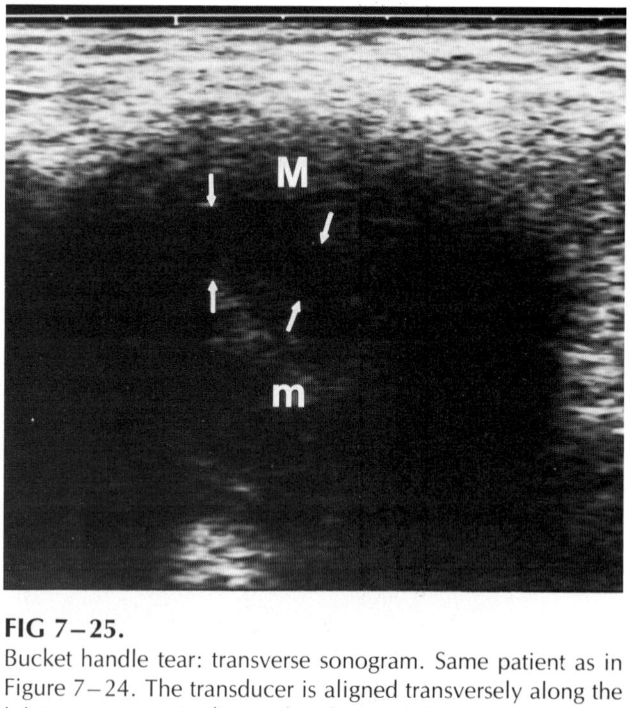


**FIG 7–25.**
Bucket handle tear: transverse sonogram. Same patient as in Figure 7–24. The transducer is aligned transversely along the joint space; anterior lies to the observer's left. A cleft *(arrows)* is seen within the meniscus *(M)* indicative of a bucket handle tear. The torn central portion of the meniscus *(m)* remains attached to the peripheral portion at the anterior and posterior extents of the meniscus.

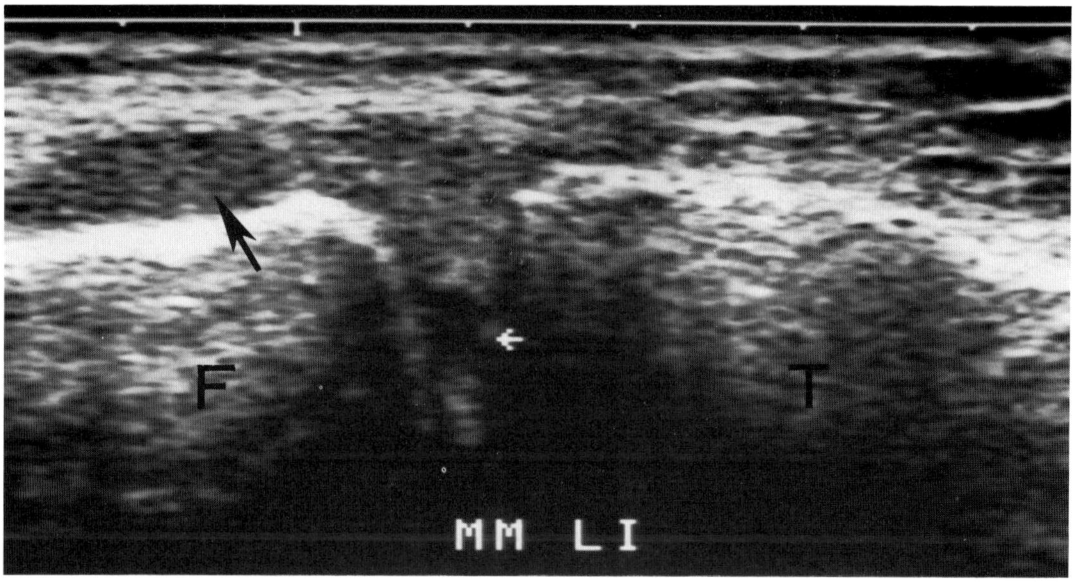


**FIG 7–26.**
Longitudinal tear of the medial meniscus: coronal sonogram. Five months after knee injury, this 32-year-old man presented with persistent pain and swelling over the medial joint space. Transducer position is identical to Figure 7–23. A discontinuity *(small arrow)* in the medial meniscus is filled with fluid. Swelling is noted within the proximal portion of the medial collateral ligament *(black arrow)*. Abbreviations: medial femoral condyle *(F)*, tibia *(T)*.

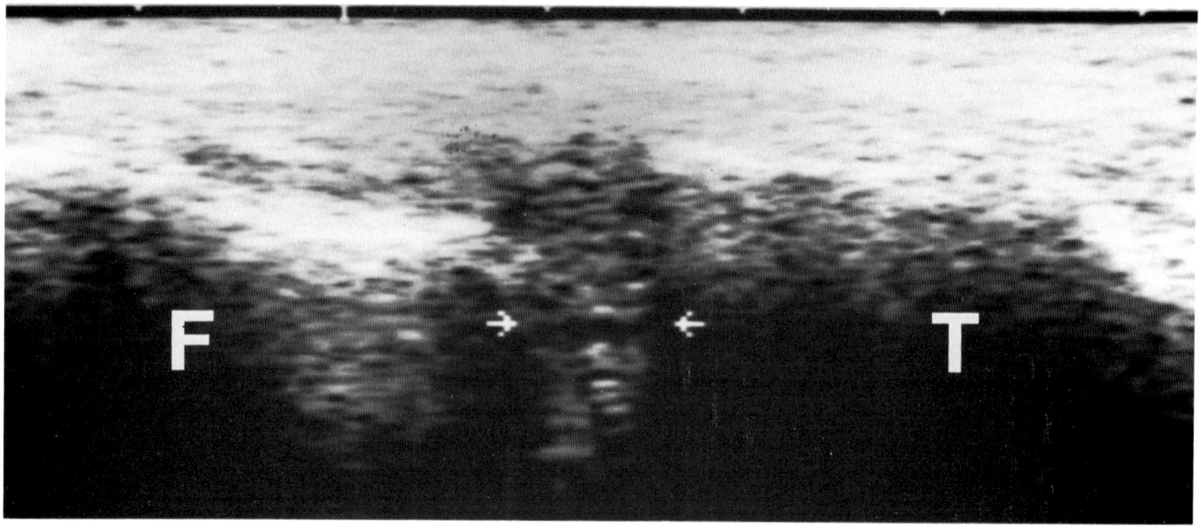


**FIG 7–27.**
Longitudinal meniscal tear: coronal sonogram. A cleft *(arrows)* is noted within the medial meniscus of this 23-year-old skier. The meniscus is diffusely decreased in echogenicity, and several highly echogenic foci are identified within the substance of the meniscus. These changes in reflectivity of the meniscus are frequently seen following intra-articular steroid injection. Abbreviations: medial femoral condyle *(F)*, tibia *(T)*.

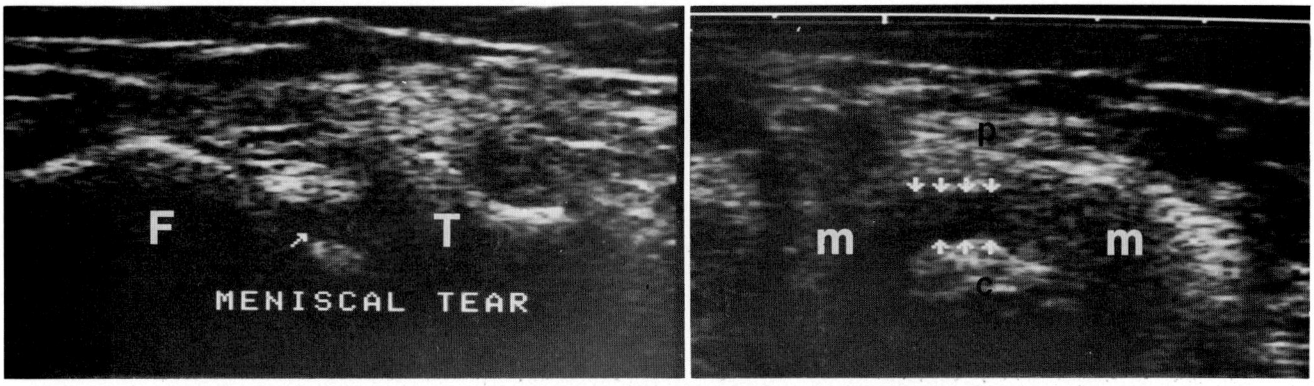


**FIG 7–28.**
**A,** longitudinal meniscal tear: coronal sonogram. The patient is a 20-year-old football player with continued pain and swelling after being hit on the knee. A longitudinal meniscal tear is seen as a hypoechoic cleft *(arrow)* in the peripheral portion of the anterior medial meniscus. Abbreviations: femur *(F)*, tibia *(T)*. **B,** longitudinal meniscal tear: transverse sonogram. Same patient as in **A.** Transducer position is identical to that in Figure 7–25. The length of the meniscal tear *(arrows)* is demonstrated, dividing the meniscus *(m)* into peripheral *(p)* and central *(c)* fragments.

sion. We have not observed hyperechoic meniscal tears in vivo. Joint fluid filling the gap in the torn menicus results in a hypoechoic appearance.

Tears of the menisci are named according to their gross pathological appearance. A tear paralleling the internal border of the meniscus is called a *longitudinal tear.* If there is separation of the central portion of the meniscus along the margins of the tear with the anterior and posterior aspects remaining attached to the peripheral meniscus, it is referred to as a *bucket handle tear. Radial tears* are oriented perpendicular to the peripheral margin of the meniscus. *Horizontal tears* extend from the superior or inferior surfaces of the meniscus to the periphery. Meniscal cysts are most frequently found in association with horizontal tears.

Meniscal cysts are multiloculated collections of mucinous material associated with the meniscal margin. The origin of these cysts is unknown but likely related to trauma. Most meniscal cysts are as-

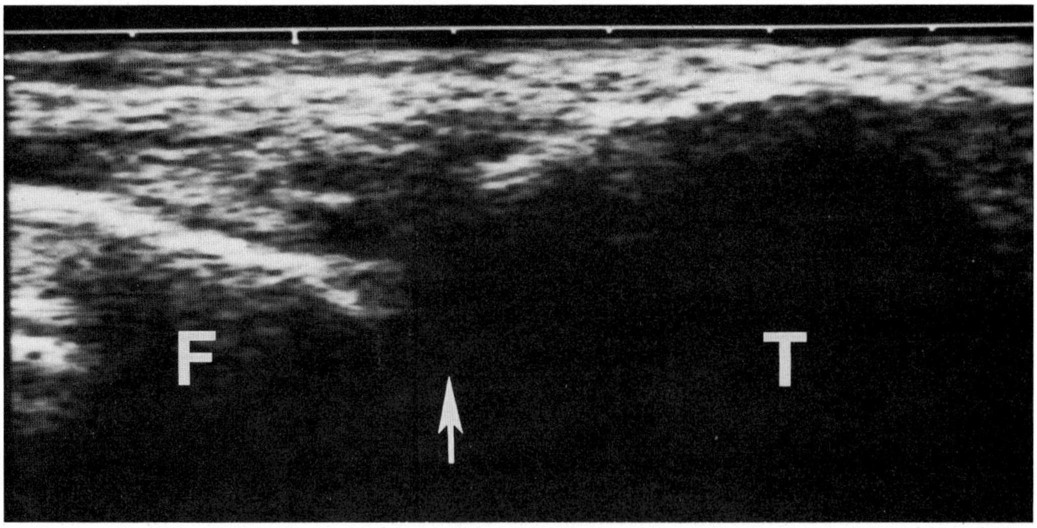


**FIG 7–29.**
Radial meniscal tear: coronal sonogram. This 52-year-old mechanic experienced severe pain on the medial side of the knee after standing up from a kneeling position. The image is obtained through the midportion of the medial meniscus. The normal triangular shape of the medial meniscus is not identified at the periphery of the joint space *(arrow)*. When the medial meniscus is not identified on coronal images, lack of sufficient valgus stress must be considered as a possible cause. Abnormal echogenicity of the medial collateral ligament is also noted in this case, providing further evidence of injury.

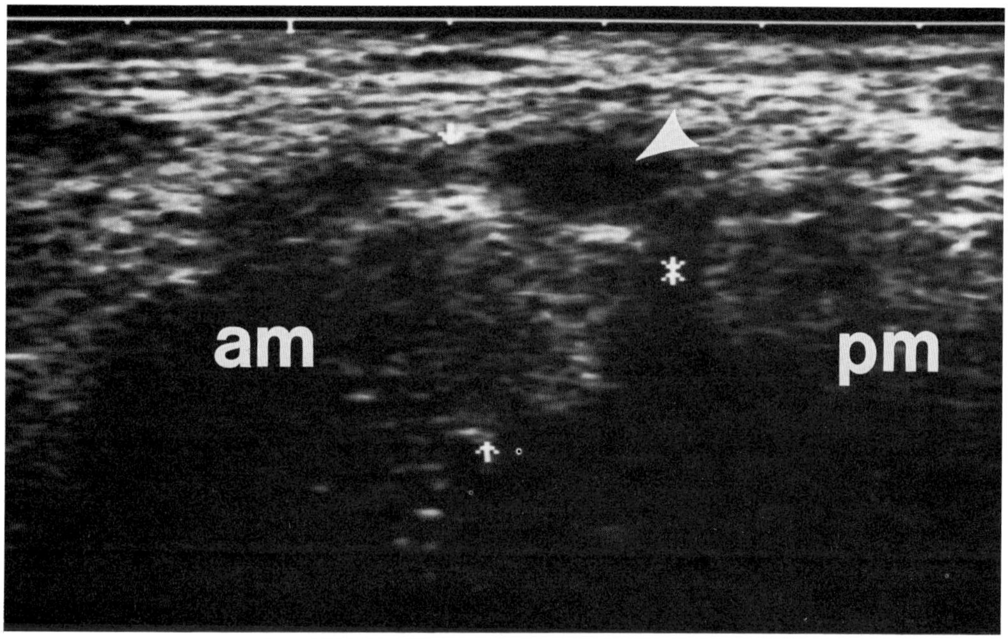


**FIG 7–30.**
Radial meniscal tear: transverse sonogram. Same patient as in Figure 7–29. The radial tear *(asterisk)* is confirmed in this plane of imaging. The tear separates the meniscus into anterior *(am)* and posterior *(pm)* fragments. *Arrows* indicate the normal dimension of the anterior meniscus. Fluid *(arrowhead)* is seen collecting between the medial collateral ligament and the peripheral margin of the meniscus.

sociated with the lateral meniscus and communicate with the joint space through a horizontal meniscal tear (Figs 7–33 and 7–34) (Resnick and Niwayama, 1988; Ferrer-Roca and Vilalta, 1980). In theory, joint motion pumps intra-articular fluid through these tears, resulting in cyst development (Figs 7–35 and 7–36). However, medial meniscal cysts are usually isolated findings, rarely associated with meniscal tears. In these cases, primary myxoid degeneration of the meniscus is thought to be responsible for cyst formation (Resnick and Niwayama, 1988). Meniscal cysts were found in 7% of excised menisci (Fig 7–37).

A decade ago, the vast majority of meniscal cysts were diagnosed at the time of surgery. Only the largest of these cysts were clinically palpable and erroneously diagnosed as soft tissue tumors, pannus, pes anserinus tendonitis, or bursitis. Today, ultrasound is the most useful modality for the detection of meniscal cysts. Preoperative sonographic diagnosis of meniscal cysts has demonstrated these lesions to be more common than previously suspected (Ferrer-Roca and Vilalta, 1980; Barrie, 1979).

Meniscal cysts are most frequently seen as hypoechoic cystic structures adjacent to the meniscus (see Fig 7–37). Occasionally, they will be anechoic. Reflectivity of the cyst contents is directly related to its viscosity. Loculations and meniscal fragments are

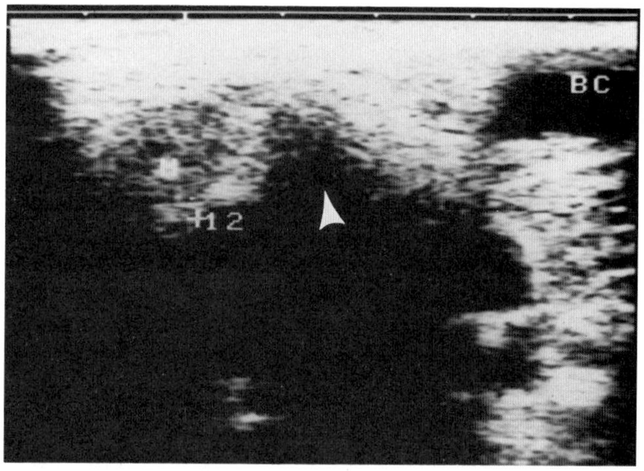


**FIG 7–31.**
Defect in the posterior medial meniscus: transverse sonogram. The patient is a 45-year-old woman who presented with pain and swelling over the medial aspect of the popliteal fossa after kicking her husband soccer style. The transducer is oriented transversely over the posteromedial aspect of the joint space. A defect *(arrowhead)* is seen in the posterior horn of the medial meniscus *(M)*. On the same image, a Baker cyst *(BC)* filled with anechoic fluid is noted. Baker's cysts are often associated with meniscal tears.

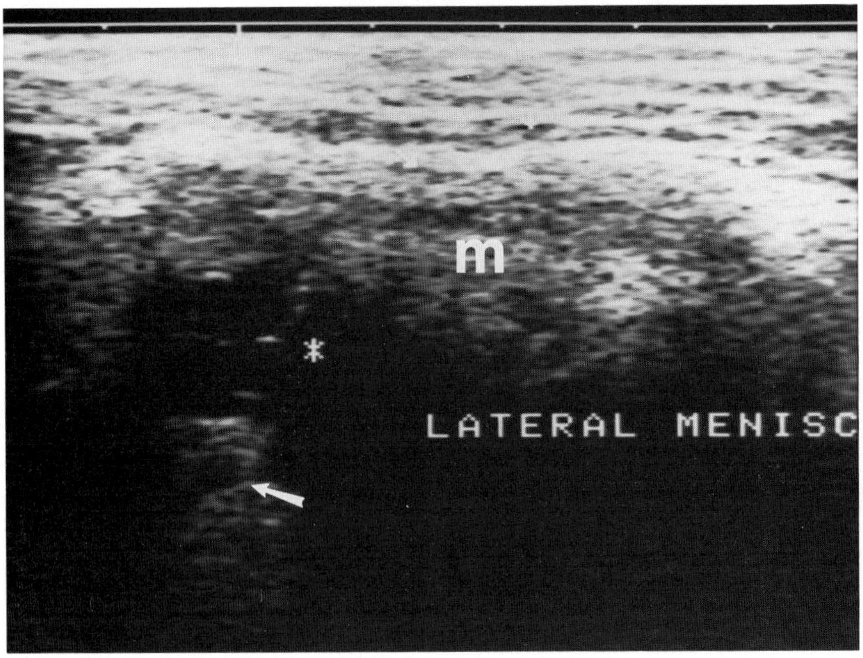


**FIG 7–32.**
Meniscal tag: transverse sonogram. A 55-year-old woman presented with persistent pain over the lateral patellofemoral joint. Shown is a transverse image of the anterior lateral meniscus. Anterior lies to the observer's left. A longitudinal gap *(asterisk)* is seen in the lateral meniscus *(m)*. There is medial displacement of a meniscal fragment or tag *(arrow)* into the joint. Arthroscopy showed no patellofemoral disease but a degenerative tear of the lateral meniscus with a free fragment.

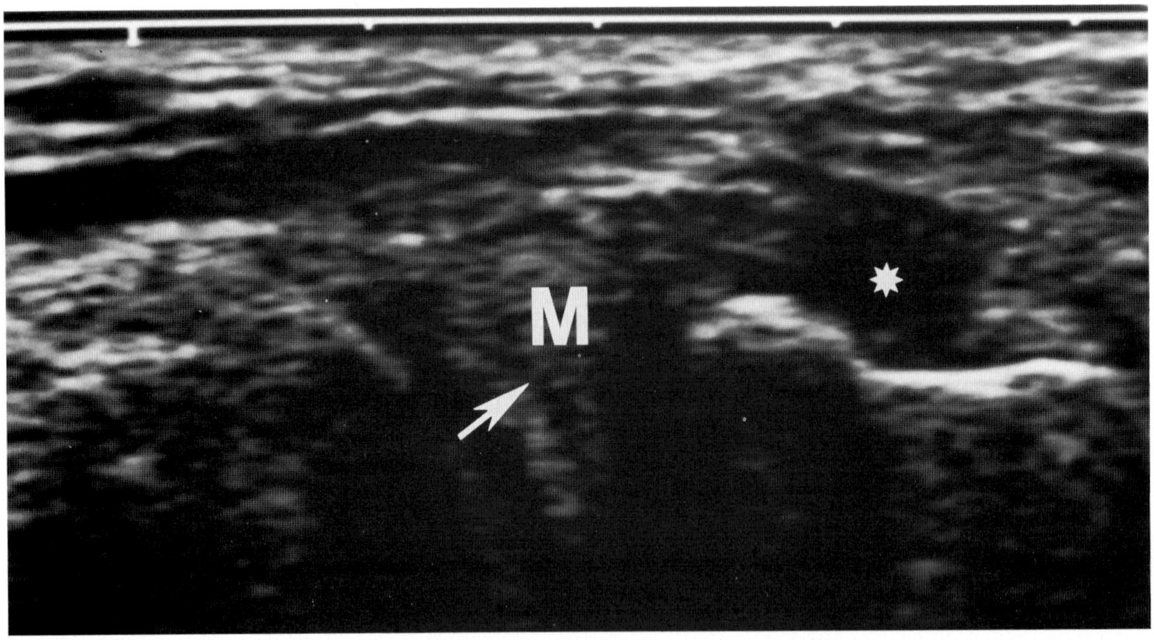


**FIG 7–33.**
Horizontal cleavage tear: coronal sonogram. After 1 year of lateral knee pain, this 42-year-old woman presented for sonographic examination. A narrow curvilinear cleft is seen extending from the inferior central surface of the lateral meniscus *(M)* to the superior peripheral surface. This was categorized as a horizontal tear during arthroscopy. A fluid collection *(*)* in the inferolateral joint recess is also noted.

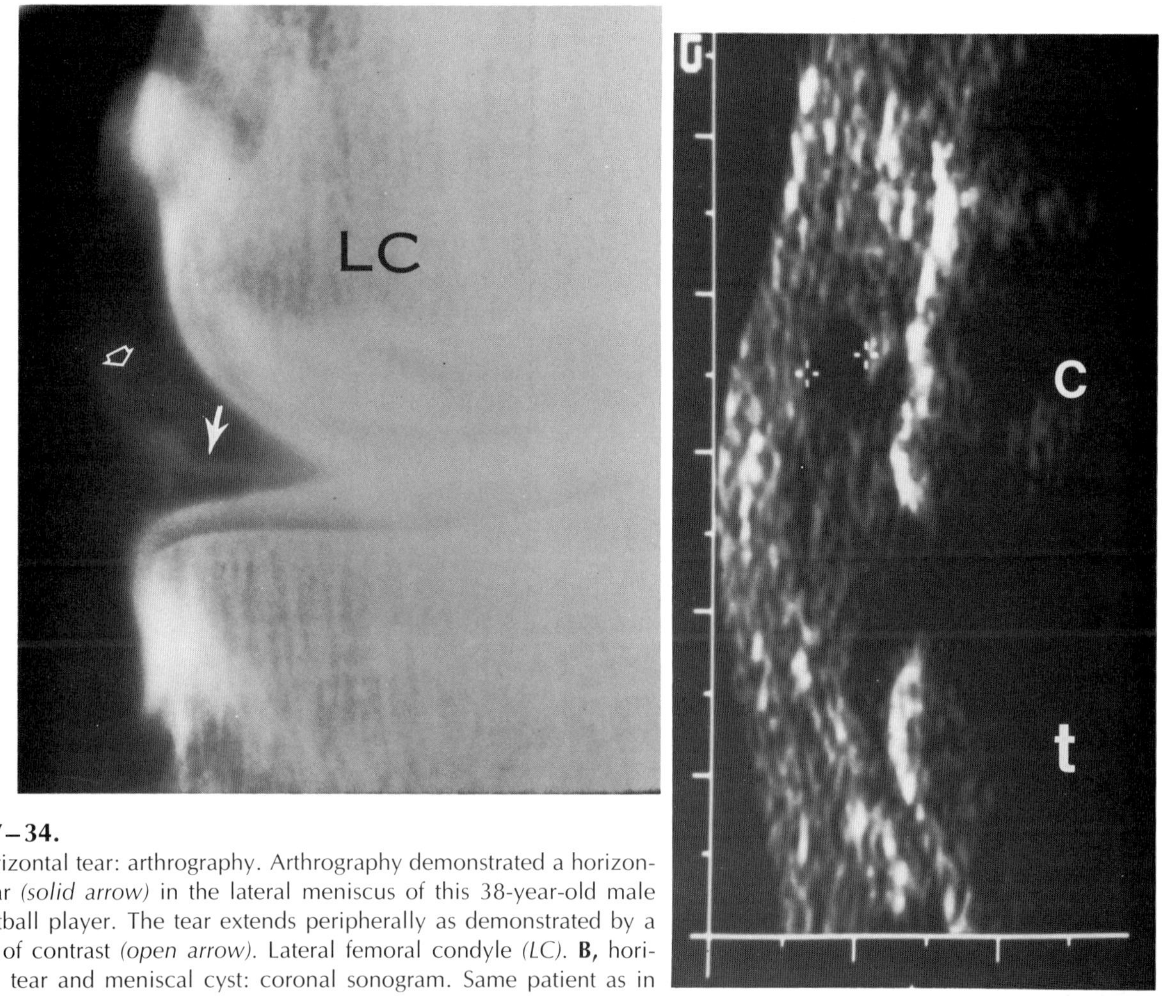


**FIG 7–34.**
**A,** horizontal tear: arthrography. Arthrography demonstrated a horizontal tear *(solid arrow)* in the lateral meniscus of this 38-year-old male basketball player. The tear extends peripherally as demonstrated by a blush of contrast *(open arrow)*. Lateral femoral condyle *(LC)*. **B,** horizontal tear and meniscal cyst: coronal sonogram. Same patient as in **A.** The transducer is positioned coronally over the lateral joint space. This image is oriented to correlate directly with the arthrogram. A small fluid collection (+) is identified adjacent to the femoral condyle. Meniscal cysts are often associated with horizontal lateral meniscal tears. Note that the meniscus is not visualized on this image due to the lack of sufficient valgus stress.

frequently present within these cysts (see Fig 7–35 and 7–36). Large meniscal cysts may erode bone under the margin of the tibial plateau, simulating an erosion of rheumatoid arthritis. Fluid within the cyst can be displaced from beneath the transducer when pressure is applied.

Degenerative changes of the menisci may be observed sonographically without associated meniscal cysts or tears. This is a frequent finding in the elderly. On sonographic images the meniscus appears swollen and decreased in echogenicity (Fig 7–38). The periphery of the meniscus bulges outward and is not compressible when pressure is applied with the transducer. Arthroscopy demonstrates swollen segments of the meniscus with myxoid degeneration.

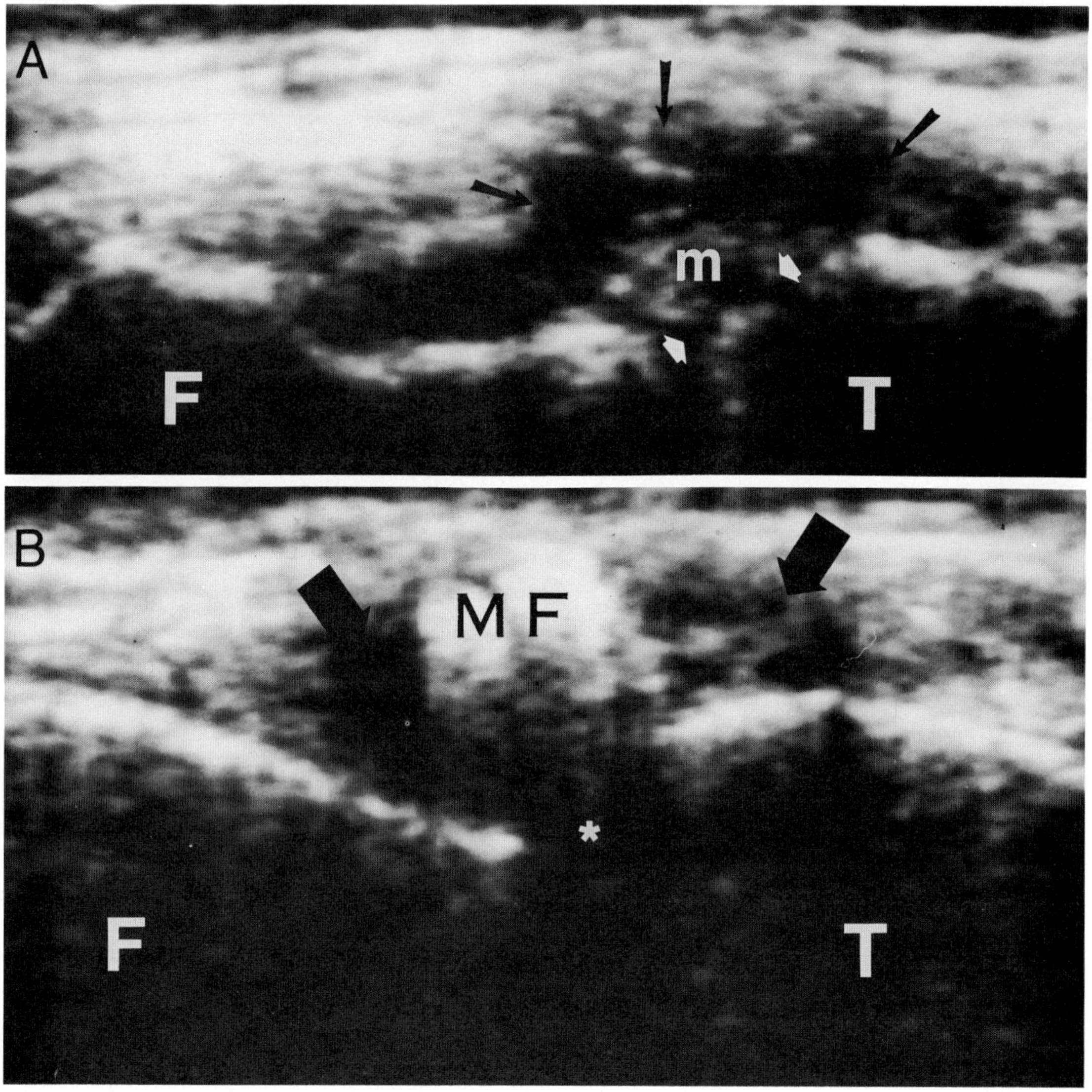


**FIG 7–35.**
**A,** lateral meniscal cyst: coronal sonogram. A 58-year-old woman presented with pain and swelling over the lateral side of the knee. Computed tomography performed at another institution demonstrated a mass in the soft tissues interpreted as a neoplasm. A cleft *(white arrows)* is noted in the lateral meniscus *(m)* contiguous with a fluid collection *(black arrows),* which surrounds the peripheral portion of the meniscus. Magnetic resonance imaging confirmed the presence of the meniscal cyst and tear. Adjacent bone marrow was normal, with no evidence of neoplasm. Abbreviations: femur *(F),* tibia *(T).* **B,** the transducer was moved more posteriorly. No meniscus is identified in the joint space *(asterisk).* Meniscal fragments *(MF)* are sequestered in the meniscal cyst *(arrows).* Abbreviations: femur *(F),* tibia *(T).*

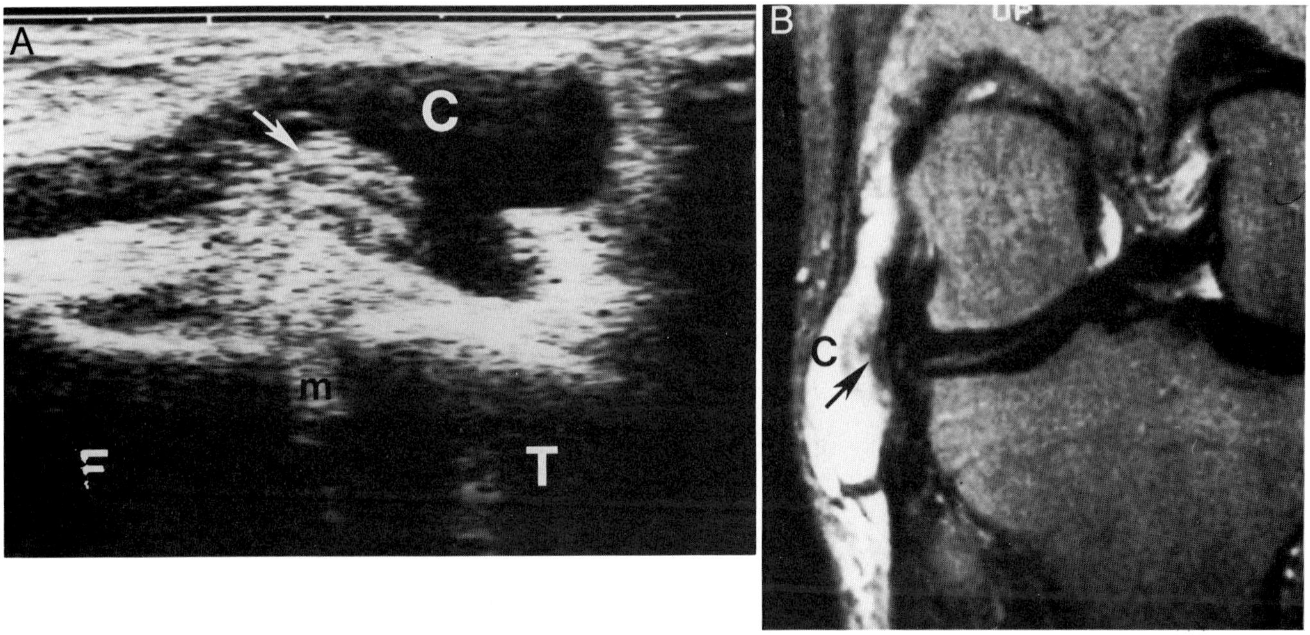


**FIG 7–36.**
Lateral meniscal cyst: coronal sonogram **(A)** and MRI **(B).** This 35-year-old man presented with pain and swelling over the posterolateral aspect of the knee. A cyst *(C)* surrounds the lateral meniscus *(m)*. The lateral portion of the meniscus bulges into the cyst *(arrow)*. These findings were confirmed with MRI and subsequently at surgery. Abbreviations: femur *(F)*, tibia *(T)*.

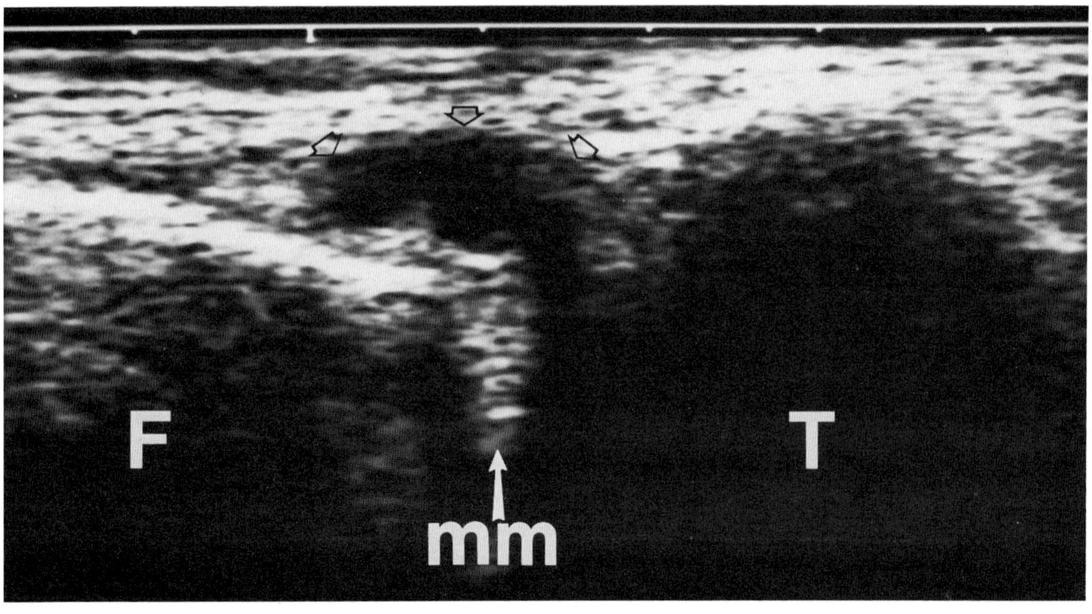


**FIG 7–37.**
Medial meniscal cyst: coronal sonogram. A small meniscal cyst *(open arrows)* associated with the medial meniscus *(mm)* is identified in this 55-year-old woman with chronic medial knee pain. This was an isolated finding; no tears were identified. Isolated meniscal cysts can be the sole cause of medial knee pain. Characteristically, meniscal cysts on the medial aspect of the knee are not associated with meniscal tears. However, lateral meniscal cysts are almost always associated with meniscal tears. Abbreviations: femur *(F)*, tibia *(T)*.

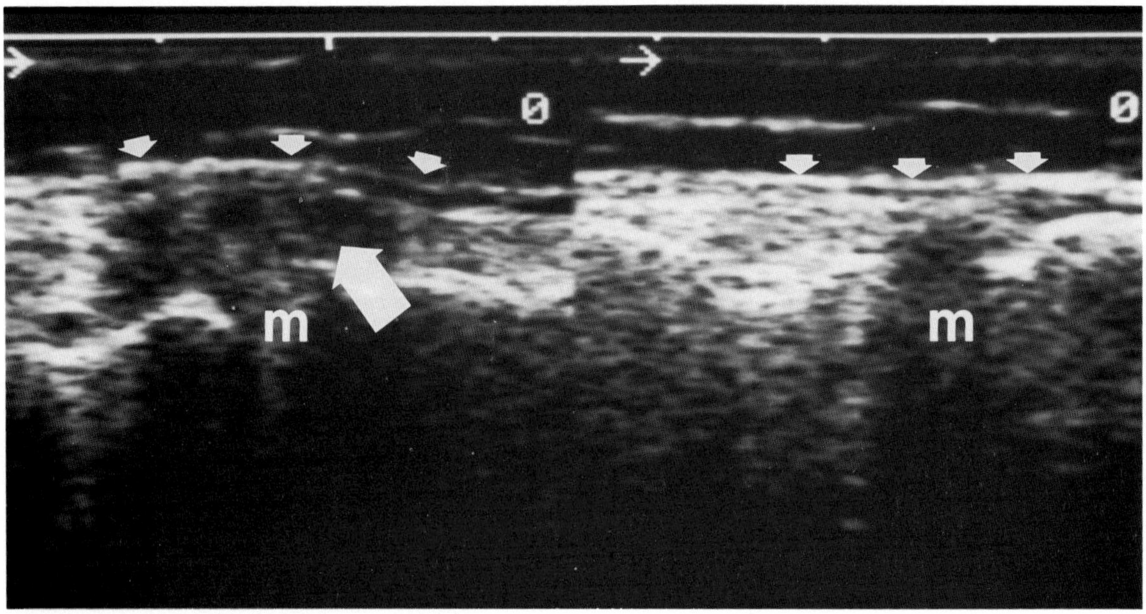


**FIG 7–38.**
Myxoid degeneration of the medial meniscus: coronal sonogram. This 50-year-old woman complained of pain over the medial aspect of the left knee. On the *left* is the left medial meniscus, and the *right* shows the right medial meniscus for comparison. The left medial meniscus *(m)* bulges medially from the joint space *(arrow)* and elevates the medial collateral ligament *(small white arrows)*. This small mass contains definite internal echos, unlike a meniscal cyst. Arthroscopy found the midsection of the meniscus to be enlarged and softened. Resection was performed, and histological analysis demonstrated myxoid degeneration.

### *Loose Bodies*

Intra-articular debris can be classified into four categories: (1) precipitation of fibrin, (2) degeneration of villi, (3) metaplasia of villi, and (4) desquamation of cartilage (Mori, 1979). Precipitation of fibrin is most frequently seen in cases of rheumatoid arthritis but may occur in any chronic inflammatory process involving the synovia (e.g., tuberculosis) (Figs 7–39 and 7–40) (Mori, 1979). In rheumatoid joints the surfaces of the villi are covered with precipitated fibrin. These fibrin collections become dislodged and migrate to the synovial recesses of the joint. Sonography easily identifies these as hyperreflective foci floating in the synovial recesses. This debris can be easily deformed by compression with the transducer (Worth et al., 1986).

Villi found in rheumatoid joints can undergo metaplasia, degeneration, and necrosis. Reduction in blood flow occurring with remission of an acute episode of inflammation will result in necrosis and shedding of the villi (Mori, 1979). Long thin villi are most susceptible to this and therefore more likely to result in loose body formation. The sonographic appearance of these loose bodies is identical to that of fibrin clots. The presence of intact synovial villi and clinical history will lead to the correct diagnosis (Figs 7–41 and 7–42).

Desquamation of articular cartilage can be a prominent feature of osteoarthritis but may also be seen in rheumatoid arthritis. Cartilage fragments can also be seen as loose bodies following intra-articular fracture or meniscal tear. Fragments of articular cartilage are often polygonal or thornlike (Fig 7–43). Like other loose bodies, they can be identified easily in the synovial recesses, but unlike fibrin clots and villi, they cannot be compressed. In addition, acoustic shadowing will be observed when calcification is present.

In rheumatoid knees treated with intra-articular steroids, punctate echogenic material can be identified. These echogenic foci correspond to calcifications that frequently result from intra-articular injection of corticosteroids (see Figs 7–39 and 7–40). Although easily identified on sonographic images, these calcifications are rarely visualized on conventional radiographs (Gilsanz and Bernstein, 1984).

## Periarticular Disease

We define periarticular disease as those processes involving the tissues surrounding the joint, specifically extra-articular and extrasynovial. Ultrasound is clearly the most elegant technique for evaluation of this type of pathology. The field of view of arthroscopy is limited to the synovial cavity. This is also true

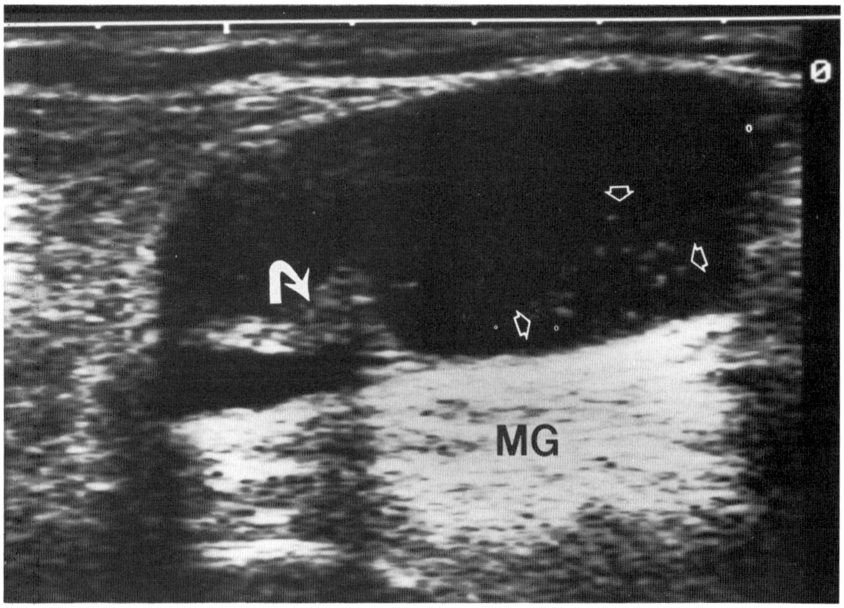


**FIG 7–39.**
Debris in rheumatoid Baker's cyst: longitudinal sonogram. This 50-year-old woman with long-standing rheumatoid arthritis was treated with intra-articular injection of steroids the day before examination. Hyperechoic strands of tissue *(curved arrow)* are common in rheumatoid Baker's cysts. Echogenic debris is noted dependently in the Baker cyst *(open arrows)*. This calcific debris is a frequent complication seen following the intra-articular injection of microcrystalline corticosteroid preparations. Abbreviations: medial gastrocnemius *(MG)*.

of arthrography. In the past, clinicians had to rely entirely on clinical examination for the diagnosis of periarticular pathology. Magnetic resonance imaging can be helpful, but it is a static examination. Ultrasound is able to demonstrate periarticular anatomy under functional conditions, thus elucidating certain types of pathology. In addition, cost, length of examination, and availability further limit the utility of MRI.

Tendon injuries, peritendonitis, bursitis, ligament tears, muscle tears, ruptures of the musculotendonous junction, and neurovascular lesions are all well evaluated using ultrasound. In later chapters all of these lesions will be discussed in detail. At this point we will confine our discussion to nerve and vascular lesions. Aneurysms, pseudoaneurysms, popliteal artery entrapment syndrome, adventitial cystic disease, venous thrombosis, and neuromas are abnormalities that may be encountered while large synovial joints are examined. Gray scale ultrasound with Doppler flow analysis is ideal for evaluation of vascular pathology. Peripheral nerves are also well evaluated using ultrasound.

### Aneurysms

Periarticular arterial aneurysms are most frequently seen adjacent to the hip and knee joints. However, these vascular lesions may be identified adjacent to any joint. Congenital aneurysms are extremely rare. The vast majority are the result of atherosclerosis. Traumatic, mycotic, syphilitic, and poststenotic aneurysms are seen less frequently (Fig 7–44).

Aneurysms adjacent to joints resulting from atherosclerotic disease are found almost exclusively in the lower extremity (Fig 7–45). Seventy percent of peripheral aneurysms are found in the popliteal artery. Fusiform aneurysms are most common in the common femoral arteries. These are usually due to extension of an infrarenal abdominal aortic aneurysm that continues into the iliac and femoral vessels. Saccular aneurysms of the common femoral arteries generally occur at the site of anastomosis of prosthetic vascular grafts. Conversely, the majority of popliteal aneurysms are saccular. They usually present as asymptomatic masses in the popliteal fossa that are not pulsatile. Therefore, on clinical examination they are difficult to distinguish from Baker's cyst (Fig 7–46). These aneurysms usually contain a large amount of thrombus. As a result, there is a high risk of distal embolization of thrombus fragments. Surgical repair of these lesions is recommended for this reason. Rupture of these aneurysms is rare. Blood flow through these lesions is very slow, in some cases too slow to detect with Doppler. The correct diagnosis is made by demonstrating continuity of the lesion with

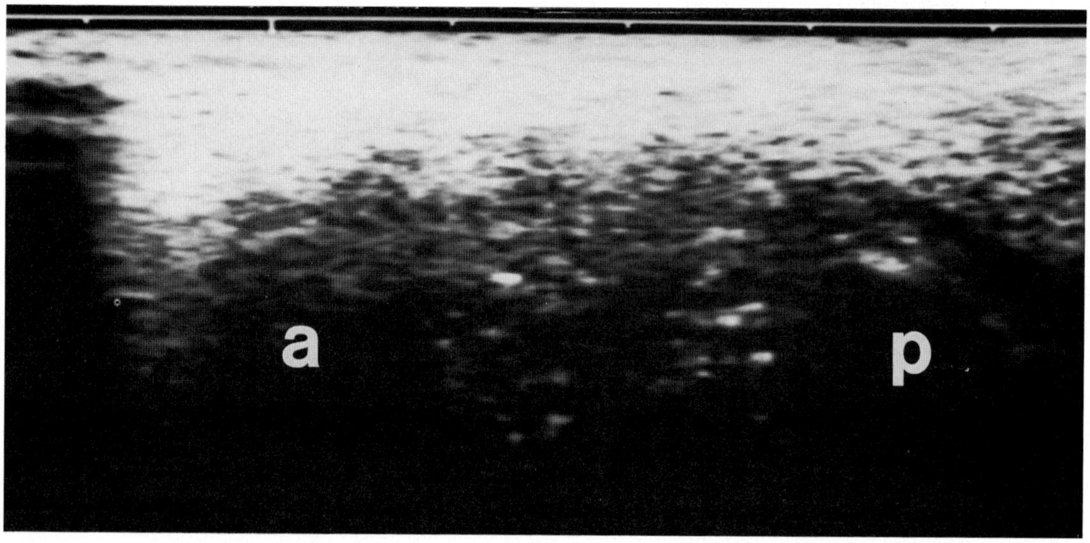


**FIG 7–40.**
Steroid-related intra-articular debris: transverse sonogram of the lateral meniscus. Same patient as in Figure 7–27. The patient received an intra-articular injection of steroids 1 week prior to the examination. Throughout the meniscus, numerous tiny highly echogenic foci are identified. These calcifications may also be observed on conventional radiographs. Abbreviations: anterior meniscus *(a)*, posterior menicus *(p)*.

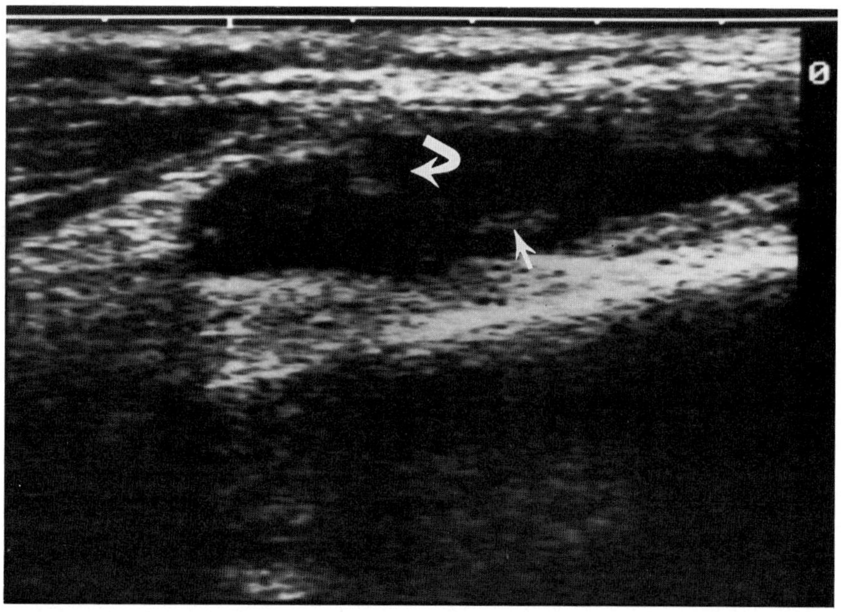

**FIG 7–41.**
Shedding of villi in rheumatoid arthritis: longitudinal sonogram. This patient is a 39-year-old woman with a 6-year history of rheumatoid arthritis and 6 months of knee involvement. She presented here with a flexion contracture of the right knee. The image presented is a longitudinal view of the suprapatellar bursa, which is distended with fluid. Irregular synovial proliferation is noted *(curved arrow)*. Free-floating echogenic masses *(arrow)* are present within the bursa. Arthroscopy demonstrated these masses to be villi that had been shed from the synovial walls. Synovectomy with debridment of this debris was performed.

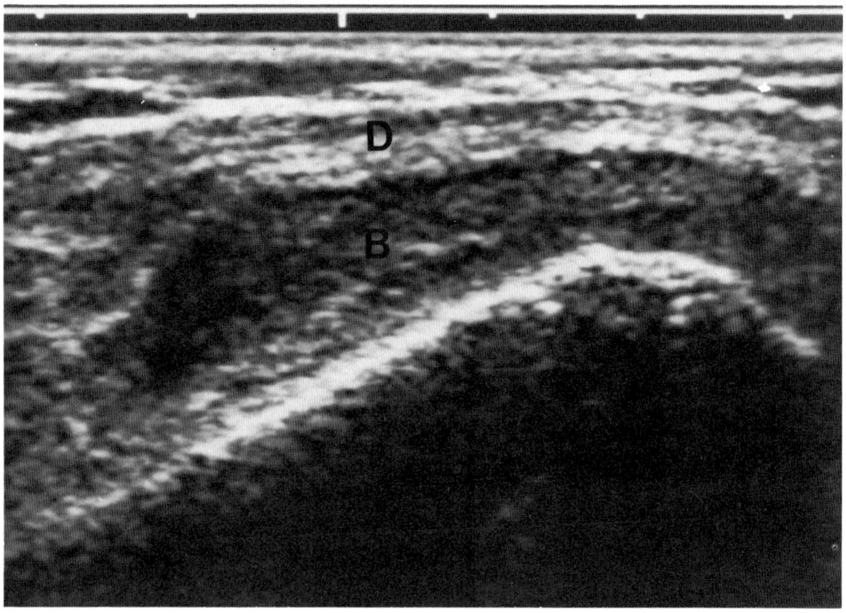


**FIG 7–42.**
Rheumatoid shoulder disease: coronal sonogram. At the time of presentation, this 53-year-old woman with a 5-year history of rheumatoid arthritis complained of increasing right shoulder pain. The transducer is positioned in the coronal plane immediately inferior to the acromion. The acromion is just out of view to the observer's right. Hypoechoic material fills the subacromial-subdeltoid bursa *(B)*. Ultrasound cannot always differentiate pannus from fibrinous debris. In this case, the sonographic appearance of the bursa explains why joint aspiration was negative despite prominent clinical swelling. Abbreviations: deltoid *(D)*.

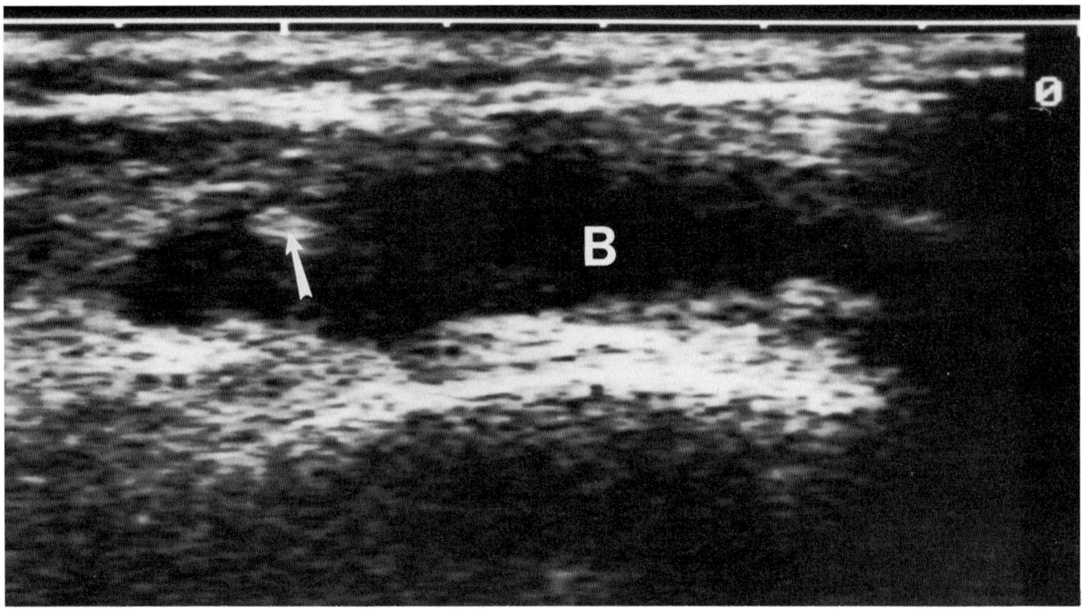


**FIG 7–43.**
Meniscal fragment after meniscal surgery: longitudinal sonogram of the suprapatellar bursa. A meniscal fragment was lost in the joint during arthroscopic surgery on the left knee of this 37-year-old man. He presented with pain and locking of the knee. This longitudinal view of the suprapatellar bursa *(B)* demonstrates a hyperechoic meniscal fragment *(arrow)* within the lumen of the fluid-filled bursa. Manipulation by the examiner clearly showed this to be a free fragment.

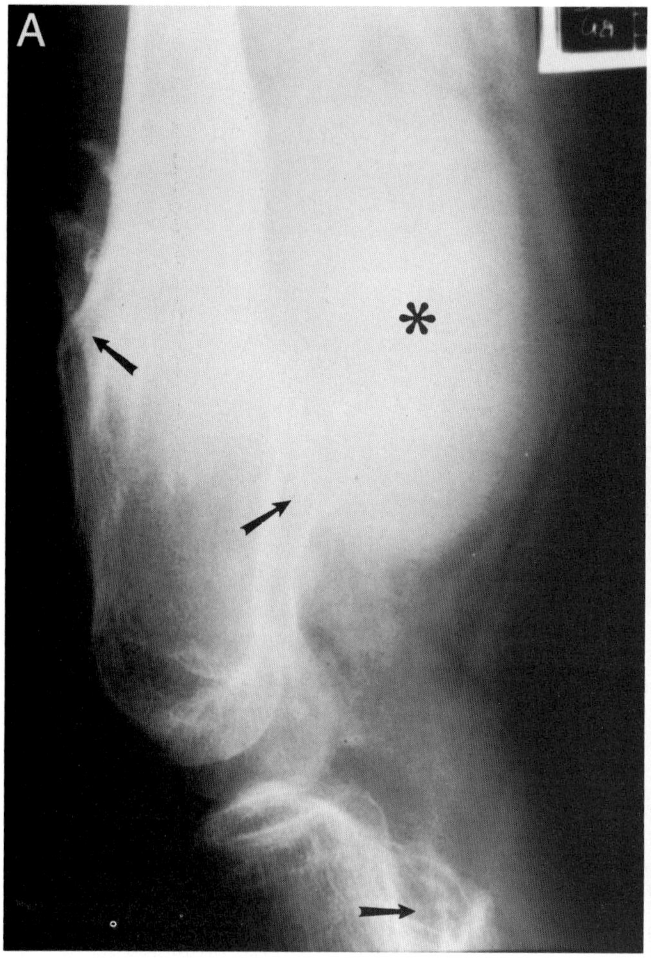


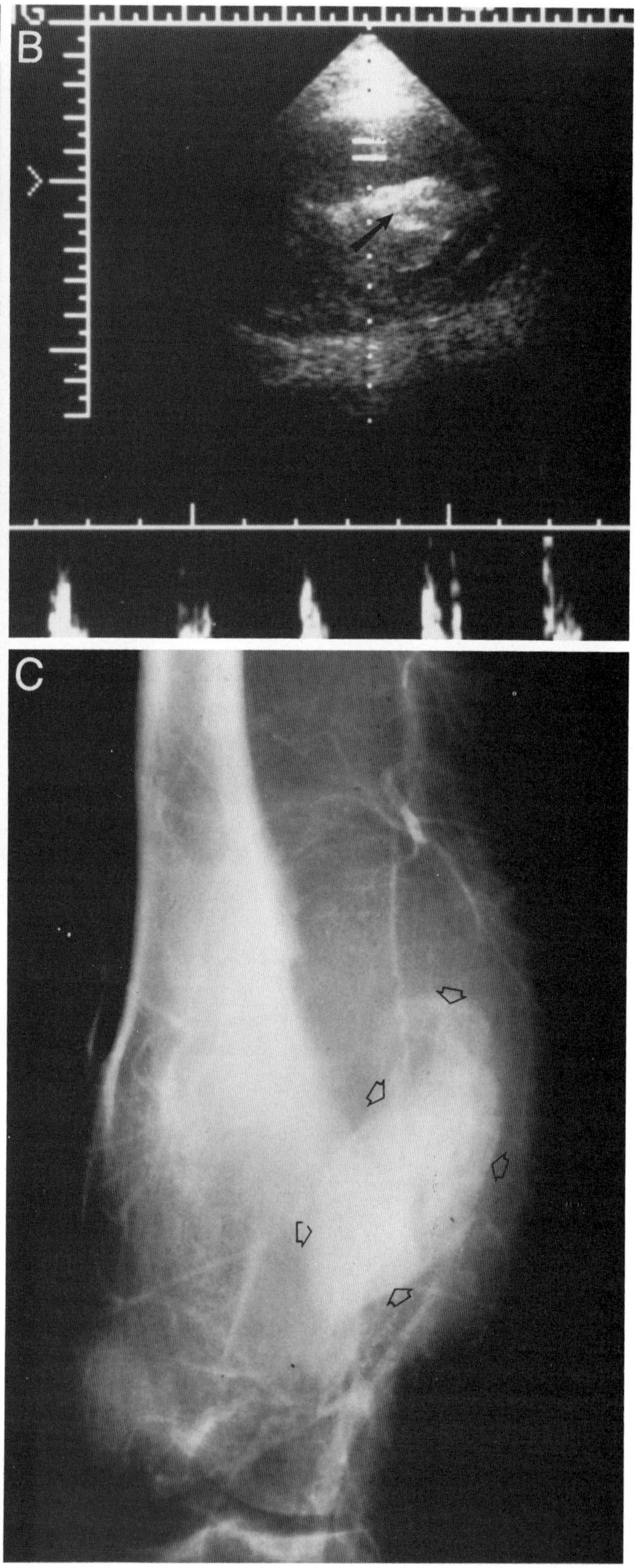


**FIG 7–44.**
**A,** popliteal mass in patient with hereditary multiple exostoses: lateral radiograph of the knee. This 17-year-old boy with hereditary multiple exostoses *(arrows)* complained of a rapidly growing mass in the popliteal fossa. The orthopedic surgeons suspected sarcomatous degeneration of an exostosis. Mass *(asterisk)* in the popliteal fossa overlies one of the exostoses and scallops the posterior cortex of the femur. A sclerotic reaction is also noted in the underlying bone. **B,** popliteal mass in patient with hereditary multiple exostoses: longitudinal sonogram with Doppler. Same patient as in **A.** The mass appears hypoechoic and continuous with the popliteal artery. An arterial Doppler signal is noted within the mass giving the diagnosis of a traumatic pseudoaneurysm. Deep to the mass the exostosis is seen as an echogenic region *(arrow).* **C,** popliteal mass in patient with hereditary multiple exostoses: femoral angiogram. Same patient as in **A** and **B.** This film of the popliteal region from the late arterial phase of a femoral angiogram demonstrates pooling of contrast material within the pseudoaneurysm *(open arrows).* The aneurysm was resected and a synthetic graft placed. Bone biopsy performed at the time of surgery demonstrated normal reactive bone.

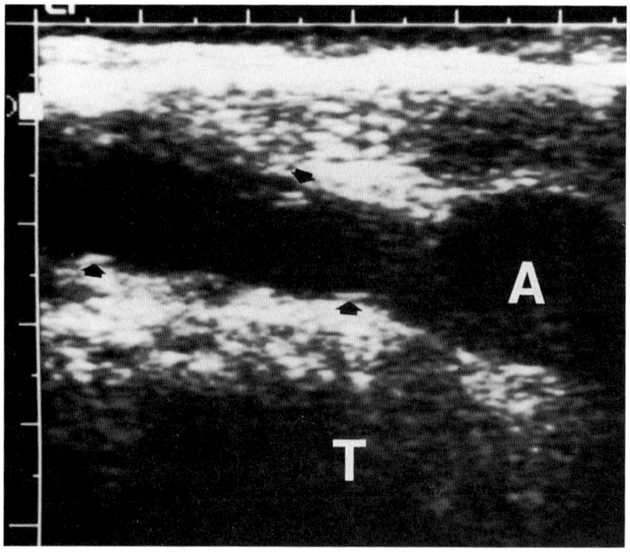


**FIG 7–45.**
Swelling of the left popliteal fossa: longitudinal sonogram. A mass was noted in the left popliteal fossa of this 62-year-old man during routine physical examination. The clinical question was whether this was joint or vascular in origin. An ectatic popliteal artery is demonstrated with frank aneurysmal dilatation distally *(A)*. Intimal calcifications are seen as hyperechoic lines *(arrows)*.

the superficial femoral artery. In most cases the distal superficial femoral artery will also be ectatic. Popliteal aneurysms are frequently associated with other atherosclerotic aneurysms. Approximately 50% of cases are bilateral. Therefore, after the diagnosis of popliteal aneurysm is made, the popliteal fossa on the opposite side, the iliac regions, and the aorta should be examined.

Poststenotic aneurysms are rare, usually found adjacent to the shoulder joint in the subclavian, axillary, and brachial arteries. They are frequently associated with cervical ribs or other type of thoracic outlet syndrome. Like popliteal aneurysms, they frequently result in distal embolization of fragments of lumenal thrombus, requiring surgical intervention. These aneurysms are almost exclusively fusiform. The definitive diagnosis can be made sonographically with the demonstration of continuity with the brachial artery and Doppler signal.

Aneurysms of the distal circulation are exceedingly rare and usually the result of trauma or septic emboli. Traumatic aneurysms may occur anywhere in the circulation. They are pseudoaneurysms, the flowing blood contained only by adventitia. As such, the risk of rupture is high, and surgical correction is indicated. Sonography does not have a role in the

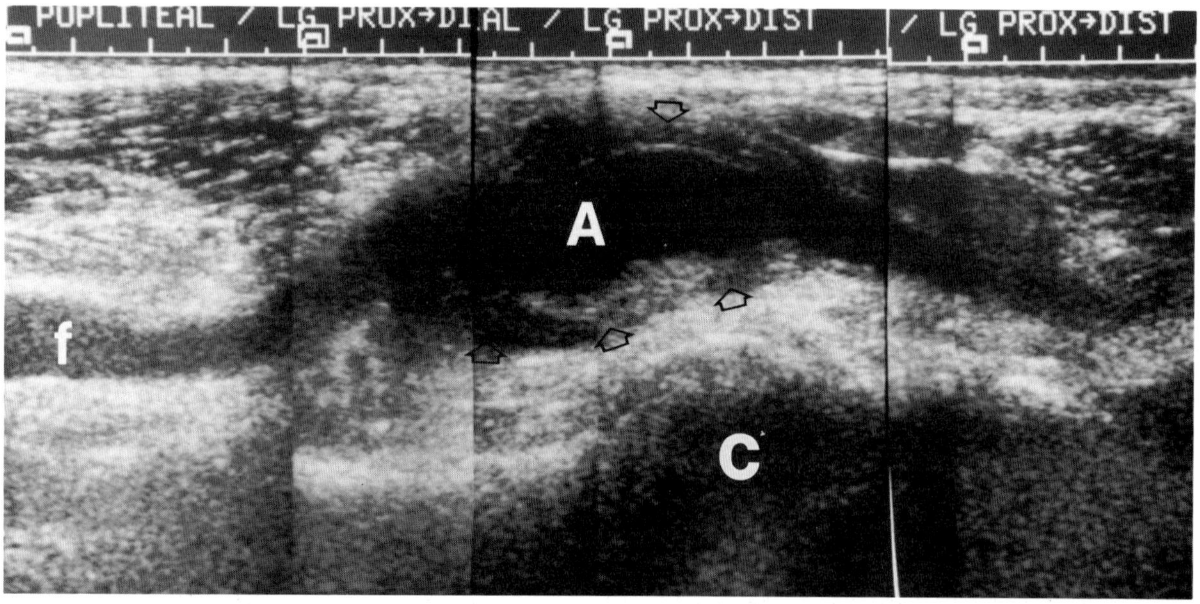


**FIG 7–46.**
Mass in the popliteal fossa: longitudinal sonogram. A mass in the right popliteal fossa of this 68-year-old man was thought to be a Baker cyst. The image displayed is a concatenation of serial longitudinal images obtained over a 16-cm length of the popliteal fossa. Aneurysmal dilatation *(A)* of the popliteal artery is noted to contain mural thrombus *(open arrows)*. When an aneurysm of the popliteal fossa is discovered, the contralateral side, common femoral arteries, and abdominal aorta must be examined. This patient had aneurysmal dilatation of the opposite popliteal, bilateral common femoral, bilateral iliac arteries, and the infrarenal aorta. Abbreviations: junction of the superficial femoral artery and popliteal *(f)*, lateral femoral condyle *(C)*.

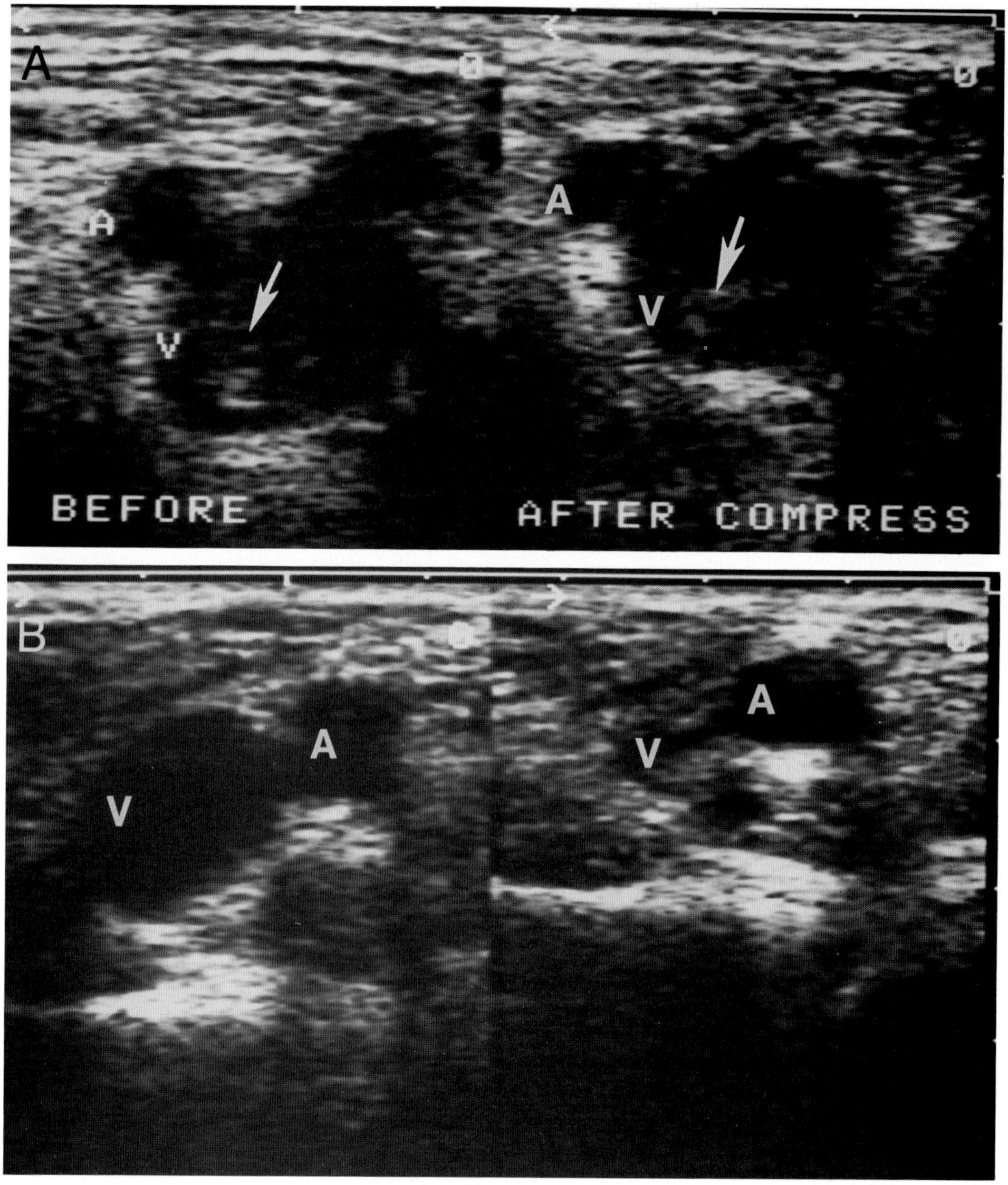


**FIG 7–47.**

**A,** deep venous thrombosis of the lower leg: transverse sonograms. This 50-year-old woman presented to the emergency room with a 3-day history of erythema and swelling of the right lower extremity. She had a 10-year history of rheumatoid arthritis, and the clinical question was deep venous thrombosis vs. pseudothrombophlebitis. Compression study was performed of the common femoral vein *(V)* at the point of junction with the saphenous. Echogenic material *(arrow)* is noted within the vein. The vein did not change in dimension when pressure was applied with the transducer. This is diagnostic for deep venous thrombosis. Abbreviations: femoral artery *(A)*. **B,** normal veins of the lower leg: transverse sonograms. Same patient and transducer position as in **A,** contralateral leg. Compression image is on the right. The common femoral vein is anechoic and fully compressible. This is the sonographic appearance of a normal vein. Abbreviations: common femoral artery *(A)*, common femoral vein *(V)*.

evaluation of acute penetrating trauma but is certainly indicated in more subtle forms of trauma. Occasionally, arteriovenous fistulas will result from penetrating trauma. If undiagnosed at initial presentation, they can cause venous aneurysms secondary to increased flow. The lower extremity is a high-resistance system. If a high degree of diastolic flow is observed proximal to the site of injury, then a fistula should be suspected.

Mycotic aneurysms most commonly are the result of bacterial endocarditis but may occur from direct intra-arterial injection in drug abusers utilizing contaminated instruments. Septic emboli lodge in the peripheral circulation resulting in aneurysm formation, usually saccular. These lesions develop and expand rapidly, requiring timely accurate diagnosis. Sonographically, these aneurysms have imaging characteristics identical to other aneurysms of the peripheral circulation. The diagnosis of mycotic aneurysm is based on clinical criteria.

### Popliteal Entrapment Syndrome

Calf claudication in a young patient without other medical problems is likely due to popliteal artery entrapment syndrome. The normal course of the popliteal artery passes between the heads of the gastrocnemius muscle. Congenital malposition of the popliteal artery medial to or within the medial head of the gastrocnemius results in compression of the artery, greatest in extension. Initially this results only in decreased pedal pulses. However, this chronic irritation of the vessel results in intimal thickening and fibrosis. If it is undiagnosed, total occlusion of the vessel will occur. Severity of the claudication is dependent on the rate at which occlusion occurs and the development of collateral flow. Diagnosis is based on identification of the popliteal artery within or medial to the medial head of the gastrocnemius muscle.

Cystic degeneration of the popliteal artery is another etiology of calf claudication. Narrowing of the popliteal artery in this condition is caused by compression from a mucoid cyst in the adventitia. It usually occurs in the middle third of the artery. Ultrasound is highly sensitive and specific for identification of these cysts, perhaps more sensitive than arteriography if filming is performed in only anteroposterior projection. Aspiration of the cyst under ultrasound guidance usually relieves symptoms. However, due to the possibility of recurrence, surgical therapy with synthetic graft placement may be desirable.

### Venous Pathology

Sonography is playing an increasing role in the diagnosis and follow-up of deep venous thrombosis. Compression gray scale imaging and Doppler flow

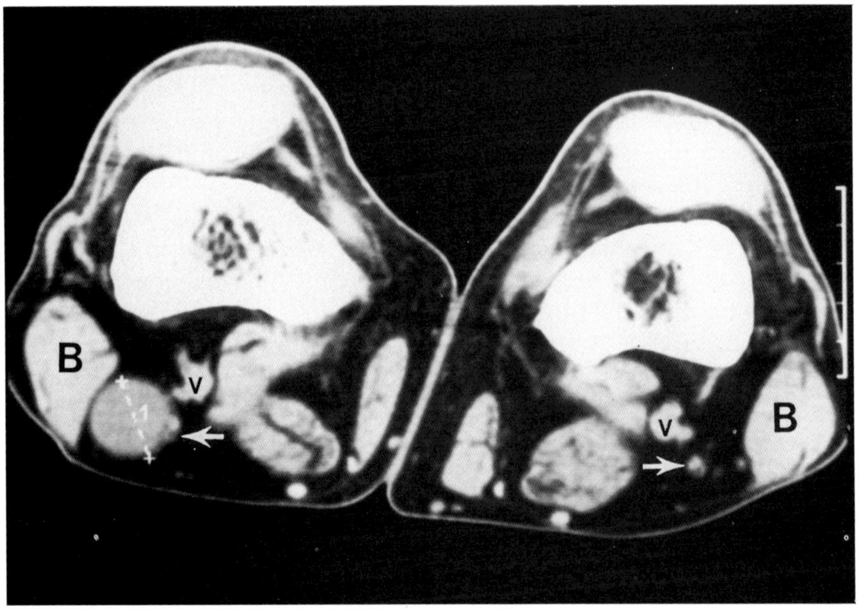


**FIG 7–48.**
Schwannoma in the right popliteal fossa: CT scan. Spinal CT was performed on this 30-year-old woman who presented with paresthesia of the right lower extremity. This examination was normal. CT of the lower extremities was performed following EMGs. A spherical mass 2.2 cm in diameter is seen arising from the lateral aspect of the tibial nerve *(arrow)*. Abbreviations: biceps muscle *(B)*, popliteal vein *(V)*.

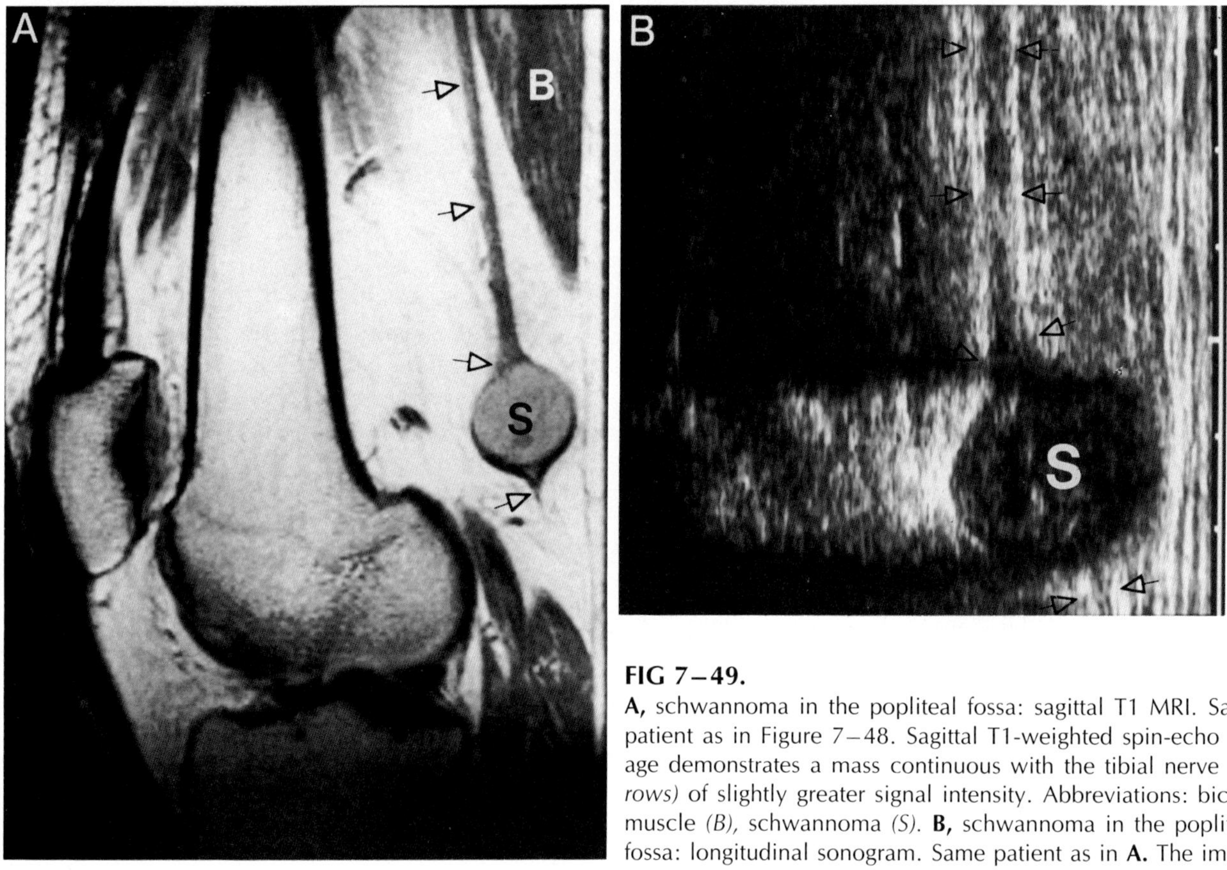


**FIG 7–49.**
**A,** schwannoma in the popliteal fossa: sagittal T1 MRI. Same patient as in Figure 7–48. Sagittal T1-weighted spin-echo image demonstrates a mass continuous with the tibial nerve *(arrows)* of slightly greater signal intensity. Abbreviations: biceps muscle *(B)*, schwannoma *(S)*. **B,** schwannoma in the popliteal fossa: longitudinal sonogram. Same patient as in **A.** The image is oriented in the same manner as the MRI examination in **A.** Ultrasound demonstrates the schwannoma *(S)* continuous with the tibial nerve *(open arrows)*. Note the increased through-transmission of sound deep to the lesion. This is a characteristic finding with schwannomas.

analysis are highly sensitive and specific for venous thrombosis (Fig 7–47) (Lensing et al., 1989). Examination of the popliteal fossa and hip should include evaluation of patency and compressibility of the popliteal and femoral veins. The vein to be examined is localized in transverse images, and then the transducer is placed directly over the vessel, scanning in its long axis. Pressure is applied with the transducer and response of the vessel observed. If the lumen of the vein cannot be completely obliterated by compression, thrombus must be present.

### *Nerve Lesions*

Normal nerves appear as linear fibrillar hyperechoic structures in close proximity to vessels. Nerves can be followed along their course in the extremities from the axilla to the wrist in the arm and ischial spine to the ankle in the leg. Benign neural tumors are generally fusiform in shape and appear hypoechoic (Figs 7–48 through 7–50). Sonography is a valuable first test when electromyography (EMG) is indicative of a peripheral nerve lesion. Its ability to quickly examine the entire extremity is another advantage over MRI.

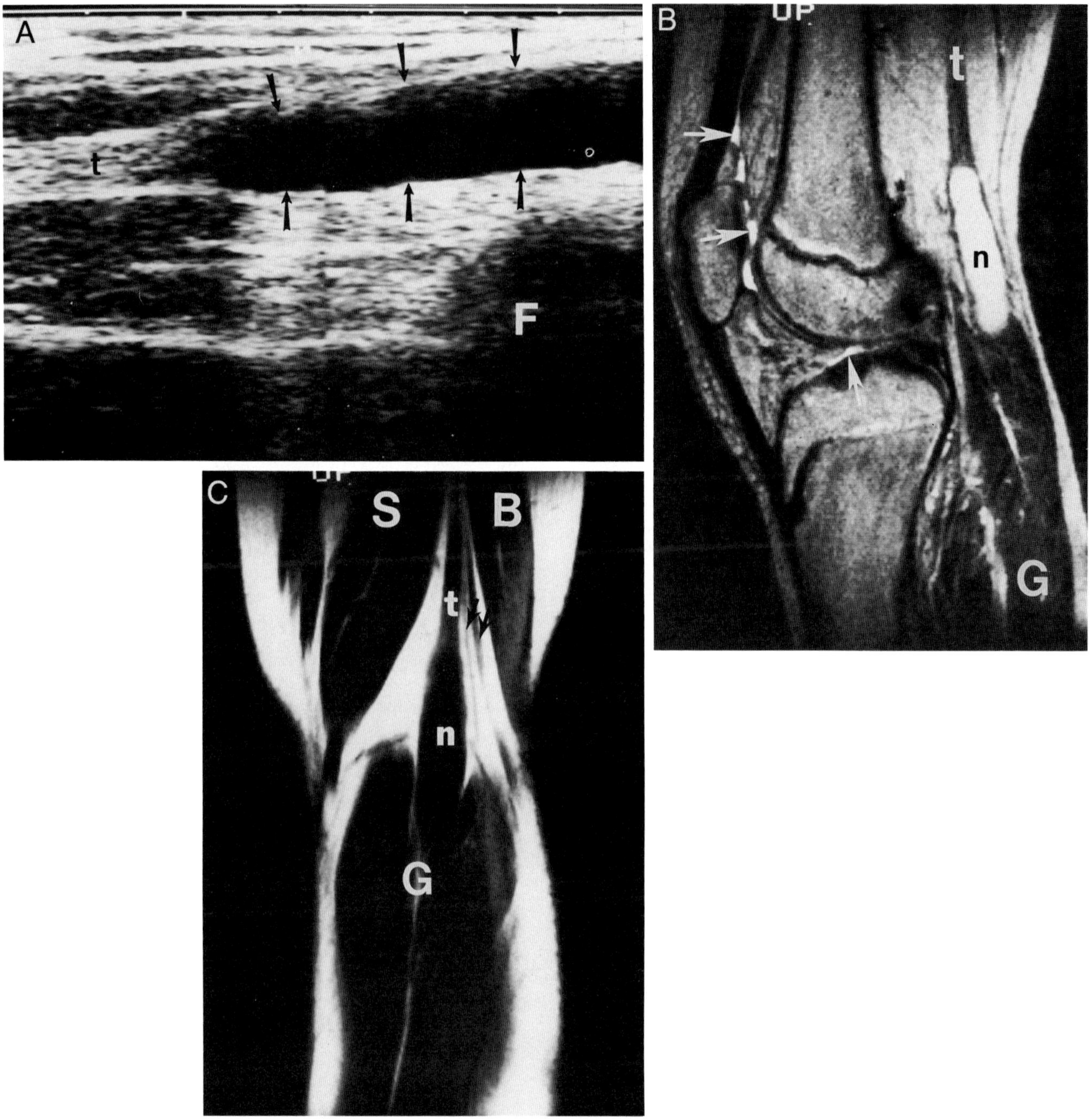


**FIG 7–50.**

**A,** neurofibroma in the popliteal fossa: longitudinal sonogram. A 12-year-old girl presented with a tingling sensation in her left leg. The CT scan of the spine was normal. Due to persistent paresthesia in the tibial nerve distribution, ultrasound examination of the sciatic nerve was performed along its entire length. Six centimeters above the joint space the normal fibrillar appearance of the tibial nerve is interrupted by a hypoechoic mass *(arrows)* centrally within the nerve. Again noted is acoustic enhancement deep to the lesion. Abbreviations: tibial nerve *(t),* femoral condyle *(F).* **B,** neurofibroma in the popliteal fossa: sagittal T2 MRI. Same patient as in **A.** The MRI examination directed by the localization provided by ultrasound reveals a hyperintense mass *(n)* within the substance of the tibial nerve *(t).* A normal amount of joint fluid *(arrows)* is present. Clinical examination was not valuable in detecting this mass, even retrospectively. Abbreviations: lateral gastrocnemius *(G).* **C,** neurofibroma in the popliteal fossa: coronal T1 MRI. Same patient as in **A** and **B.** Coronal T1-weighted spin-echo image of the posterior popliteal fossa demonstrates fusiform enlargement *(n)* of the tibial nerve *(t).* A biopsy of this low signal intensity mass demonstrated a neurofibroma. Abbreviations: biceps muscle *(B),* gastrocnemius *(G),* semimembranosus *(S).*

## REFERENCES

Adam R, Hendry GMA, Moss J, et al: Arthrosonography of the irritable hip in childhood: A review of 1 year's experience. *Br J Radiol* 1986; 59:205–208.

Aisen AM, McCune WJ, McGuire A, et al: Sonographic evaluation of the cartilage of the knee. *Radiology* 1984; 153:781–784.

Barrie HJ: The pathogenesis and significance of meniscal cysts. *J Bone Joint Surg (Br)* 1979; 61:184.

Bloom W, Fawcett DW: *A Textbook of Histology*, ed 10. Philadelphia, WB Saunders Co, 1975.

Cooperberg PL, Tsang L, Truelove L, et al: Gray scale ultrasound in the evaluation of rheumatoid arthritis of the knee. *Radiology* 1978; 126:759–763.

Dörr U: The painful hip joint. Diagnostic possibilities offered by sonography. *Fortschr Geb Röntgenstr Nuklearmed Erganzungsband* 1988; 148:487–491.

Ferrer-Roca O, Vilalta C: Lesions of the meniscus. II: Horizontal cleavages and lateral cysts. *Clin Orthop* 1980; 146:301.

Gielen JL, van Holsbeeck MT, Hauglustaine D, et al: Growing bone cysts in long-term hemodialysis. *Skeletal Radiol* 1990; 19:43–49.

Gilsanz V, Bernstein BH: Joint calcification following intraarticular corticosteroid therapy. *Radiology* 1984, 151:647–649.

Hammer M, Mielke H, Wagener P, et al: Sonography and NMR imaging in rheumatoid gonarthritis. *Scand J Rheumatol* 1986; 15:157–164.

Lensing AWA, Prandoni P, Brandjes D, et al: Detection of deep-vein thrombosis by real-time B-mode ultrasonography. *N Engl J Med* 1989; 320:342–345.

Marchal GJ, van Holsbeeck MT, Raes M, et al: Ultrasonography in transient synovitis of the hip in children. *Radiology* 1987; 162:825–828.

Mori Y: Debris observed by arthroscopy of the knee. *Orthop Clin North Am* 1979; 10:559–563.

Peck J: Ultrasound of the painful hip in children. *Br J Radiol* 1986; 59:205–208.

Resnick D, Niwayama G: *Diagnosis of Bone and Joint Disorders*. ed 2. Philadelphia, WB Saunders Co, 1988.

Richardson ML, Selby B, Montana MA, et al: Ultrasonography of the knee. *Radiol Clin North Am* 1988; 26:63–75.

Selby B, Richardson ML, Montana MA, et al: High resolution sonography of the menisci of the knee. *Invest Radiol* 1986; 21:332–335.

Taveras JM, Ferrucci JT: *Radiology: Diagnosis—Imaging—Intervention*, vol 5. Philadelphia, J.B. Lippincott, 1987, chap. 1.

van Holsbeeck M, Gielen JL, Martens M, et al: Sonography of the medial collateral ligament [abstract]. *Radiology* 1988a; 169:178.

van Holsbeeck M, van Holsbeeck K, Gevers G, et al: Staging and follow-up of rheumatoid arthritis of the knee. Comparison of sonography, thermography, and clinical assessment. *J Ultrasound Med* 1988b; 7:561–566.

Vandeputte L, Mulier JC, Mulier F: Transient synovitis of the hip joint in children. *Acta Orthop Belg* 1971; 37:186–193.

Wingstrand H: Transient synovitis of the hip in the child. *Acta Orthop Scand* 1986; 57(suppl):45–46.

Worth WD, Hermann E, Meudt R, et al: Stellenwert der Arthrosonographie in die Beurteilung der Exsudativen und proliferativen Synovialitis. *Z Rheumatol* 1986; 45:263.

# 8

# Pediatric Musculoskeletal and Spinal Sonography

Michael A. DiPietro, M.D.*

Many of the recently developed applications for real-time sonography of the musculoskeletal system (Harcke et al., 1988) have been in children, some exclusively so. Cases of congenital dislocated hip (CDH) can be missed on the neonatal clinical examination. Radiography is also of limited usefulness at that age. The price of late detection and treatment is the lower likelihood of a successful reduction and a normal hip. Sonography clearly depicts the unossified hip and has thereby found widespread use for detecting CDH, including cases that were normal clinically. Its extreme sensitivity has increased our awareness of normal developmental maturation of the hip and of the possibility of overtreatment of "early CDH." The majority of this chapter addresses CDH, including background information learned from large, European, presonographic, clinical screening programs in an attempt to put sonography's role into perspective in this complicated problem.

Other uses of sonography include distinguishing joint dislocations from displaced epiphyses (e.g., displaced Salter-Harris fractures), detecting joint effusions, and delineating the extent of midline or near midline soft tissues masses on the back with possible occult tethered spinal cord.

## CONGENITAL DISLOCATED HIP

### Background

Congenital dislocated hip is "an anomaly of the hip joint, present at birth, in which the head of the femur is, or may be, partially or completely dislocated from the acetabulum" (Dunn, 1969). It is regarded as a spectrum of instability: subluxable, subluxed, dislocatable, or dislocated hips with varying degrees of deformity (i.e., dysplasia) of the acetabulum. However, cases of bilateral CDH of variable severity from side to side support the notion of it being a spectrum of a single pathological entity rather than several entities (Dunn, 1969).

Congenital dislocated hip is a congenital deformation, a late in utero alteration of a previously normally formed part, by persistent gentle forces. There is a special category of CDH, teratological, in which the dislocation had occurred early in utero secondary to a problem such as myelodysplasia or arthrogryposis and which accounts for only 2% of cases. In contrast to the more common form of CDH, in the teratological form, adaptive changes in the hips are advanced at birth and are often obvious clinically and radiologically.

There are mechanical and physiological causes for CDH (Dunn, 1976b; Hensinger, 1979). The mechanical cause of deformation is constraint and compression of the fetus in utero. In utero spatial restriction of fetal leg movements can occur with oligohydramnios, first pregnancy (tight maternal uterine and abdominal muscles), and breech presentation (30%–50% of CDH). Other deformations also caused by in utero constraint such as sternocleidomastoid torticollis and metatarsus adductus have an increased association of CDH. The physiological cause of CDH is laxity of the hip joint and the influence of maternal hormones, which is thought to explain the 6:1 female/male incidence of CDH and its 20% familial incidence (familial sensitivity?).

*Dr. DiPietro is Associate Professor, Section of Pediatric Radiology, Department of Radiology; University of Michigan Hospitals, Ann Arbor, Michigan.

Near the time of birth, the joint capsule is elastic and stretched, resulting in a tendency for the femoral head to dislocate. However, the shape of the joint and the soft tissues are nearly normal, so that if the femoral head is kept in the acetabulum, the joint capsule will return to a normal configuration in a few weeks (Hensinger, 1979). The dislocatable unstable hip is treated by splinting with reliable, but not excessive, constraint until it can be held firmly by its own connective tissue, which had been stretched prior to birth. Ninety-six percent of CDHs become radiologically and physiologically normal when they are treated early (Hensinger, 1979). Were the hip not successfully reduced, adaptive changes in the acetabulum (steep roof, pseudoacetabulum formation), femoral head (small, flattened, delayed ossification), ligaments, and muscles (contractures) would occur; and with more deformity to reverse, obtaining a satisfactory reduction would be difficult.

A clinical study, based on physical examination of 7,742 newborns, revealed an increased frequency of breech-delivered babies in the group with unstable or dislocated vs. normal hips (17.3% vs 4.4%) (Barlow, 1962). One and one-half percent of all the newborns had dislocatable or dislocated hips soon after birth. However, 60% of those became stable by 1 week and 90% by 2 months with or without treatment. These patients also had normal 1-year follow-up physical examinations and radiographs yielding a final CDH incidence of 0.15% of live births, similar to that reported in other series.

Most unstable newborn hips (80% [Dunn 1976b]) resolve spontaneously, although this is unlikely after 2 months (Barlow, 1962). Adduction, extension, and immobility of the hips reduces the natural rate of resolution of dislocatable hips, whereas partial abduction and partial flexion encourages resolution. The possibility of spontaneous resolution raises controversy about treating some Graf type 2 hips (physiologically immature; see later discussion) (Berman et al., 1986). On the other hand, treatment, if eventually needed, is more difficult after 2 months due to the development of adductor tightness. Barlow (1962) recommends treatment in the first week.

An autopsy study performed on babies who died during labor or soon after birth suggested that the crucial pathology of CDH is the labrum (Dunn, 1969, 1976a). Babies between 13 and 40 weeks' gestation with nonsubluxable hips showed little difference in general morphology over this wide gestational range. In those babies (27–44 weeks' gestation) with dislocatable hips, the crucial pathology was shown to be the labrum, which was everted and stretched posteriorly and superiorly to variable degrees. When dislocation was partial, the hip was restrained by the capsule and ligamentum teres, which, however, were stretched when instability and dislocation were worse. In addition, the labrum was inverted, the acetabulum shallow, and the femoral head small, pitted, discolored, and less spherical in the markedly unstable dislocated hips. The number of right and left abnormal hips was the same, although in clinical studies, an incidence of 20% right and 25% bilateral CDH has been cited (Hensinger, 1979).

What is CDH? Subluxable hips, dislocatable or dislocated hips, acetabular dysplasia with or without joint laxity, physiological immaturity of the acetabulum, physiological laxity of the capsule in the newborn? One problem with evaluating the effect of screening and treating CDH has been the variability and indistinctness of the condition under study. "Distinctions between congenital dislocation, acetabular dysplasia, and normality are blurred in the mists of radiological images" (Catterall, 1984). Reports concerning the management of CDH in infants less than 1 year seldom strictly define the criteria for diagnosis, yet CDH covers a broad spectrum between normality and severe dysplasia (Catterall, 1984).

Even with clinical screening programs there has been no decrease in the incidence of late-discovered cases of CDH (Castelein and Sauter, 1988; Place et al., 1978). It is best to reduce CDH before 6 months (MacEwen and Zembo, 1987); however, early dysplasia may escape clinical diagnosis (Langer, 1987). The incidence of CDH is 0.1%, whereas the incidence of unstable hips is as high as 2%. Therefore, most unstable hips are not CDH. However, there is no guarantee that the unstable hip that later becomes stable will not develop CDH. Accordingly, in one Swedish report, all unstable hips were treated (Hansson et al., 1983).

The incidence of CDH with clinical screening programs is reported between 0.25% and 2% of live births, whereas before such programs it was 0.07% to 0.15% of live births (Place et al., 1978). The rate of late diagnosed cases was no different, and the apparent increase in the incidence of CDH could merely be a difference in classification, including many cases that might have recovered spontaneously without treatment (Place et al., 1978). A reported increased incidence from 0.15% (prescreen) to 0.18% (with screen) could have also been due to the detection of cases that formerly had gone unnoticed until adulthood (MacKenzie and Wilson, 1981). One might accept the high false positive rate, with the risk of overtreatment, if the false negative rate (CDH diagnosed late, missed in newborn exam) dropped. Yet the reported incidence of late diagnosed (missed?) CDH was 0.078% of live births where all had newborn examinations (Place et al., 1978). The screening clinical examination that tests for increased joint laxity has failed to decrease the occurrence of late CDH, which is harder to treat. Were some of the clinical screening examinations that failed to predict subsequent CDH (i.e., misses) truly negative (i.e., stable hips) and the instability developed later? The assumption that infant CDH is merely a progression from the unstable neonatal hip is unproved (Berman and Klenerman, 1986). Yet clinical screening programs focus on joint instability assuming that hip dysplasia, with or without dislocation, is a progression from neonatal instability (Castelein and Sauter, 1988). In another series, ten infants with dysplastic or dislocated hips all had normal, stable neonatal clinical examinations (Davies and Walker, 1984).

Acetabular dysplasia might be secondary to failure to maintain concentric reduction of the femoral head due to joint laxity. Yet it might be a distinct entity that is not detectable without joint laxity or femoral head displacement until the child bears weight (MacKenzie and Wilson, 1981).

If one treats early, there exists the possibility of needless and potentially dangerous overtreatment. However, if treatment is postponed until later, some cases might be missed or lost and treatment can be more difficult. Those who oppose immediate splinting cite the spontaneous resolution rate of Barlow (1962) together with the risk of iatrogenic avascular necrosis. Is the resolved "normal" hip, as determined on physical examination, truly completely normal (Clarke, 1986)? There is a need to know the natural history of hip instability in the first weeks of life (Clarke, 1986). Sonography confirmed Barlow's (1962) clinical impression of spontaneous resolution, probably due to increased muscular activity and capsular tightening. The risk of avascular necrosis following splinting is 2.5% in one study (Saies et al., 1988). However, if extreme abduction is avoided during splinting, and if it is done early, before contracture, when little force is needed, the risk of avascular necrosis is minimized (Hansson et al., 1983; Tredwell and Davis, 1989).

## Sonography

In recent years, various investigators have stressed particular aspects of observing the infant hip sonographically, including the pioneering work of Graf (1983) and early reports of Novick et al. (1983) and Harcke et al. (1984).

Many sonographers now use a combined approach. In essence, the three features noted are:

1. The position of the femoral head within the acetabulum.
2. Any change in position of the femoral head within the acetabulum with stress (Barlow maneuver). This is the dynamic examination.
3. The steepness of the acetabular roof with quantification of the acetabular angles and an assessment of the degree of maturity or dysplasia of the acetabulum.

In other words, the hip is assessed with respect to where the femoral head is, how much it moves with stress, and how well formed the acetabulum is to contain the femoral head.

### *Technique*

With respect to dislocation, infant hips are studied from their lateral aspect. The transducer is placed lateral to the femoral head over the greater trochanter. I prefer to use a linear array transducer of 5 MHz or greater, whichever will allow adequate penetration of the tissues. The hips are routinely studied from this lateral approach in two planes, coronal and transverse, with respect to the bony pelvis. When using this method, one quickly realizes that the coronal images are similar to viewing anteroposterior radiographs (Fig 8–1) and that the transverse images are similar to viewing axial computed tomography (CT) scans of the pelvis (Fig 8–2). However, since scanners show the transducer surface on the top of the viewing screen, the top is therefore lateral (Fig 8–3). The sonographer learning this technique will experience a period of acclimation to the anatomy displayed in this orientation.

### *Findings*

In the coronal projection, the femoral head should fit within the acetabulum as a ball within a socket. A small ossific nucleus might be seen, but in the young infant the femoral head is usually unossified and exhibits a speckled appearance of low echogenicity. The femoral head should sit within the deepest portion of the socket (acetabulum) (Fig 8–3). Superior and lateral displacement is easily seen (Fig 8–4). Posterior displacement is inferred from the coronal view when the plane in which one sees the femoral head in its largest diameter (the midplane of the femoral head) is in the more posterior portion of the acetabulum and not within the deepest portion of the socket. One can also identify the triradiate cartilage of the acetabulum by the through transmission of the

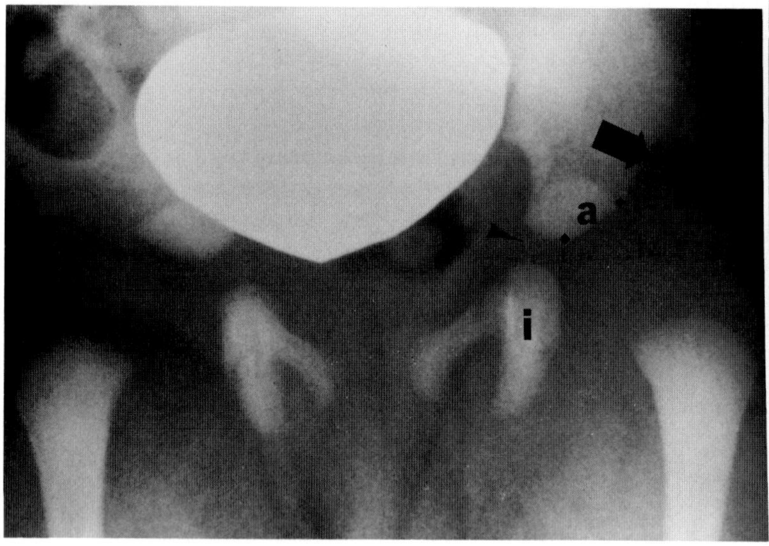


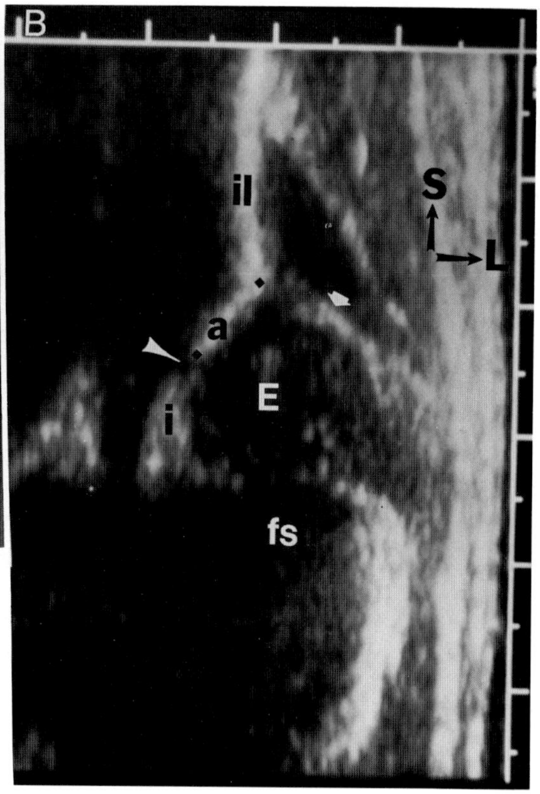


**FIG 8–1.**
**A,** anteroposterior radiograph shows the steep left acetabular roof *(a).* **B,** the coronal sonogram of the same hip viewed vertically to correspond to the anteroposterior radiograph also shows the steep acetabular roof *(a).* Although the sonogram is more akin to an anteroposterior tomogram of the hip, the acetabular roof is similarly seen. The lateral ilium *(large arrowhead)* on the radiograph roughly corresponds to the lateral bony margin ofthe ilium *(il)* on the sonogram. The triradiate cartilage *(small arrowheads* and *ischium)* are shown. Abbreviations: unossified femoral head *(E),* femoral shaft *(fs),* labrum *(small broad arrowhead).*

ultrasound traversing the unossified femoral head and the unossified medial wall of the acetabulum (i.e., triradiate cartilage) (see Fig 8–4). Ossification of the femoral head with age prevents adequate sonographic assessment of the acetabulum (Fig 8–5). Sonography is limited between 6 and 12 months and is often useless after 1 year. Magnetic resonance imaging (MRI) is a reasonable alternative to arthrography for complicated and older dislocated hips (Johnson et al., 1988).

The sonographic anatomy of the hip correlates well with the autopsy findings in comparative cadaver studies (Nichols et al., 1986; Yousefzadeh and Ramilo, 1987; Keller, 1985; Dunn, 1969). The articular cartilage is markedly hypoechoic hyaline cartilage; the unossified acetabular roof and the unossified femoral head, both of which will eventually ossify, are of mixed but mostly low echogenicity. Fibrocartilage of the true labrum is echogenic (Yousefzadeh and Ramilo, 1987) (Fig 8–6).

In the transverse view, the femoral head is normally centered over a somewhat wide, open V. The triradiate cartilage separates the two limbs of the V, and again, through transmission will identify the triradiate cartilage as it did on the coronal view. The posterior limb of the V is the ischium and the anterior limb the pubis. This is similar to the anatomy seen on axial CT images if one were to rotate the CT scan 90 degrees clockwise or counterclockwise depending on the hip being viewed (see Fig 8–2). Only one limb of the V might be seen, because the ossified femoral neck obscures the pubic ramus if the transverse view is obtained with the hip flexed (Fig 8–7). However, centering of the femoral head over the triradiate cartilage will verify its normal position in this plane. One readily sees posterior and lateral displacement in this view (see Fig 8–7).

A stress examination of the hips to assess for joint laxity, subluxability, or dislocatability can be combined with the real-time sonographic evaluation (Berman and Hollingdale, 1987). Mobility of the flexed hip with force applied in an attempt to push it posteriorly (i.e., Barlow's [1962] maneuver) can reveal up to 6 mm of posterior displacement in a normal newborn (Keller et al., 1988b). The amount of displacement decreases gradually to about 1 to 2 mm by 5 months (Keller, 1990).

The dynamic examination of the hips looking for stability is performed in the transverse plane with the hips flexed. During the scan, a Barlow maneuver

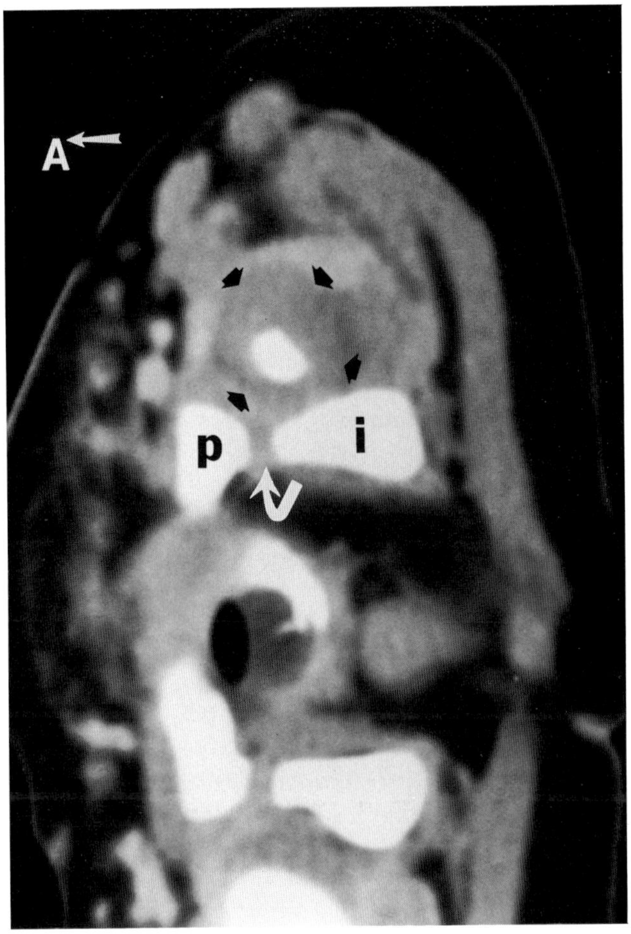


**FIG 8–2.**
Computed tomography scan of the pelvis turned on its side to correspond to the transverse sonographic view. *Straight arrow* points anteriorly *(A)*. The femoral head *(arrowheads)* with its ossified nucleus lies within the acetabulum. Abbreviations: pubis *(p)*, ischium *(i)*, triradiate cartilage *(curved arrow)*.

in which force is applied to the flexed hip in an attempt to force it posteriorly, moving the femur like a piston, is performed. Posterior subluxation or dislocation of the femur with stress signifies a positive maneuver. However, instead of relying on the feel of the hip dislocating, one can see and quantify the amount of motion sonographically. The use of a split screen is useful for this part of the study (Figs 8–8 through 8–10). On one half of the screen, a nonstress transverse view is obtained. With that image frozen, it is matched on the other half of the screen before applying stress. When stress is applied, using the nonstress view as a reference assures that the scanning plane has not changed with the performance of the maneuver. We can then compare the position of the femoral head to the ischium and triradiate cartilage on both views and measure the amount of posterior movement that has occurred (see Fig 8–10).

Joint mobility is not the only important factor in the development of CDH, although it is what is assessed clinically. Much of the literature on sonography of CDH has pertained to standardizing and quantifying the steepness and developmental maturity of the acetabular roof (Fig 8–11). The acetabular shape plays an important role in CDH, and it, like mobility, undergoes a normal period of maturation. In one study, 79 of 82 (96%) hips with steep acetabular roofs on sonography became normal. However, three remained dysplastic, requiring treatment in spite of being stable clinically and sonographically (Castelein and Sauter, 1988).

Frank dislocation with the femoral head behind the acetabulum along the ilium is rare. The hip is more commonly subluxed at the level of the acetabular rim (Clarke, 1986) (Fig 8–12). Containment of the femoral head and coverage by the bony and cartilaginous portions of the acetabular roof are easily determined sonographically. The bony roof should be long and flat (noninclined) enough so that with its cartilaginous portion, it contains the femoral head when force is applied to it (i.e., stability) (Zieger et al., 1986). The center of the femoral head is normally medial to the lateral bony acetabular rim in infants less than 3 months old. Forty-five percent of newborns have an immature pattern with the femoral head's center lateral to the bony rim under the cartilaginous acetabular roof. However, only 10% of infants have this pattern persisting between 5 and 16 weeks of age (Zieger and Schulz 1986). Zieger described a "vector line" to assess osseous acetabular coverage (Zieger, 1986; Zieger and Schulz, 1986).

Bony coverage of the femoral head, as determined sonographically, was compared with the radiographic acetabular index (Morin et al., 1985). All hips with more than 58% of the femoral head covered by the bony acetabulum had normal radiographs. All hips with less than 33% coverage had abnormal radiographs with increased indices (i.e., roof too steep). However, those with intermediate coverage had variable acetabular indices.

The inclination (steepness) of the acetabular roof can be measured sonographically (alpha angle of Graf [1984]) or radiographically (acetabular index) (see Fig 8–6). Correlation between these methods is poor (Morin et al., 1985), most likely due to the lack of common reference points between a tomographic (ultrasound) and nontomographic (x-ray) technique. However, the sonographic alpha angle and the radiographic acetabular index are inversely

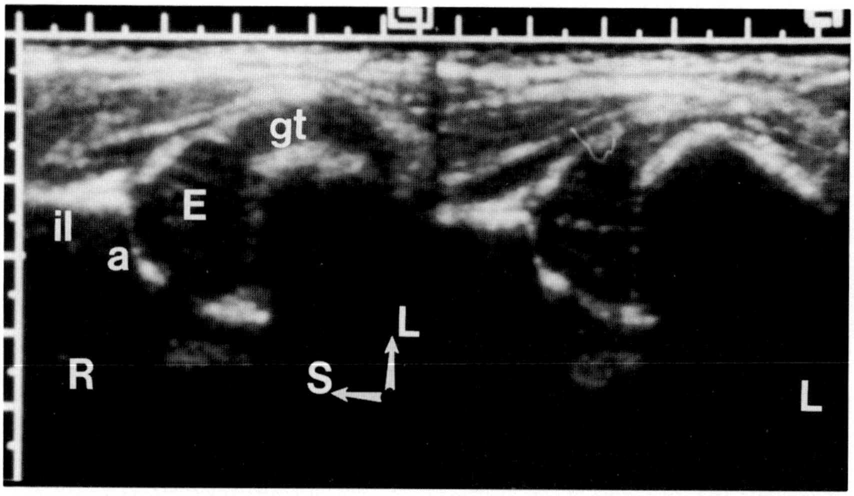


**FIG 8–3.**
Split screen coronal views of right and left hips in a 5-month-old. The hips are oriented horizontally as they would be seen on the viewing screen. The *arrows* point superiorly *(S)* and laterally *(L)*. The unossified femoral epiphysis *(E)* is normally positioned within a well-formed acetabulum *(a)*. The lateral bony margin of the ilium *(il)* and the unossified greater trochanter *(gt)* are shown.

related and are, if only conceptually, grossly complementary.

The dysplastic acetabulum, associated with CDH, is shallow and has a steep acetabular roof. The acetabular angle as measured on plain film is the angle between lines drawn tangent to the acetabular roof and a horizontal reference (Hilgenreiner's) line drawn through both triradiate cartilages. A normal acetabular angle by this method is less than 30 degrees. As the acetabular roof becomes steeper, the acetabular angle as measured by this method also increases. A similar measurement is made sonographically. However, since the reference baseline through the triradiate cartilage (Hilgenreiner's line) cannot be determined sonographically, angle measurement is standardized in another way (Graf, 1984). The coronal view is obtained in a specified standard plane. The bony acetabular promontory and the acetabular labrum should be in view, and the lateral aspect of the ilium should appear straight and horizon-

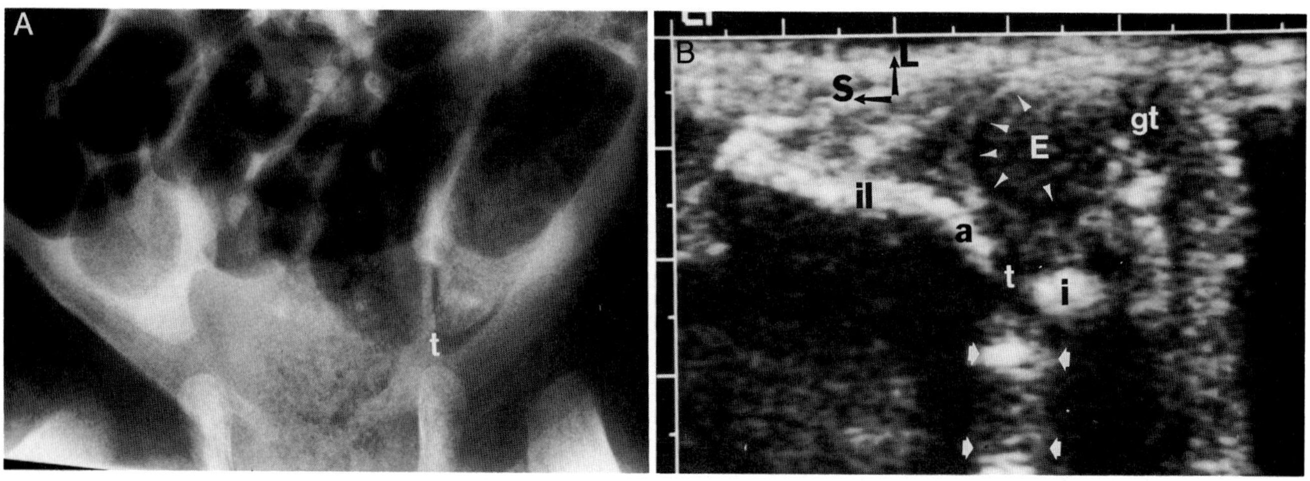


**FIG 8–4.**
**A,** anteroposterior radiograph of the pelvis and, **B,** coronal sonogram (oriented horizontally as it would be seen on the viewing screen) demonstrate a dysplastic left acetabulum with a superiorly and laterally dislocated left hip. The findings are obvious on the radiograph. Sonogram shows the unossified femoral epiphysis *(E,* outlined by *small arrowheads)* and the greater trochanter *(gt)* displaced laterally and superiorly. Triradiate cartilage *(t)* with through transmission (between *broad arrowheads*) on the sonogram. Ilium *(il)*, acetabular roof *(a)*, ischium *(i)*. The right acetabulum is also dysplastic. *Arrows* point superiorly *(S)* and laterally *(L)*.

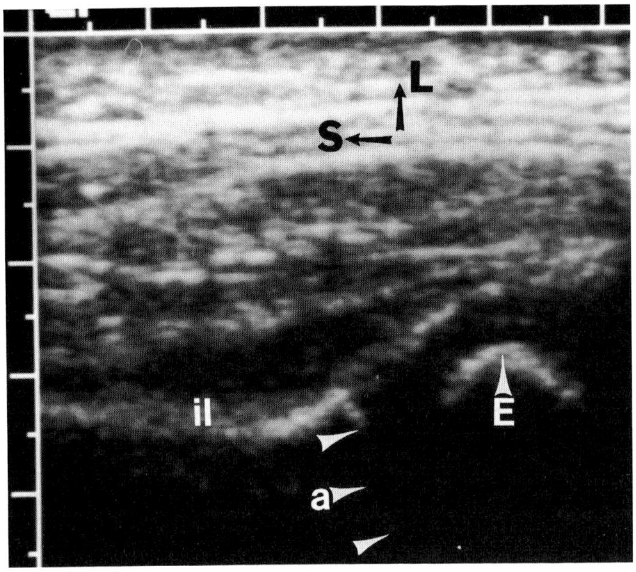


**FIG 8–5.**
Coronal projection as shown on the viewing screen. Ossification is too progressed in this 15-month-old to allow good visualization of the acetabulum *(a)*. The epiphysis is shadowed by its ossified lateral margin *(E)*. Abbreviations: ilium *(il)*. *Arrows* point superiorly *(S)* and laterally *(L)*.

tal on the viewing screen, not curved (Graf, 1987) (see Fig 8–3). For this determination, the patient is often on his or her side with the hip slightly flexed and slightly internally rotated. The sonographic reference baseline is the straight line extension of the lateral aspect of the ilium, which grossly approaches the perpendicular to Hilgreiner's radiographic line. The alpha angle is measured between the baseline and a line drawn from the triradiate cartilage to the lateral aspect of the bony acetabular roof (see Fig 8–6). The less mature or dysplastic the acetabulum is, the steeper is its roof, and the lesser the alpha angle becomes (see Fig 8–11). The alpha angle is important diagnostically because it represents acetabular roof ossification, which promotes joint stability (Graf, 1984).

The less used beta angle is measured between a line drawn from the lateral aspect of the labrum to the lateral bony margin of the acetabulum and the same lateral iliac baseline (see Fig 8–11). The beta angle provides an assessment of eversion of the cartilaginous acetabular roof and, hence, superior femoral head displacement.

Eighty-eight percent sensitivity and 97% specificity for sonography (compared with radiography and clinical examination) were described in one study of 122 children (mean age 22 weeks) who were referred for evaluation (Clarke et al., 1985). Another report of a large screening study cited a 100% sensitivity and 98% specificity of sonography against radiography (Zieger, 1986). Interpretation of such data is difficult since there is no gold standard against which to test sonography. There was a 94% interoperator agreement in the assignment of hip type. Assignment was based on a global visual analysis of the hip, considering the acetabular roof inclination and femoral head position, rather than a single angle measurement. A 4 to 6 degree interobserver variation for determining alpha was also noted (Zieger, 1986).

The continued occurrence of late CDH missed on the neonatal clinical screening examination questions the validity of the clinical test and the premise that the abnormality is present and detectable at birth (Clarke, 1986). Real-time sonography shows not only the position of the unossified femoral head and its displacement from the acetabulum with stress, but also the developmental anatomy or dysplasia of the acetabulum, and is therefore not as limited as the clinical screening examination. Sonography has identified new categories of possible abnormality (Clarke, 1986). It has shown that the progression from clinical instability to stability is not an assurance of reduction and normality. Sonography confirms the abnormality of the clinically unstable hip and aids in establishing its severity, which varies from case to case. Both hips are often abnormal even if only one is clinically suspect.

The false negative clinical examination compared to sonography has been revealed in several studies. Although all 9 of 20 newborns with unstable hips who underwent immediate splinting became normal on follow-up clinical examinations, sonograms, and radiographs at 4 months, the other 11 who were followed by only an orthopedist fared differently. Five stabilized spontaneously, and two remained unstable. Four were thought to be normal clinically (false negative) but were found to be abnormal on sonography (Clarke, 1986). Eighty-two hips in 307 unselected newborns had abnormal sonograms (alpha less than 60 degrees or sonographic instability) with normal clinical examinations. However, 3 of the 82 were definitely dysplastic, whereas 79 reverted to normal sonographically and clinically by 12 weeks (Castelein and Sauter, 1988). Paradoxically, all the hips that developed persistent dysplasia had initial alpha angles between 50 and 60 degrees, whereas all hips (seven) with alpha angles between 40 and 50 degrees became normal (Castelein and Sauter, 1988). Four of 69 neonates at risk had CDH sonographically with normal clinical examinations in another study (Walter et al., 1989).

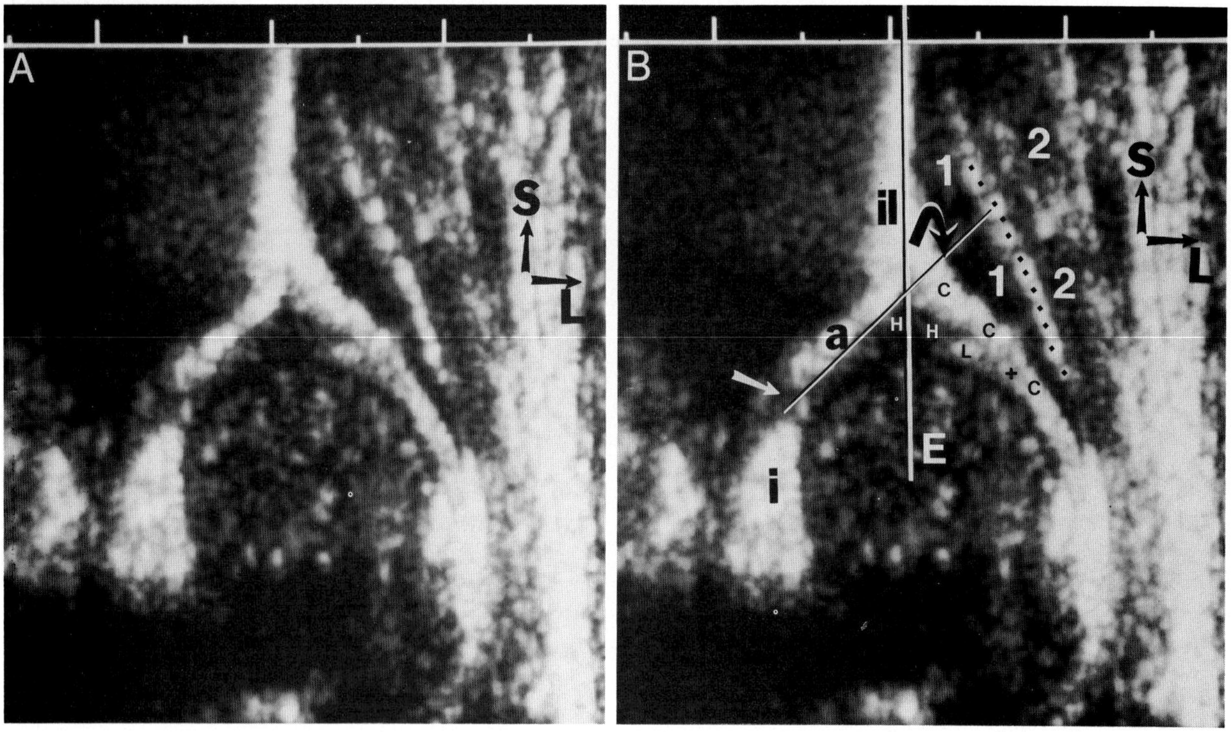


**FIG 8–6.**
Coronal views oriented vertically similar to a frontal radiograph of a left hip are shown without **(A)** and with **(B)** labels. *Arrows* denote superior *(S)* and lateral *(L)*. The unossified epiphysis *(E)* is well seated within the acetabulum *(a)*. Although slightly less than one half of the femoral head is covered by the bony acetabulum as noted by extending a straight line from the lateral iliac margin *(il)*, it is well covered by the lateral, unossified acetabular roof, composed of hypoechoic hyaline cartilage *(H)*, which will later ossify. The lateral margin of the acetabular roof *(L)* is echogenic and represents fibrocartilage of the acetabular labrum. Epiphysis *(E)*, ilium *(il)*, bony acetabular roof *(a)*, unossified acetabular roof (hyaline cartilage) *(H)*, labrum *(L)*, ischium *(i)*, triradiate cartilage *(straight arrow)*, joint capsule *(C)*, gluteus minimus *(1)*, gluteus medius *(2)*. *Dots* outline gluteal intermuscular septum. The alpha angle of Graf *(curved arrow)* is drawn by the intersection of the line parallel to the straight iliac projection *(il)* and the line formed along the bony acetabular roof. It measures the degree of steepness of the bony acetabular roof. The alpha angle of 46 degrees indicates a Graf type 2C hip. The beta angle (not shown), which measures the superior deviation of the cartilaginous roof, is drawn between the lateral aspect of the cartilaginous roof (+) and the intersection of the iliac and acetabular roof lines.

Another series (Clarke et al., 1989) revealed 3 cases of late CDH (10–12 months) among 4,169 low-risk newborns with normal clinical examinations but no sonograms. The remaining 448 newborns born that year had abnormal clinical examinations or were at risk for CDH and underwent screening sonograms. Three hundred fifty had normal sonograms and remained normal in follow-up. Seventeen were abnormal sonographically and were treated although 5 had normal clinical exams (false negative clinical exam). The remaining 81 with normal sonograms became normal without treatment, and in 17 this was in spite of major hip displacement (Clarke et al., 1989).

The palpable and sometimes almost audible "clunk" is the hip relocating with the Ortolani maneuver, a clinical sign of hip dislocatability. In contrast, many soft tissue, bone, and cartilage structures can produce a more delicate "click" with movement of the normal infant hip (Hensinger, 1979). Sonography of 29 neonates with clicks on physical examination revealed normal sonograms in 19 and minor displacements in 10, all of which became normal in 3 to 12 weeks (Clarke, 1986). In another study, dislocation and relocation noted clinically (clunks) in infant hips correlated with increased mobility on sonography, whereas palpable clicks occurred with either normal or increased mobility (Keller et al., 1988a). The findings of some large-scale sonographic screening programs have been reported. Sonography of 11,000 newborns revealed a 1.5% abnormality rate (Schulz and Zieger, 1986). Sonograms obtained for clinical indications were abnormal in 17% in another series (Zieger and Schulz, 1987). Many of the "abnormals,"

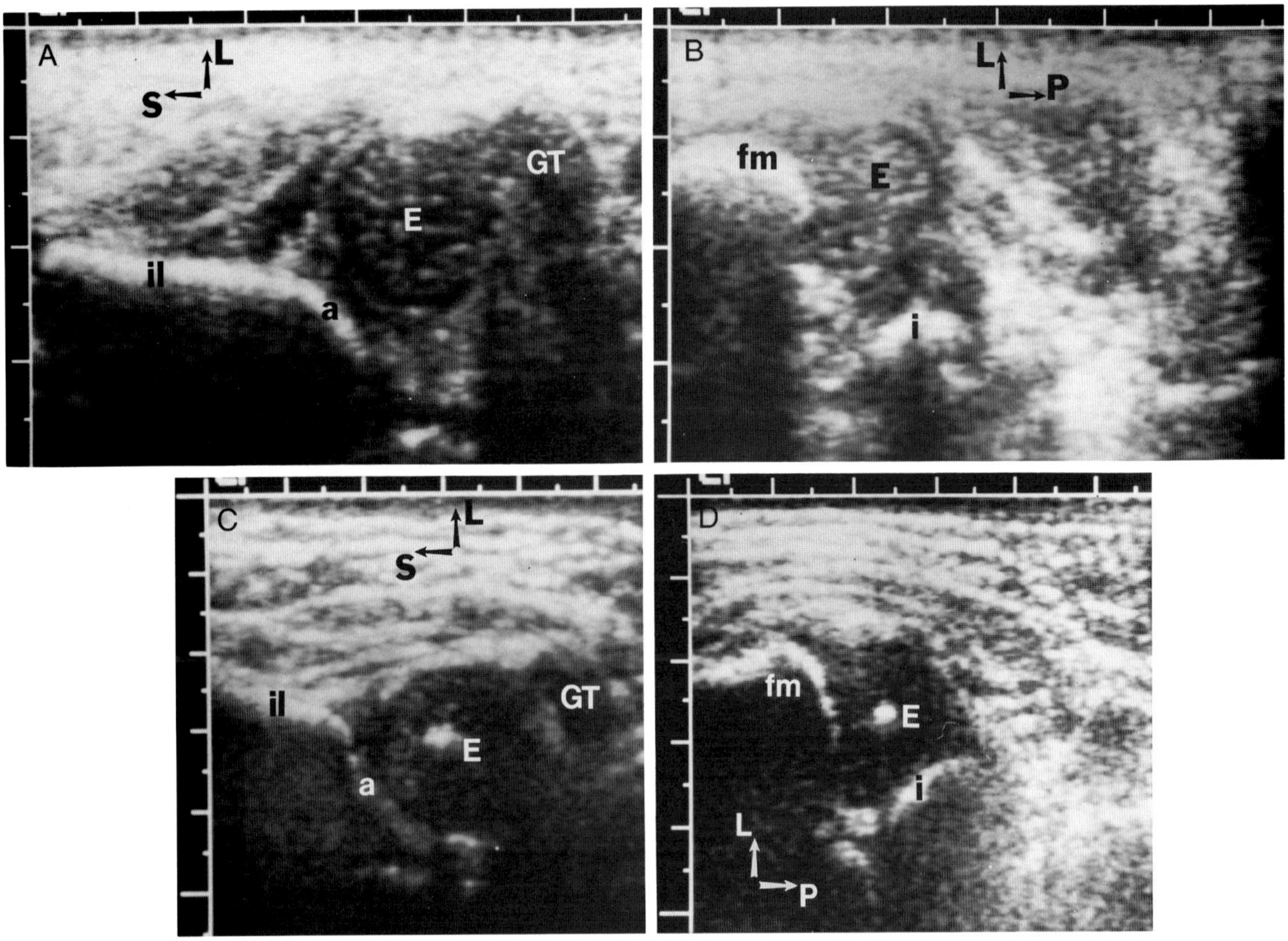


**FIG 8–7.**
**A** and **B,** coronal and transverse views as seen on the viewing screen show a subluxed left hip at 3 days of age (Graf type 3). Orientation arrows point laterally *(L)*, superiorly *(S)*, and posteriorly *(P)*. The epiphysis *(E)* is displaced superiorly, laterally, and posteriorly. Ilium *(il)*, acetabulum *(a)*, femoral metaphysis *(fm)*, ischium *(i)*, greater trochanter *(GT)*. **C** and **D,** follow-up views at 4 months following fixation in a harness show the normally positioned hip within a better-formed acetabulum. The ossific nucleus of the femoral epiphysis *(E)* is seen.

however, were Graf types 1B (a variant with a normal alpha angle and a shorter cartilaginous roof) (Graf, 1987) and 2A and were not treated.

Since only 70% of dysplastic hips will heal if they are unrecognized by 3 months, every newborn was screened sonographically for CDH at one center, and the results in 1,000 newborns were reported (Szoke et al., 1988). Forty-eight percent of the hips were Graf type 1B, 50% type 2A, 1.5% type 2C, 0.3% type 3, and 0.15% type 4 (Szoke et al., 1988). Although approximately 10% of type 2A hips can get worse, none deteriorated in this series. No late diagnosed cases of CDH were reported; however, the authors did not directly address this issue.

Another sonographic screening study of 615 unselected newborns revealed 84.7% with type 1 hips and (according to the least developed hip) 13% type 2A, 2% type 2C, and 0.3% type 3 (Exner, 1988). All hips with alpha angles between 50 and 59 degrees were normal without treatment in 4 to 12 weeks.

Graf 1B and 2A type hips are common in preterm and small for gestational age neonates (Zieger and Hilpert, 1987). Until 3 months, 85% of infants' hips are type 2 (Graf, 1984). The Graf method with its types and subtypes is complicated and intimidating, especially to novices of this classification system. Graf (1984), who introduced, developed, and advocates the sonographic quantitation of acetabular morphology through standard angle measurements, stated, "Inexperienced investigators can use the angle data to gain skill in applying this method, but the use of angles should not be overemphasized." However,

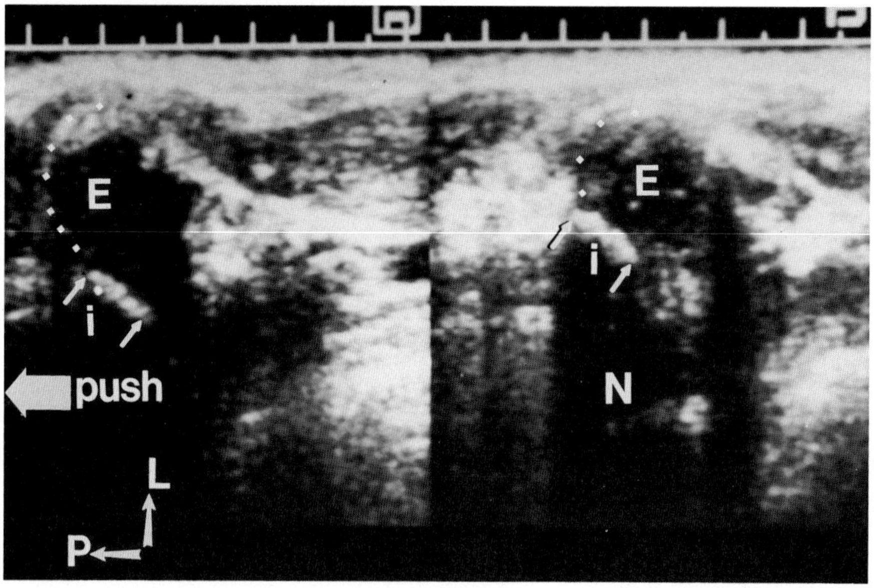


**FIG 8–8.**
Transverse views of a 2-day-old shown with split screen and in horizontal orientation as they would be viewed directly on the viewing screen. Note the rather marked posterior displacement of the unossified femoral head (*E*, outlined by *dots*) on the ischium (*i*, demarcated with *arrows*) with stress. *Arrows* indicate posterior (*P*) and lateral (*L*) directions. Abbreviations: nonstress, neutral view (*N*). Note how, with stress, the ossified metaphysis crosses the triradiate cartilage, thereby preventing the through transmission that was obvious on the neutral view.

such standardization will be useful in comparing clinical studies, because the clinical entity, CDH, will be better defined than it has been in presonographic series. The types, by defining the abnormality (together with the results of stress test), improve the interpretability of studies on the natural history and therapy of CDH. A difficulty in comparing many studies of clinical screening has been the lack of, or vagueness of, a definition of CDH, as criticized by Catterall (1984).

Exner (1988) designed a transparent template that is placed over the coronal view and is used to quickly

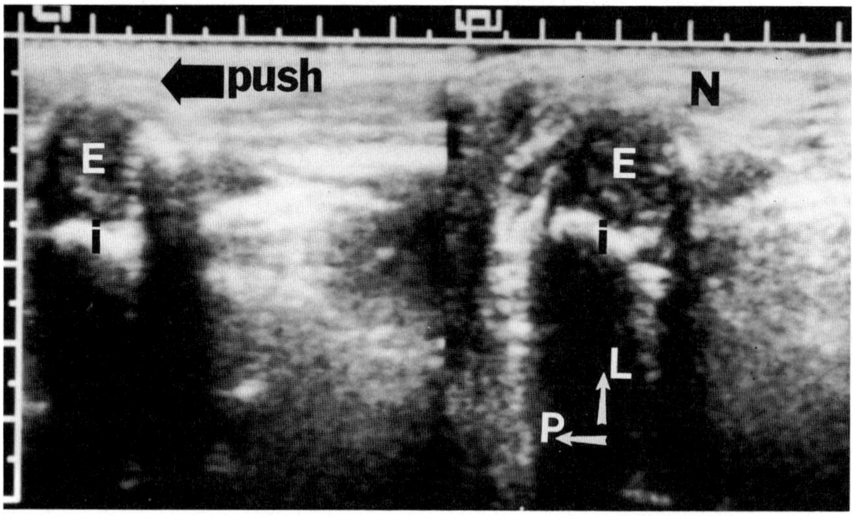


**FIG 8–9.**
Transverse views of the hip of a 2-week-old using a split screen (viewed horizontally) at rest (*N*, neutral) and with Barlow's maneuver. *Arrows* indicate lateral (*L*) and posterior (*P*) directions. The unossified femoral head (*E*) is displaced posteriorly 4 mm from the ischium (*i*) with stress.

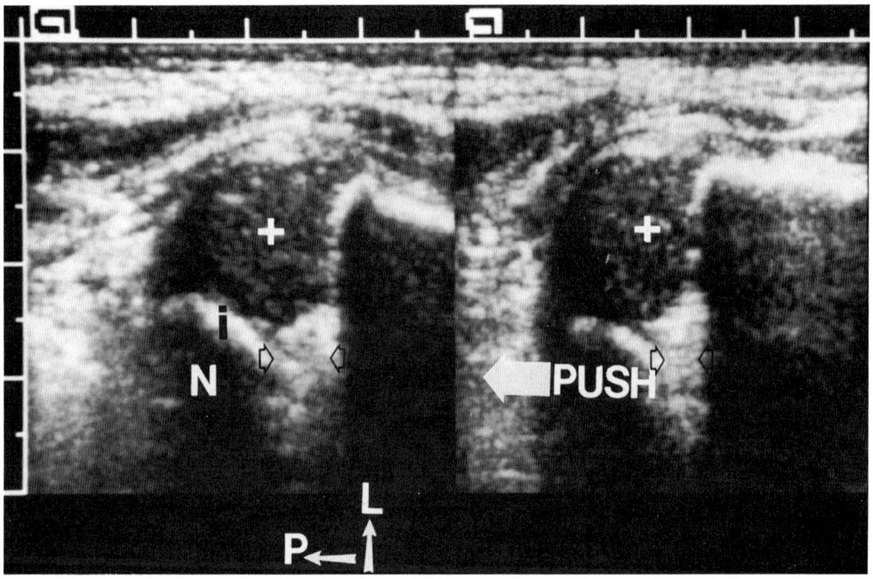


**FIG 8–10.**
Split screen showing the hip in transverse projection oriented horizontally as it would be seen on the viewing screen. The arrows point laterally *(L)* and posterior *(P)*. The left view is without stress *(N)*, and the right view is with stress *(PUSH)*. The femoral head is displaced 2 mm posterior on the ischium *(i)* with the stress maneuver *(Barlow)*. Displacement is best appreciated in real time during the examination; however, it can be measured by either comparing the center of the femoral head (+) with the ischium or by comparing the distance between the anterior edge of the ischium and the metaphysis *(between arrowheads)*.

determine whether the alpha angle is more than 55 degrees and the beta angle less than 72 degrees, which is normal. Alpha 50 degrees is also noted on the template, because a follow-up sonogram is recommended in 4 to 6 weeks to check for maturation. Hips with alpha angles less than 50 degrees are considered dysplastic and are treated. The beta angle is used to distinguish the dysplastic types (2D-4); it increases with superior dislocation of the femoral head and elevation of the cartilaginous acetabular roof.

In addition, another measurement has been proposed, delta angle, to measure the concavity of the acetabular fossa (Schulz and Zieger, 1986; Zieger et al., 1986; Zieger and Schulz, 1987), which is measured between the iliac baseline and a tangent drawn to the ossified acetabular roof near its rim. The use of this angle is not widespread, and it is not clear to me exactly how the tangent is drawn to produce this angle. Yet, ROC analysis found the delta angle to be the best discriminator between normal and dysplasia, followed by Graf's alpha angle (Zieger and Schulz, 1986; Zieger, 1986). The acetabular rim, important for forming the delta angle, is always looked at, at least qualitatively, as to its shape and ossification (Zieger and Schulz, 1987).

A simplified approach to the Graf types is (Table 8–1): the acetabulum with an alpha angle greater than 60 degrees (probably even 55 degrees) is normal; 50 to 59 degrees is followed up in 4 to 6 weeks or possibly treated if older than 3 months; less than 50 degrees is treated. Reduction can be checked sonographically with the hip in a Pavlik harness (Boal and Schwentker, 1985; Grissom et al., 1988) or in a spica cast, although in the latter case, CT (Hernandez, 1984) is probably easier and doesn't require cutting a hole in an important part of the cast to get standard coronal and transverse views.

In conclusion, who should undergo sonography for CDH? Should newborn sonographic screening be universal? The detection of occult cases using sonography in infants with normal clinical examinations has been demonstrated. This could be the way to reduce the incidence of "late CDH" that has not diminished with the adoption of large clinical screening programs. On the other hand, the high incidence of physiological immature hips will warrant a large number of follow-up sonograms to check interval maturation of these hips, which will occur in the vast majority of cases. Otherwise, overtreatment could result in some cases of avascular necrosis.

Should newborn sonographic screening be limited to high-risk patients or those with abnormal or equivocal clinical examinations? Since the clinical exam will select the hips to be studied, clinically si-

**TABLE 8–1.**
Simplified Synopsis of Graf Sonographic Hip Types*†

| Type | Description | Alpha Angle (Degrees) | Beta Angle (Degrees) | Comment | Treatment |
|---|---|---|---|---|---|
| 1 | Normal | *>60†* | | Should not dislocate in absence of a neuromuscular imbalance with altered biomechanics | None |
| 2A | Physiological immature *<3 mo.* | *50–59* | | | Observe until matures to type 1 |
| 2B | Delayed ossification *>3 mo.* | 50–59 | | | Abduction orthosis |
| 2C | Very deficient bony acetabulum but femoral head still concentric | *43–49* | <77 | At this stage beta angle becomes important | Abduction orthosis |
| 2D | Femoral head *subluxed* | 43–49 | >77 | Increased beta angle signifies an elevated, everted labrum and subluxation | Reduction |
| 3 | *Dislocated* | <43 | >77 | | Reduction |
| 4 | *Severe* dysplasia/ dislocation | *Not measurable;* flat, shallow, bony acetabulum | | *Labrum inverted; interposed* between femoral head and ilium | Possible open reduction |

*Adapted from Donaldson JS: Pediatric musculoskeletal US, in Poznanski AK, Kirkpatrick JA Jr (eds): *Diagnostic Categorical Course in Pediatric Radiology.* Oak Brook, Ill, Radiological Society of North America Publications, 1989, pp 77–88; and Graf R: *Guide to Sonography of the Infant Hip.* New York, Thieme Medical Publishers, 1987.
†Distinguishing feature of each type is in italics.

lent CDH will never be studied sonographically and will continue to be missed, whereas many hips studied for clicks or laxity will be normal.

The question as to who should be studied in sonographic screening programs for CDH is complicated. Yet the fact remains that sonography detects cases of CDH that clinical examination misses.

## ROTATIONAL DEFORMITIES

Although sonographic assessment of lower extremity rotational deformities, femoral anteversion (Moulton and Upadhyay, 1982; Phillips et al., 1985; Terjesen and Anda, 1987), and tibial torsion (Joseph et al., 1987) have been described, CT seems to be easier, more precise, and more reproducible and therefore remains the method of choice (Hernandez 1983, Hernandez et al., 1981). The sonographic technique for rotational measurements is presented in chapter 9.

## EPIPHYSEAL ALIGNMENT

Unossified epiphyses are invisible radiographically. However, as already shown in the hip, they are well demonstrated sonographically—a fact that can be used to particular advantage in infants. Radiographic malalignment of the ossified humeral shaft with the glenoid or with the ossified radial or ulnar shafts following a difficult, traumatic delivery, could indicate shoulder or elbow dislocation, respectively. Yet, in children the physis is relatively weak compared with ligaments and tendons, in which case separations of the epiphyses from their metaphyses commonly occur. Distinguishing dislocation of the entire bone, including its unossified epiphysis from a displaced Salter-Harris fracture of the epiphysis (i.e., slipped epiphysis), is often impossible on plain radiography in the neonate and young infant, although at this age the latter should always be suspected. Formerly, arthrography was required to outline the epiphysis, showing its relationship to the joint and the shaft (White et al., 1987). Now, sonography can often achieve the same result.

One must utilize multiple projections when scanning. An offset (water bag or gel pad) might be helpful when the structures are very superficial and in the transducer's near field. Remaining oriented by looking not only at the screen but at the extremity and the transducer with frequent comparisons with the normal side, similarly positioned, is crucial. In my experience, "being there" was very important to

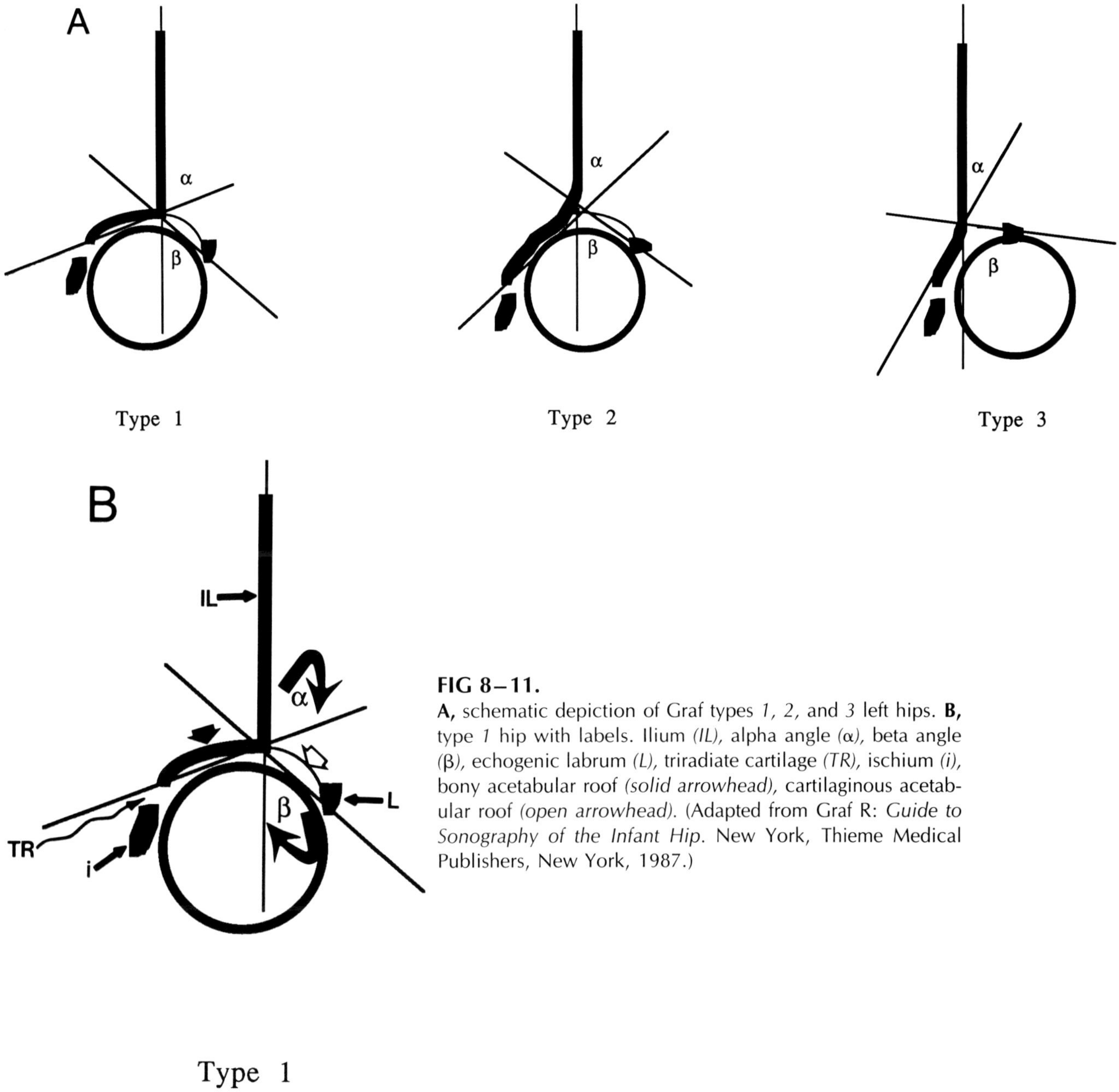


**FIG 8–11.**
**A,** schematic depiction of Graf types *1*, *2*, and *3* left hips. **B,** type *1* hip with labels. Ilium *(IL)*, alpha angle *(α)*, beta angle *(β)*, echogenic labrum *(L)*, triradiate cartilage *(TR)*, ischium *(i)*, bony acetabular roof *(solid arrowhead)*, cartilaginous acetabular roof *(open arrowhead)*. (Adapted from Graf R: *Guide to Sonography of the Infant Hip*. New York, Thieme Medical Publishers, New York, 1987.)

establish optimal, interpretable views. Because these injuries are uncommon, one will not develop a large experience as with CDH or hip effusions. A little hands-on experimentation might be needed in each case.

A slipped proximal humeral epiphysis is well demonstrated in the coronal plane from a lateral approach (Zieger et al., 1987a). A dislocated distal humeral condylar fracture is seen in the longitudinal projection from a posterior approach (Fig 8–13) (Dias et al., 1988). Notice how the displaced condylar fracture (see Fig 8–13) and the congenital elbow dislocation (Fig 8–14) associated with Larsen's syndrome appear similar on plain radiography since the epiphyses are unossified. Yet sonography, obtained from the posterior aspect of the fractured elbow, showed the humeral condyles aligned with the proximal forearm epiphyses and shafts, all of which were displaced posterior to the humeral shaft (see Fig 8–13). However, the epiphyses remain aligned with their shafts in congenital dislocations (Figs 8–14 and 8–15).

Slipped capital femoral epiphyses, a cause for the painful hip in older children, may be noted on sonography (Dörr et al., 1988). It is readily detected on

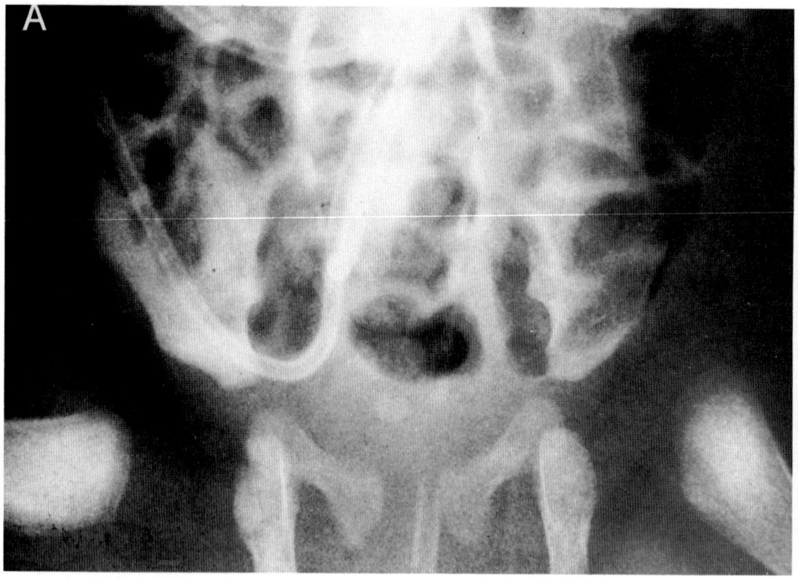


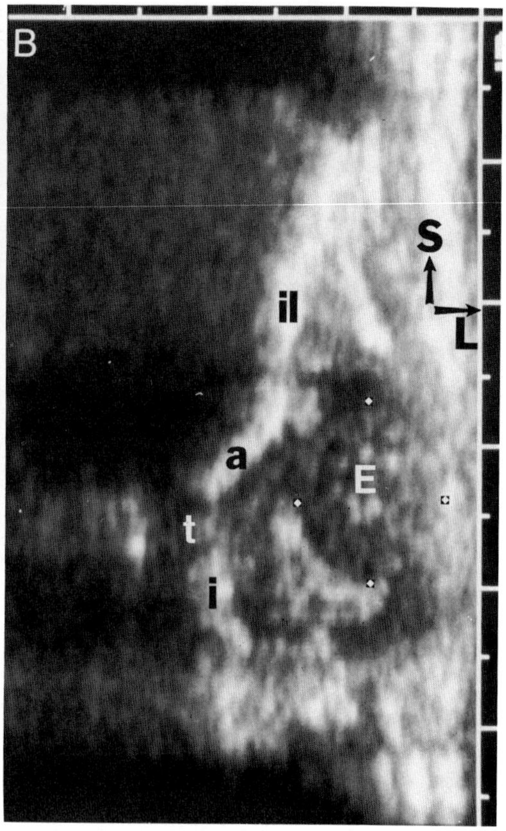


**FIG 8–12.**
**A,** dislocated left hip in neonate with myelomeningocele (note lumbosacral dysraphism and ventriculoperitoneal shunt tubing). The dislocated left hip and dysplastic left acetabulum are obvious on the plain radiograph. **B,** the corresponding coronal sonogram depicted vertically to match the radiograph shows the markedly dislocated unossified femoral epiphysis *(E,* surrounded by *dots).* The short, steep, acetabular roof *(a)* with shallow acetabulum is noted. Abbreviations: ischium *(i),* triradiate cartilage with through-transmission *(t),* ilium *(il).* Arrows point superiorly *(S)* and laterally *(L).*

plain film, and most likely the sonogram would have been obtained looking for joint effusion as a cause of the hip pain.

The ability of sonography to show unossified bones has been used to check alignment of the navicular with the talus in infants with congenital vertical talus (Schlesinger et al., 1989). The unossified navicular is visualized by scanning sagittally along the dorsum of the foot with the distal tibia, talus, and navicular in view (Fig 8–16).

## PAINFUL EXTREMITY

The site of abnormality is often not clinically apparent in the young child with a painful extremity. The child might merely be unwilling to move it or have it moved. Joint effusion, especially of the hip, is often a consideration that is difficult to exclude clinically or by plain film radiography. Joint aspiration under fluoroscopy with a postaspirate arthrogram to document intra-articular needle position, especially in cases of a dry tap, has been the standard way to exclude hip joint effusion. However, the experience with sonography of joints (especially hip) has increased so that it, instead of aspiration and arthrography, can serve to noninvasively diagnose or exclude joint effusion (Adam et al., 1986; Alexander et al., 1989).

In contrast to scanning the hip from a lateral approach as is done for CDH, the joint is studied from an anterior (ventral) approach in longitudinal and transverse planes because fluid tends to accumulate anterior to the femoral neck in the inferior, caudal recess of the joint (Fig 8–17). One scans in the sagittal plane over the extended hip and rotates the transducer slightly to be parallel with the long axis of the femoral neck. External rotation of the hip will bring the medial, most inferior recess of the joint space into the scanning plane, aiding in the detection of small effusions (Fig 8–18).

Effusions as small as 1 ml can be detected (Marchal et al., 1987). Three criteria have been used to detect hip effusion:

1. An absolute measurement where the anterior joint capsule is displaced 3 mm or greater anterior to the femoral neck at a point halfway between the physis and the caudal insertion of the joint capsule on the femoral neck.
2. Distention of the joint capsule 2 mm more than the asymptomatic, presumably normal side. It is

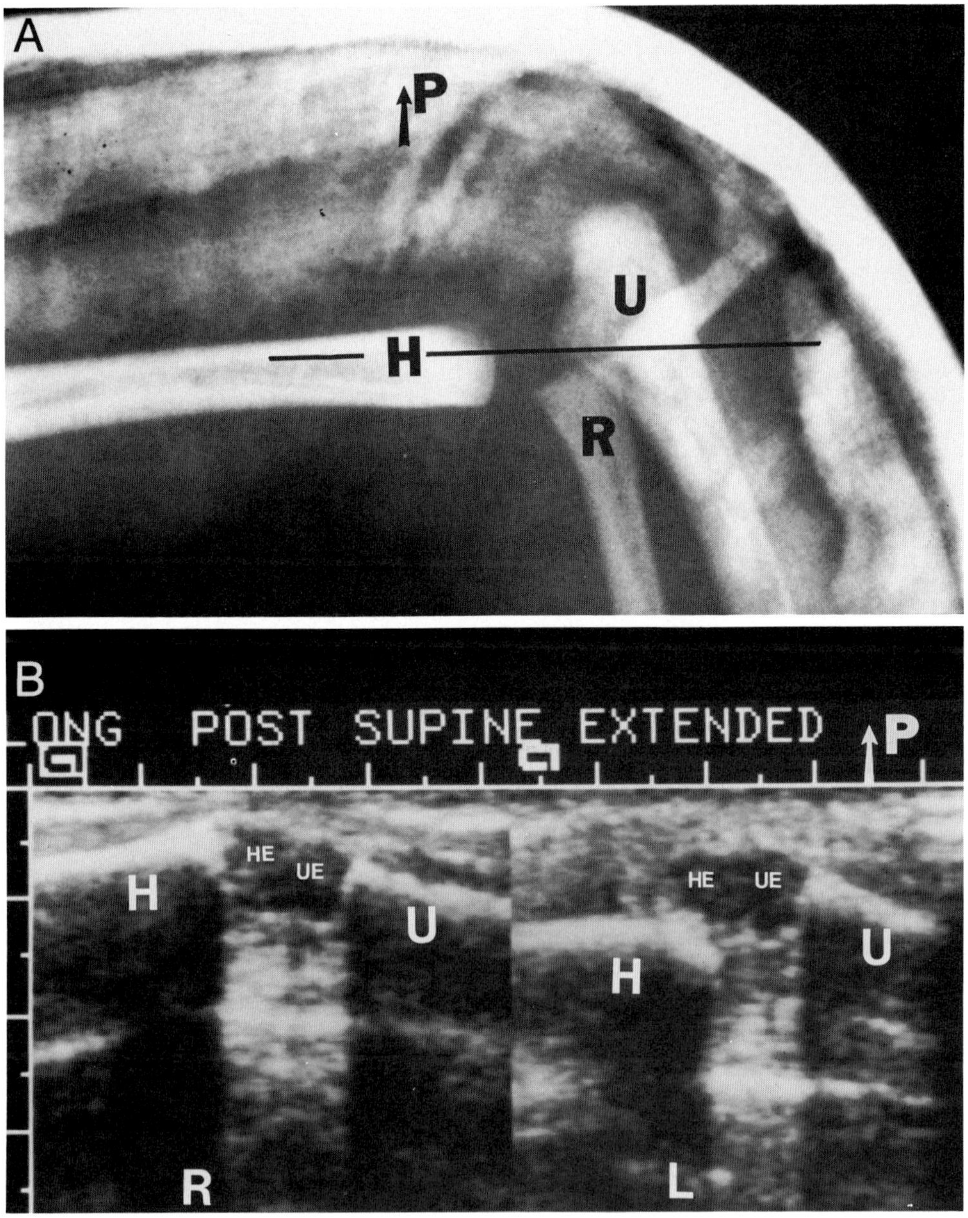


**FIG 8–13.**
**A,** lateral radiograph of this newborn's left elbow in a posterior splint shows obvious elbow malalignment. The *arrow* points posteriorly. The radius *(R)* and ulna *(U)* are posterior to the humerus *(H)*. Status of the unossified elbow epiphyses is unknown from the radiograph. **B,** sagittal views of both elbows were obtained from their posterior aspects and are shown on the split screen in a projection matching the radiograph. The right hypoechoic distal humeral *(HE)* and proximal ulnar *(UE)* epiphyses appear aligned with their shafts (i.e., normal). Note the through transmission through the cartilaginous epiphyses. The distal left humeral epiphysis is separated from its shaft, indicating that this is not an elbow dislocation but that it is a fracture with displacement of the distal humeral epiphysis, which has maintained its relationship with the radius and ulna.

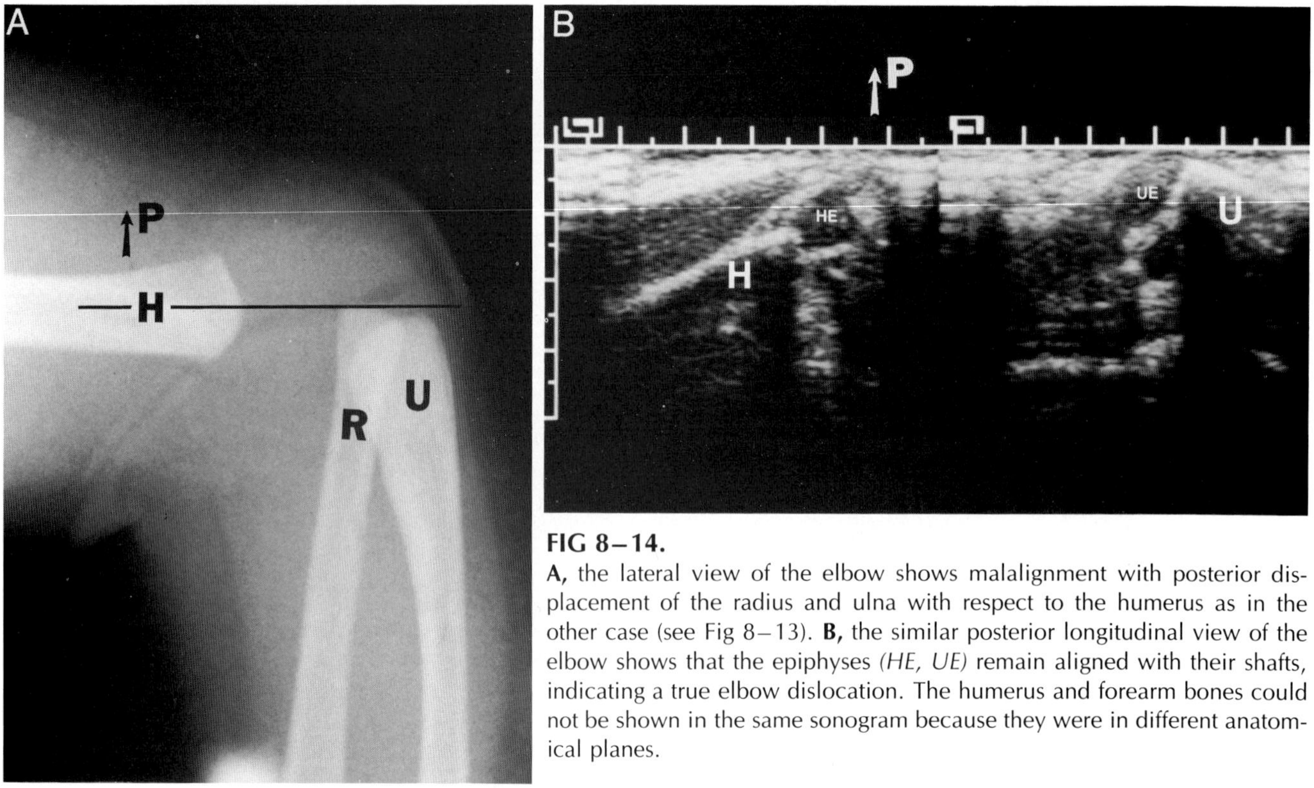


FIG 8–14.
**A,** the lateral view of the elbow shows malalignment with posterior displacement of the radius and ulna with respect to the humerus as in the other case (see Fig 8–13). **B,** the similar posterior longitudinal view of the elbow shows that the epiphyses *(HE, UE)* remain aligned with their shafts, indicating a true elbow dislocation. The humerus and forearm bones could not be shown in the same sonogram because they were in different anatomical planes.

important, therefore, that the scanning planes and the positions of the hips (i.e., extension, rotation) be identical.

3. Simple observation of the shape of the joint capsule. It is normally concave anteriorly, parallel to the curvature of the femoral neck. If it is convex anteriorly, this indicates bulging of the joint capsule, which in most cases is due to an effusion.

The anterior bulge of the joint capsule is best seen with the hip extended. When the hip is flexed, the fluid is redistributed. The anterior capsule is not as

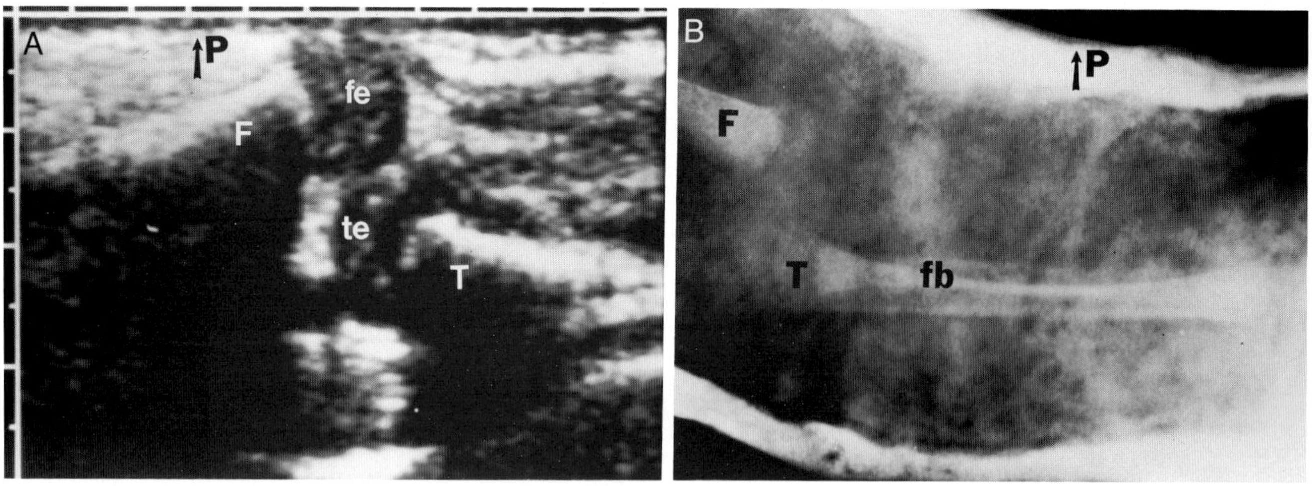


FIG 8–15.
**A,** longitudinal sonogram obtained from the posterior aspect of the knee and in the same orientation as the plain radiograph shows that the unossified distal femoral epiphysis *(fe)* and proximal tibial epiphysis *(te)* remain with their respective shafts, indicating a true dislocation. **B,** the lateral radiograph obtained through a plaster cast shows a dislocated knee in this neonate. The *arrow* points posteriorly *(P)*. The tibia *(T)* and fibula *(fb)* are displaced anteriorly with respect to the femur *(F)*.

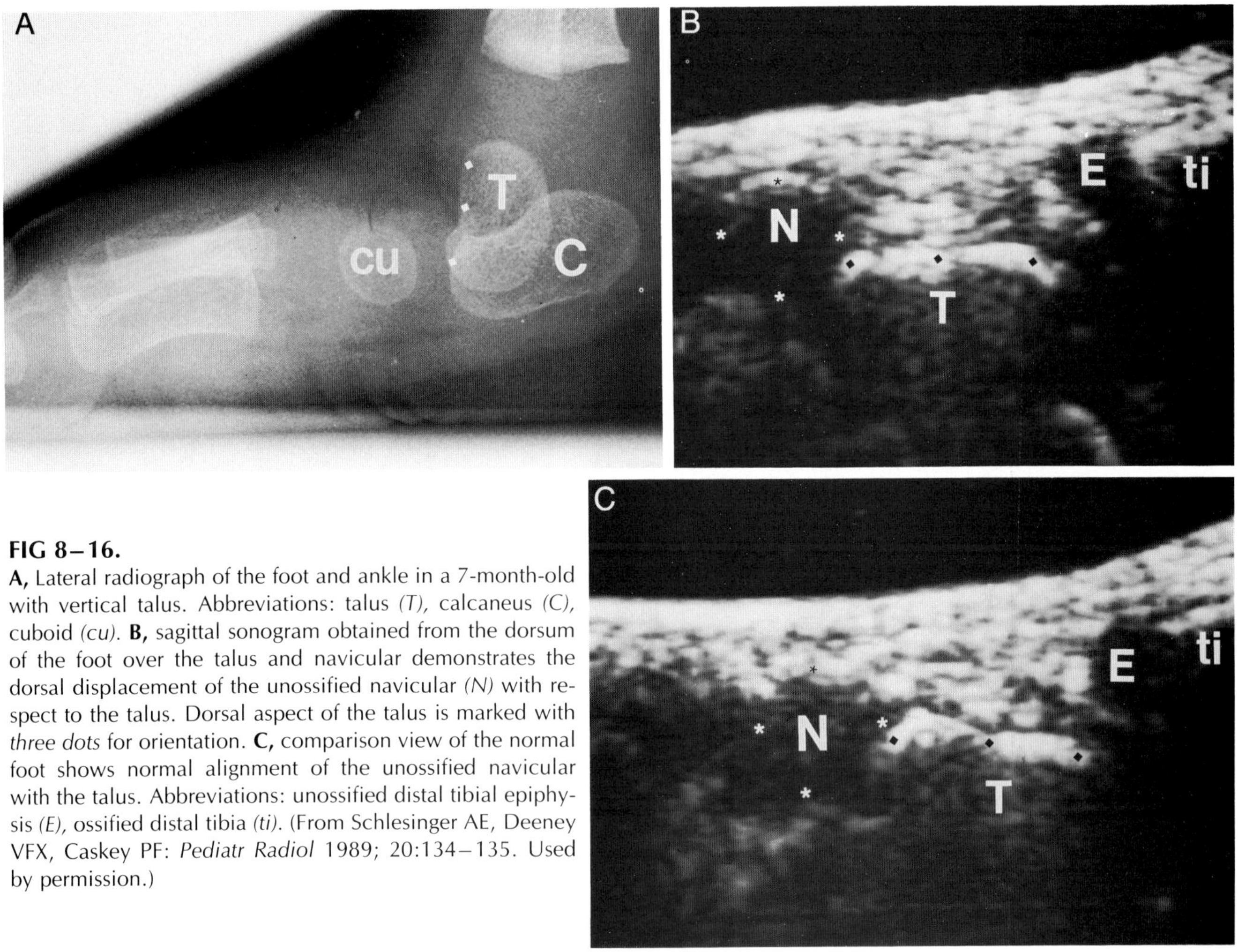


**FIG 8–16.**
**A,** Lateral radiograph of the foot and ankle in a 7-month-old with vertical talus. Abbreviations: talus *(T)*, calcaneus *(C)*, cuboid *(cu)*. **B,** sagittal sonogram obtained from the dorsum of the foot over the talus and navicular demonstrates the dorsal displacement of the unossified navicular *(N)* with respect to the talus. Dorsal aspect of the talus is marked with *three dots* for orientation. **C,** comparison view of the normal foot shows normal alignment of the unossified navicular with the talus. Abbreviations: unossified distal tibial epiphysis *(E)*, ossified distal tibia *(ti)*. (From Schlesinger AE, Deeney VFX, Caskey PF: *Pediatr Radiol* 1989; 20:134–135. Used by permission.)

distended, and hence, the fluid is not as apparent sonographically (Marchal et al., 1987). An irritable hip might be flexed and difficult to extend, precluding adequate sonography. The ability of sonography to determine the nature of the effusate and to distinguish septic arthritis from transient synovitis is not completely determined. Increased echogenicity of the fluid (difficult to ascertain in effusions less than 2–3 mL) and a thick articular capsule (more than 2 mm) suggests septic arthritis (Dörr et al., 1988; Zieger et al., 1987b). However, the main value of sonography is to confirm or exclude the presence of a joint effusion. A dry aspirate of a hip in which an effusion was noted sonographically indicates that the hip joint was missed. A study of 61 hip sonograms and arthrocenteses in 48 patients (6 months–13 years) yielded 4 hips with normal sonograms from which fluid was aspirated (i.e., false negative sonogram). In all cases the fluid was less than 1.0 mL (Foster et al., 1989).

Effusions in other joints, such as knees or ankles, can be detected sonographically, although they are more often apparent clinically than are hip effusions (Fig 8–19 through 8–21). Sonography can differentiate an effusion from periarticular soft tissue swelling.

Legg-Calvé-Perthes disease and slipped capital femoral epiphysis have been detected sonographically (Suzuki et al., 1987; Dörr et al., 1988) although the diagnoses are usually made radiographically. The irregular, fissured femoral head might be visible in the former and the misaligned femoral head in the latter. An effusion can be seen with either diagnosis (Dörr et al., 1988). Effusion might be seen with early Perthes disease, and follow-up is therefore needed if pain persists.

Osteomyelitis has been detected sonographically (Abiri et al., 1989) as a small focal fluid collection directly in contact with bone. Occult fractures have also been detected (Graif et al., 1988). Hip sonography on a 2-month-old who wouldn't move her left leg excluded a hip effusion. However, sonography of the

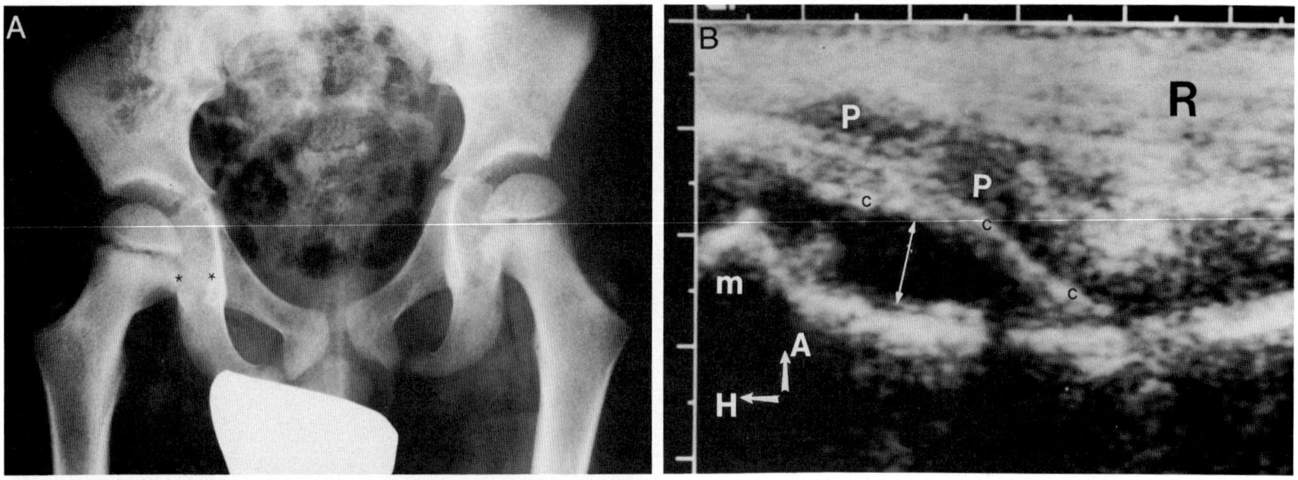


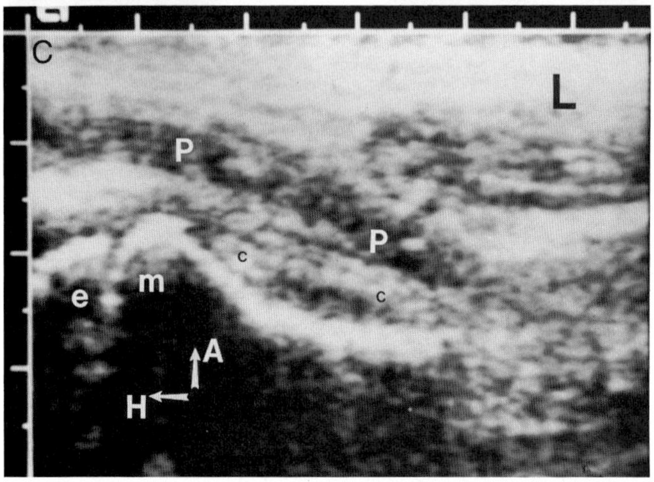


**FIG 8–17.**
**A,** the anteroposterior radiograph demonstrates slight widening of the medial aspect of the right hip joint in this 5-year-old with a painful right hip. **B,** sagittal view obtained over the femoral head and neck shows the marked right hip effusion *(double-headed arrow),* which causes bulging of the joint capsule *(C).* **C,** the normal left hip is shown for comparison. Psoas *(P),* femoral epiphysis *(e),* femoral metaphysis *(m). Arrows* point anteriorly *(A)* and toward the head *(H).*

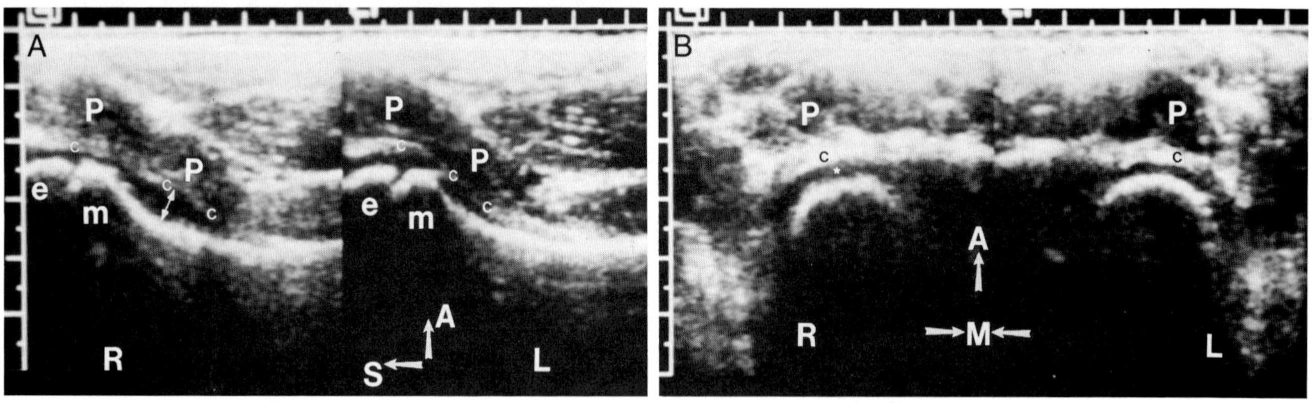


**FIG 8–18.**
Sagittal **(A)** and transverse **(B)** views of both hips viewed on split screens demonstrating a small right hip effusion (*double-headed arrow* on sagittal view and *star* on transverse view). One milliliter of clear fluid was aspirated. *Arrows* point anteriorly *(A),* superiorly *(S),* and medially *(M).* Abbreviations: joint capsule *(C),* psoas *(P),* femoral epiphysis *(e),* femoral metaphysis *(m),* right *(R),* left *(L).*

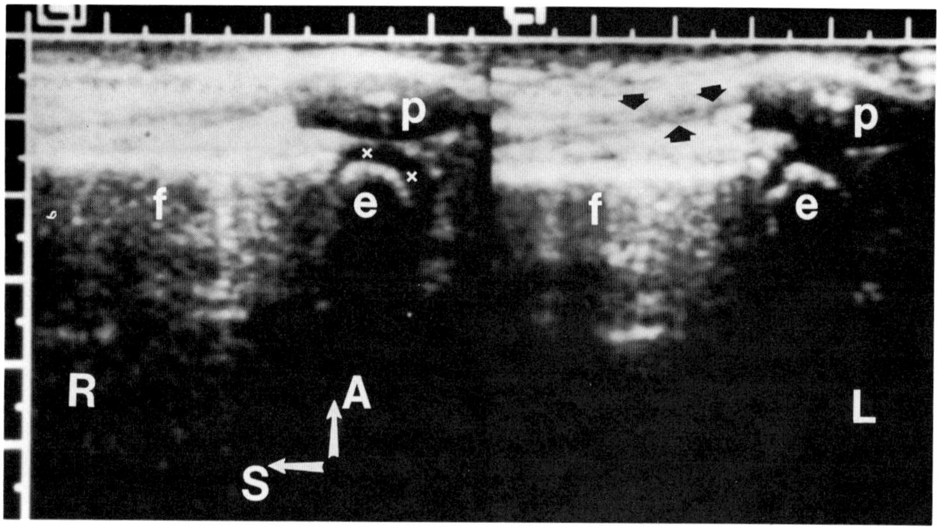


**FIG 8–19.**
Sagittal views of both distal femurs *(f)* viewed on a split screen (*R,* right; *L,* left). Patella with small central ossific nucleus *(p),* ossified distal femoral epiphysis *(e)* with hypoechoic articular cartilage *(x).* The quadriceps femoris tendon is seen inserting into the patella *(arrowheads). Arrows* point anteriorly *(A)* and superiorly *(S).*

rest of the leg showed a metaphyseal (Salter-Harris II) corner fracture of the distal tibia, which was confirmed by radiography and was later shown to be due to child abuse (Fig 8–22).

When the cause of a limp or irritable extremity cannot be localized, as is often in the case in a young child, radionuclide bone scan is useful to detect the focus of the problem (Alexander et al., 1988).

Sonography has been used to measure the thickness of muscle and subcutaneous tissue (Heckmatt et al., 1988) and to evaluate skeletal muscle in children with various types of muscular dystrophies. Its ability to recognize involved muscle groups can be used to select sites for needle muscle biopsy (Heckmatt and Dubowitz, 1987).

## SOFT TISSUE MASSES AND THE SPINAL CANAL

A child presenting with a soft tissue mass or cutaneous abnormality such as a hairy patch, midline

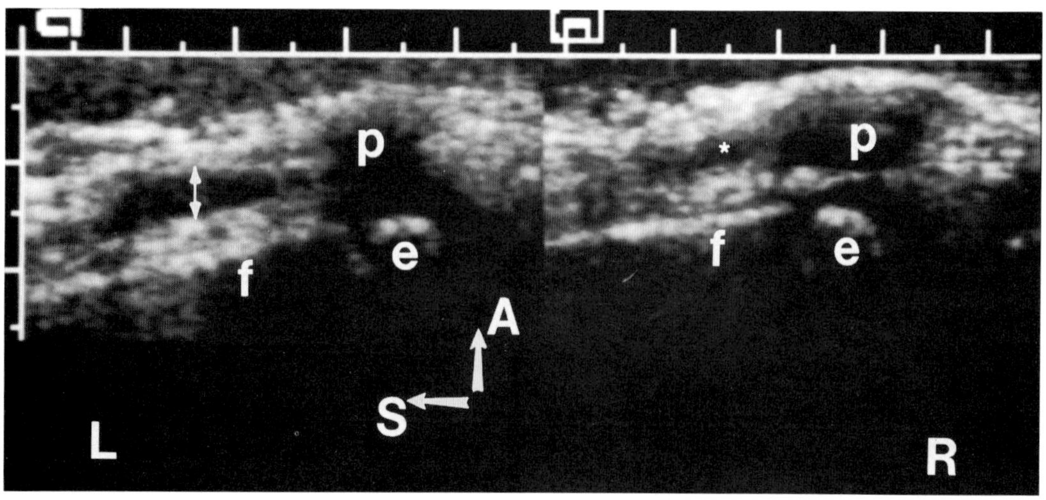


**FIG 8–20.**
Sagittal views of both distal femurs on a split screen. There is a broad, hypoechoic *(double-headed arrows)* band in the left suprapatellar region, indicating a left knee effusion. There is a much smaller similar finding in the right knee, which is either quadriceps femoris tendon or a tiny right knee effusion. Patella *(p),* femoral shaft *(f),* distal femoral epiphysis *(e). Arrows* point anteriorly *(A)* and superiorly *(S).*

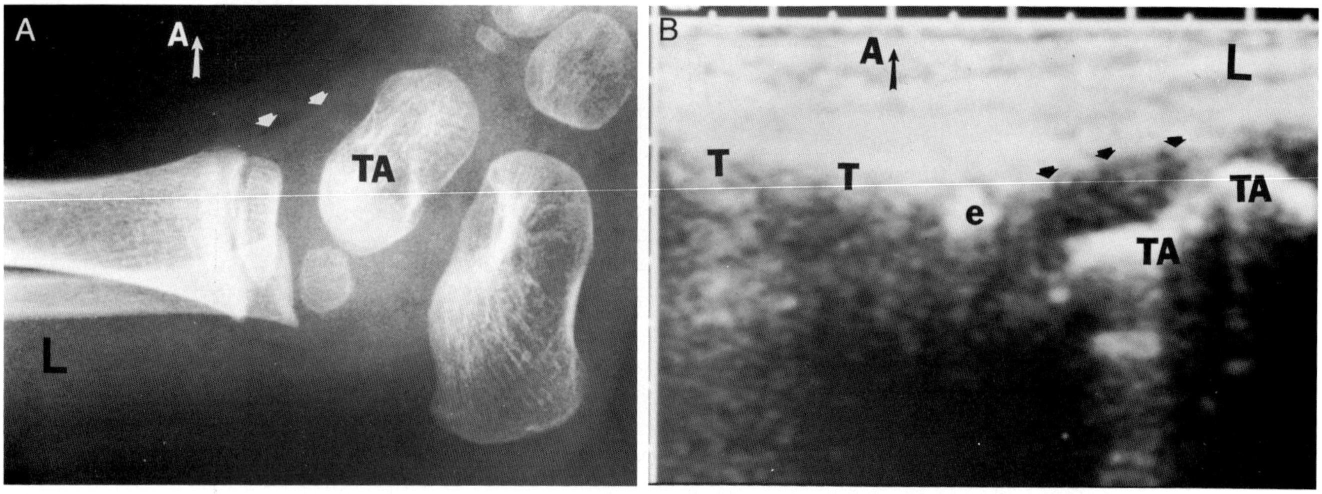


**FIG 8–21.**
**A,** lateral radiograph of the left ankle rotated to match the sonogram (*arrow* points anteriorly). Note the prominent deep soft tissues in the ankle joint *(arrowheads),* anterior to the talus *(TA)* and consistent with an ankle joint effusion. **B,** the sagittal sonogram of the left ankle demonstrates the joint effusion *(arrowheads)* anterior to the talus *(TA).* Abbreviations: distal tibial shaft *(T),* distal tibial epiphysis *(e). Arrow* points anteriorly.

dimple, or hemangioma on the back might indeed have an associated tethered spinal cord (Albright et al., 1989; Hall et al., 1981; Storrs and Walker, 1987). This is especially true with subcutaneous lipomas (Emery and Lendon, 1969; Lhowe et al., 1987). One should know how to evaluate the spinal canal as well as the mass, and therefore this section is included in this chapter.

The soft tissue mass or hemangioma can be seen sonographically (Glasier et al., 1987; Yamaguchi et al., 1987) (Fig 8–23). Although in most cases the sonography does not appear to be specific as to the tissue type of the soft tissue mass, the depth and extent of the mass can often be determined sonographically (Figs 8–24 and 8–25). It is important to see if the mass extends into the spinal canal and to determine the level of the conus medullaris. In cases of a hairy patch, hemangioma, or dimple, the spinal cord can be tethered with a low conus medullaris without seeing any direct continuation from the cutaneous abnormality.

The spinal canal is easily seen sonographically in

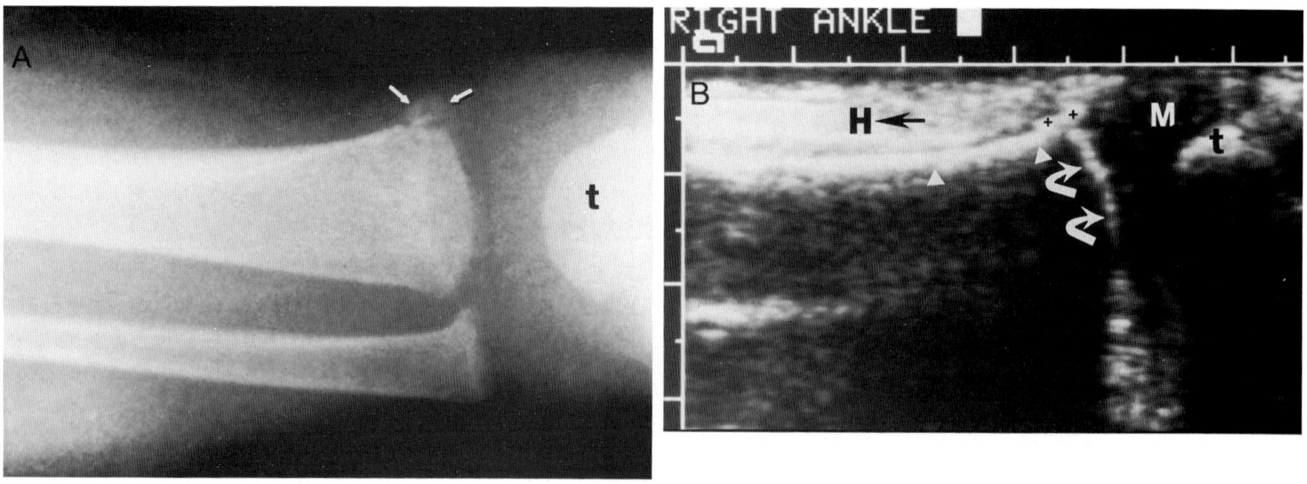


**FIG 8–22.**
**A,** longitudinal (coronal) sonogram of the distal tibia demonstrating a corner metaphyseal fracture *(arrows)* in an abused 2-month-old. **B,** the plain radiograph is oriented similarly for comparison. Abbreviations: unossified medial malleolus *(M),* talus *(t)* distal tibial metaphysis *(curved arrow),* distal medial tibial shaft *(arrowheads).* The *arrow* points toward the patient's head.

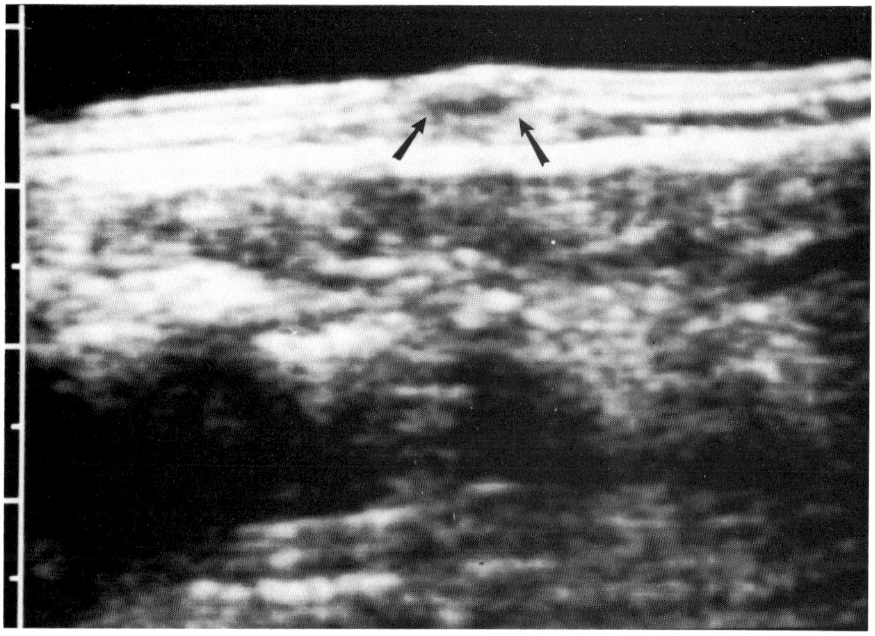

**FIG 8–23.**
Sonogram of a 7-mm superficial hemangioma *(arrows)* on a 10-month-old's back.

the newborn and young infant (DiPietro and Venes, 1988; Gusnard et al., 1986; Kawahara et al., 1987; Naidich et al., 1984, 1986; Raghavendra et al., 1983; Rubin et al., 1988; Scheible et al., 1983). However, in contrast to what one might initially think, older infants and children can also be evaluated sonographically (DiPietro and Venes, 1988; Rubin et al., 1988). Increased ossification and thickness of the soft tissues precludes obtaining the broad panoramic view in the older child compared with that obtained in the neonate and young infant. However, with careful technique, the spinal canal can be visualized between bony elements, and the level of the conus medullaris can be determined (Fig 8–26).

### Technique

The patient is examined prone, often over a pillow or blanket roll that flexes the spine and separates the posterior spinal elements to allow better access to

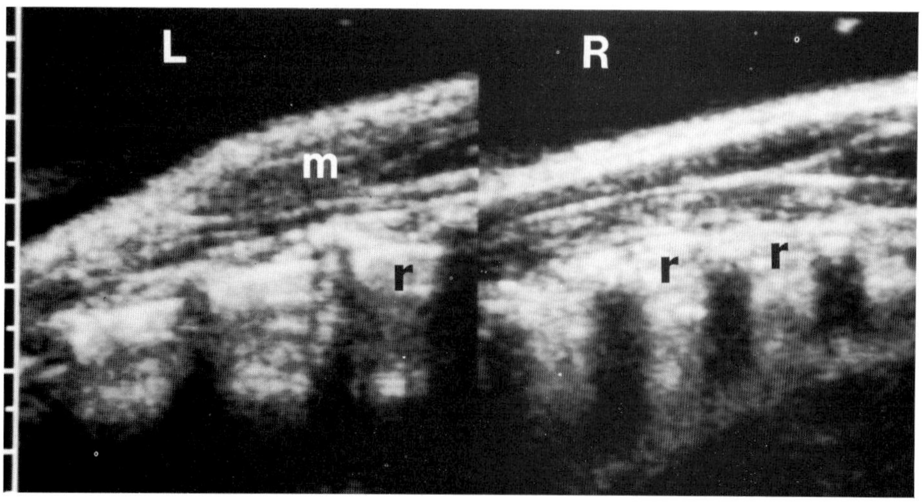


**FIG 8–24.**
Longitudinal sonogram of a lipoma in the upper left back. It was found to be well circumscribed at surgery. The normal comparable right side is shown for comparison on the split screen. *r* denotes some of the ribs.

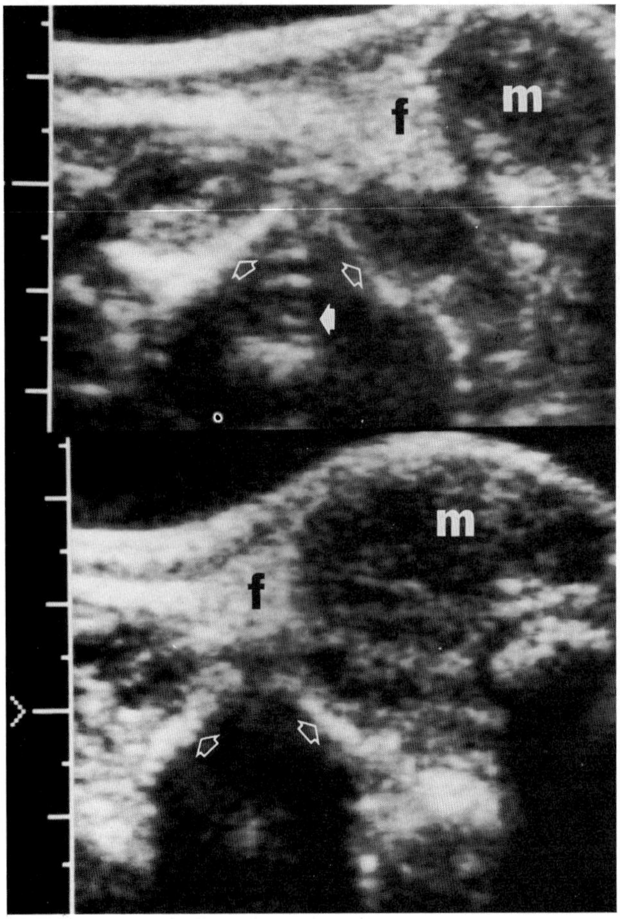


**FIG 8–25.**
Transverse sonograms obtained over the upper back showing a 6 by 3 cm lobulated mass medial to the right scapula in a 5-day-old. The mass *(m)* is deep in the soft tissues but was separate from paraspinal musculature, and there was no evidence of invasion of the spinal canal. At surgery and pathology, it was found to be a hemangiopericytoma. Abbreviations: fat *(f)*, thoracic lamina *(open arrowheads)*, spinal cord *(closed arrowhead)*.

the spinal canal. The child can likewise be studied in the decubitus posture. Midline scanning succeeds in the infant as the posterior spinal elements are incompletely ossified. In the child, a slightly lateral, interlaminar approach with medial angulation toward the spinal canal is needed (DiPietro and Venes, 1988). With the transducer parasagittal, off the midline but angled medially, the sonographic beam passes between lamina and into the spinal canal (see Fig 8–26).

It is probably easiest to start scanning over the sacrum because one can identify the sacral elements that appear like steps as they ascend posteriorly and hence toward the transducer (Fig 8–27). The anechoic, fluid-filled thecal sac is identified. Often the caudal extent of the sac is at S2 (Koroshetz and Taveras, 1986) (see Fig 8–27,C). While scanning in the sagittal plane is continued, the transducer is moved more cephalad into the lumbar canal (see Fig 8–27,B). Scanning is continued up into the thoracic canal (see Fig 8–27,A), although the upper and middle thoracic canal are more difficult to visualize as the child gets older due to the marked overlap of the posterior thoracic spinal elements. The spinal cord is identified as a hypoechoic tubular structure with an echogenic line in its midportion (see Fig 8–27,A). This line is believed to be the central canal, although there has been some controversy in the literature in this regard. The caudal extent of the spinal cord, the conus medullaris, has the appearance of the tip of a carrot in the sagittal view (see Fig 8–27,B). Echogenic strands are seen extending caudally from it, the roots of the cauda equina (see Figs 8–27,B, C, G, and H). Even in the older child, when visualization of the spinal canal is only partial, one may infer the position of the conus when at one level the spinal cord is clearly seen and when at the next level only roots of the cauda equina are present (see Fig 8–26).

In the transverse plane, the spinal cord appears round. The upper lumbar cord is larger than the middle and low thoracic cord (see Figs 8–27,D and E). As one continues to scan caudally, one encounters the tapering of the spinal cord, the conus medullaris, which sometimes assumes a somewhat crescent shape (see Figs 8–27,F and G). The tip of the conus medullaris should be no lower than the inferior aspect of L2 (Wilson and Prince, 1989) (see Fig 8–27,C). In my experience, the level of the conus medullaris has often been either clearly normal, in the high lumbar canal near the thoracolumbar junction, at the level of the origin of the lowest ribs, or clearly abnormal, in the low lumbar or sacral canal. If the patient is cooperative, one can count vertebral bodies and estimate the level of the conus. The split screen is sometimes useful (see Fig 8–27,C). Any question as to the level of the conus when it is near the midlumbar canal is resolved by marking the level of the conus on the skin and then obtaining a radiograph with a small marker (e.g., BB) at that site. The vertebral level of the conus can then be determined from the plain film.

A sinus tract can sometimes be seen using a standoff pad (Fig 8–28). A subcutaneous hump on the lower back, usually a lipoma, can be seen extending into the spinal canal with an associated low conus medullaris, confirming an associated tethered spinal cord (Figs 8–29 and 8–30). This is often

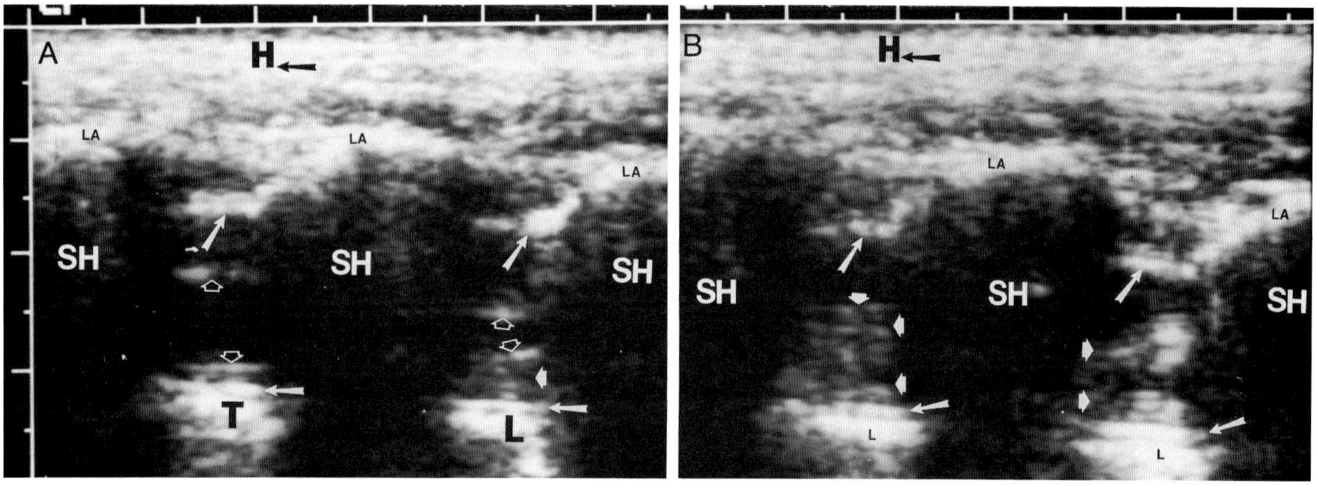


**FIG 8–26.**
Sagittal views of the low thoracic to midlumbar spinal canal in a 9-year-old. **A,** normal tapering of the spinal cord *(open arrowheads),* indicating the position of the conus medullaris at the thoracolumbar junction. **B,** below that level, the cord is no longer seen. Only echogenic strands of the cauda equina *(closed arrowheads)* are visible. Though ossification precludes obtaining a panoramic view as in small infants, the lowest extent of the spinal cord can be determined from observing the spinal canal between the extensive shadowing *(SH)* from laminae *(LA). Arrows* denote dura. Abbreviations: thoracic vertebra *(T),* lumbar vertebra *(L).* Patient's head is toward the viewer's left.

readily apparent with careful scanning. Intraspinal extension of a paraspinal mass can occasionally be seen sonographically (Patel, 1985), although not to the extent or as easily as it is on CT or MRI (Fig 8–31).

## ILIZAROV LEG LENGTHENING

A recently described use for sonography is to evaluate for new bone growth in patients who have recently undergone an Ilizarov leg lengthening procedure (Young et al., 1990). This procedure, named for its developer, Dr. G.A. Ilizarov of the Soviet Union, has been recently introduced in the United States, although it has been used in Europe and was originally described more than 30 years ago. A corticotomy (not osteotomy) is performed while leaving the periosteum intact. Fine, tensile wires that have been inserted through the bone are attached to an external fixator ring, and the two fragments are slowly and gradually stretched apart at a rate of approximately 1 mm daily. The patient is ambulatory throughout this because the external device provides support. As new bone forms along the interzone, it can be seen sonographically as thin, echogenic strands bridging the gap between 4 and 7 weeks before they are visible radiographically (Fig 8–32). Documentation of this early visualization is important, because if the distraction is too rapid, new bone will not form, and if it is too slow, the gap will ossify too quickly, before the desired lengthening will have been attained. We have found the 5-MHz linear array transducer to be most useful, although higher frequencies might even be better. A sector scanner, with its smaller head, might be needed early, before much distraction has occurred and the fixator rings are still relatively close together.

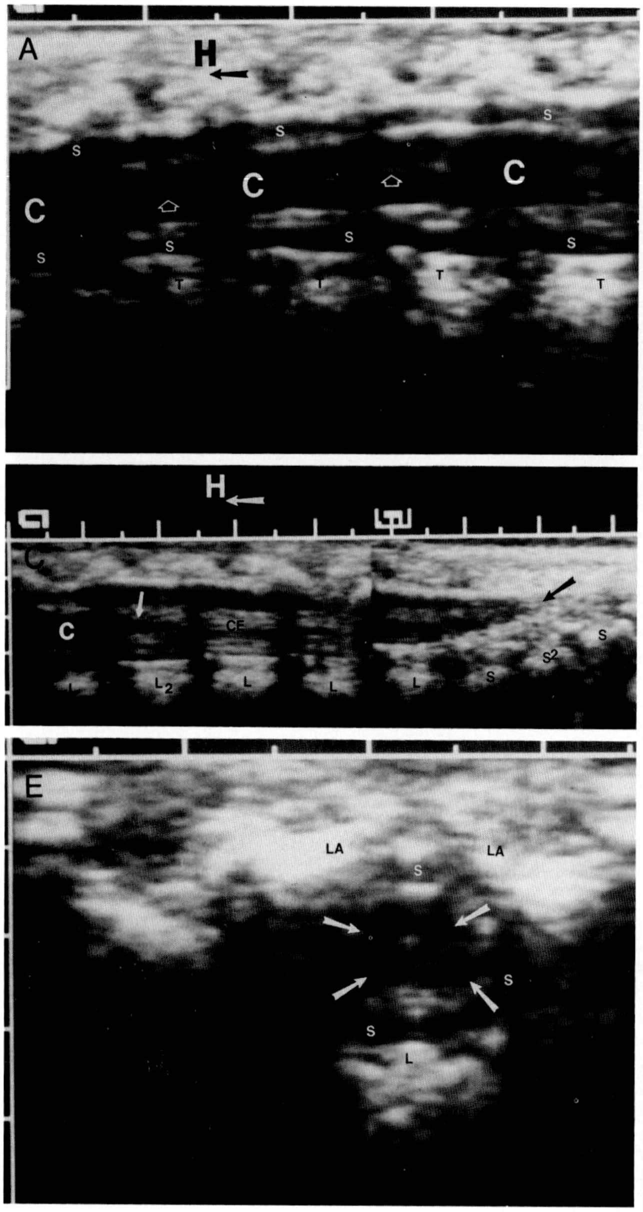


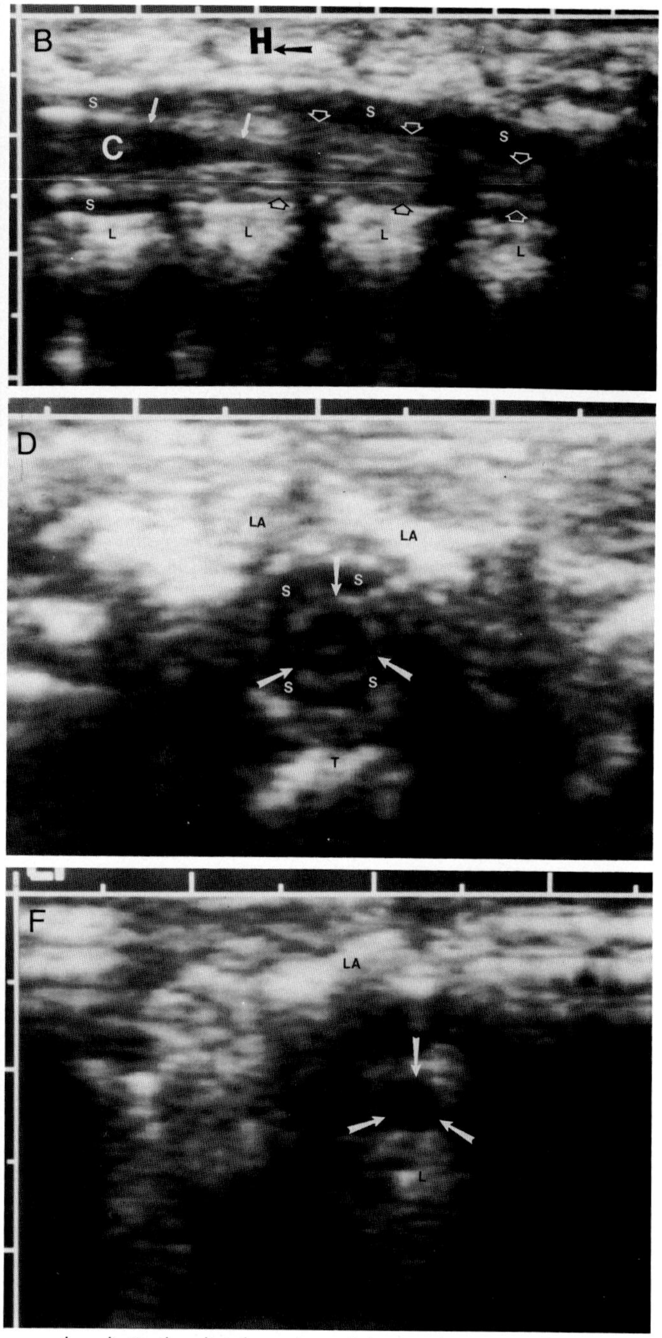


**FIG 8–27.**
**A–C,** intraspinal sonography of a normal 5-week-old. Sagittal views of the back with the patient's head toward the viewer's left. **A,** the hypoechoic middle to lower thoracic and lumbar spinal cord *(C)* is shown with the echogenic central canal *(open arrowhead)*. **B,** the distal portion of the cord tapers *(white arrows)* forming the conus medullaris. Longitudinal echogenic nerve roots *(open arrowheads, CE)* extend through the remainder of the spinal canal as the cauda equina. **C,** the split screen enables one to accurately depict most the lumbosacral canal in one picture. The most caudal portion of the thecal sac *(black arrow)* often corresponds to S2. Abbreviations: subarachnoid space *(white S)*, thoracic vertebra *(T)*, lumbar vertebra *(L)*, sacral segments *(black S)*. **D–H,** transverse views of the spinal canal at various levels. **D** and **E,** the hypoechoic spinal cord is outlined *(straight arrows)*. In **E,** the spinal cord is thickest at the high lumbar level. **F** and **G,** in the upper lumbar canal, the cord *(arrows)* tapers at the conus medullaris *(open arrowheads)* surrounded by echogenic nerve roots forming the cauda equina *(CE)*. **H,** the cauda equina extends into the lower lumbosacral canal appearing as echogenic clusters *(wavy arrows)*. Abbreviations: thoracic vertebra *(T)*, lumbar vertebra *(L)*, lamina *(LA)*, subarachnoid space *(S)*, the margins of the low lumbar canal *(closed arrowheads)*.

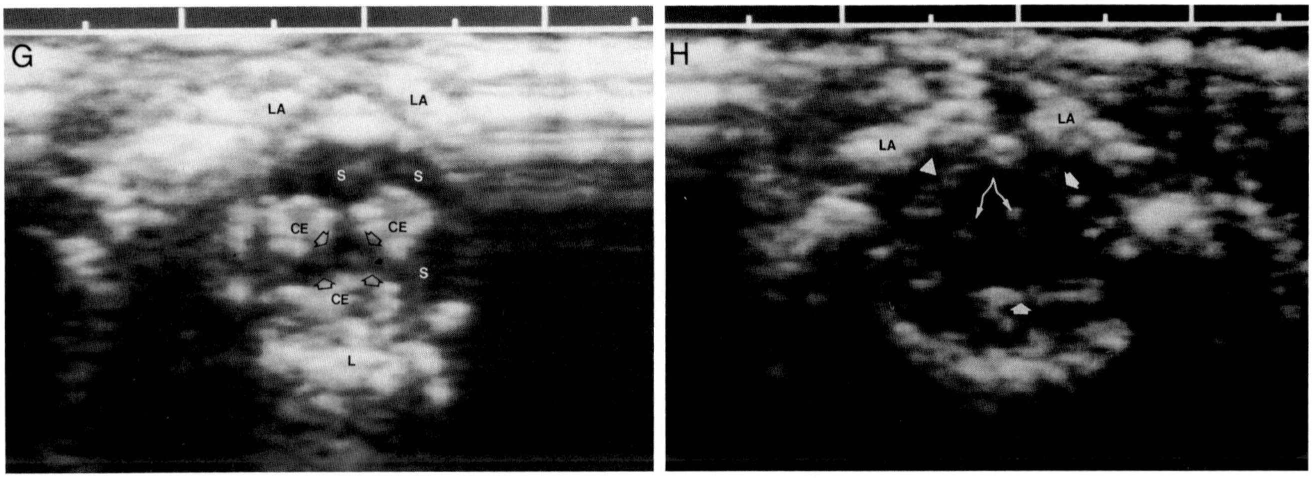


**FIG 8–27 (cont.).**

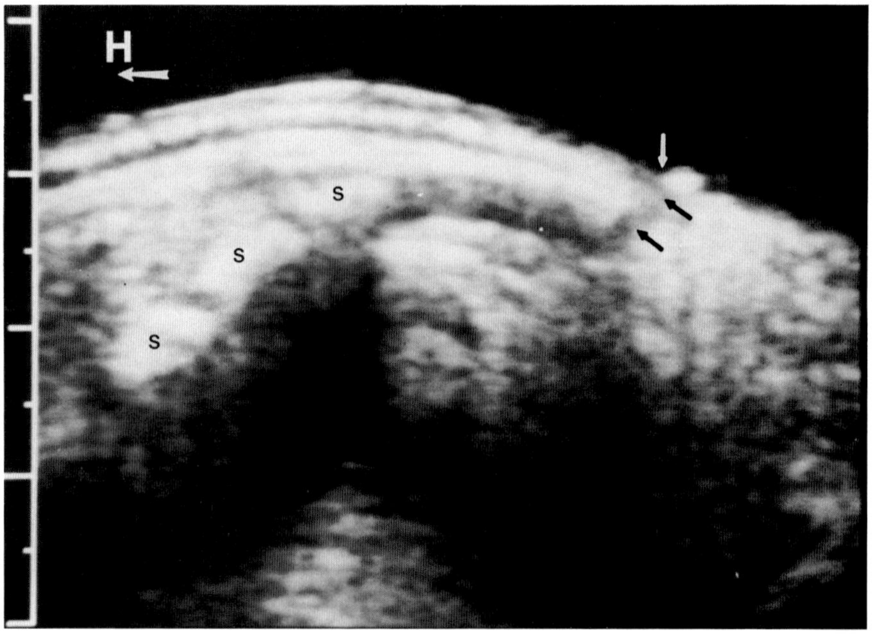


**FIG 8–28.**
Sagittal sonogram obtained over the lower portion of the sacrum *(S)* showing a very low pilonidal sinus *(arrows)*, far below the sacrum and spinal canal. Patient's head is toward the viewer's left.

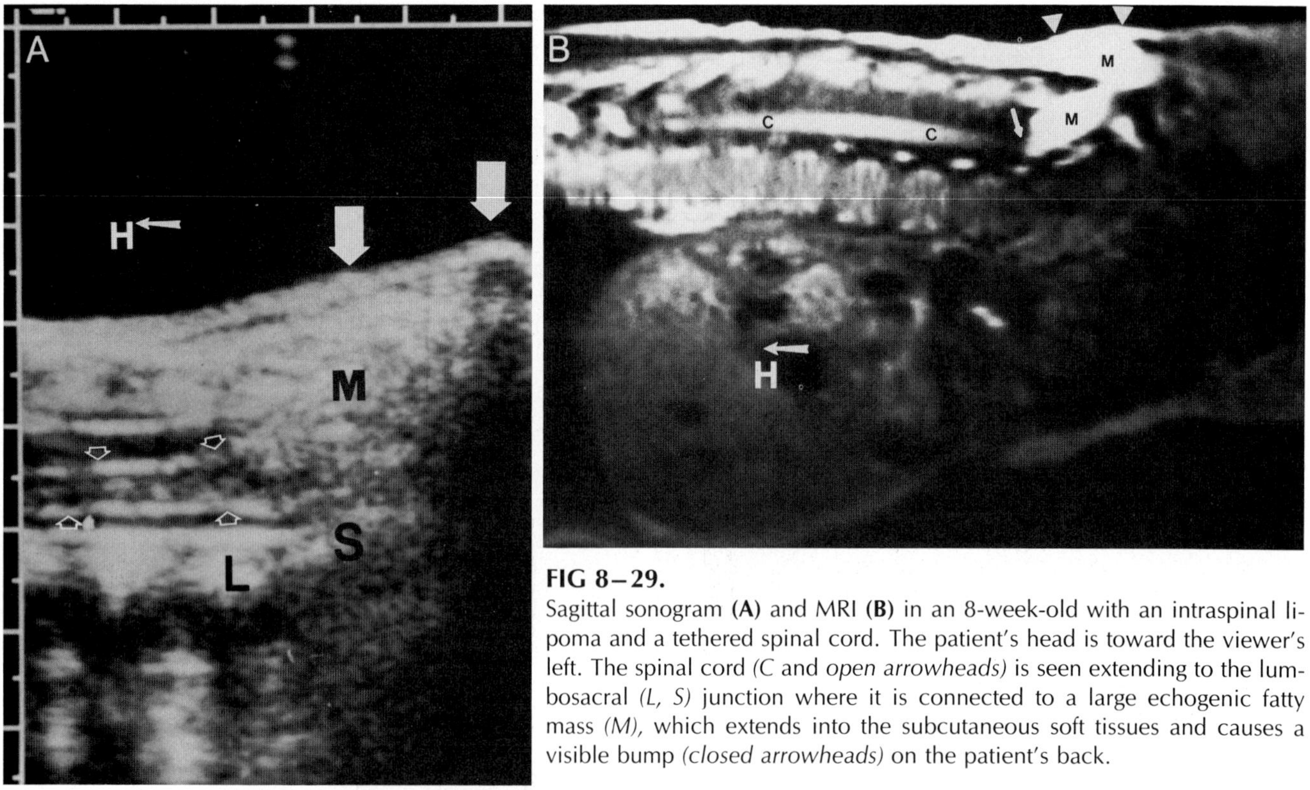


**FIG 8–29.**
Sagittal sonogram **(A)** and MRI **(B)** in an 8-week-old with an intraspinal lipoma and a tethered spinal cord. The patient's head is toward the viewer's left. The spinal cord *(C* and *open arrowheads)* is seen extending to the lumbosacral *(L, S)* junction where it is connected to a large echogenic fatty mass *(M),* which extends into the subcutaneous soft tissues and causes a visible bump *(closed arrowheads)* on the patient's back.

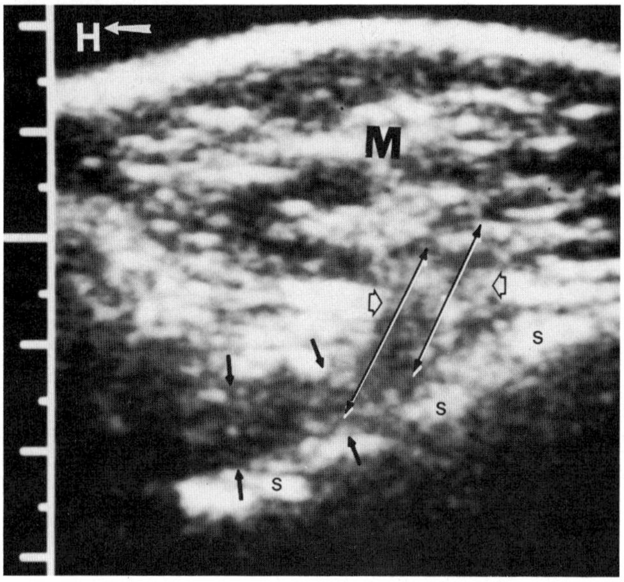


**FIG 8–30.**
Sagittal sonogram obtained over the sacrum where a subcutaneous mass (lipoma) *(M)* is noted to extend into the low sacral canal *(arrows)*. In real time the echogenic mass was noted to pulsate in and out of the spinal canal *(double-headed arrows)* through the dorsal sacral bony defect *(open arrowheads)*. Abbreviations: sacral elements *(S)*. The patient's head is toward the viewer's left.

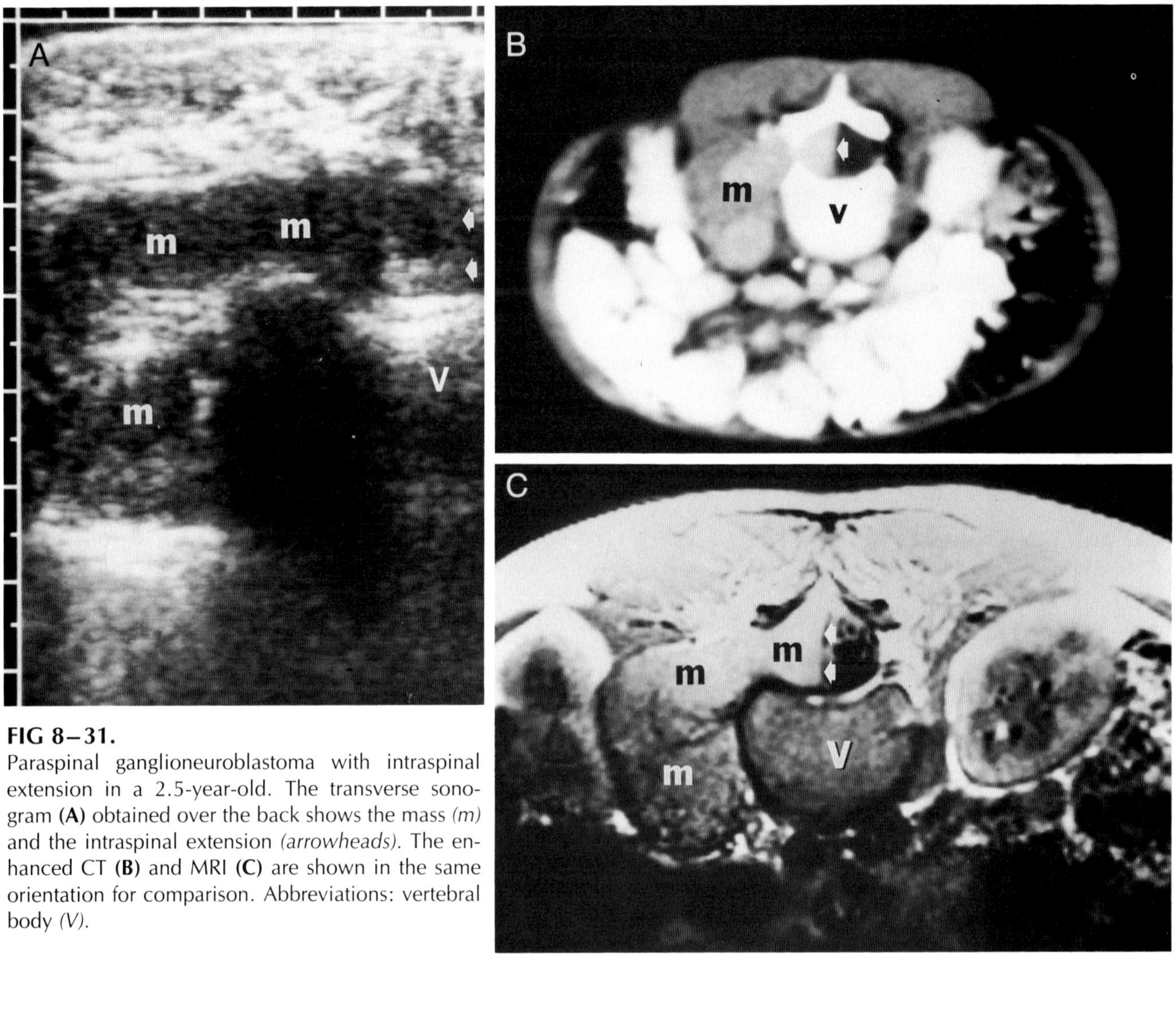


**FIG 8–31.**
Paraspinal ganglioneuroblastoma with intraspinal extension in a 2.5-year-old. The transverse sonogram **(A)** obtained over the back shows the mass *(m)* and the intraspinal extension *(arrowheads)*. The enhanced CT **(B)** and MRI **(C)** are shown in the same orientation for comparison. Abbreviations: vertebral body *(V)*.

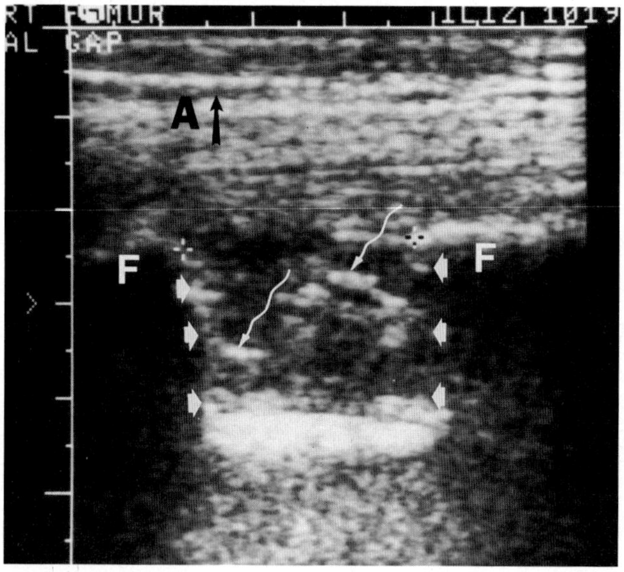


**FIG 8–32.**
Longitudinal view of the femoral corticotomy gap with approximately 2.6 cm of distraction between the femoral fragments *(F)*. Broad echogenic strands within the corticotomy *(wavy arrows)* indicate new bone in the interzone. Broad echogenic strands anterior (superficial) to the femur are muscle bundles.

## REFERENCES


Abiri MM, Kirpekar M, Ablow RC: Osteomyelitis: Detection with US. *Radiology* 1989; 172:509–511.

Adam R, Hendry GMA, Moss J, et al: Arthrosonography of the irritable hip in childhood: A review of 1 year's experience. *Br J Radiol* 1986; 59:205–208.

Albright AL, Gartner JC, Wiener ES: Lumbar cutaneous hemangiomas as indicators of tethered spinal cords. *Pediatrics* 1989; 83:977–980.

Alexander JE, Seibert JJ, Aronson J, et al: A protocol of plain radiographs, hip ultrasound, and triple phase bone scans in the evaluation of the painful pediatric hip. *Clin Pediatr* 1988; 27:175–181.

Alexander JE, Seibert JJ, Glasier CM, et al: High-resolution hip ultrasound in the limping child. *J Clin Ultrasound* 1989; 17:19–24.

Barlow TG: Early diagnosis and treatment of congenital dislocation of the hip. *J Bone Joint Surg (Br)* 1962; 44:292–301.

Berman L, Catterall A, Meire HB: Ultrasound of the hip: A review of the applications of a new technique. *Br J Radiol* 1986; 59:13–17.

Berman L, Hollingdale J: The ultrasound appearance of positive hip instability tests. *Clin Radiol* 1987; 38:117–118.

Berman L, Klenerman L: Ultrasound screening for hip abnormalities: Preliminary findings in 1001 neonates. *Br Med J* 1986; 293:719–722.

Boal DKB, Schwentker EP: The infant hip: Assessment with real-time US. *Radiology* 1985; 157:667–672.

Castelein RM, Sauter AJM: Ultrasound screening for congenital dysplasia of the hip in newborns: Its value. *J Pediatr Orthop* 1988; 8:666–670.

Catterall A: What is congenital dislocation of the hip [editorial]? *J Bone Joint Surg (Br)* 1984; 66:469–470.

Clarke NMP: Sonographic clarification of the problems of neonatal hip instability. *J Pediatr Orthop* 1986; 6:527–532.

Clarke NMP, Clegg J, Al-Chalabi AN: Ultrasound screening of hips at risk for CDH. Failure to reduce the incidence of late cases. *J Bone Joint Surg (Br)* 1989; 71:9–12.

Clarke NMP, Harcke HT, McHugh P, et al: Real-time ultrasound in the diagnosis of congenital dislocation and dysplasia of the hip. *J Bone Joint Surg (Br)* 1985; 67:406–412.

Davies SJM, Walker G: Problems in the early recognition of hip dysplasia. *J Bone Joint Surg (Br)* 1984; 66:479–484.

Dias JJ, Lamont AC, Jones JM: Ultrasonic diagnosis of neonatal separation of the distal humeral epiphysis. *J Bone Joint Surg (Br)* 1988; 70:825–828.

DiPietro MA, Venes JL: Real-time sonography of the pediatric spinal cord: Horizons and limits. *Concepts Pediatr Neurosurg* 1988; 8:120–132.

Donaldson JS: Pediatric musculoskeletal US, in Poznanski AK, Kirkpatrick JA Jr (eds): *Diagnostic Categorical Course in Pediatric Radiology.* Oak Brook, Ill, Radiological Society of North America Publications, 1989, pp 77–88.

Dörr U, Zieger M, Hauke H: Ultrasonography of the painful hip: Prospective studies in 204 patients. *Pediatr Radiol* 1988; 19:36–40.

Dunn PM: Congenital dislocation of the hip (CDH): Necropsy studies at birth. *Proc R Soc Med* 1969; 62:1035–1037.

Dunn PM: The anatomy and pathology of congenital dislocation of the hip. *Clin Orthop* 1976a; 119:23–27.

Dunn PM: Perinatal observations on the etiology of congenital dislocation of the hip. *Clin Orthop* 1976b; 119:11–22.

Emery JL, Lendon RG: Lipomas of the cauda equina and other fatty tumours related to neurospinal dysraphism. *Dev Med Child Neurol* 1969; 20(suppl):62–70.

Exner GU: Ultrasound screening for hip dysplasia in neonates. *J Pediatr Orthop* 1988; 8:656–660.

Foster JL, Babcock DS, Oestreich AE, et al: Hip pain in children: Comparison of sonography and arthrocentesis in detecting hip joint effusion [abstract]. *AJR* 1989; 153:201.

Glasier CM, Seibert JJ, Williamson SL, et al: High resolution ultrasound characterization of soft tissue masses in children. *Pediatr Radiol* 1987; 17:233–237.

Graf R: New possibilities for the diagnosis of congenital hip joint dislocation by ultrasonography. *J Pediatr Orthop* 1983; 3:354–359.

Graf R: Fundamentals of sonographic diagnosis of infant hip dysplasia. *J Pediatr Orthop* 1984; 4:735–740.

Graf R: *Guide to Sonography of the Infant Hip.* New York, Thieme Medical Publishers, 1987.

Graif M, Stahl-Kent V, Ben-Ami T, et al: Sonographic detection of occult bone fractures. *Pediatr Radiol* 1988; 18:383–385.

Grissom LE, Harcke HT, Kumar SJ, et al: Ultrasound evaluation of hip position in the Pavlik harness. *J Ultrasound Med* 1988; 7:1–6.

Gusnard DA, Naidich TP, Yousefzadeh DK, et al: Ultrasonic anatomy of the normal neonatal and infant spine. correlation with cryomicrotome sections and CT. *Neuroradiology* 1986; 28:493–511.

Hall DE, Udvarhelyi GB, Altman J: Lumbosacral skin lesions as markers of occult spinal dysraphism. *JAMA* 1981; 246:2606–2608.

Hansson G, Nachemson A, Palmen K: Screening of children with congenital dislocation of the hip joint on the maternity wards in Sweden. *J Pediatr Orthop* 1983; 3:271–279.

Harcke HT, Clarke NMP, Lee MS, et al: Examination of the infant hip with real-time ultrasonography. *J Ultrasound Med* 1984; 3:131–137.

Harcke HT, Grissom LE, Finkelstein MS: Evaluation of the musculoskeletal system with sonography. *AJR* 1988; 150:1253–1261.

Heckmatt JZ, Dubowitz V: Ultrasound imaging and directed needle biopsy in the diagnosis of selective involvement in muscle disease. *J Child Neurol* 1987; 2:205–213.

Heckmatt JZ, Pier N, Dubowitz V: Measurement of quadriceps muscle thickness and subcutaneous tissue thickness in normal children by real-time ultrasound imaging. *JCU* 1988; 16:171–176.

Hensinger RN: Congenital dislocation of the hip. *Clin Symp* 1979; 31(1).

Hernandez RJ: Evaluation of congenital hip dysplasia and tibial torsion by computed tomography. *J Comput Tomogr* 1983; 7:101–108.

Hernandez RJ: Concentric reduction of the dislocated hip. Computed-tomographic evaluation. *Radiology* 1984; 150:266–268.

Hernandez RJ, Tachdjian MO, Poznanski AK, et al: CT determination of femoral torsion. *AJR* 1981; 137:97–101.

Johnson ND, Wood BP, Jackman KV: Complex infantile and congenital hip dislocation: Assessment with MR imaging. *Radiology* 1988; 168:151–156.

Joseph B, Carver RA, Bell MJ, et al: Measurement of tibial torsion by ultrasound. *J Pediatr Orthop* 1987; 7:317–323.

Kawahara H, Andou Y, Takashima S, et al: Normal development of the spinal cord in neonates and infants seen on ultrasonography. *Neuroradiology* 1987; 29:50–52.

Keller MS: Personal communication, 1990.

Keller MS, Chawla HS: Sonographic delineation of the neonatal acetabular labrum. *J Ultrasound Med* 1985; 4:501–502.

Keller MS, Rattner Z, Weltin GG: Quantitation of instability in the clinically abnormal newborn and infant hip [abstract]. *AJR* 1988a; 151:623.

Keller MS, Weltin GG, Rattner Z, et al: Normal instability of the hip in the neonate: US standards. *Radiology* 1988b; 169:733–736.

Koroshetz AM, Taveras JM: Anatomy of the vertebrae and spinal cord (chapter 101), in Taveras JM (ed): *Radiology-Diagnosis-Imaging-Intervention.* Philadelphia, JB Lippincott Co, 1986, p 5.

Langer R: Ultrasonic investigation of the hip in newborns in the diagnosis of congenital hip dislocation: Classification and results of a screening program. *Skeletal Radiol* 1987; 16:275–279.

Lhowe D, Ehrlich MG, Chapman PH, et al: Congenital intraspinal lipomas: Clinical presentation and response to treatment. *J Pediatr Orthop* 1987; 7:531–537.

MacEwen GD, Zembo MM: Current trends in the treatment of congenital dislocation of the hip. *Orthopedics* 1987; 10:1663–1669.

MacKenzie IG, Wilson JG: Problems encountered in the early diagnosis and management of congenital dislocation of the hip. *J Bone Joint Surg (Br)* 1981; 63:38–42.

Marchal GJ, Van Holsbeeck MT, Raes M, et al: Transient synovitis of the hip in children: Role of US. *Radiology* 1987; 162:825–828.

Morin C, Harcke HT, MacEwen GD: The infant hip: Real-time US assessment of acetabular development. *Radiology* 1985; 157:673–677.

Moulton A, Upadhyay SS: A direct method of measuring femoral anteversion using ultrasound. *J Bone Joint Surg (Br)* 1982; 64:469–472.

Naidich TP, Fernbach SK, McLone DG, et al: Sonography of the caudal spine and back: Congenital anomalies in children. *AJNR* 1984; 5:221–234.

Naidich TP, Radkowski MA, Britton J: Real-time sonographic display of caudal spinal anomalies. *Neuroradiology* 1986; 28:512–527.

Nichols GW, Schwentker EP, Boal DKB: Correlation of anatomy and ultrasonographic images in the infant hip: an experimental cadaver study. *J Pediatr Orthop* 1986; 6:410–415.

Novick G, Ghelman B, Schneider M: Sonography of the neonatal and infant hip. *AJR* 1983; 141:639–645.

Patel RB: Sonographic diagnosis of intraspinal neuroblastoma. *JCU* 1985; 13:565–569.

Phillips HO, Greene WB, Guilford WB, et al: Measurement of femoral torsion: Comparison of standard roentgenographic techniques with ultrasound. *J Pediatr Orthop* 1985; 5:546–549.

Place MJ, Parkin DM, Fitton JM: Effectiveness of neonatal screening for congenital dislocation of the hip. *Lancet* 1978; 2:249–251.

Raghavendra BN, Epstein FJ, Pinto RS, et al: The tethered spinal cord: Diagnosis by high-resolution real-time ultrasound. *Radiology* 1983; 149:123–128.

Rubin JM, DiPietro MA, Chandler WF, et al: Spinal ultrasonography: Intraoperative and pediatric applications. *Radiol Clin North Am* 1988; 26:1–27.

Saies AD, Foster BK, Lequesne GW: The value of a new ultrasound stress test in assessment and treatment of clinically detected hip instability. *J Pediatr Orthop* 1988; 8:436–441.

Scheible W, James HE, Leopold GR, et al: Occult spinal dysraphism in infants: Screening with high-resolution real-time ultrasound. *Radiology* 1983; 146:743–746.

Schlesinger AE, Deeney VFX, Caskey PF: Sonography of the nonossified tarsal navicular cartilage in an infant with congenital vertical talus. *Pediatr Radiol* 1989; 20:134–135.

Schulz RD, Zieger M: The present standard of ultrasonography in newborn and young infant hips. *Ann Radiol* 1986; 29:681–685.

Storrs BB, Walker ML: Sacral dermal sinus: Occult sacral masses discovered by routine ultrasound. *Concepts Pediatr Neurosurg* 1987; 7:172–178.

Suzuki S, Awaya G, Okada Y, et al: Examination by ultrasound of Legg-Calvé-Perthes disease. *Clin Orthop* 1987; 220:130–136.

Szöke N, Kühl L, Heinrichs J: Ultrasound examination in the diagnosis of congenital hip dysplasia of newborns. *J Pediatr Orthop* 1988; 8:12–16.

Terjesen T, Anda S: Femoral anteversion in children measured by ultrasound. *Acta Orthop Scand* 1987; 58:403–407.

Tredwell SJ, Davis LA: Prospective study of congenital dislocation of the hip. *J Pediatr Orthop* 1989; 9:386–390.

Walter RS, Donaldson JS, Davis CL, et al: Ultrasound screening of high risk newborns for congenital dysplasia of the hip [abstract]. *AJR* 1989; 153:201.

White SJ, Blane CE, DiPietro MA, et al: Arthrography in evaluation of birth injuries of the shoulder. *J Can Assoc Radiol* 1987; 38:113–115.

Wilson DA, Prince JR: MR imaging determination of the location of the normal conus medullaris throughout childhood. *AJR* 1989; 152:1029–1032.

Yamaguchi M, Takeuchi S, Matsuo S: Ultrasonic evaluation of pediatric superficial masses. *JCU* 1987; 15:107–113.

Young JWR, Kostrubiak IS, Resnik CS, et al: Sonographic evaluation of bone production at the distraction site in Ilizarov limb-lengthening procedures. *AJR* 1990; 154:125–128.

Yousefzadeh DK, Ramilo JL: Normal hip in children: Correlation of US with anatomic and cryomicrotome sections. *Radiology* 1987; 165:647–655.

Zieger M: Ultrasound of the infant hip. Part 2: Validity of the method. *Pediatr Radiol* 1986; 16:488–492.

Zieger M, Dörr U: Pediatric spinal sonography. Part I: Anatomy and examination technique. *Pediatr Radiol* 1988; 18:9–13.

Zieger M, Dörr U, Schulz RD: Sonography of slipped humeral epiphysis due to birth injury. *Pediatr Radiol* 1987a; 17:425–426.

Zieger M, Hilpert S: Ultrasonography of the infant hip: Part IV. Normal development in the newborn and preterm neonate. *Pediatr Radiol* 1987; 17:470–473.

Zieger M, Hilpert S, Schulz RD: Ultrasound of the infant hip: Part I. Basic principles. *Pediatr Radiol* 1986; 16:483–487.

Zieger M, Schulz RD: Method and results of ultrasound in hip studies. *Ann Radiol (Paris)* 1986; 29:383–386.

Zieger M, Schulz RD: Ultrasonography of the infant hip: Part III. Clinical application. *Pediatr Radiol* 1987; 17:226–232.

Zieger MM, Dörr U, Schulz RD: Ultrasonography of hip joint effusions. *Skeletal Radiol* 1987b; 16:607–611.

Zieger MM, Dörr U, Schulz RD: Pediatric spinal sonography: Part II: Malformations and mass lesions. *Pediatr Radiol* 1988; 18:105–111.

# 9

# Sonography of the Dermis, Hypodermis, Periosteum, and Bone

## EPIDERMIS, DERMIS, AND HYPODERMIS

The epidermis and dermis combined are referred to as the *corium*, or more commonly, the *skin*. Stratified squamous epithelium forms the epidermis, which is usually on the order of 100 μm thick. Highly cornified regions of the skin are the exception. Epidermis of the palms of the hands and soles of the feet may be up to 2 mm thick. The dermis is the primary supporting structure of the skin, containing vessels, nerves, lymphatics, glands, hair follicles, and a thick network of collagen fibers (Bloom and Fawcett, 1980). It typically measures 1 to 3 mm thick. The subcutaneous tissue, hypodermis, consists of loose connective tissue and fat cells (Fig 9–1). Clearly, the thickness of the hypodermis varies greatly depending on the region of the body and the body habitus of the patient. Blood vessels and nerves traverse the hypodermis on their way to the dermis. High-frequency transducers (10, 7.5, and 5 MHz) currently available are capable of distinguishing the skin from the hypodermis. A standoff pad is required for this examination. The skin appears as a uniform hyperechoic layer of variable thickness as described earlier. A distinction between epidermis and dermis cannot be made on sonographic images. Hypodermis has a much less echogenic appearance, containing some linear and curvilinear echogenic septa. These echogenic septa correspond to the fibrous interconnections supporting the hypodermis (see Fig 9–1) (Vincent, 1988). Closely associated with this network of fibers are blood vessels, nerves, and lymphatics (Fig 9–2).

Pathology of the skin is well evaluated by clinical examination, and biopsy is easily performed. Ultrasound currently has no role in clinical evaluation of lesions confined to the skin but may someday be of value in follow-up of systemic diseases affecting the skin, such as scleroderma. A current research application is measurement of the thickness of the corium following full-thickness burn (Fig 9–3). Ultrasound's ability to distinguish the skin from the hypodermis and underlying muscle also make it valuable in anthropomorphic studies. Measurement of the subcutaneous fat layer is easily performed using sonography and is more reproducible than other methods. This parameter is applied in several formulas to calculate lean body mass, a valuable indicator of the level of conditioning in athletes (Fried, 1986; Chiba et al., 1989).

Lesions of the subcutaneous tissues are well evaluated with sonography. Edema in the hypodermis results in markedly increased echogenicity (Figs 9–4 and 9–5) (Vincent, 1988). The sonographic appearance of edema in the hypodermis is the inverse of that normally seen. Fibrous septa now appear hypoechoic relative to the surrounding echogenic fat. This is likely due to minute fat-fluid interfaces, which are the result of excess interstitial fluid. Echogenicity of the septa may, in fact, be slightly decreased due to distended lymphatics.

Generalized edema of the subcutaneous tissues is seen in cellulitis, panniculitis, lymphedema, venous insufficiency, subcutaneous hemorrhage (Fig 9–6), and ruptured Baker's cyst (Fig 9–7). Diffuse swelling seen in lymphedema involves the hypodermis and

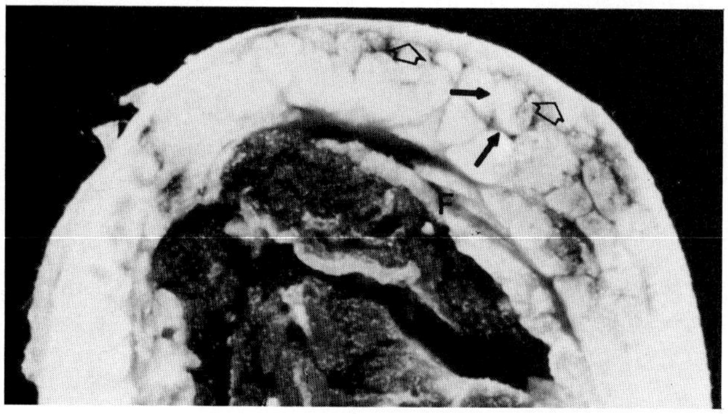


**FIG 9–1.**
Subcutaneous fat: transverse anatomical section. A loose fibrous connective tissue network *(black arrows)* provides a structural matrix for the subcutaneous fat. This connective tissue network anchors the skin *(open arrows)* to the underlying fascia *(F)* while allowing a moderate degree of motion between the layers.

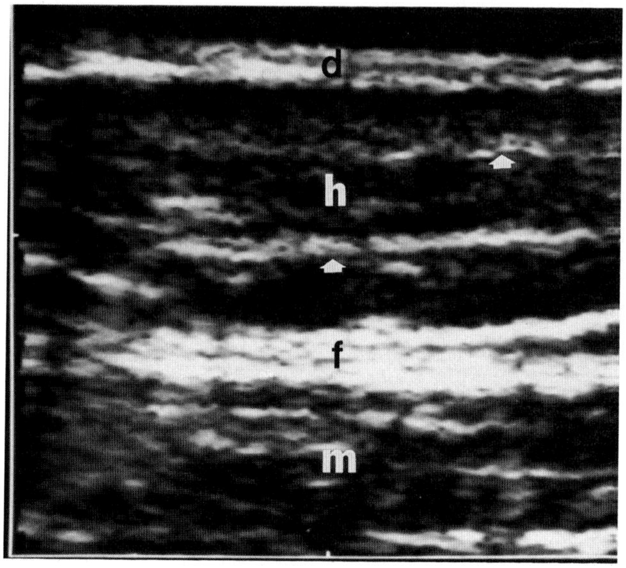


**FIG 9–2.**
Normal skin and subcutaneous tissues: longitudinal sonogram of the thigh. This longitudinal image obtained of a moderately obese 25-year-old man demonstrates the epidermis and dermis as a single hyperechoic layer *(d)*. Its thickness varies considerably depending on location. Subcutaneous fat appears hypoechoic relative to the collagen-rich dermis. The subcutaneous fat appears less echogenic as the ratio of fat to lean body mass increases. Fibrous septa *(arrows)* within this layer are hyperreflective relative to the fat lobules. As they become more widely separated by larger fat lobules, echogenicity of the fat decreases. Abbreviations: hypodermis/subcutaneous fat *(h)*, fascia *(f)*, muscle *(m)*.

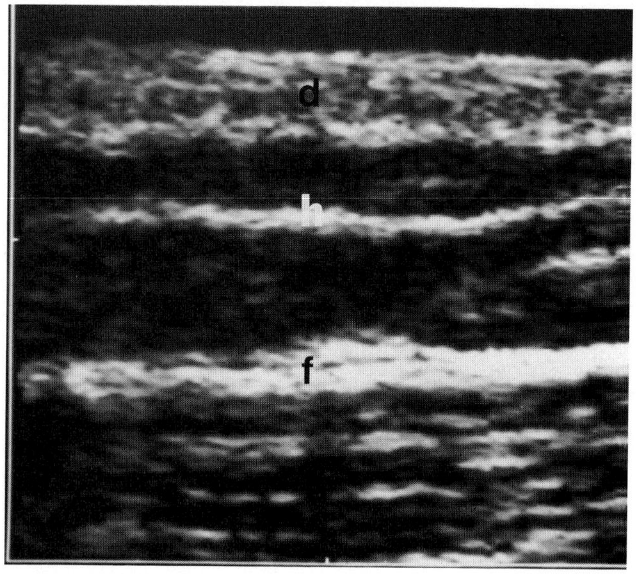


**FIG 9–3.**
Skin changes following a burn wound: longitudinal sonogram. A 20-year-old man 3 months following a large second-degree burn of the anterior thigh. Transducer position is identical to that in Figure 9–2. Fibrous scarring of the dermis *(d)* results in marked thickening of this hyperechoic layer. In this case, it is increased in thickness approximately fourfold. The degree of scarring is a function of the depth of burn, complication of healing by infection, and genetic factors. Abbreviations: hypodermis *(h)*, fascia *(f)*.

muscles equally (Yeh and Rabinowitz, 1982). Edema seen in thrombophlebitis tends to be greater in muscles, which appear hypoechoic. Cellulitis can have a nodular tumor-like appearance in intravenous (IV) drug abusers. However, in other patients with cellulitis, a gradual transition from abnormal to normal will be observed (Yeh and Rabinowitz, 1982; Sandler et al., 1984; Gitschlag et al., 1982). Subcutaneous abscess is visualized as a hypoechoic fluid collection bounded by increased echogenicity of the surrounding soft tissues. Debris is commonly identified within the fluid. Foreign bodies are often the inciting agent responsible for subcutaneous abscess. When wooden fragments are entrapped within the subcutaneous tissues, the ensuing inflammatory reaction will decompose the piece of wood. The echogenic wood splinter becomes progressively less echogenic until it finally disappears within the inflammatory collection. Some foreign bodies are not the result of accident but are surgically introduced. Vascular grafts (Fig 9–8) and orthopedic implants may serve as a nidus for infection. Regardless of the cause, sonography is a valuable tool in the diagnosis of inflammatory lesions of the hypodermis.

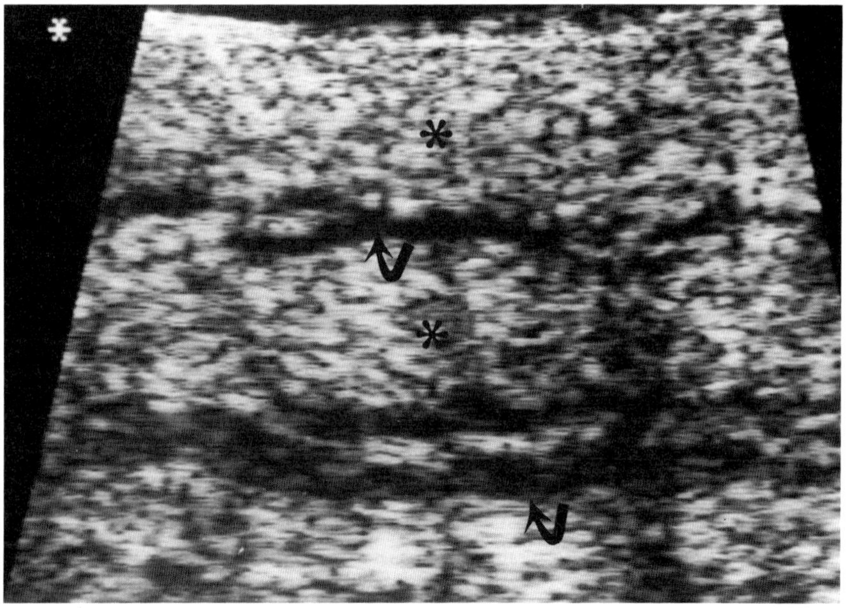

**FIG 9–4.**
Marked edema of the hypodermis: longitudinal sonogram of the thigh. This 42-year-old woman presented with deep venous thrombosis of the right leg. The transducer is positioned over the posterior thigh just above the popliteal fossa. Compare this image with the normal appearance of the hypodermis shown in Figure 9–2. The dermis can no longer be distinguished from the hypodermis. A complete reversal in the echogenicity of structures within the hypodermis is observed. The fibrous septa *(curved arrows)* containing distended lymphatics appear hypoechoic relative to the markedly echogenic fat lobules *(asterisks)*. Increased reflectivity of the fat lobules may be attributed to microscopic fat-fluid interfaces resulting from edema.

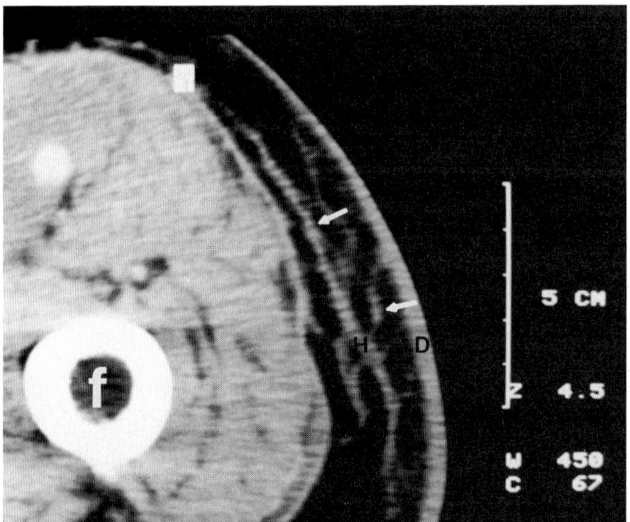


**FIG 9–5.**
Edema of the hypodermis: CT of the thigh. Marked edema of the hypodermis *(H)* is seen in this patient with a mycotic aneurysm of the femoral artery. Thickened septa *(arrows)* are seen throughout the subcutaneous fat. Normally these septa are almost imperceptible. Note that the thickest septa run parallel to the skin *(D)* and fascia. Abbreviations: femur *(f)*.

## PERIOSTEUM

Thickening of the periosteum is a nonspecific finding seen in trauma, fracture, osteotomy, infection, tumor, Paget's disease, venous stasis, hypertrophic pulmonary osteoarthropathy, metabolic diseases, and endocrinopathies (Resnick and Niwayama, 1988). Sonography will clearly demonstrate periosteal thickening, but in the majority of cases conventional radiography is the best method to survey the periosteum. Two specific exceptions to this statement are the diagnosis of osteomyelitis and stress fracture. In these cases, ultrasound is superior in the detection of periosteal reaction, well before mineralization becomes evident by radiography.

Acute osteomyelitis is recognized on sonographic images by elevation of the periosteum by a hypoechoic layer of purulent material (Fig 9–9) (Abiri et al., 1988). This appearance is identical to that of subperiosteal hemorrhage, but the clinical history will usually clinch the diagnosis (Fig 9–10). The absence of soft tissue between the fluid and cortical bone helps to differentiate osteomyelitis from other fluid collections (Abiri et al., 1988). The diagnosis of acute osteomyelitis by ultrasound with aspiration and cul-

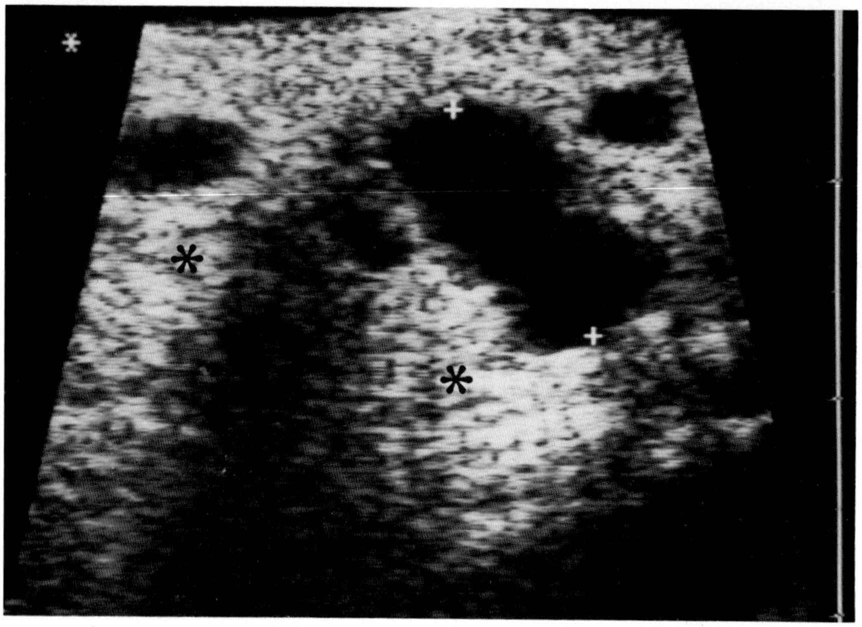

**FIG 9–6.**
Subcutaneous hemorrhage: transverse sonogram. The patient is a 28-year-old volleyball player who was kicked in the thigh during a match. He presented with redness and induration of the anterior thigh. This transverse image of the upper thigh demonstrates multiple hypoechoic fluid collections indicative of hematoma (+). The surrounding fat appears markedly hyperechoic *(asterisks)*. No septa are identified, likely due to rupture.

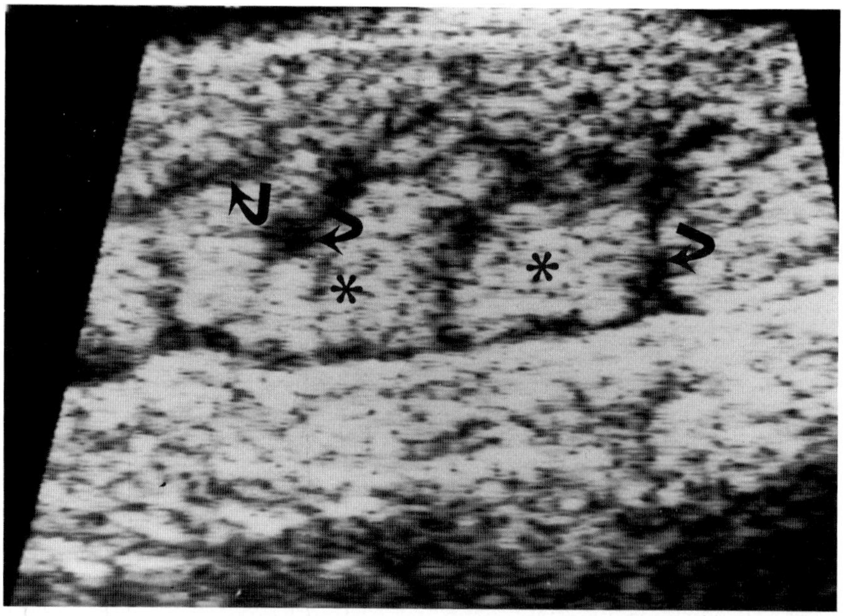

**FIG 9–7.**
Edema of the hypodermis: longitudinal sonogram of the posterior calf. In this patient with recently diagnosed rheumatoid disease, marked edema of the hypodermis is noted following rupture of a Baker's cyst. Again, reversal of the normal echo texture of the hypodermis is noted. Fat lobules *(asterisks)* appear markedly hyperechoic, and the connective tissue septa *(curved arrows)* are hypoechoic. In the calf, septa running perpendicular to the fascia and skin are more prominent. This appearance of the edematous hypodermis is always seen in cases of ruptured Baker's cyst.

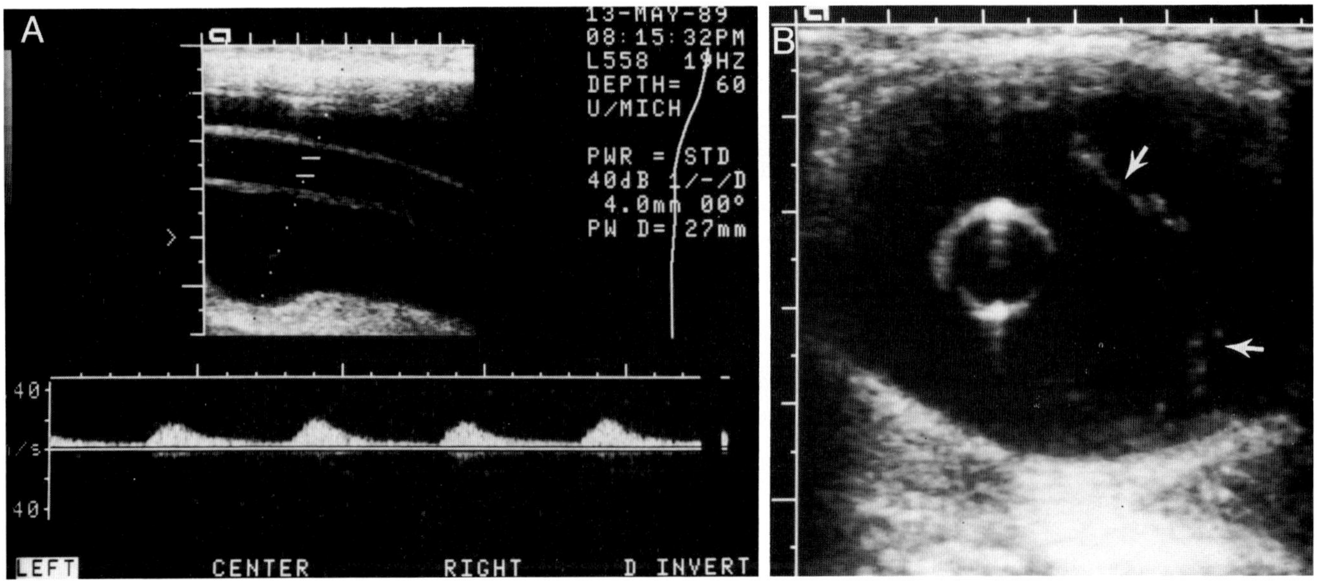


**FIG 9–8.**
**A,** extra-anatomical bypass graft infection: longitudinal sonogram. This 64-year-old man with bilateral axillary-femoral bypass grafts presented with systemic sepsis and induration over the right graft. A large fluid collection is noted surrounding the graft, which is demonstrated to be patent by Doppler. The surrounding subcutaneous fat is diffusely increased in echogenicity, and debris is seen floating within the collection. **B,** extra-anatomical bypass graft infection: transverse sonogram. Same patient as in **A.** Once again, clumps of debris *(arrows)* are identified within this inflammatory collection. Aspiration was performed under ultrasound guidance, and cultures yielded *Streptococcus viridans.* Note the reverberation artifact off of the anterior and posterior walls of this Gore-Tex graft.

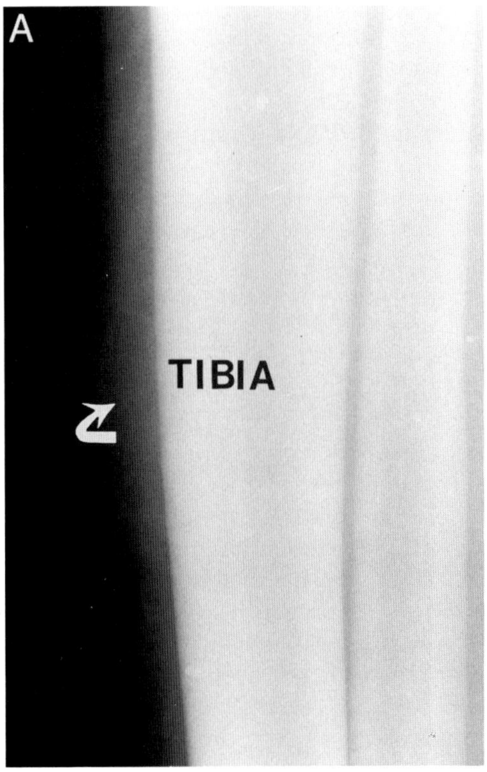


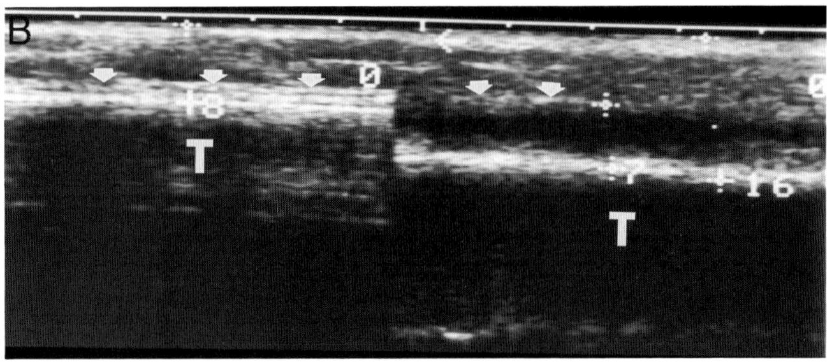


**FIG 9–9.**
**A,** acute osteomyelitis-lateral radiograph. A 25-year-old diabetic man who presented with 2 days of fever and pain over the midtibia. No osseous changes are present. Some edema *(curved arrow)* of the soft tissues overlying the tibia is noted. **B,** acute osteomyelitis: longitudinal sonogram of the anterior tibia. Same patient as in **A.** The transducer is positioned over the anteromedial surface of the tibia. Side to side comparison is provided in this split screen image, asymptomatic left tibia on the *left.* A normal relationship of the hyperechoic fascia/periosteum *(arrows)* to the tibial cortex is seen on the *left.* A hypoechoic fluid collection elevates the fascia/periosteum off of the tibial cortex on the *right.* Soft tissue overlying the right tibia is twice as thick as the normal left side. Abbreviations: tibia *(T).*

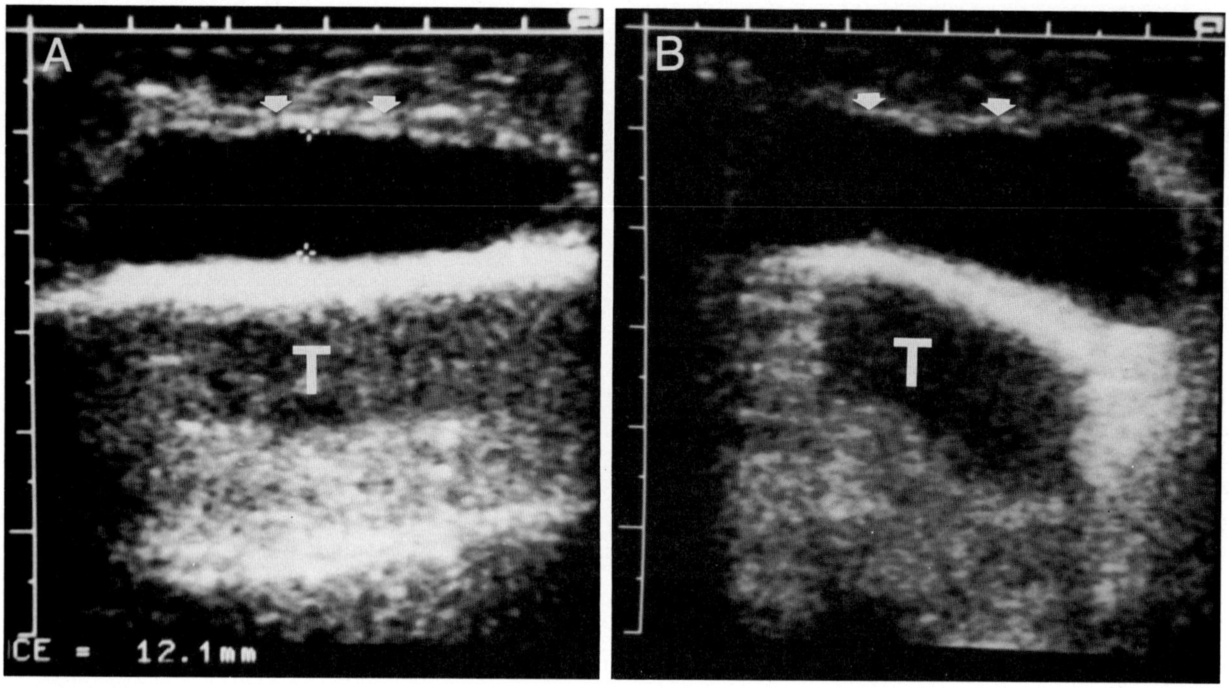


**FIG 9–10.**

**A,** subperiosteal hematoma: longitudinal sonogram. This 22-year-old soccer player received a blow to the anterior tibia. He presented with continued pain and swelling 5 days after injury. The transducer is positioned over the midtibia along its length. Anechoic hematoma elevates the periosteum *(arrows)*. Abbreviations: tibia *(T)*. **B,** subperiosteal hematoma: transverse sonogram. Same patient as in **A.** The hematoma that elevates the periosteum *(arrows)* tracks along the anteromedial cortex of the tibia *(T)*.

ture of the offending organism often expedites initiation of appropriate therapy. Results of ultrasound examination are usually positive 24 hours following the onset of fever in cases of osteomyelitis. Results of radionuclide scintigraphy using technetium 99m–methylene diphosphonate become positive after approximately 48 to 72 hours. Galium 67 scintigraphy will take up to 72 hours just to obtain adequate images, thus delaying the diagnosis by up to 5 days. Radiographic changes indicative of osteomyelitis may not become evident for 1 week or more.

In chronic osteomyelitis, sonography is not only valuable in making the diagnosis but can assess involvement of the adjacent soft tissues and guide placement of a percutaneous drainage catheter (van Sonnenberg et al., 1987). Unlike acute osteomyelitis, changes in the surrounding soft tissues are often observed in chronic osteomyelitis. Soft tissue abscess related to chronic osteomyelitis is identified as a hypoechoic or anechoic fluid collection. These collections are noted to extend around the bony contours in more than 70% of cases (van Sonnenberg et al., 1987).

After sonographic localization of the abscess, a needle can be placed into the collection using sonographic guidance. A percutaneous drainage catheter is then placed using the Seldinger technique. The collection is aspirated dry and the cavity irrigated with sterile saline until the fluid returned is clear. It is of the utmost importance that the patient be given IV antibiotics immediately prior to placement of a percutaneous drainage catheter, because this procedure will result in bacteremia. This approach is very efficient because diagnosis and therapy are accomplished in the same session (van Sonnenberg et al., 1987). Further imaging studies, such as computed tomography (CT), magnetic resonance imaging (MRI), or radionuclide scintigraphy are not necessary.

The periosteal reaction seen in other disease processes differs slightly from that seen in osteomyelitis. A very thin hyperechoic line defines the soft tissue–periosteal interface. The thickened periosteum appears less echogenic, measuring several millimeters thick. These findings have been most helpful in the diagnosis of stress fractures and hairline fractures (Fig 9–11). Stress fractures are most frequently seen in the lower extremities of runners. The clinical symptoms of pain related to exercise, associated with local tenderness, are rather nonspecific. Ultrasound will make the diagnosis by identification of focal pe-

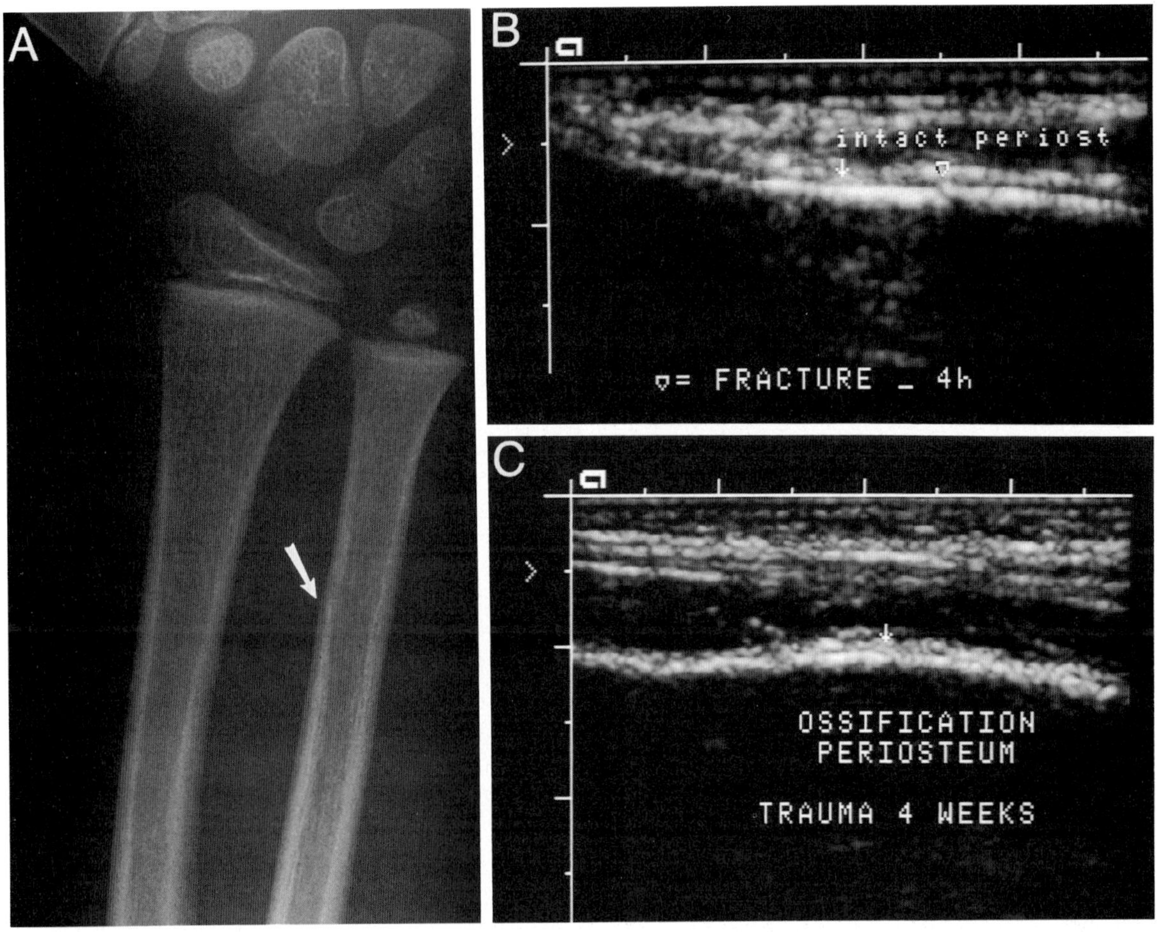


**FIG 9–11.**
**A,** acute hairline fracture of the ulna: anteroposterior radiograph. A minute cortical discontinuity *(arrow)* is noted in the distal ulna of this 7-year-old boy following a playground injury. **B,** acute hairline fracture of the ulna: longitudinal sonogram. Same patient as in **A,** 4 hours after trauma. The transducer is placed of the distal ulna along its length. Hematoma slightly elevates the ulnar periosteum *(arrow)* adjacent to the site of cortical discontinuity *(arrowhead)*. **C,** 4½ weeks after injury, the cortical discontinuity is no longer recognized, and periosteal reaction *(arrow)* is observed as a homogeneous hyperechoic layer overlying the cortex.

riosteal thickening in the symptomatic region (Fig 9–12). The examination can be performed at the time of the patient's initial presentation, weeks before radiographic changes are apparent. At the same time, other soft tissue lesions can be excluded as the source of pain.

The differential diagnosis of periosteal thickening in the tibia includes the fascial thickening seen in shin splints. Fascial thickening in shin splints will be limited to the site of attachment of the fascia cruris. The ability to differentiate these two entities is one of the principle advantages of ultrasound over radionuclide scintigraphy in the diagnosis of stress fractures.

Other areas in which sonographic evaluation of the periosteum has proved of value include the demonstration of a favorable response during Ilizarov bone lengthening and follow-up of Paget's disease. Evaluation of periosteal thickening of the tibia (Fig 9–13) in Paget's disease has been shown to be an indicator of disease activity (Maldague and Malghem, 1987). The evaluation of the Ilizarov bone lengthening technique was discussed in the previous chapter on pediatric musculoskeletal sonography.

## BONE

The bone–soft tissue interface is highly reflective, seen as a bright line with acoustic shadowing deep to the interface. The inability to image the medullary cavity of bone using pulse-echo ultrasound has led to the misconception that it is not well suited for the evaluation of bone. However, the high reflectivity of cortical bone and the tomographic nature of ultra-

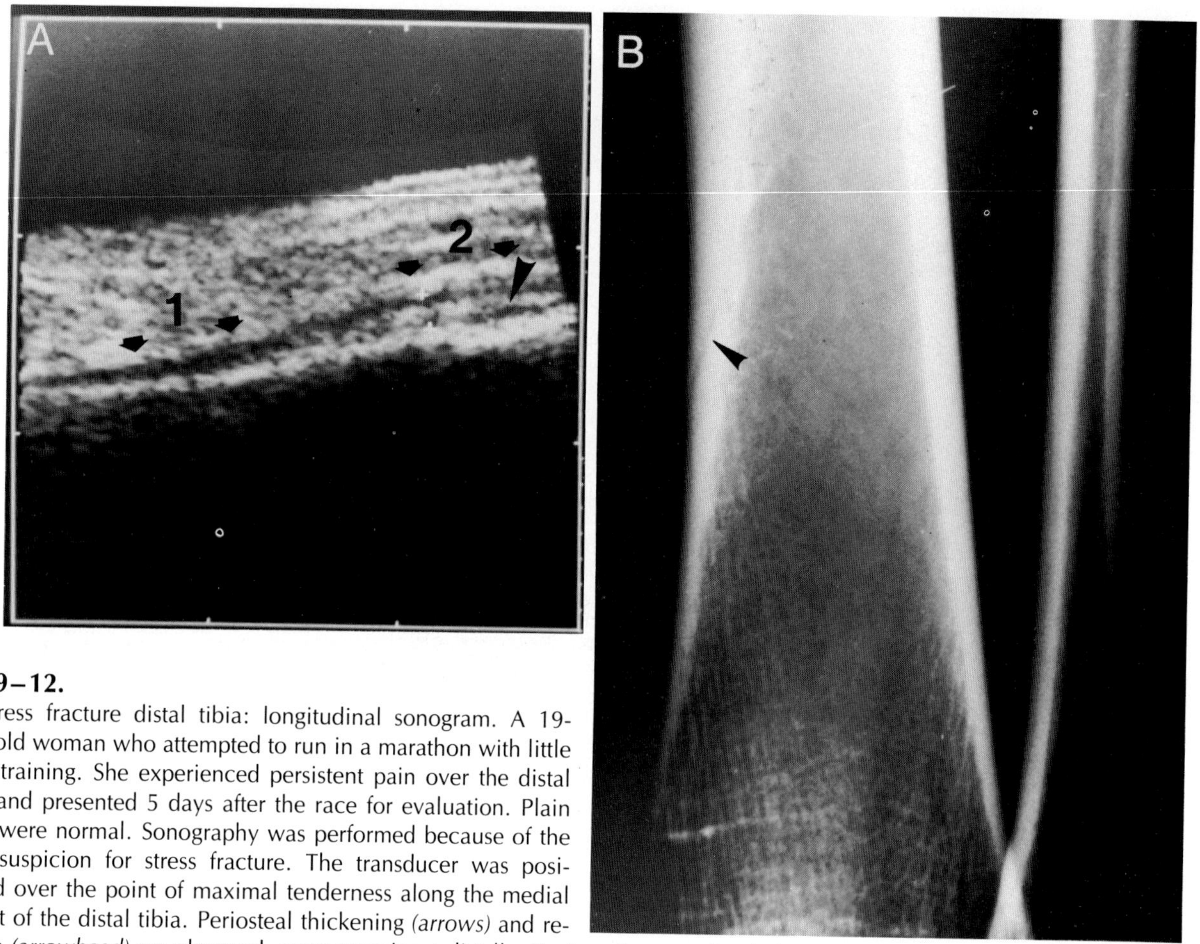


**FIG 9–12.**
**A,** Stress fracture distal tibia: longitudinal sonogram. A 19-year-old woman who attempted to run in a marathon with little prior training. She experienced persistent pain over the distal tibia and presented 5 days after the race for evaluation. Plain films were normal. Sonography was performed because of the high suspicion for stress fracture. The transducer was positioned over the point of maximal tenderness along the medial aspect of the distal tibia. Periosteal thickening *(arrows)* and reaction *(arrowhead)* are observed, more prominent distally. **B,** stress fracture distal tibia: anteroposterior radiograph. Same patient as in **A,** 8 days after injury. A tiny cortical discontinuity *(arrowhead)* is identified in the medial cortex of the distal tibia in the region of abnormality observed with ultrasound. This was not evident on the prior radiographic examination. Stress fractures in the tibia are usually oblique, making differentiation from vascular canals difficult.

sound imaging make it ideal for evaluation of bony contours (Figs 9–14 and 9–15). In all musculoskeletal ultrasound examinations, the identification of grooves, fossae, tuberosities, trochanters, and epicondyles guides the study. For example, we distinguish the subscapularis and supraspinatus tendons of the rotator cuff by identification of the bicipital groove. Pathological changes in the bony contours are equally well recognized. Marginal erosions and synovial inclusions found in rheumatoid disease are more easily seen sonographically than with conventional radiography (van Holsbeeck et al., 1988).

Ultrasound has also proved valuable in the detection of Hill-Sachs lesions (Graf and Schuler, 1988). These posttraumatic defects are often hard to visualize on conventional radiographs. The large number of special projections advocated for radiographic visualization of these lesions indicates the difficulty of this problem. Angulated internal rotation views, Hermodsson's tangential projection, Stryker's, Didiee, and West Point views are all helpful (Resnick and Niwayama, 1988). However, they are often difficult to obtain and interpret. Ultrasound will quickly and clearly demonstrate the contours of the entire circumference of the humeral head. Hill-Sachs deformity is recognized as a defect in the cortical contour of the posterolateral surface of the humeral head. The lesion is the result of anterior shoulder dislocation with the humeral head impacting on the anteroinferior aspect of the glenoid. The typical patient is a young active adult. Identification of Hill-Sachs lesions is important because it indicates a predisposition for recurrent luxation.

In addition to evaluation of the sequellae of dislocation, ultrasound can be valuable in demonstrating subluxations at joints that are difficult to evaluate radiographically. One example is the symphysis pubis

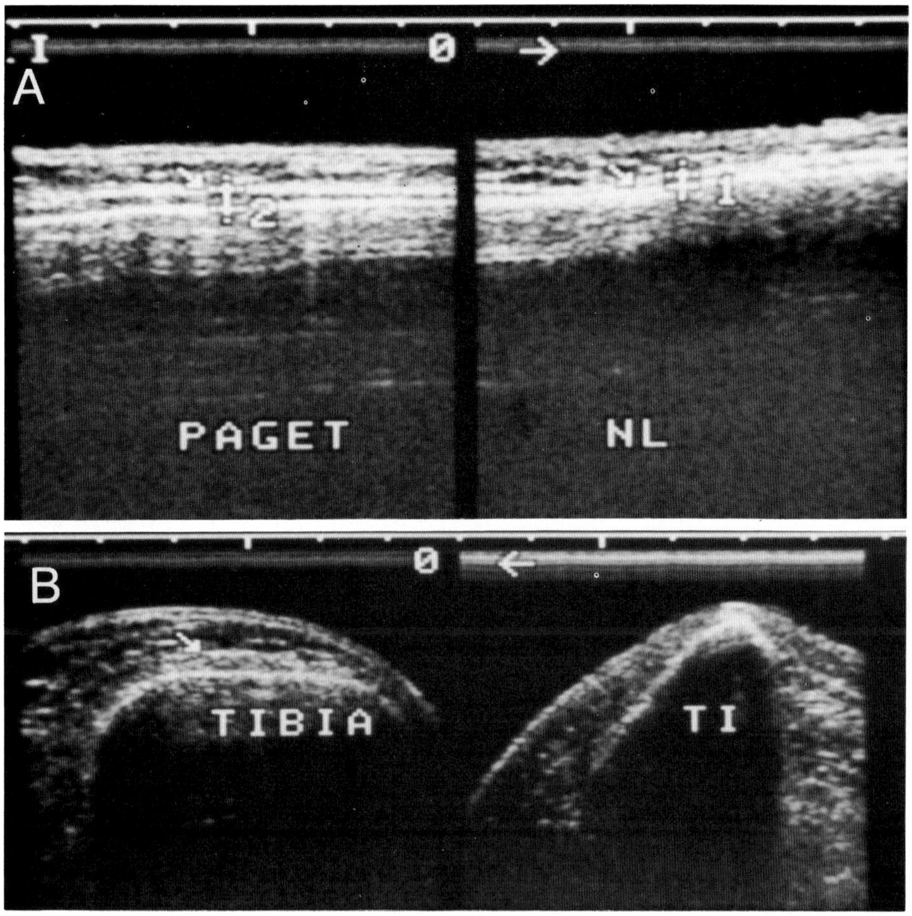


**FIG 9–13.**
**A,** periosteal thickening in Paget's disease: longitudinal sonogram of the tibia. The patient is a 70-year-old man with Paget's disease involving the right tibia. Side by side comparison is provided with these longitudinal images of the midshaft of the tibia. Periosteum of the involved tibia is twice as thick as the normal side. **B,** periosteal thickening in Paget's disease: transverse sonogram of the tibia. Same patient as in **A.** These transverse images were obtained at the same level as the longitudinal images. A normal inverted V shape of the tibia *(TI)* is seen on the *right.* On the *left,* the tibia affected by Paget's disease is grossly deformed and surrounded by a markedly thickened periosteum *(arrow).*

(Fig 9–16). Disruption of the pubic symphysis is a relatively common injury in soccer players due to the torsional stress induced while forcefully kicking. The ensuing pelvic pain is usually investigated with an anteroposterior radiograph, which is invariably unremarkable. Real-time sonographic examination quickly and easily demonstrates the abnormal movement.

Fractures occult on conventional radiographs can often be detected with sonography (see Fig 9–11). High-frequency transducers are extremely important in these cases. The technique of "sonographic palpation" is valuable for localizing the region of interest. Often a cortical discontinuity can be identified. Observation of adjacent subperiosteal hemorrhage will confirm the diagnosis. Several nonimaging ultrasound techniques have also been used for the evaluation of occult fracture. Transmission ultrasound with a power output of 3 W/cm$^2$ has been successfully used to detect occult fractures of the scaphoid (Shenouda and England, 1987). Bone mineralization and architecture can also be evaluated using transmission ultrasound (Heaney et al., 1989). Transmission ultrasound can also be used to produce cross-sectional images in a technique called *orthographic transmission ultrasound* (Carson et al., 1977; Hentz et al., 1984, 1987). In this technique, the attenuation of the sound beam is measured, rather than reflection, as in pulse-echo imaging. The disadvantage of this technique is that it currently requires a water path, thus limiting its use to the extremities. Differences in attenuation of sound in various soft tissues is greater

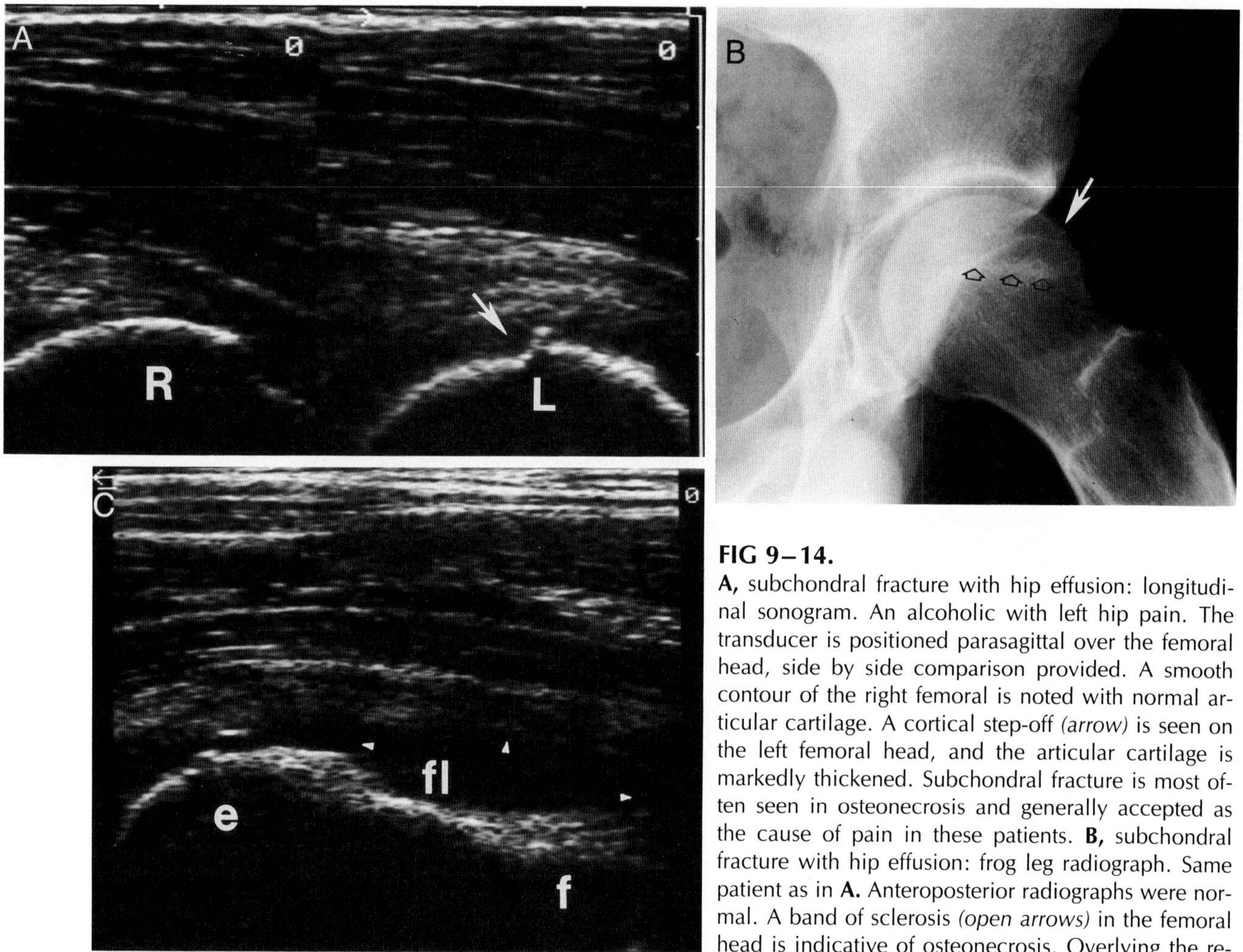


**FIG 9–14.**
**A,** subchondral fracture with hip effusion: longitudinal sonogram. An alcoholic with left hip pain. The transducer is positioned parasagittal over the femoral head, side by side comparison provided. A smooth contour of the right femoral is noted with normal articular cartilage. A cortical step-off *(arrow)* is seen on the left femoral head, and the articular cartilage is markedly thickened. Subchondral fracture is most often seen in osteonecrosis and generally accepted as the cause of pain in these patients. **B,** subchondral fracture with hip effusion: frog leg radiograph. Same patient as in **A.** Anteroposterior radiographs were normal. A band of sclerosis *(open arrows)* in the femoral head is indicative of osteonecrosis. Overlying the region of osteonecrosis is a segment of cortical irregularity *(white arrow).* **C,** subchondral fracture with hip effusion: longitudinal sonogram. Same patient as in **A** and **B.** The transducer is aligned over the left femoral neck *(f).* An anechoic fluid collection *(fl)* elevates the anterior joint capsule *(arrowheads).* Effusions are commonly found in cases of osteonecrosis. The irregular epiphysis *(e)* is again noted.

than the differences in attenuation of x-rays. Therefore, greater soft tissue contrast resolution is afforded by transmission ultrasound. In vitro studies of cadaver hands have demonstrated image quality similar to that of MRI. Clinical transmission ultrasound equipment is not generally available at present but hopefully will become widely available in the near future.

Sonographic evaluation of bone in children is often more valuable than conventional radiography because cartilage is visualized on ultrasound images. Ossification of the growth cartilage can be detected sonographically well before it is evident on conventional radiographs (Graf and Schuler, 1988; Harcke et al., 1986). As discussed in the chapter on pediatric applications, ultrasound is ideal for the evaluation of congenital hip dysplasia. It provides both morphological and functional information about the developing acetabulum and femoral head (Graf and Schuler, 1988; Harcke et al., 1988). Furthermore, repeat ultrasound examinations performed to follow growth and maturation of the hip do not carry the risks associated with ionizing radiation.

Changes in the cartilaginous hip and ossification center are also seen in Legg-Calvé-Perthes disease. Until recently, arthrography (Fig 9–17) had been the principle means of evaluating deformity of the developing femoral epiphysis. These examinations require deep sedation or general anesthesia, in addition to the risks normally associated with this invasive procedure (Suzuki et al., 1987). Ultrasound provides the same, if not more, information. Normal epiphyseal cartilage has a homogeneous hypoechoic appearance. The smooth spherical appearance of the devel-

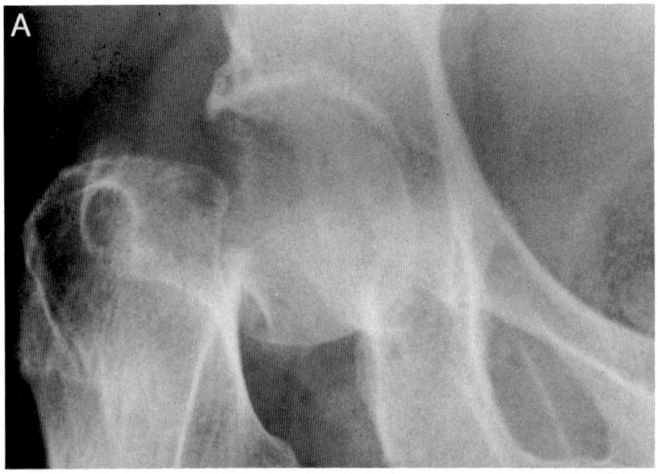


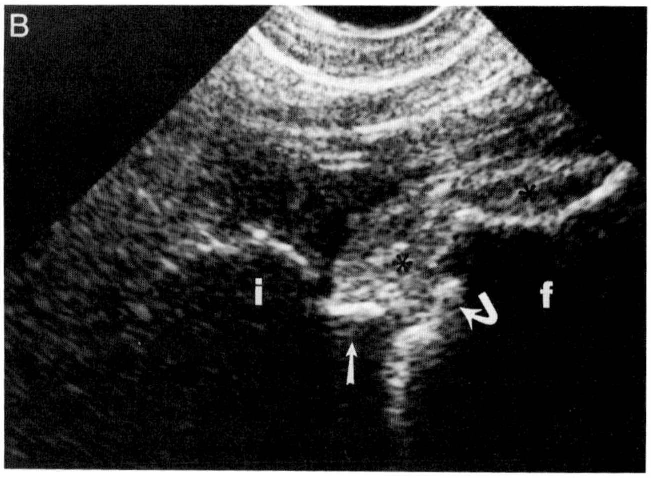


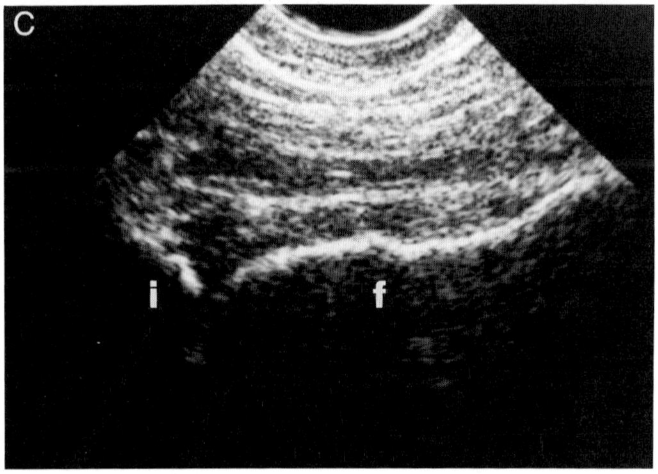


**FIG 9–15.**
**A,** femoral neck fracture: anteroposterior radiograph. This 68-year-old female nursing home patient fell while getting out of bed. A grossly displaced fracture of the femoral neck is evident. **B,** femoral neck fracture: longitudinal sonogram. Same patient as in **A.** The fracture *(curved arrow)* with associated capsular hematoma *(asterisks)* shows marked anterior displacement of the distal femoral neck *(f)* relative to the femoral head *(arrow)*. Abbreviations: ilium *(i)*. **C,** femoral neck fracture: longitudinal sonogram of the asymptomatic femur. Same patient as in **A** and **B.** A normal smooth contour of the junction of the femoral head and neck is observed on the asymptomatic left side. The iliofemoral ligament is straight and not displaced. Abbreviations: proximal femur *(f)*, ilium *(i)*.

oping femoral head is easily demonstrated. Ossification is identified as a stippled echogenic region centrally within the cartilage. As growth continues, the ratio of the volume of the ossification center to the volume of unossified cartilage increases.

In addition to the detection of deformity responsible for hinge abduction, ultrasound can demonstrate deformity secondary to necrosis by measurement of the echo index (Suzuki et al., 1987). Coronal images are obtained through the midportion of the acetabulum with the hip in 35 degrees of adduction. The midportion of the acetabulum can be recognized by the straight linear appearance of the iliac wing. The greatest height of the femoral epiphysis is measured from the proximal extent of the growth plate (Fig 9–18, line *A*). A second measurement of the radius of the femoral head is obtained along the growth plate (see Figure 9–18, line *B*). The echo index is obtained by dividing the height *(A)* by the radius *(B)*. A normal femoral head should be spherical, in other words, echo index equals 1.0. This value can be compared with the opposite side using the echo quotient, which is the echo index of the right hip divided by the echo index of the left hip. These data are valuable indicators of the degree of deformity of the femoral head and can be used in the follow-up of Legg-Calvé-Perthes disease. Orthopedic surgeons base their treatment on the degree of deformity of the cartilage and the presence of hinge abduction. The ultimate goal of therapy is maintenance of the normal spherical shape of the femoral head. Echo index and echo quotient are good indicators of the success of treatment over time.

## MEASUREMENT TO DETECT ABNORMAL DEVELOPMENT OF THE EXTREMITIES

During the last half century, many investigators have tried to develop reproducible methods for measurement of the axis of skeletal elements. Among these are retroversion of the humeral head, anteversion of the femur, and tibial torsion. Ultrasound has been proposed as a quick method to provide a rough

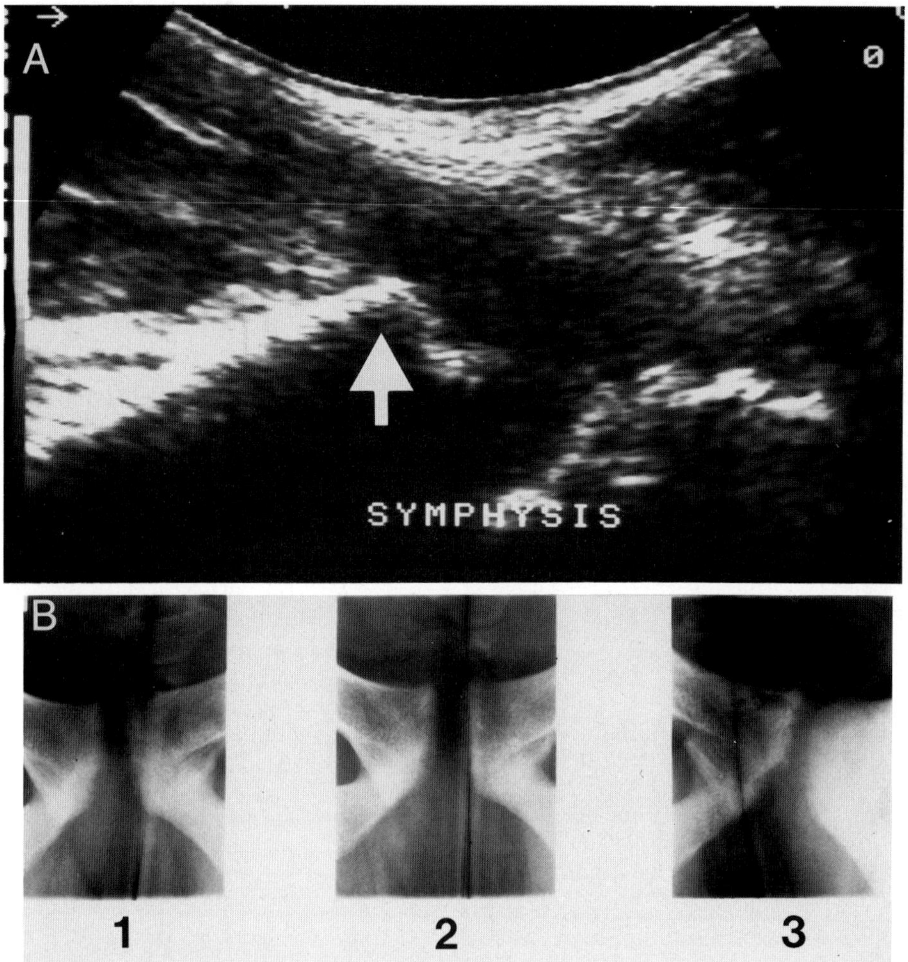


**FIG 9–16.**
**A,** instability of the pubic symphysis: transverse sonogram. This 23-year-old professional soccer player was referred with the clinical diagnosis of adductor tendonitis. The transducer is positioned transversely over the symphysis pubis. Anterior displacement *(arrow)* of the right pubic bone is observed relative to the left. This is a common injury in soccer players due to forceful kicking. **B,** instability of the pubic symphysis: upright differential weight-bearing views. Same patient as in **A.** This series of radiographs was obtained with weight bearing on the left leg *(1),* both legs *(2),* and right leg *(3).* Subchondral sclerosis and erosion of the inferior aspect of the symphysis are present. This technique is relatively time consuming when compared with real-time ultrasound examination. In addition, ultrasound will demonstrate the degree of anteroposterior motion, which is not evident on these radiographs.

measure of these parameters. No studies have carefully addressed the question of relative accuracy of these methods, but estimation of the error in sonographic measurement may be as high as 20%. Therefore, we believe that the role of ultrasound is simply that of a rough screening tool. Its value in surgical planning and follow-up is very limited. Computed tomography provides the most accurate means of calculating these parameters (Hernandez et al., 1981; Berman, 1986).

### Humeral Retroversion

The axis through the bicipital groove of the humeral head and the axis of the humeral epicondyles form an angle of approximately 30 degrees in normal individuals. This angle is referred to as the angle of humeral retroversion. Decreased retroversion of the humeral head predisposes the patient to anterior dislocation of the shoulder (Graf and Schuler, 1988). Pathological retroversion can be corrected by a rotational osteotomy, which increases the degree of retroversion.

Ultrasound measurement (Figs 9–19 and 9–20) of

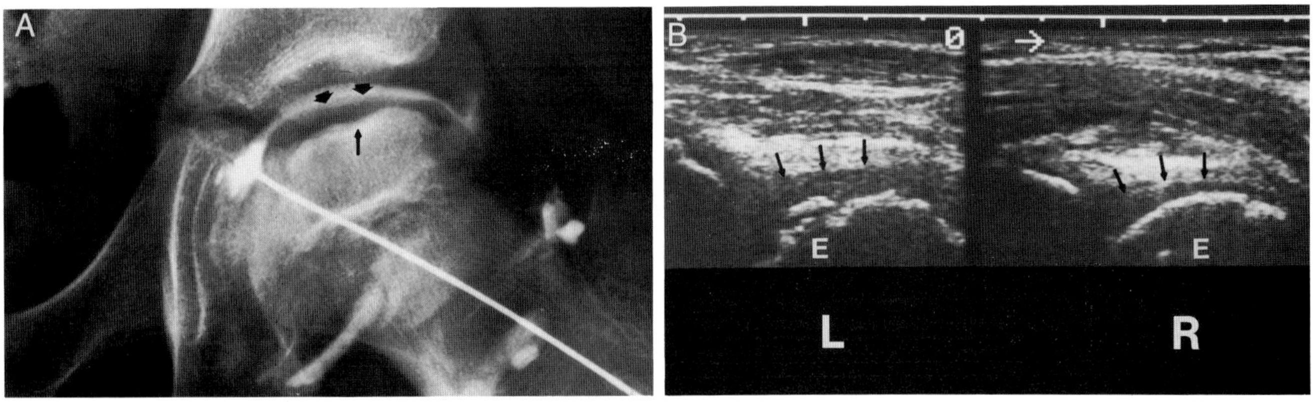


**FIG 9–17.**
**A,** Legg-Calvé-Perthes disease: arthrogram. An 8-year-old boy with known Legg-Calvé-Perthes disease and painful hip motion. Clinical examination demonstrated possible hinge abduction. The arthrogram shows flattening of the lateral femoral head covered by intact cartilage. The density of the epiphysis is increased. **B,** Legg-Calvé-Perthes disease: coronal sonogram. Same patient as in **A.** The transducer is positioned over the lateral aspect of the femoral head with the hip in 10 degrees of adduction. The involved left hip is irregularly flattened, and the overlying cartilage *(arrows)* is thickened. No contour deformities of cartilage are observed, thus excluding hinge abduction. Information provided by sonographic examination is identical to that provided by arthrography. Advantages of ultrasound are the noninvasive nature of the study and the ability to compare with the asymptomatic side.

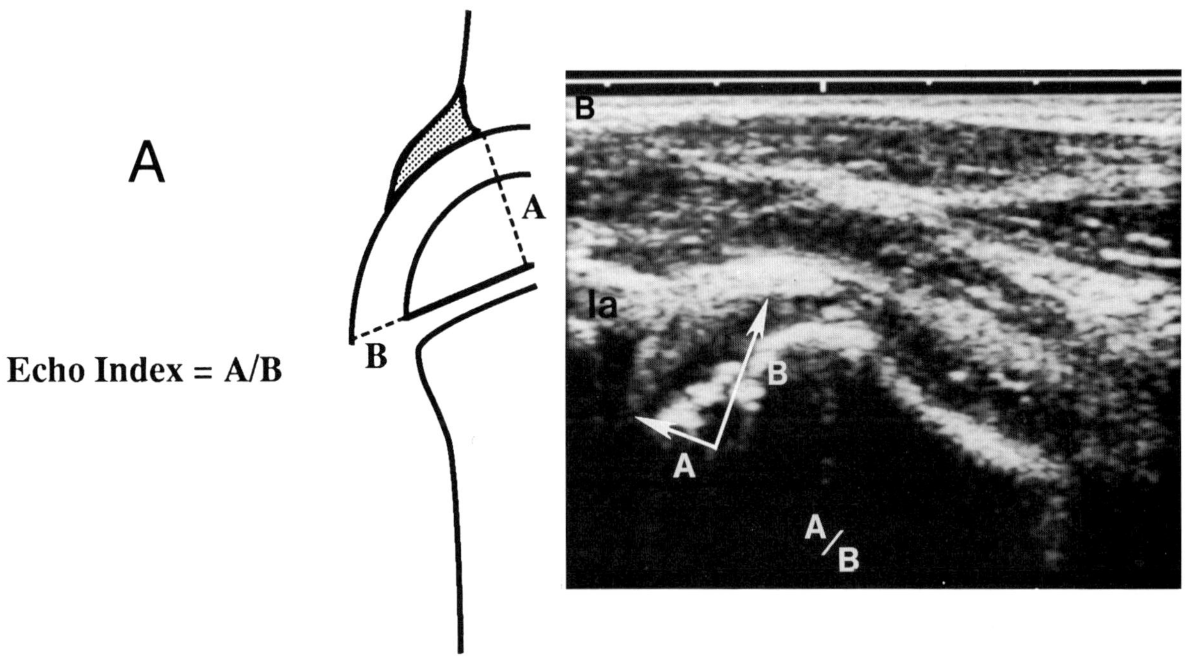


**FIG 9–18.**
**A,** schematic diagram of echo index measurement. **B,** echo index of the femoral epiphysis: coronal sonogram. This 9-year-old boy was diagnosed 3 years ago with Legg-Calvé-Perthes disease. This coronal image obtained with the hip in adduction demonstrates a flattened irregular femoral epiphysis. A poor prognosis is associated with flattening of the femoral head. The echo index is a valuable indicator of the degree of deformity of the femoral epiphysis, which can be used in follow-up. Abbreviations: labrum *(la)*.

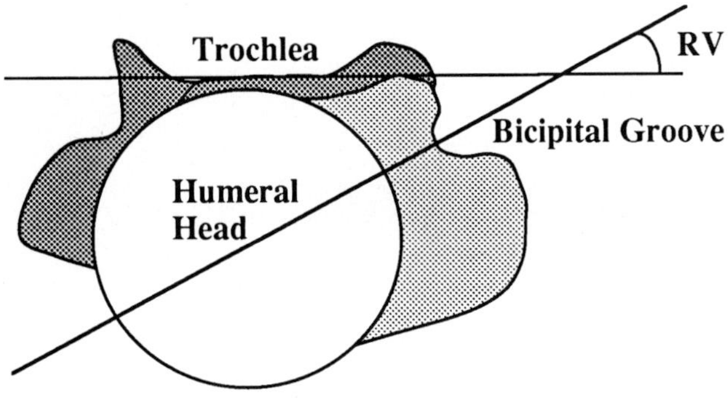


**FIG 9–19.**
Measurement of humeral retroversion: schematic representation.

humeral retroversion is performed with the patient supine and the arm in the anatomical position. The greatest depth of the bicipital groove is identified on transverse images with the transducer face parallel with the surface of the examination table. A small carpenter's level attached to the back of the transducer will greatly help in making reproducible images and accurate measurements. The second transverse image obtained is the distal humerus at the level of the trochlea and capitellum. Once again, the transducer face is kept parallel to the examination table. On the first image, a line is drawn from the center of the humeral head through the center of the bicipital groove. This line should be perpendicular to the transverse ligament. A line parallel to the plane of the anterior surface of the trochlea is then drawn on the second image. When these images are aligned along their border rules, the intersection of these lines yields the angle of humeral retroversion.

### Femoral Anteversion

The measurement of femoral anteversion is a problem similar to humeral retroversion. Excessive femoral anteversion will result in a toe-in deformity. Two radiographic methods are available. In a technique developed by Rippstein (1955), two radiographs are taken in special projections defined by a standard leg positioning device. A modified device was designed by Elsasser and Walker (Dihlmann, 1985). This method yields only an estimate of femoral neck-shaft angles. Computed tomography is without question the most accurate means of evaluating femoral anteversion (Hernandez et al., 1981). Several transverse images are obtained through the femoral neck and distal femur. Lines are drawn along the plane of the femoral neck and femoral condyles, yielding the angle of femoral anteversion. The leg must be stabilized to prevent movement during the examination. The range of normal femoral anteversion is 15 to 20 degrees.

Anteversion of the femoral neck is measured sonographically in a manner similar to humeral retroversion (Figs 9–21 and 9–22). The patient is supine with the lower leg flexed 90 degrees, hanging dependently over the end of the examination table. This allows stable reproducible positioning of the femur. A trans-

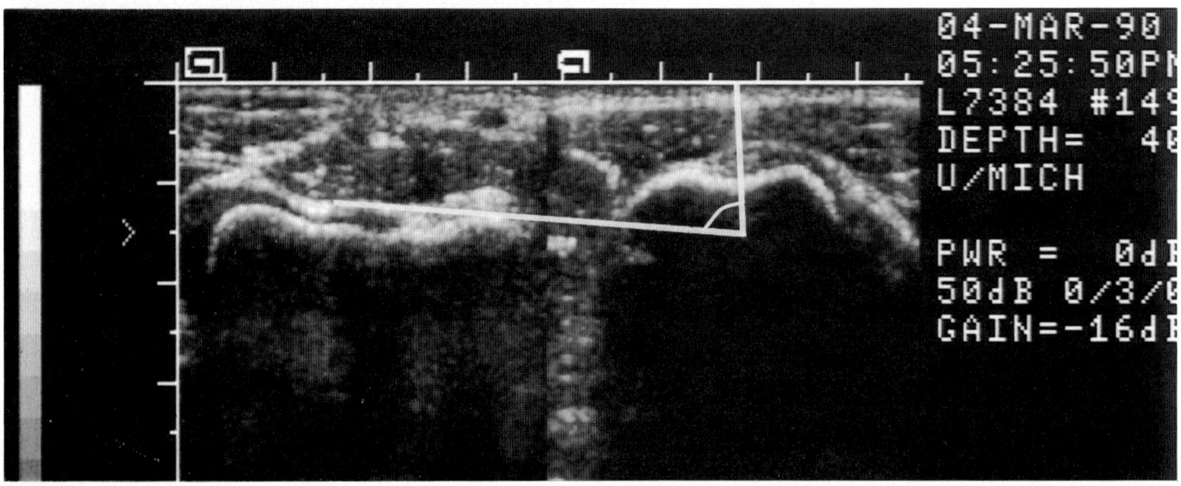


**FIG 9–20.**
Sonographic measurement of humeral retroversion. These images were obtained in a normal 7-year-old boy. A line is drawn from the center of the humeral head bisecting the bicipital groove. Another line is drawn parallel to the trochlea. The intersection of these lines yields the angle of humeral retroversion, as indicated. Sonographic measurement of humeral retroversion tends to overestimate the true angle. This method of measurement provides only a rough estimation of humeral retroversion.

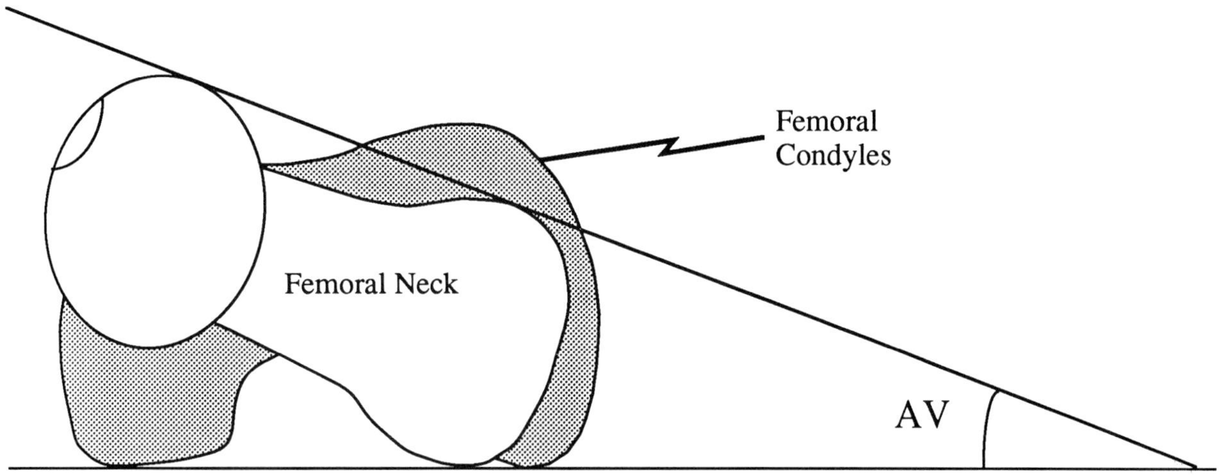


**FIG 9–21.**
Measurement of femoral anteversion: schematic representation.

verse image showing both the femoral head and greater trochanter is obtained with the transducer face parallel to the examination table. The axis of the femoral condyles is assumed to be parallel to the examination table when the patient is positioned as described earlier. The angle of femoral anteversion is obtained by drawing a line tangential to both the anterior surface of the femoral head and the greater trochanter. Its intersection with the border rule defining the face of the transducer yields the angle of femoral anteversion. This technique provides an estimate of femoral anteversion that will demonstrate gross pathology, but CT is the method of choice for surgical planning and follow-up (Hernandez et al., 1981; Berman, 1986).

### Tibial Torsion

Lateral torsion of the tibia is responsible for the normal alignment of the foot. Normal tibial torsion is

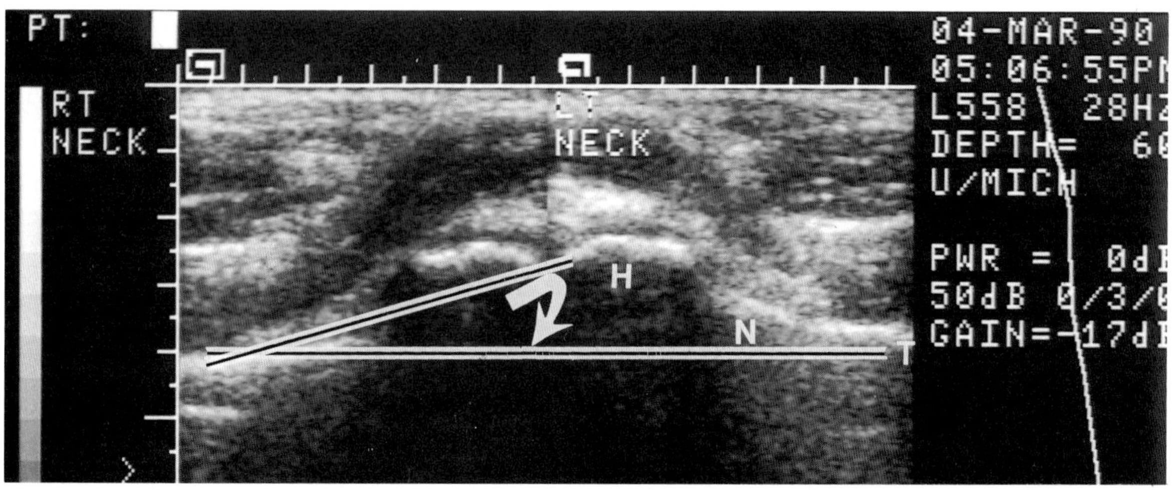


**FIG 9–22.**
Sonographic measurement of femoral anteversion. Same normal 7-year-old boy as in Figure 9–20. This split screen image demonstrates both femoral necks, patient's right at the observer's left. Measurement of the angle of femoral anteversion is demonstrated on the right femur. A line is drawn parallel to the femoral neck. Its intersection with a line parallel with the border rule at the top of the image defines the angle of femoral anteversion *(curved arrow)*. Femoral anteversion in this case measured 13 degrees, within the normal range of 10 to 15 degrees for a well-developed child. The angle of femoral anteversion in an infant is approximately 35 degrees. This rapidly decreases as the child develops. Abbreviations: femoral head *(H)*, femoral neck *(N)*, greater trochanter *(T)*.

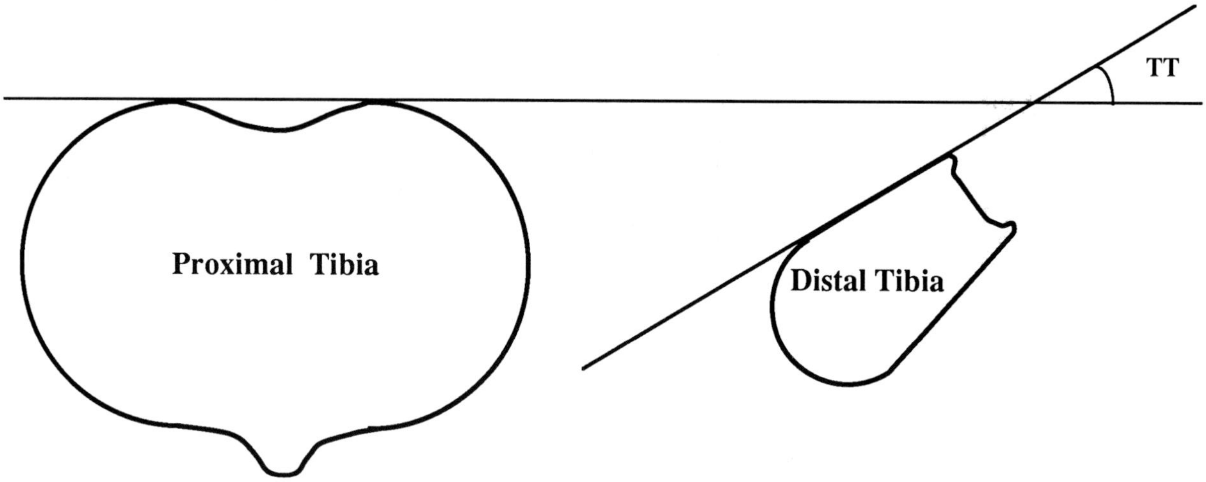


**FIG 9–23.**
Measurement of tibial torsion: schematic representation.

approximately 20 degrees. Tibial torsion may be increased, decreased, absent, or reversed. Abnormal tibial torsion is the most common cause of toe-in deformity in children. In the majority of cases, abnormal tibial torsion will be spontaneously corrected during growth. However, in those cases in which deformity persists, measurement of tibial torsion and femoral anteversion are required to establish the site of abnormality. Corrective osteotomy can then be performed. Excessive tibial torsion is seen in patients with cerebral palsy and spina bifida (Joseph et al., 1987). Surgical correction in these cases depends on the general condition of the patient and the severity of the deformity.

Sonographic measurement (Figs 9–23 and 9–24) is performed with the child lying prone on the examination table with the feet extending just beyond the end of the table. The knee and ankle can be stabilized with sand bags. Transverse images of the proximal and distal tibia are obtained with the transducer face parallel to the examination table. The proximal image is obtained at the level of the tibial condyles, just below the joint space. Imaging of the distal tibia is done at the level of the tibiofibular joint. Figure

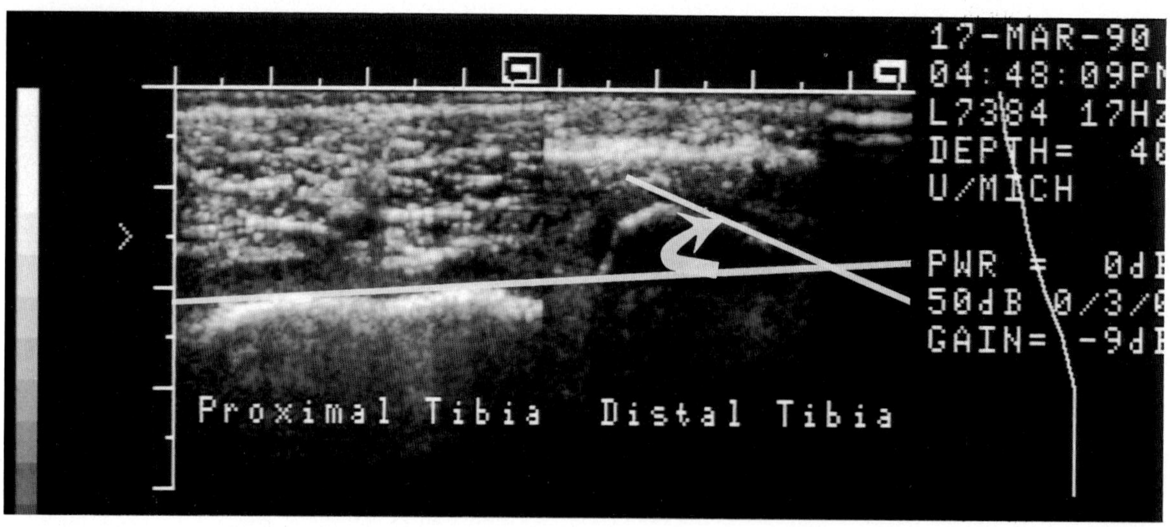


**FIG 9–24.**
Sonographic measurement of tibial torsion. Same normal 7-year-old boy as in Figure 9–22, prone with the feet extending freely over the end of the examination table. The proximal transverse image is obtained at the level of the tibial condyles and the distal image at the level of the tibiofibular joint. Lines are drawn parallel to the posterior surfaces of the bone. Their intersection defines the angle of tibial torsion.

9–24 demonstrates the calculation of tibial torsion. Accuracy of these sonographic measurements are comparable with those obtained using CT (Joseph et al., 1987). The advantage of ultrasound in these cases is the lack of ionizing radiation. Therefore, ultrasound is preferred for long-term follow-up.

## TUMORS

Ultrasound has been used to examine both soft tissue (Fig 9–25) and bone tumors, but clearly MRI and CT provide greater information about the extent of tumor involvement. Magnetic resonance imaging and CT demonstrate the full extent of local invasion, in addition to evaluating metastatic disease. Examination of neoplasia with ultrasound is limited to the soft tissue component of the mass and cortical contours of the involved bone. It can clearly demonstrate the relationship of a mass to adjacent nerves, vessels, and bone (Vincent, 1988). Medullary involvement of bone is not evaluated. Assessment of metastatic disease is also limited, principally to parenchymal organs. However, sonography may play a role in targeting CT or MRI examinations.

Sonography is particularly well suited for evaluating peripheral nerve lesions (Reuter et al., 1982; Chinn, 1982). It provides the capability of examining a peripheral nerve along its entire length in a few minutes, allowing targeted CT or MRI examination of the region of interest. This technique saves considerable time on the CT or MR scanner, eliminating a screening examination. When electromyographic studies are indicative of a peripheral nerve lesion, ultrasound can often serve as the initial screening examination.

Normal nerves are recognized as hyperechoic cords with a fibrillar structure, adjacent to vessels (Fornage, 1988). These are easily distinguished from tendons by lack of motion during movement of the extremity. Flexion and extension maneuvers will readily distinguish nerves from tendons during real-time examination. Most tumors of peripheral nerves are fusiform in shape and decreased in echogenicity relative to normal nerve tissue. A distinction between neurofibromas and schwannomas cannot be made sonographically; both appear hypoechoic and demonstrate increased through transmission. In neurofibromatosis, multiple neurofibromas are observed in peripheral nerves, and it is not unusual to identify them in muscular branches (Fig 9–26). A similar appearance has been reported in leprosy (Fornage and Noret, 1987). Sarcomatous degeneration of neurofibromas may occur, resulting in rapid growth of the tumor with ill-defined borders and irregular margins (Lattes, 1982; Lange et al., 1987; Reading et al., 1988). However, poor margination of a lesion is not a reliable sign of malignancy, because malignant lesions may have a pseudocapsule (Fig 9–27).

Lipomas of the subcutaneous tissues usually appear hyperechoic and are always soft pliable masses. Other benign tumors, such as hemangiomas (Figs

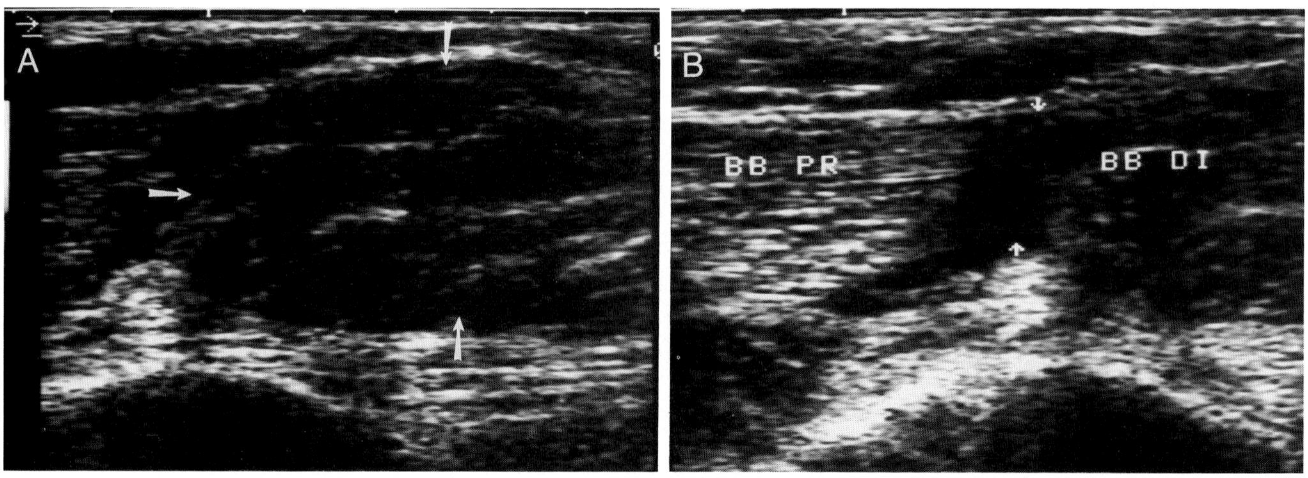


**FIG 9–25.**
**A,** upper arm pseudotumor: longitudinal sonogram. This 60-year-old woman with a mass in anterior upper left arm was sent for sonographic examination to evaluate the extent of presumed soft tissue sarcoma. The mass *(arrows)* has an appearance similar to muscle with visualization of fibroadipose septa. **B,** the transducer was moved proximal to the mass, and a rupture of the proximal biceps muscle is demonstrated *(arrows)*. Proximal *(BB PR)* and distal *(BB DI)* ends of the ruptured biceps are retracted. The distal end was more masslike and clinically palpable, thus called a neoplasm by clinicians.

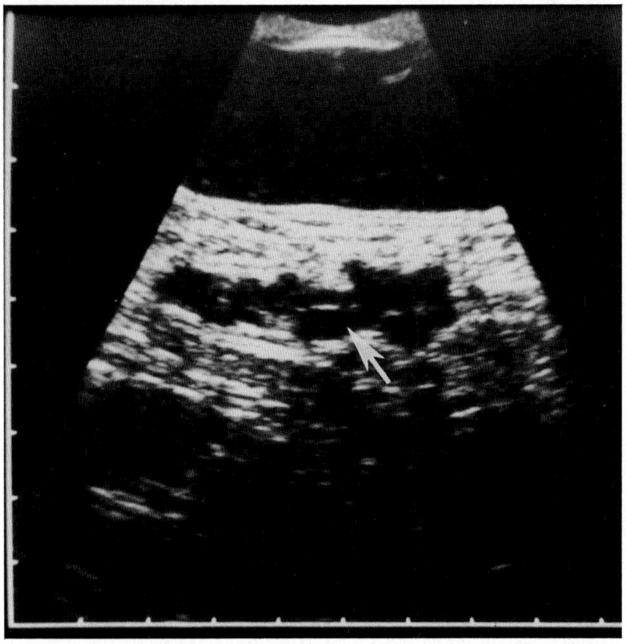

**FIG 9–26.**
Cluster of small neurofibromas: longitudinal sonogram. A mass was palpated in the right gluteal region of this 30-year-old woman with neurofibromatosis. The transducer is positioned along the length of the gluteus maximus muscle. A cluster of neurofibromas *(arrow)* is seen as a lobular mass overlying the gluteus.

9–28 and 9–29), can also be evaluated with ultrasound. Color Doppler studies may be valuable in identifying feeding and draining vessels, but flow within the lesion may be too slow to detect with Doppler. Compression imaging will demonstrate these lesions to be fluid-filled vessels. Occasionally, phleboliths may be identified within these lesions. In general, masses containing calcifications require further imaging investigation with radiographs, CT, or MRI. Intramuscular calcifications or calcifications adjacent to bone without associated soft tissue mass, as described in chapter 3, can safely be called myositis ossificans (Fig 9–30) based solely on sonographic appearance.

When a mass is of indeterminate nature but is not vascular, ultrasound-guided biopsy can easily be performed (Reading et al., 1988). The technique employed is essentially the same as that used for the aspiration of fluid collections. Sonographic biopsy is safe, effective, and cost effective. It will usually accelerate the process of diagnosis by avoiding the scheduling problems associated with CT-guided biopsy.

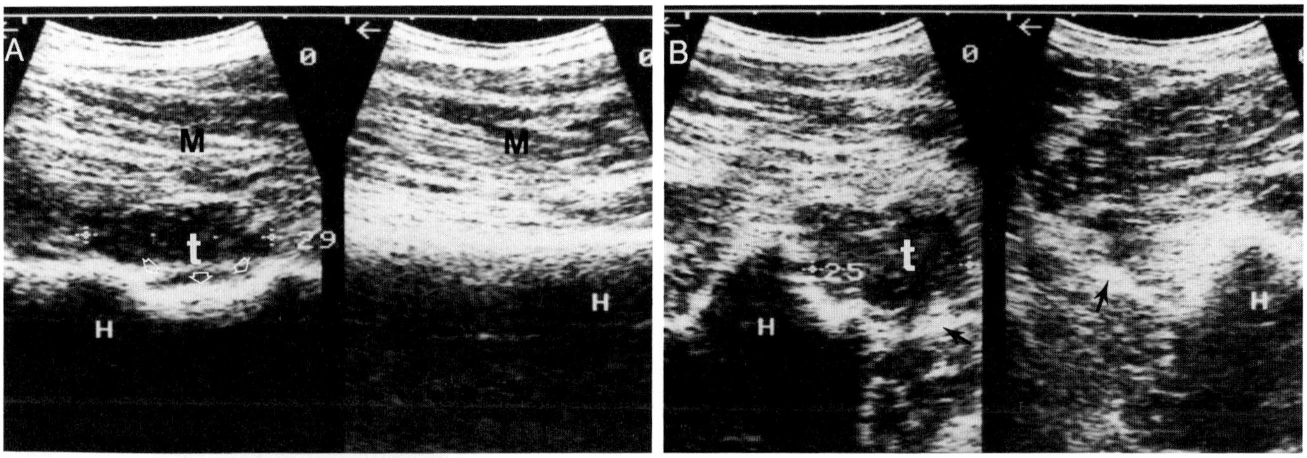


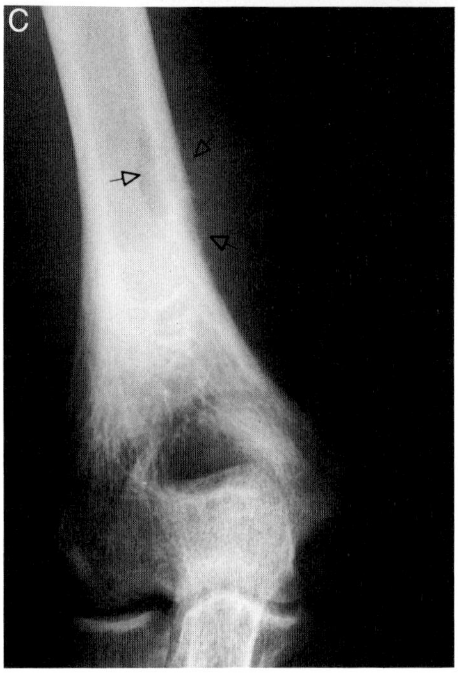


**FIG 9–27.**
**A,** fibrosarcoma above the elbow: longitudinal sonogram. A 55-year-old man who presented with a firm mass at the medial aspect of the right distal upper arm. An ovoid hypoechoic mass *(t)* 2.9 cm in diameter is noted deep to the brachial muscle *(M)* on the *left.* A comparable view of the normal left arm is seen on the *right.* Erosion of the underlying humerus *(open arrows)* is also observed, consistent with a malignant tumor. Note that the superficial margin of the tumor is quite distinct, indicative of pseudocapsule formation. **B,** fibrosarcoma above the elbow: transverse sonogram. Same patient as in **A.** The transducer is rotated 90 degrees; the right arm is once again on the left side of the split screen images. The medial aspects of each arm are at the center of the figure. Immediately anterior to the ulnar nerve *(arrow)* is the previously noted fibrosarcoma *(t)* of the right upper arm. Again invasion of the medial aspect of the humerus is noted. A normal nondisplaced hyperechoic ulnar nerve *(arrow)* is seen in the left arm. Abbreviations: humerus *(H).* **C,** fibrosarcoma above the elbow: anteroposterior radiograph. Same patient as in **A** and **B.** Osseous destruction *(open arrows)* at the medial aspect of the distal humerus is associated with irregular periosteal reaction. The soft tissues are swollen, but the extent of the mass cannot be appreciated on this film.

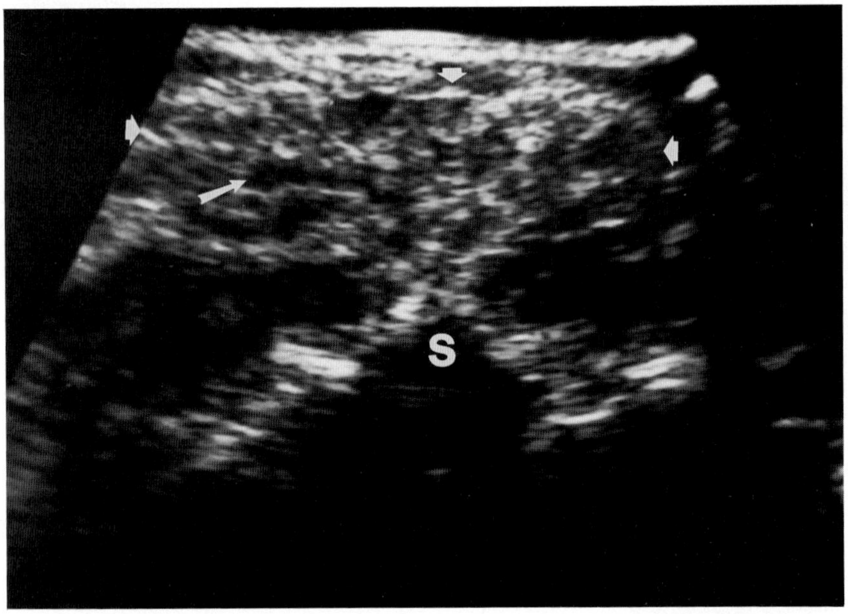


**FIG 9–28.**
Capillary hemangioma of the posterior neck: transverse sonogram. The patient is a 2-year old-boy with swelling of the posterior neck centrally. This transverse image directly over the spinous processes *(S)* demonstrates a large mass *(small arrows)* of mixed echogenicity. It is predominantly hyperechoic relative to the surrounding muscle. Surgical excision demonstrated this to be a capillary hemangioma. Hypoechoic spaces *(large arrow)* within the mass represent venous lakes. Intramuscular hemangiomas and cavernous hemangiomas would be hypoechoic relative to the surrounding muscle, unlike this capillary hemangioma.

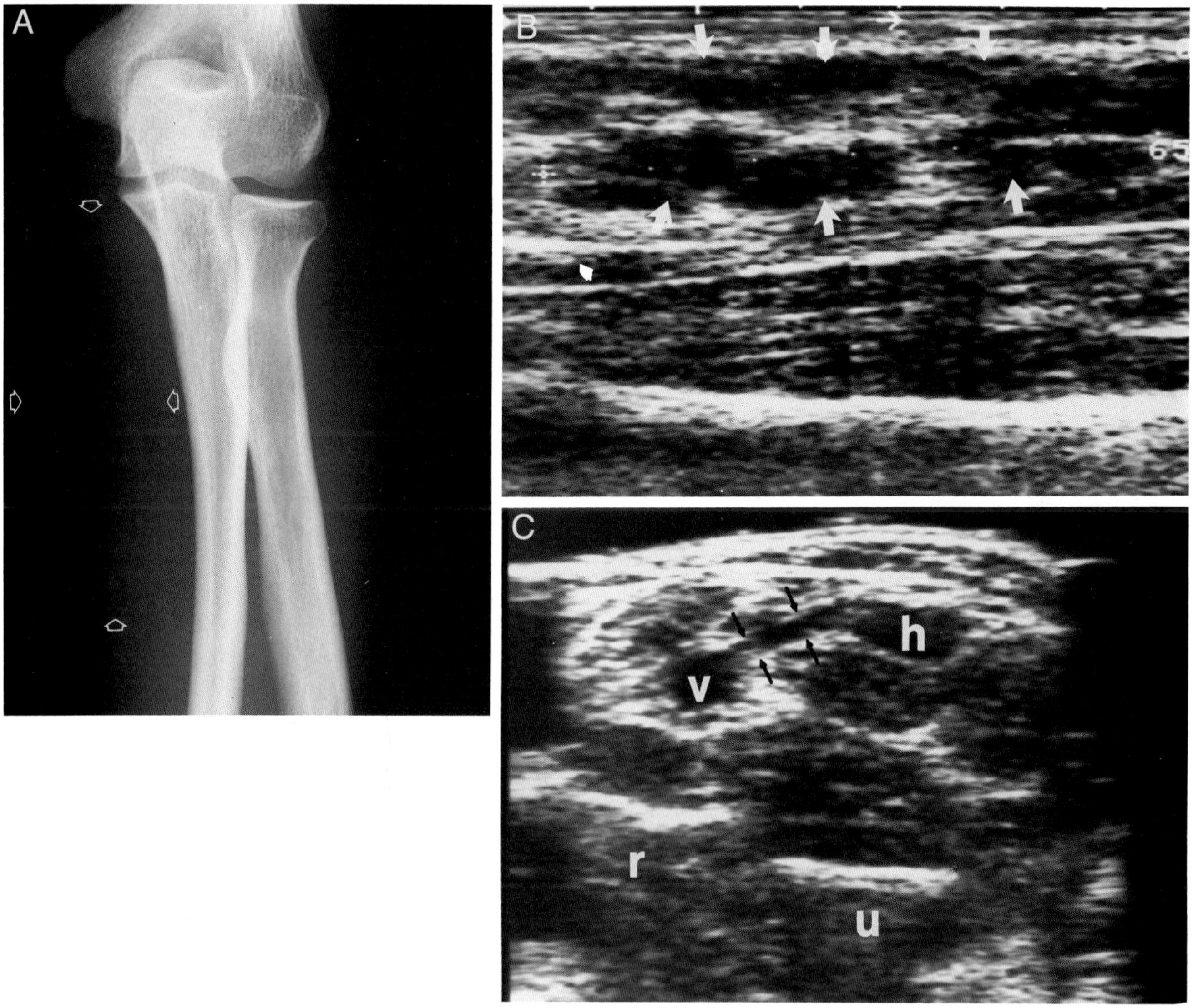


**FIG 9–29.**

**A,** intramuscular angiomatosis: anteroposterior radiograph of the forearm. A large mass is present in the medial aspect of the left forearm of this 10-year-old boy. Multiple phleboliths *(open arrows)* are present in the soft tissues of the medial aspect of the forearm. Phleboliths may be seen in angiomatosis, hemangiomas, and lymphangiomas. **B,** intramuscular angiomatosis: longitudinal sonogram. Same patient as in **A.** Serpiginous vessels *(arrows)* are observed extending deep into the substance of the flexor muscles. At the time of surgery, this lesion had an infiltrative appearance. However, it is a hamartoma. **C,** intramuscular angiomatosis: transverse sonogram. Same patient as in **A** and **B.** The transducer was rotated to image transversely. A large draining vein *(arrows)* extends from the lesion *(h)* to a normal vein *(v)*. Color Doppler demonstrated sluggish flow within the lesion and a normal venous signal in the vein. This image demonstrates the lesion to be limited to the superficial flexor compartment of the forearm. Abbreviations: radius *(r)*, ulna *(u)*.

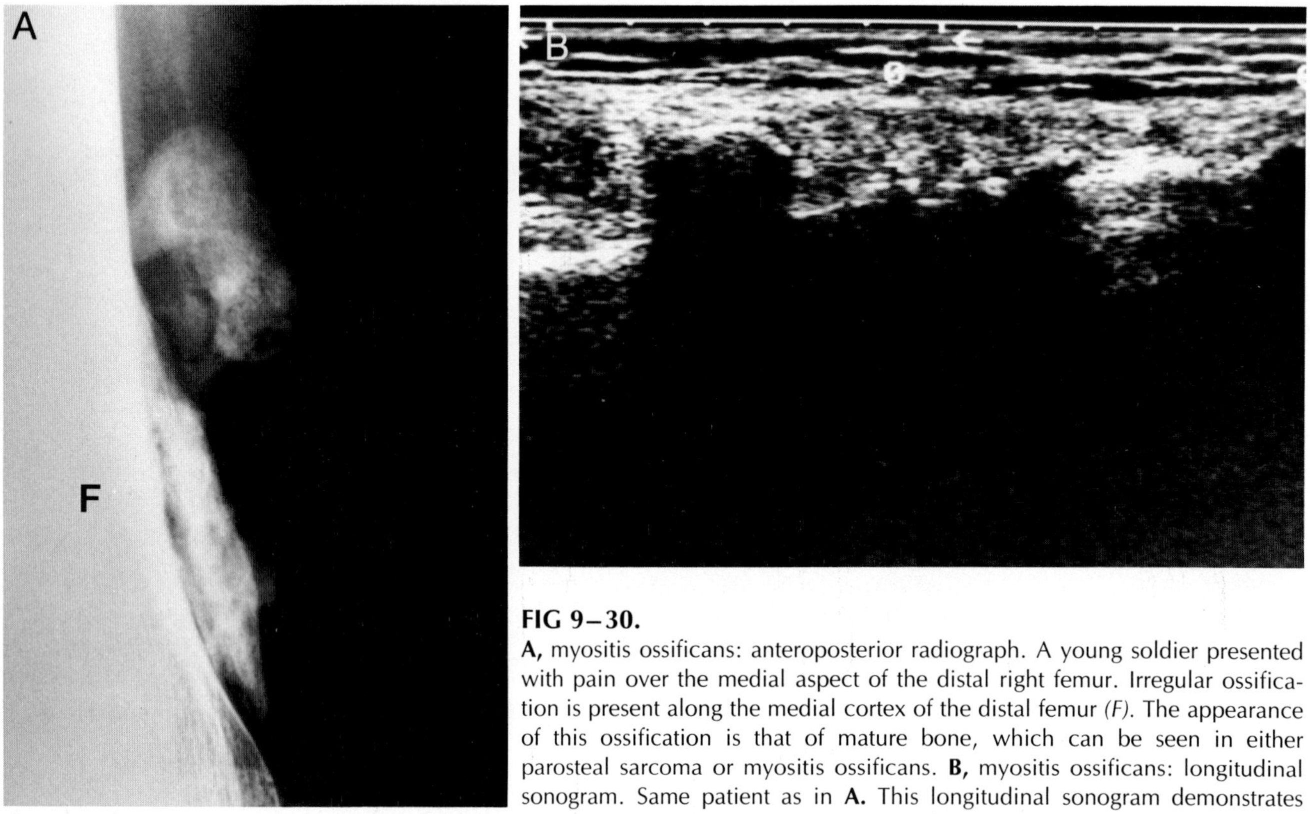


**FIG 9–30.**
**A,** myositis ossificans: anteroposterior radiograph. A young soldier presented with pain over the medial aspect of the distal right femur. Irregular ossification is present along the medial cortex of the distal femur *(F)*. The appearance of this ossification is that of mature bone, which can be seen in either parosteal sarcoma or myositis ossificans. **B,** myositis ossificans: longitudinal sonogram. Same patient as in **A.** This longitudinal sonogram demonstrates that no soft tissue mass is present associated with the regions of ossification. This finding with the history of chronic trauma is consistent with the diagnosis of myositis ossificans. Despite the sonographic finding, biopsy was performed and results interpreted as myositis ossificans.

## REFERENCES

Abiri MM, Kirpekar M, Ablow RC: Osteomyelitis: Detection with ultrasound (work in progress). *Radiology* 1988; 169:795–797.

Berman M: Ultrasound of the hip: A review of the applications of a new technique. *Br J Radiol* 1986; 59:13–17.

Bloom W, Fawcett DW: *A Textbook of Histology.* ed 11. Philadelphia, WB Saunders Co, 1980.

Carson PL, Oughton TV, Hendee WR, et al: Imaging soft tissue through bone with ultrasound transmission tomography by reconstruction. *Med Phys* 1977; 4:302–309.

Chiba T, Lloyd DA, Bowen A, et al: Ultrasonography as a method of nutritional assessment. *J Parenter Enter Nutr* 1989; 13:529–534.

Chinn DH: Unusual ultrasonographic appearance of a solid schwannoma. *J Clin Ultrasound* 1982; 10:243–245.

Dihlmann W: *Joints and Vertebral Connections.* New York, Georg Thieme Verlag, 1985.

Fornage BD, Nerot C: Sonographic diagnosis of tuberculoid leprosy. *J Ultrasound Med* 1987; 6:105–107.

Fornage BD: Peripheral nerves of the extremities: Imaging with ultrasound. *Radiology* 1988; 167:179–182.

Fried AM, Coughlin K, Griffen WO: The sonographic fat/muscle ratio. *Invest Radiol* 1986; 21:71–75.

Gitschlag KF, Sandler MA, Madrazo BL, et al: Disease in the femoral triangle: Sonographic appearance. *AJR* 1982; 139:515–519.

Graf R, Schuler P: *Sonographie am Stütz und Bewegungsapparat bei Erwachsenen und Kindern.* Weinheim, Edition Medizin VCH, 1988.

Harcke HT, Lee MS, Sinning L, et al: Ossification center of the infant hip: Sonographic and radiographic correlation. *AJR* 1986; 147:317–321.

Harcke HT, Grissom LE, Finkelstein MS: Evaluation of the musculoskeletal system with sonography. *AJR* 1988; 150:1253–1261.

Heaney RP, Avioli LV, Chesnut CH, et al: Osteoporotic bone fragility. Detection by ultrasound transmission velocity. *JAMA* 1989; 261:2986–2990.

Hentz VR, Marich KW, Dev P: Preliminary study of the upper limb with the use of ultrasound transmission imaging. *J Hand Surg (Am)* 1984; 9:188.

Hentz VR, Green PS, Arditi M: Imaging studies of the cadaver hand using transmission ultrasound. *Skeletal Radiol* 1987; 16:474–480.

Hernandez RJ, et al: CT determination of femoral torsion. *AJR* 1981; 137:97–101.

Joseph B, Carver RA, Bell MJ, et al: Measurement of tibial torsion by ultrasound. *J Pediatr Orthop* 1987; 7:317–323.

Lange TA, Austin CW, Seibert JJ, et al: Ultrasound imaging as a screening study for malignant soft-tissue tumors. *J Bone Joint Surg* 1987; 69A:100–105.

Lattes R: *Tumors of the Soft Tissues.* Washington DC, Armed Forces Institute of Pathology, 1982.

Maldague B, Malghem J: Dynamic radiologic patterns of Paget's disease of bone. *Clin Orthop* 1987; 217:127–151.

Reading CC, Charboneau JW, James EM, et al: Sonographically guided percutaneous biopsy of small (3 cm or less) masses. *AJR* 1988; 151:189–192.

Resnick D, Niwayama G: Diagnosis of bone and joint disorders. Philadelphia, WB Saunders Co, 1988.

Reuter KL, et al: Ultrasonography of a plexiform neurofibroma of the popliteal fossa. *J Ultrasound Med* 1982; 1:209–211.

Rippstein J: Zur Bestimmung der Antetorsion des Schenkelhalses mittels zweier Rögntgenaufnahmen. *Z Orthop* 1955; 86:345.

Sandler MA, Alpern MB, Madrazo BL, et al: Inflammatory lesions of the groin: Ultrasonic evaluation. *Radiology* 1984; 151:747–750.

Shenouda NA, England JPS: Ultrasound in the diagnosis of scaphoid fractures. *J Hand Surg* 1987; 220:130–136.

Suzuki S, Awaya G, Okada Y, et al: Examination by ultrasound of Legg-Calvé-Perthes disease. *Clin Orthop* 1987; 220:130–136.

van Holsbeeck M, van Holsbeeck K, Gevers G, et al: Staging and follow-up of rheumatoid arthritis of the knee. Comparison of sonography, thermography, and clinical assessment. *J Ultrasound Med* 1988; 7:561–566.

vanSonnenberg E, Wittich GR, Casola G, et al: Sonography of thigh abscess: Detection, diagnosis and drainage. *AJR* 1987; 149:769–772.

Vincent LM: Ultrasound of soft tissue abnormalities of the extremities. *Radiol Clin North Am* 1988; 26:131–144.

Yeh HC, Rabinowitz JG: Ultrasonography of the extremities and pelvic girdle and correlation with computed tomography. *Radiology* 1982; 143:519–525.

# 10

# Sonography of Rheumatoid Disease

Early diagnosis, assessment of inflammation and detection of complications of rheumatoid disease are problems that can be addressed using ultrasound. The use of noninvasive techniques is especially important in patients with rheumatoid disease. Joint aspiration and arthrography are extremely painful when performed on an inflamed joint. In addition, these procedures are associated with significant risk of infection in this population due to immune suppression from chronic steroid therapy. Arthrography, arthroscopy, and joint aspiration must be kept to an absolute minimum in these patients (Figs 10–1 and 10–2) (Moore et al., 1975).

Over the past 20 years, many noninvasive techniques have been proposed to assess articular involvement of rheumatoid disease. Among them are radioisotope scanning and thermography. These techniques have been of limited benefit. Most recently, magnetic resonance imaging (MRI) and sonography have come to the forefront in the noninvasive evaluation of rheumatoid disease. Given ideal conditions, both of these modalities are equally well suited for the diagnosis and follow-up of rheumatoid disease. However, as we are all aware, ideal conditions rarely exist. Sonography has a tremendous advantage over MRI in that the examination can be performed quite rapidly, within approximately 15 minutes. Rheumatoid patients cannot tolerate lying motionless on a hard table for the 60 to 90 minutes required for an MRI examination. In addition, during sonographic examination the patient may move other extremities relatively freely, and it does not require sedation. Cost and availability factors also strongly favor ultrasound.

## DIAGNOSIS

The initial diagnosis of rheumatoid disease usually requires the correlation of clinical examination, laboratory tests, and radiographic findings (Beecker, 1959; Ropes et al., 1959). A definitive diagnosis of rheumatoid arthritis can be made when marginal erosions are identified in the characteristic distribution on conventional radiographs. Marginal erosions are the result of chronic synovial inflammation. In the early stages of disease, a magnifying lens proves helpful in identifying tiny erosions. However, detection of small erosions may be significantly delayed because they are obscured on the radiograph due to technical factors, most often patient positioning. As a result, diagnosis may be delayed several months, and joint destruction continues untreated.

Ultrasound has proved to be extremely valuable in the early diagnosis of rheumatoid disease. It is exquisitely sensitive in detecting increased intra-articular fluid and is capable of detecting synovial inflammation before marginal erosions are present (Fig 10–3) (Marchal et al., 1987; van Holsbeeck et al., 1988; Gielen et al., 1990). In early stages of active rheumatoid disease, cartilage edema is present, which can be detected sonographically as cartilage thickening. Erosions are more easily detected using ultrasound (Fig 10–4 through 10–6). A comparison of sonography with conventional radiography in the detection of erosions in the shoulder demonstrated that we could detect 27% more erosions using ultrasound.

Increased intra-articular fluid is a nonspecific finding indicative of joint pathology. Although it is nonspecific, joint effusion remains a valuable indi-

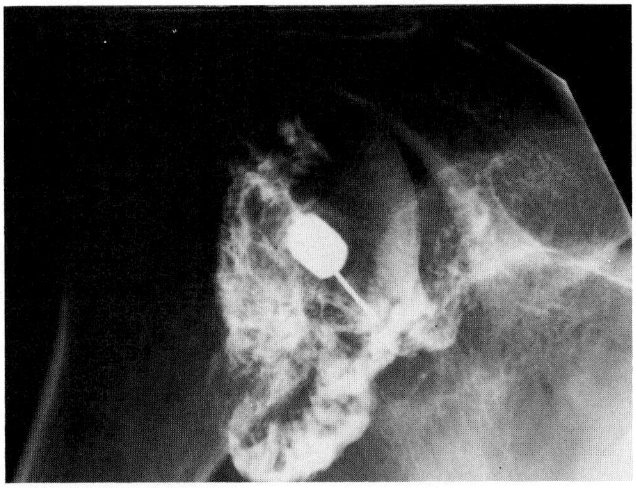

**FIG 10–1.**
Early rheumatoid arthritis: shoulder arthrogram. This single contrast arthrogram was performed on a 56-year-old woman with a 4-month history of polyarticular arthritis, predominantly in the right shoulder. Numerous filling defects indicative of synovial proliferation are noted. The rotator cuff is intact. Arthrography performed in patients with synovial inflammation is extremely painful and usually a prolonged procedure. In addition, these invasive procedures are associated with an increased risk of joint infection. Invasive procedures must be kept to a minimum in these immune compromised patients.

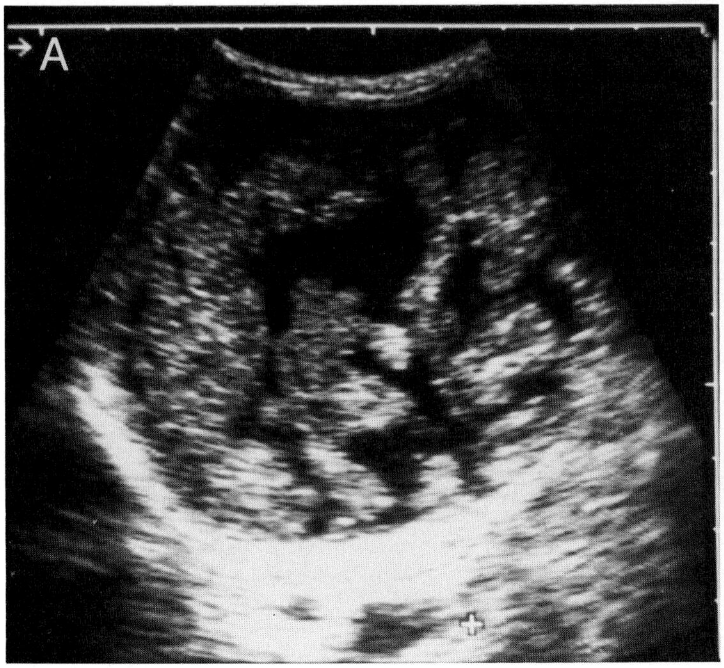


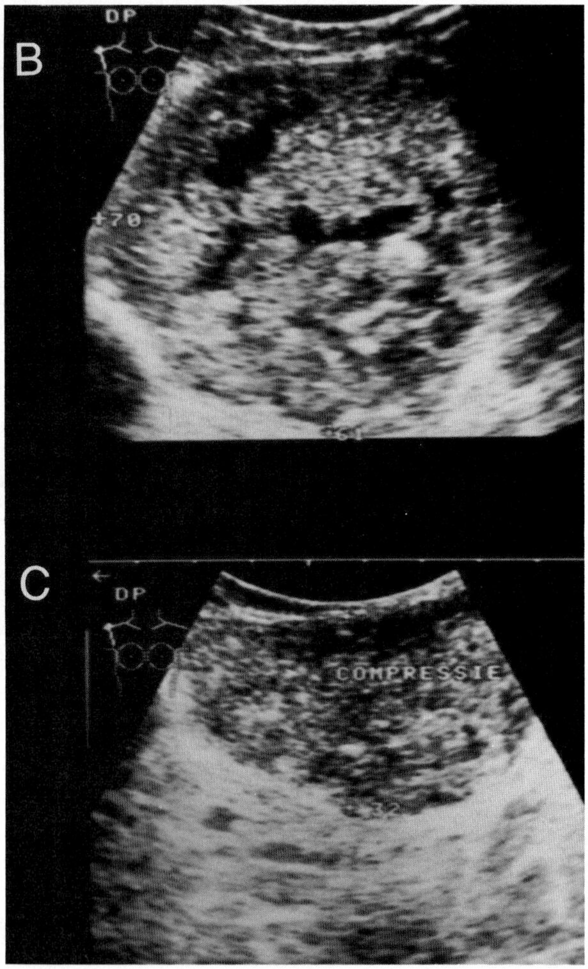


**FIG 10–2.**
**A,** rheumatoid arthritis of the shoulder: transverse sonogram of the axillary recess of the shoulder. Same patient as in Figure 10–1. An enlarged axillary recess filled with pannus is seen on this transverse image. Villous synovial proliferation is evident within the fluid-filled joint. Sonography provides an excellent evaluation of the joint noninvasively. In addition to evaluation of the synovium, the rotator cuff and biceps tendon are also well evaluated during the same session. **B and C,** chronic inflammation of the axillary recess of the shoulder: transverse sonograms. Same patient as in **A.** Another advantage of sonography over arthrography is the ability to accurately quantitate synovial proliferation. These two images demonstrate the necessity of compression with the transducer to accurately measure total synovial thickness. Note the marked change in anteroposterior dimension of the axillary recess when fluid is expressed by compression.

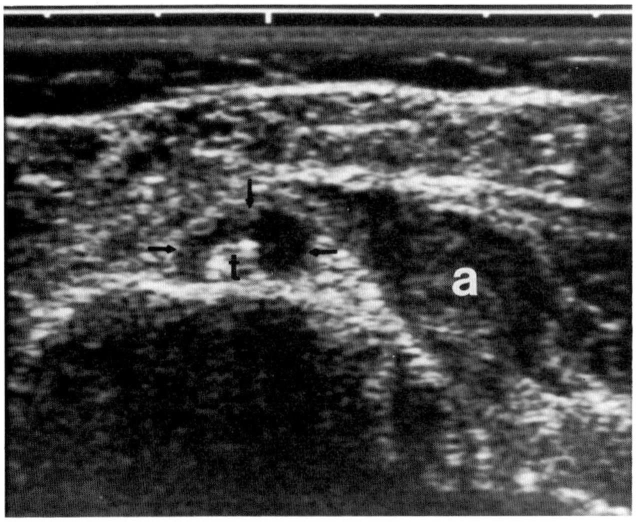


**FIG 10–3.**
Rheumatoid arthritis of the shoulder: transverse sonogram of the bicipital groove. Hypoechoic fluid and synovial proliferation *(arrows)* surround the long biceps tendon *(t)* and distend an enlarged axillary recess in this 53-year-old woman with a 2-year history of rheumatoid arthritis. An enlarged axillary recess *(a)* is pathognomonic for chronic joint disease. Often the diagnosis of rheumatoid disease can be made sonographically by the observation of an enlarged axillary recess with synovial proliferation several months prior to the development of marginal erosions.

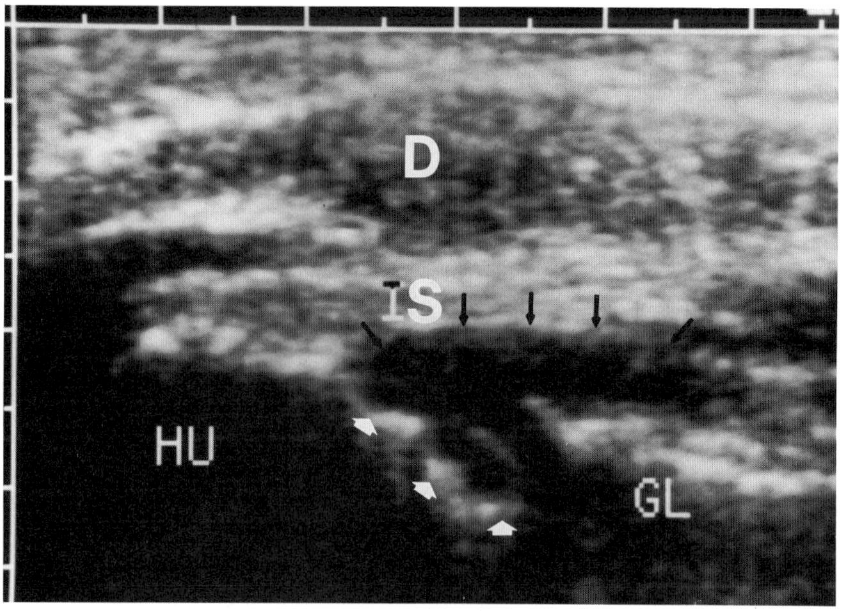


**FIG 10–4.**
Effusion in the shoulder joint: transverse sonogram of the posterior glenohumeral joint. The patient is a 39-year-old woman who developed symmetrical joint pain during pregnancy. Conventional radiographs taken after delivery were normal. Sonographic examination of the shoulder demonstrated a joint effusion with marginal erosions *(white arrows)* adjacent to the insertion of the infraspinatus tendon *(IS)*. The best location to detect small effusions within the shoulder is the posterior synovial recess *(black arrows)* between the glenohumeral joint and the infraspinatus tendon. A fluid film greater than 2 mm thick at this location is indicative of joint effusion. The first considerations when effusion is detected in an adult without other findings are joint infection and rheumatoid arthritis. Aspiration is then indicated to make the definitive diagnosis. The observation of erosions in this case made the diagnosis of rheumatoid arthritis. Abbreviations: deltoid *(D)*, glenoid *(GL)*, humerus *(HU)*.

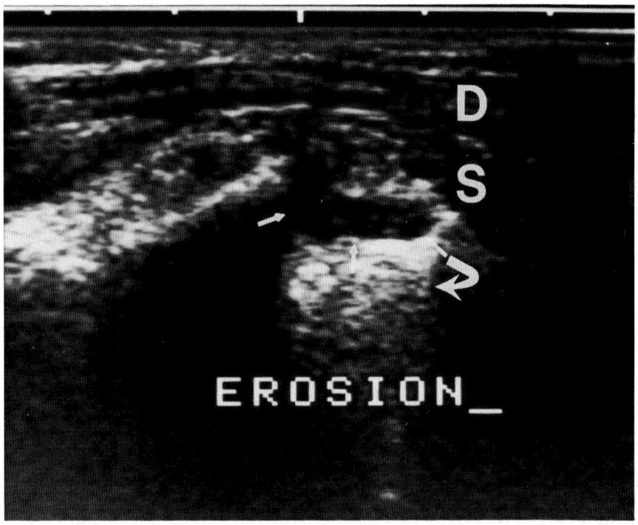


**FIG 10–5.**
Humeral head erosion: transverse sonogram. This woman has a long history of rheumatoid disease. The transducer is positioned transversely over the lateral aspect of the shoulder just inferior to the acromion. A large erosion *(small arrows)* is observed deforming the humeral head. The appearance of increase through transmission *(curved arrow)* is frequently seen deep to large erosions. Rheumatoid erosions contain hypoechoic pannus. In combination with an absence of cortical bone in the depth of the lesion, imaging of a small amount of trabecular bone is possible. Ultrasound has been demonstrated to be more sensitive than conventional radiography in the detection of marginal erosions and Hill-Sachs lesions. Abbreviations: deltoid *(D)*, supraspinatus *(S)*.

cator of active joint disease. Ultrasound has been shown to be unquestionably the best method for detection of increased intra-articular fluid (Adam et al., 1986; Peck, 1986; Marchal et al., 1987). Effusions as small as 1 mL can be detected with ultrasound (Marchal et al., 1987). Careful comparison with the contralateral joint is often helpful, but symmetrical disease may be misleading. The film of synovial fluid separating the leaves of the synovia in the joint recesses should not exceed 3 mm (Fig 10–7).

Inflammation of the synovial membrane in rheumatoid arthritis is observed as irregular thickening. Fat surrounding the inflamed synovia is increased in echogenicity. Therefore, the thickened synovial membrane has a hypoechoic appearance. Measurement of synovial thickness must be made with firm compression applied with the transducer. This expresses synovial fluid from the joint recess, allowing a reproducible measure of synovial thickness. In addition to a very irregular contour of the synovial membrane,

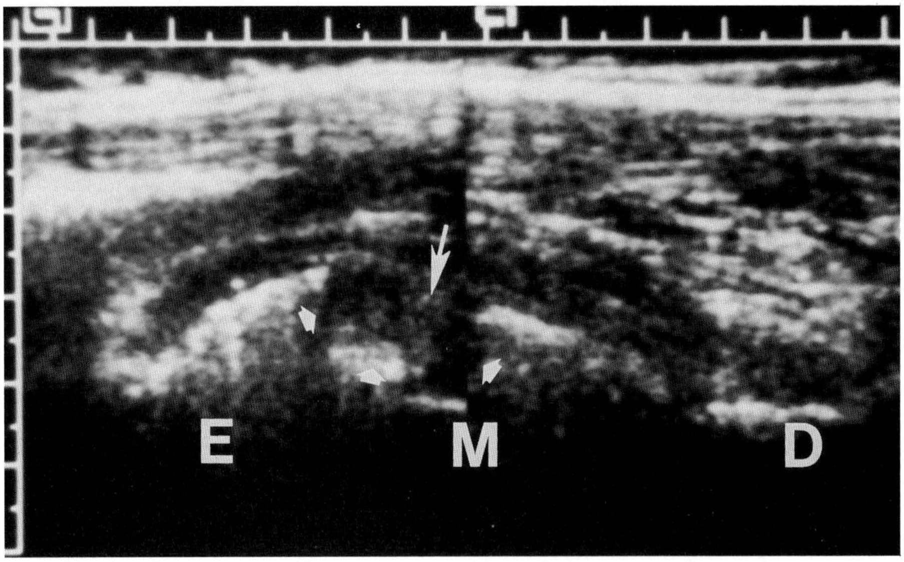


**FIG 10–6.**
Rheumatoid arthritis of the hip: longitudinal sonogram. Over the past 7 months, this 61-year-old woman experienced pain and stiffness in the right hip, which was worse in the morning. Conventional radiographs demonstrated a normal joint space and a lucency in the femoral neck of unknown significance. Blind aspiration of the hip yielded a "dry tap." The transducer is positioned lateral to the femoral artery in the axis of the femoral neck. A large marginal erosion *(small arrows)* is seen in the proximal femoral neck. Hypoechoic pannus *(large arrow)* fills the erosion. This examination made the initial diagnosis of rheumatoid involvement of the hip. During the following 6 months she developed a symmetrical polyarthritis. Abbreviations: diaphysis *(D)*, epiphysis *(E)*, metaphysis *(M)*.

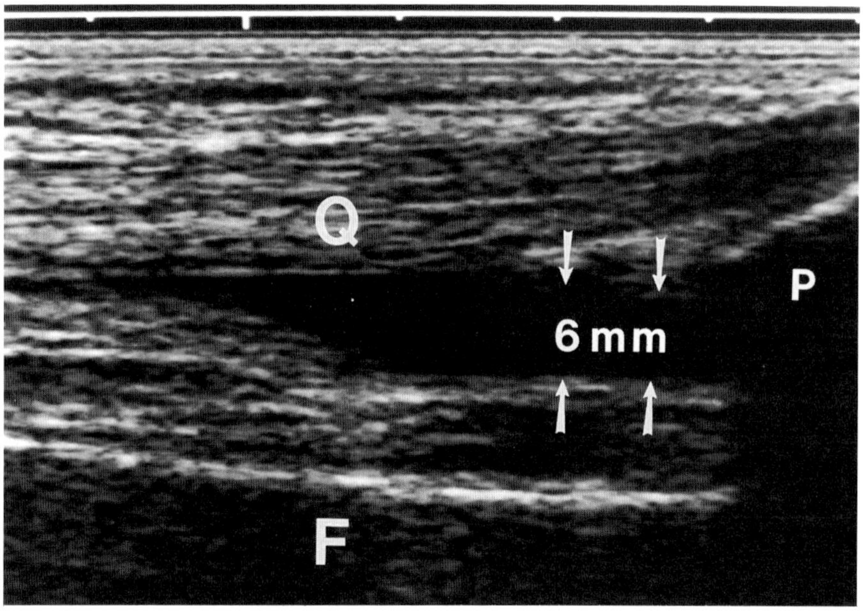


**FIG 10–7.**
Knee effusion in chondromalacia: longitudinal sonogram. A 17-year-old girl with bilateral patellar subluxation and recurrent swelling of the knees presented for sonographic evaluation. In this standard longitudinal view of the suprapatellar bursa, an anechoic fluid collection is identified within the bursa *(arrows)*. The normal fluid film within the suprapatellar bursa should not exceed 3 mm. Sonography is very sensitive in detection of joint effusion. Increases in joint fluid as small as 1 mm can be detected. Effusion is a nonspecific finding seen in a wide variety of pathology, in this case chondromalacia patellae. Abbreviations: femur *(F)*, patella *(P)*, quadriceps *(Q)*.

synechia (Fig 10–8) between the synovial walls can be observed in late rheumatoid disease (van Holsbeeck et al., 1988). Pannus found in affected joints has the appearance of hypoechoic soft tissue masses, a pseudotumor appearance (Fig 10–9) (Cooperberg et al., 1978). Often it has a villous configuration continuous with the synovial walls. These masses can be differentiated from aggregations of proteinaceous material within the joint by their broad-based attachment to the synovial walls (van Holsbeeck et al., 1988; Worth et al., 1986).

Cartilage edema in early stages of rheumatoid arthritis can be detected sonographically by thickening of the articular cartilage. Electronic calipers make these measurements quite easy (Aisen et al., 1984). Normal weight-bearing cartilage ranges from 1.2 to 1.9 mm thick (Aisen et al., 1984; Richardson et al., 1988). Measurements of cartilage thickness greater than 2.5 mm are generally considered pathological. Chronic inflammation of the cartilage will result in permanent damage to the articular surface. This is observed sonographically as blurring of the articular surface. Continued destruction of the cartilage due to rheumatoid disease will be seen as pitting of the articular surface and measurable thinning of the cartilage.

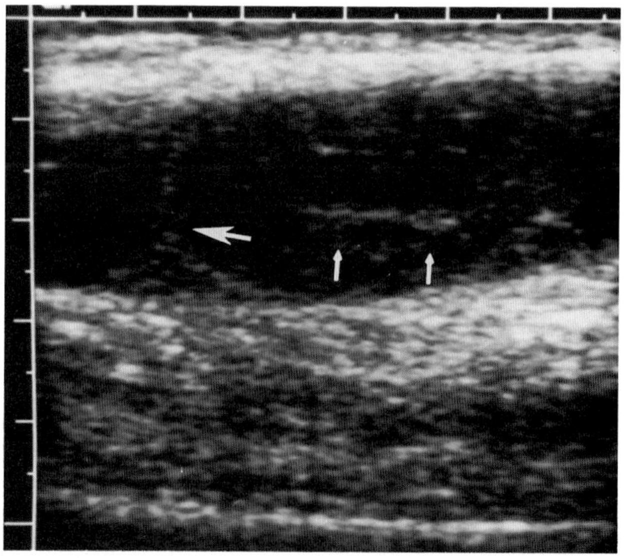

**FIG 10–8.**
Rheumatoid arthritis of the knee: longitudinal sonogram of the popliteal fossa. A large Baker's cyst is noted in the popliteal fossa. The presence of synechiae *(large arrow)*, synovial proliferation, and debris *(small arrows)* are characteristic findings in chronic synovitis.

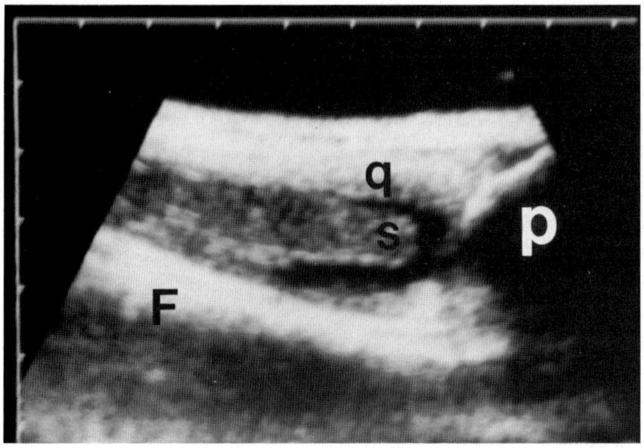


**FIG 10–9.**
Diagnosis of rheumatoid knee synovitis: suprapatellar approach. The suprapatellar bursa provides a sonographic window through which we can evaluate for joint effusion, synovial thickening, and intra-articular loose bodies. This bursa is not obscured by bone, making it ideal for ultrasound examination. In this patient with rheumatoid arthritis, a large synovial polyp *(s)* surrounded by fluid is seen extending down from the superior extent of the bursa. Abbreviations: femur *(F)*, patellar base *(p)*, quadriceps tendon *(q)*.

Marginal erosions are identified as crater-like defects in the bony contours along the edges of the articular cartilage. Pannus is recognized as hypoechoic soft tissue filling these cortical defects. As stated earlier, sonography is more sensitive than conventional radiography in detection of these erosions. In more advanced disease intraepiphyseal "cysts" can be identified. Like marginal erosions, they contain hypoechoic pannus.

### Differential Diagnosis

Uncomplicated joint effusion, characterized by anechoic intra-articular fluid, can be seen in a wide variety of joint pathology (see Fig 10–7). Traumatic disorders such as osteochondritis dissecans, chondromalacia patellae, patellar subluxation, medial plica syndrome, and meniscal tears are among the numerous possible etiologies. Calcium pyrophosphate dihydrate deposition disease (CPPD) is also associated with uncomplicated effusion. All of these entities are easily distinguished sonographically from rheumatoid arthritis by the absence of irregular synovial thickening.

Synovial proliferation (see Fig 10–9) is characteristic of the inflammatory arthritides and chronic infection. In clinical practice, the differential diagnosis of rheumatoid arthritis most frequently includes psoriatic arthritis and Reiter's disease. Psoriatic arthritis and Reiter's disease are rarely manifested by villous synovial proliferation. However, the absence of villi attached to the irregularly thickened synovium is not a reliable means of differentiating these entities from rheumatoid arthritis. The definitive diagnosis in these cases depends on radiographic findings, laboratory data, and clinical presentation. Pigmented villonodular synovitis is another proliferative disorder involving the synovial membrane that cannot be differentiated solely by sonographic criteria. The definitive diagnosis can sometimes be made with conventional radiography, but surgical biopsy is often required to exclude synovial sarcoma. A few cases of amyloid deposition in the synovium have been detected sonographically in hemodialysis patients. It produces a slightly nodular appearance in the synovium, but no villi are present.

Chronic infection is a rare cause of joint effusion in combination with synovial proliferation. Tuberculosis, brucellosis, syphilis, and fungal infection all may produce sonographic findings similar to the inflammatory arthritides. Radiographic findings are also quite similar to rheumatoid arthritis. Clinical history and laboratory findings will be suggestive of infection, but definitive diagnosis depends on aspiration of the joint and isolation of the causative organism. The presence of anechoic fluid within the joint generally reduces suspicion for infection. However, the lack of echos within the fluid does not absolutely exclude infection. One case of joint sepsis in which the fluid was clearly anechoic has been reported (Marchal et al., 1987). Whenever infection is suspected, joint aspiration is mandatory.

Ultrasound guidance has several advantages over fluoroscopy in joint aspiration. The major advantage is that ultrasound has the ability to localize fluid collections. If aspiration is performed using fluoroscopy, a "dry tap" does not exclude infection (Fig 10–10). Debris within the joint may block flow through the needle, or loculations may inhibit aspiration. Sonographic guidance can direct the needle to the largest fluid-filled joint recess, avoiding vital structures. Duration of the procedure is usually not significantly longer than a routine ultrasound examination (McGahan et al., 1987).

Several small changes in aspiration technique will facilitate sonographic guidance. A linear array transducer placed within a sterile surgical glove is essential. If sterile ultrasound gel is not available, povidone-iodine or chlorhexidine gluconate skin prep solution can be substituted. Radial arrays and sector scanners make the procedure more difficult because of the varying angle of incidence of the sound beam

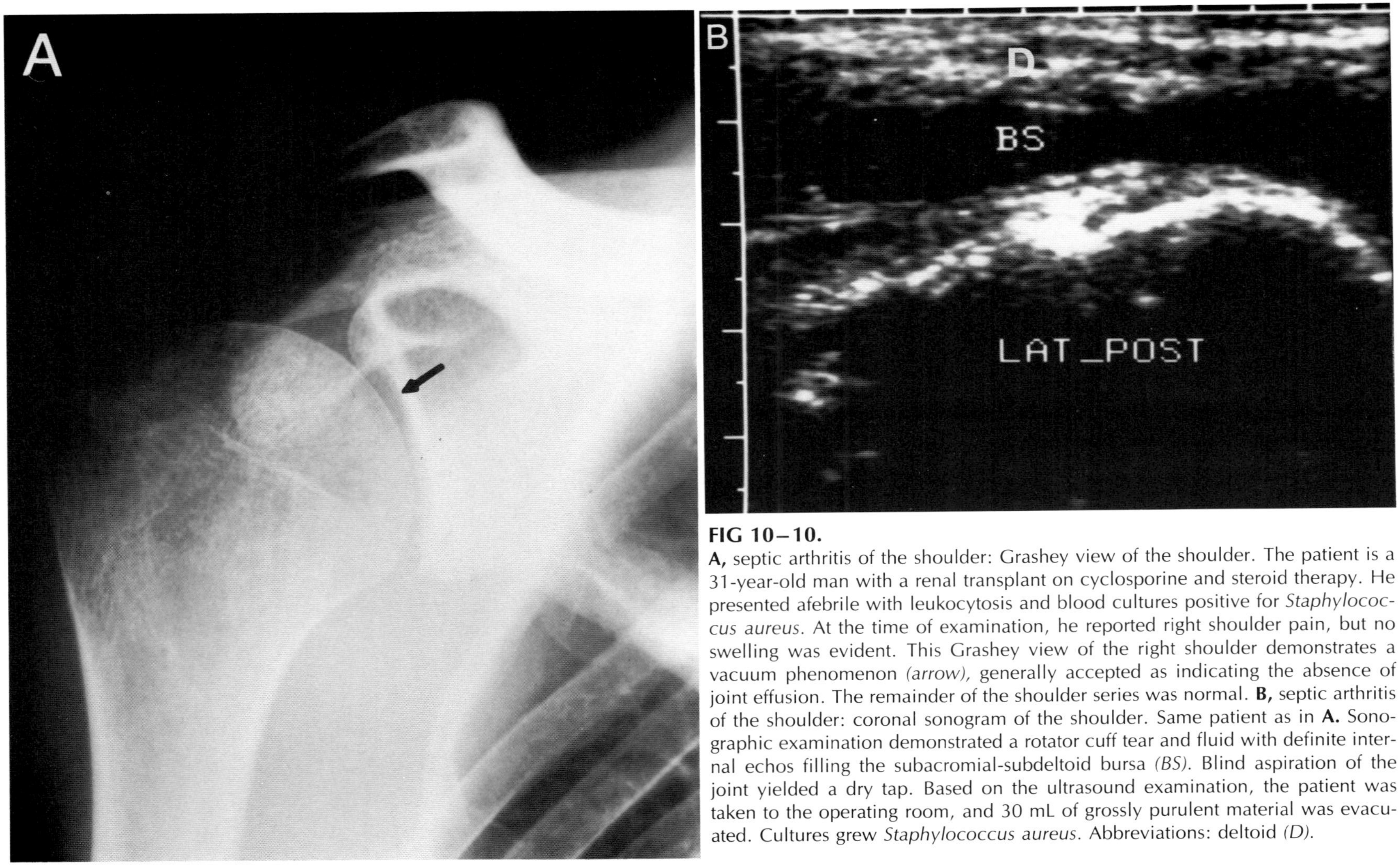


**FIG 10–10.**
**A,** septic arthritis of the shoulder: Grashey view of the shoulder. The patient is a 31-year-old man with a renal transplant on cyclosporine and steroid therapy. He presented afebrile with leukocytosis and blood cultures positive for *Staphylococcus aureus*. At the time of examination, he reported right shoulder pain, but no swelling was evident. This Grashey view of the right shoulder demonstrates a vacuum phenomenon *(arrow)*, generally accepted as indicating the absence of joint effusion. The remainder of the shoulder series was normal. **B,** septic arthritis of the shoulder: coronal sonogram of the shoulder. Same patient as in **A.** Sonographic examination demonstrated a rotator cuff tear and fluid with definite internal echos filling the subacromial-subdeltoid bursa *(BS)*. Blind aspiration of the joint yielded a dry tap. Based on the ultrasound examination, the patient was taken to the operating room, and 30 mL of grossly purulent material was evacuated. Cultures grew *Staphylococcus aureus*. Abbreviations: deltoid *(D)*.

with the needle. The needle is best visualized when the incident sound beam is perpendicular to the shaft of the needle. A 20-gauge spinal needle is routinely used for these procedures. Needles with textured surfaces are available for sonographic procedures and allow better visualization of the needle. Another trick to identifying the location of the needle tip is removal and replacement of the stylet. This will pump a tiny amount of air through the needle, producing a small bubble at the tip. The echogenic gas with shadowing is easily detected. However, this trick cannot be used more than three or four times during the same procedure. Visualization of landmarks will be obscured by shadows produced by the bubbles if this trick is used excessively. Aspiration technique is otherwise unchanged from a fluoroscopic approach.

## FOLLOW-UP OF RHEUMATOID ARTHRITIS

Previously, there were no noninvasive techniques to accurately evaluate the activity of rheumatoid arthritis in large synovial joints. Fortunately, clinical examination of small superficial joints, such as the interphalangeal joints, is a reliable means of evaluating disease activity in rheumatoid patients. Resolution of swelling in the interphalangeal joints is a reliable indicator of healing. However, it is not possible to evaluate inflammation in deep-seated joints by clinical examination. Arthrotomy and biopsy were the only accurate means available for deep-seated joints, but these procedures are associated with significant morbidity in this patient population. Laboratory data and conventional radiography are of little help (DeSilva et al., 1986). Magnetic resonance imaging has been proposed as a noninvasive means of assessing joint inflammation (Hammer et al., 1986). These investigators noted a significant change in T2 characteristics of the synovium after therapy, but no prospective studies of follow-up over time are presently available. Also, the time required to perform the examination is usually intolerable to these patients. Thermography has shown some promise; however, it is also a difficult examination to perform. Radioisotope scanning has proved to be of questionable value. Ultrasound has demonstrated accuracy and reliability in the evaluation of joint inflammation associated with rheumatoid arthritis (van Holsbeeck et al., 1988; Cooperberg et al., 1978). It is an examination that can be performed quickly and without discomfort.

Adoption of a standardized technique for sonographic examination is essential. When evaluating the knee, we obtain sagittal images of the suprapatellar bursa with the patient supine and the knee flexed 30 degrees. The transducer is positioned along the length of the rectus femoris tendon with its inferior extent touching the superior border of the patella. Two sets of images are obtained. The first set is obtained with firm pressure applied with the transducer (Fig 10–11,C), compressing the bursa between the transducer and the femur. This image allows reproducible measurement of total synovial thickness, a sum of the thickness of the anterior and posterior walls of the bursa. The second set of images requires two examiners. One examiner positions the transducer while the other compresses the lateral joint recesses. Compression of the lateral recesses forces fluid into the suprapatellar bursa. These images are obtained with minimal transducer pressure and provide a reliable estimation of the quantity of joint fluid (see Fig 10–11,A and B).

Acute exacerbations of rheumatoid disease are accompanied by accumulation of increased intra-articular fluid and synovial thickening. Instituting appropriate therapy results in marked reduction in joint fluid within 24 hours, and approximately one half of patients will show complete resolution of joint effusion within 10 days (van Holsbeeck et al., 1988). Reduction in synovial thickness occurs almost as rapidly. A 36% decrease in mean synovial thickness was observed 10 days after initiating steroid therapy. The rate of response has been shown to be dose related (see Fig 10–11,D).

The knee is used as an example because it is a large synovial joint frequently involved by rheumatoid disease. These techniques can easily be applied to any large synovial joint. Evaluation of rheumatoid disease in the shoulder is best performed by imaging the posterior synovial recess due to accessibility. The transducer is oriented transversely, perpendicular to the plane of the glenoid (see Fig 10–4). Measurement at the elbow, hip, and ankle are obtained using an anterior approach. At the elbow, images are obtained with the transducer positioned in the long axis of the radius over the radial head and the arm in extension. Examination of the hip is done in the same way an examination for transient synovitis is performed (see Fig 10–6). The transducer is placed along the length of the femoral neck. Longitudinal images of the ankle are obtained over the anterior aspect of the tibiotalar joint.

## COMPLICATIONS OF RHEUMATOID DISEASE

### Infection

Most patients afflicted with rheumatoid arthritis are chronically immune suppressed due to steroid

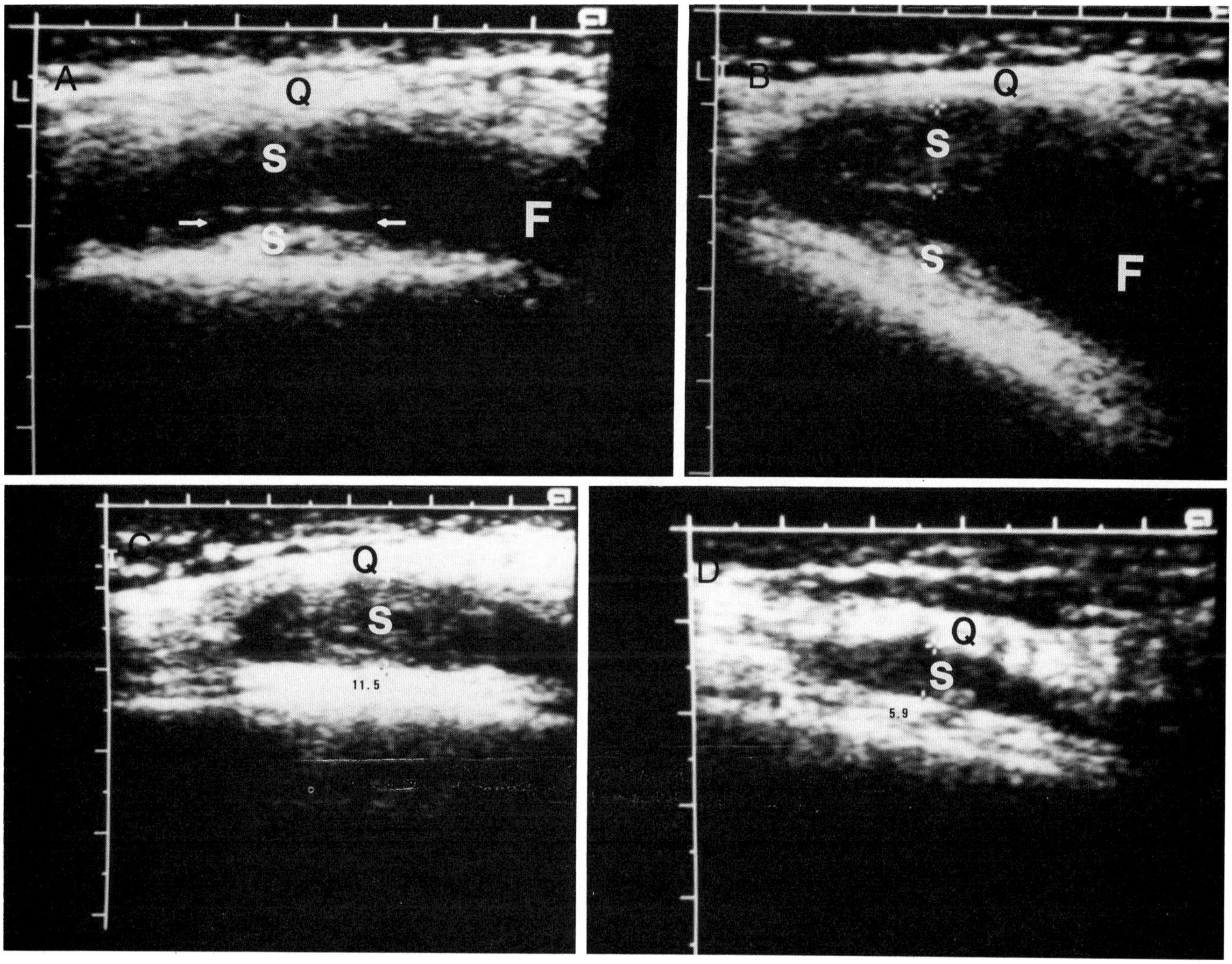


**FIG 10–11.**
**A,** follow-up of rheumatoid synovitis: suprapatellar scanning. New onset of left knee pain was reported by this patient with a history of rheumatoid arthritis for more than 10 years. This longitudinal image of the suprapatellar bursa demonstrates a small amount of fluid *(arrows)* within the bursa and marked synovial thickening *(S)*. The fluid collection *(F)* increases in dimension immediately above the patella. Abbreviations: quadriceps tendon *(Q)*. **B,** accurate reproducible estimation of the magnitude of joint effusion requires two examiners. One examiner applies pressure to the lateral joint recesses, thus forcing fluid *(F)* into the suprapatellar bursa. The other examiner positions the transducer. This has proved to be the best method for following the magnitude of joint effusion. Abbreviations: synovial thickening *(s)*, quadriceps tendon *(Q)*. **C,** the transducer has not been moved, but pressure has been applied with the transducer. All fluid has been expressed from the bursa. Pannus remains hypoechoic relative to the quadriceps tendon *(Q)* and surrounding fat, despite firm compression. Total synovial thickness measured on compression images is an accurate measure of synovial proliferation and can be used to follow disease activity over time. In this case, total synovial thickness prior to therapy measures 11.5 mm. Abbreviations: synovium *(S)*. **D,** 10 days after intra-articular injection of a long-acting corticosteroid preparation. Firm pressure is again applied to make an accurate measurement of the total synovial thickness (TST). A favorable response to therapy is indicated in this case by a decrease in TST to 5.9 mm. Abbreviations: quadriceps *(Q)*, synovium *(S)*. (From van Holsbeeck M, van Holsbeeck K, Gevers G, et al: *J Ultrasound Med* 1988; 7:561–566. Used by permission.)

therapy. They are predisposed to develop joint infection, which can rapidly lead to systemic sepsis if unrecognized. Ultrasound can be very valuable in the early diagnosis of joint and bursal infection in patients with rheumatoid arthritis. Prompt initiation of therapy with intravenous (IV) antibiotics and, if indicated, surgical drainage is of the utmost importance. In our experience, more than 70% of infected joints in rheumatoid patients were shoulders. This may be due to the frequent use of intra-articular steroid injection to treat shoulder pain in these patients.

Joint and bursa infection are characterized by

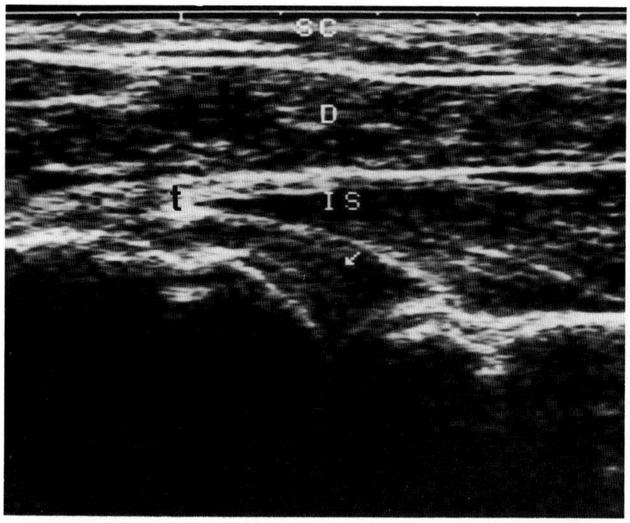


**FIG 10–12.**
Infected shoulder: transverse sonogram. A 50-year-old woman with known rheumatoid arthritis presented with increased pain and swelling of the right shoulder after intra-articular steroid injection. The distended posterior synovial recess *(arrow)* is difficult to identify due to the isoechoic nature of the fluid within the joint. The infraspinatus muscle *(IS)* is elevated by this fluid collection. Normally, this muscle is in direct contact with the posterior glenoid and humerus. Evaluation of joint effusion in the shoulder is difficult from an anterior approach due to humeral retroversion and surrounding bony structures. A posterior approach is much more sensitive in detecting joint effusion in the shoulder. Abbreviations: deltoid *(D)*, infraspinatus tendon *(t)*, subcutaneous fat *(SC)*.

rapid accumulation of fluid (Fig 10–12) within the infected joint or bursa. When acute swelling of an isolated joint is detected in a rheumatoid patient, infection must be suspected. A sonographic sign of infection in a rheumatoid joint is the marked increase in intra-articular fluid with an absence of significant increase in synovial membrane thickness. In an exacerbation of rheumatoid disease, the joint effusion will be in proportion to increased synovial membrane thickness. In addition, infected fluid will have definitive internal echos in the majority of cases. The most commonly isolated organism is *Staphylococcus aureus* (Resnick and Niwayama, 1988). When gas-producing organisms are present, gas bubbles identified as echogenic foci with shadowing may be observed. Despite these signs, the decision to aspirate the joint remains based on clinical criteria. The consequences of untreated joint infection in these patients are rapidly devastating. In addition to aiding in the early diagnosis of infection, ultrasound can provide assistance in aspiration of the involved joint or bursa, as described earlier.

## Baker's Cyst

Baker's cysts are frequently found in rheumatoid knees (Figs 10–13 and 10–14) as a result of increased intra-articular fluid. Sonography is very valuable in the evaluation popliteal cysts and their complications. This complication of rheumatoid disease is discussed in detail in Chapter 7.

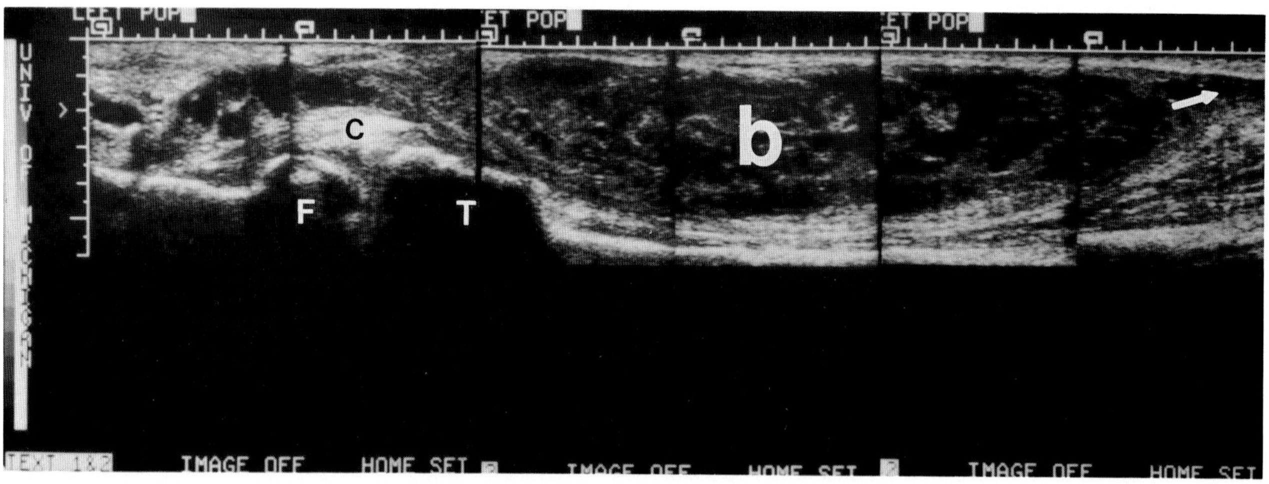


**FIG 10–13.**
Ruptured Baker's cyst in rheumatoid arthritis: longitudinal sonogram. This 50-year-old man has a 4-month history of symmetrical polyarticular rheumatoid disease. He presents to the emergency room with acute swelling of the left calf. Results of contrast venography were negative. Ultrasound was then performed, revealing a huge rheumatoid Baker's cyst *(b)* measuring approximately 35 cm long. This huge mass is filled with hypoechoic pannus. Surrounding soft tissues are abnormally increased in echogenicity. Note that the inferior extent of the cyst appears pointed *(arrow)*, a typical finding in ruptured Baker's cysts. Abbreviations: femoral condyle *(F)*, posterior joint capsule *(C)*, tibial condyle *(T)*.

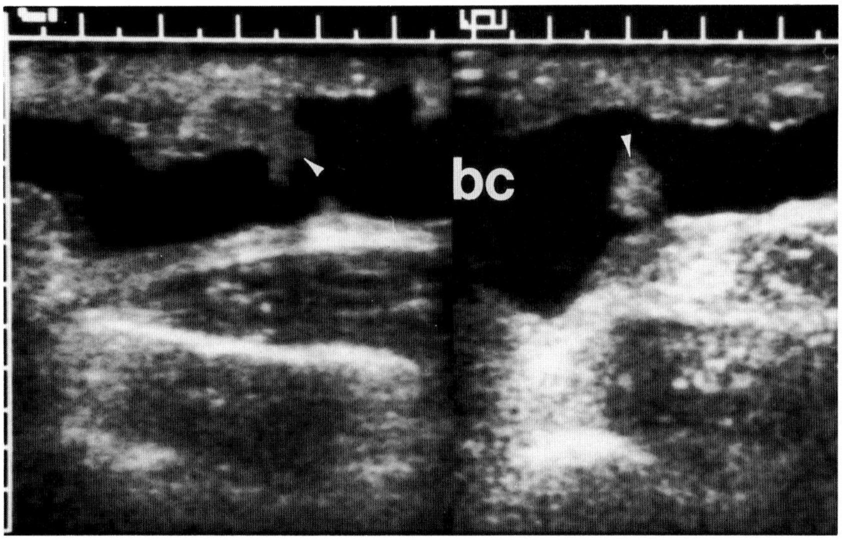


**FIG 10–14.**
Baker's cyst in amyloid arthropathy: longitudinal sonogram. This patient has been receiving chronic hemodialysis for more than 10 years. The transducer is positioned over the medial aspect of the popliteal fossa in the long axis of the leg. A Baker's cyst *(bc)* filled with anechoic fluid is noted with irregular thickening of the synovial walls. This nodular appearance *(arrowheads)* is commonly seen in patients with amyloidosis secondary to hemodialysis.

## Periarticular Complications of Rheumatoid Disease

To this point we have considered rheumatoid disease as an articular process. However, inflammation is not restricted to articular structures. Periarticular soft tissues are inevitably involved in this inflammatory process. Sinus tract formation, ligamentous laxity, muscle atrophy, muscle rupture, and other soft tissue atrophy may be seen as a result of corticosteroid therapy (Resnick and Niwayama, 1988).

Tendon ruptures and tenosynovitis are common in rheumatoid patients. The synovial sheaths surrounding tendons are affected by rheumatoid disease in the same manner as the synovia of joints and bursae. Sonography demonstrates the characteristic irregular synovial thickening seen in joints and bursae (Fig 10–15 and 10–16). Tendon ruptures likely result from synovial inflammation with associated release of proteolytic enzymes. Inflammation within the tendon itself will also cause significant weakening of the tendon. The catabolic effect of corticosteroids on mesenchymal tissues may also contribute to rupture. Supporting structures of the tendons, the vincula tendinum, may be similarly involved. This will allow tendon dislocation, tendon entrapment, and further deterioration in function.

Clinical manifestations of tendon rupture and destruction of their supporting structures are most familiar in the hands. Mallet fingers, Boutonnière deformity, swan neck deformity, ulnar drift, and extensor tendon subluxations are all too familiar manifestations of rheumatoid arthritis in the hands (Swanson and Swanson, 1973; Smith and Kaplan, 1967). Ultrasound is extremely valuable in identifying sites of rupture, allowing the surgeon to better plan reconstructive procedures.

Tendon ruptures are not limited to the hands. The supraspinatus, patellar, and Achilles tendons are also frequent sites of rupture (Resnick and Niwayama, 1988). In patients with active rheumatoid disease for greater than 5 years, 32% will have rotator cuff tears. These lesions are often bilateral (Fig 10–17).

Rheumatoid nodules are shown in Figure 10–18.

## Complications of Intra-articular Steroid Therapy

In addition to an increased incidence of joint infection, intra-articular steroid injection can lead to destructive changes resembling a neuropathic joint. The pathogenesis of this Charcot-like joint destruction is unknown but is likely multifactorial. Osteonecrosis, steroid-induced osteoporosis, intra-articular crystal deposition, steroid-induced insensitivity to pain, and direct steroid effect on cartilage all have been implicated as causative factors (Gray and Gottlieb, 1983; Miller and Restifo, 1966; Chandler et al., 1959). In these cases, sonography demonstrates large joint effusions containing small fragments of cartilage and bone. These loose bodies are admixed with calcium hydroxyapatite crystals, which are commonly found in steroid-treated joints. Sonographic images will also demonstrate associated ligamentous and tendinous lesions.

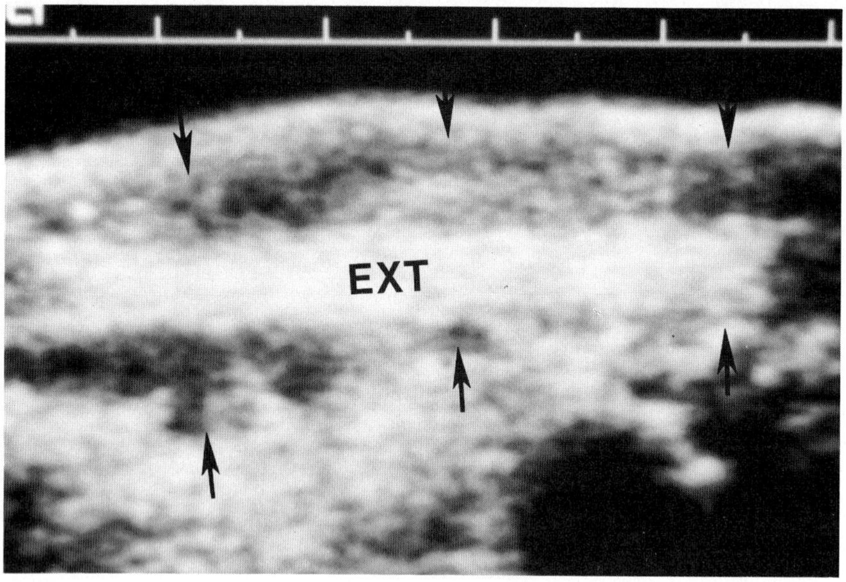


**FIG 10–15.**
Rheumatoid inflammation of the synovial sheath of the extensor pollicis longus tendon: longitudinal sonogram. Swelling over the dorsum of the right hand was noted in this 65-year-old man with rheumatoid arthritis predominating in the hands and feet. The transducer is positioned along the length of the extensor pollicis longus tendon *(EXT)*. The synovial sheath *(arrows)* surrounding the tendon is markedly thickened with an irregular appearance. Compare this appearance with that of a normal tendon sheath in Figure 10–16.

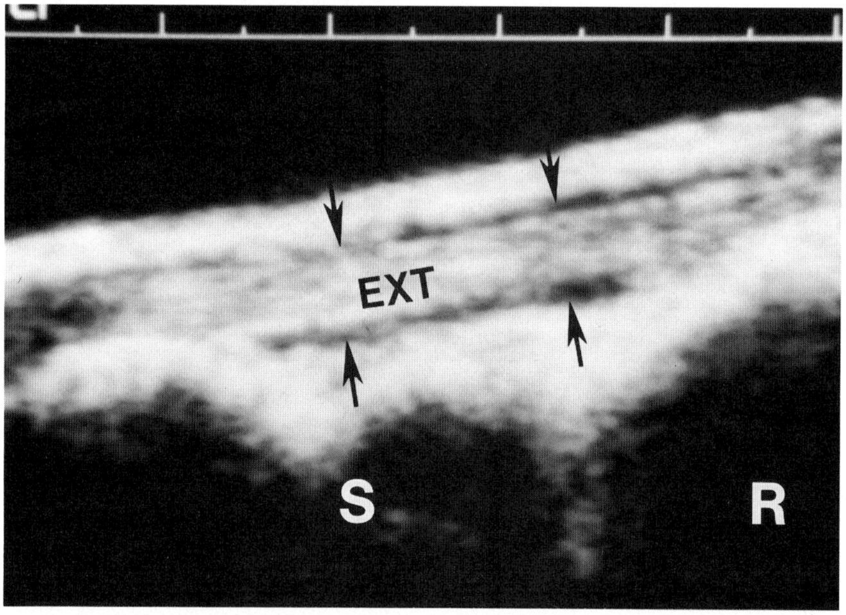


**FIG 10–16.**
Normal tendon sheath of the extensor pollicis longus tendon: longitudinal sonogram. A tendon sheath *(arrows)* of a normal 25-year-old man is shown in this longitudinal sonogram. Normal tendon sheaths are no more than 2 to 3 mm thick. Abbreviations: extensor pollicis longus *(EXT)*, scaphoid *(S)*, distal radius *(R)*.

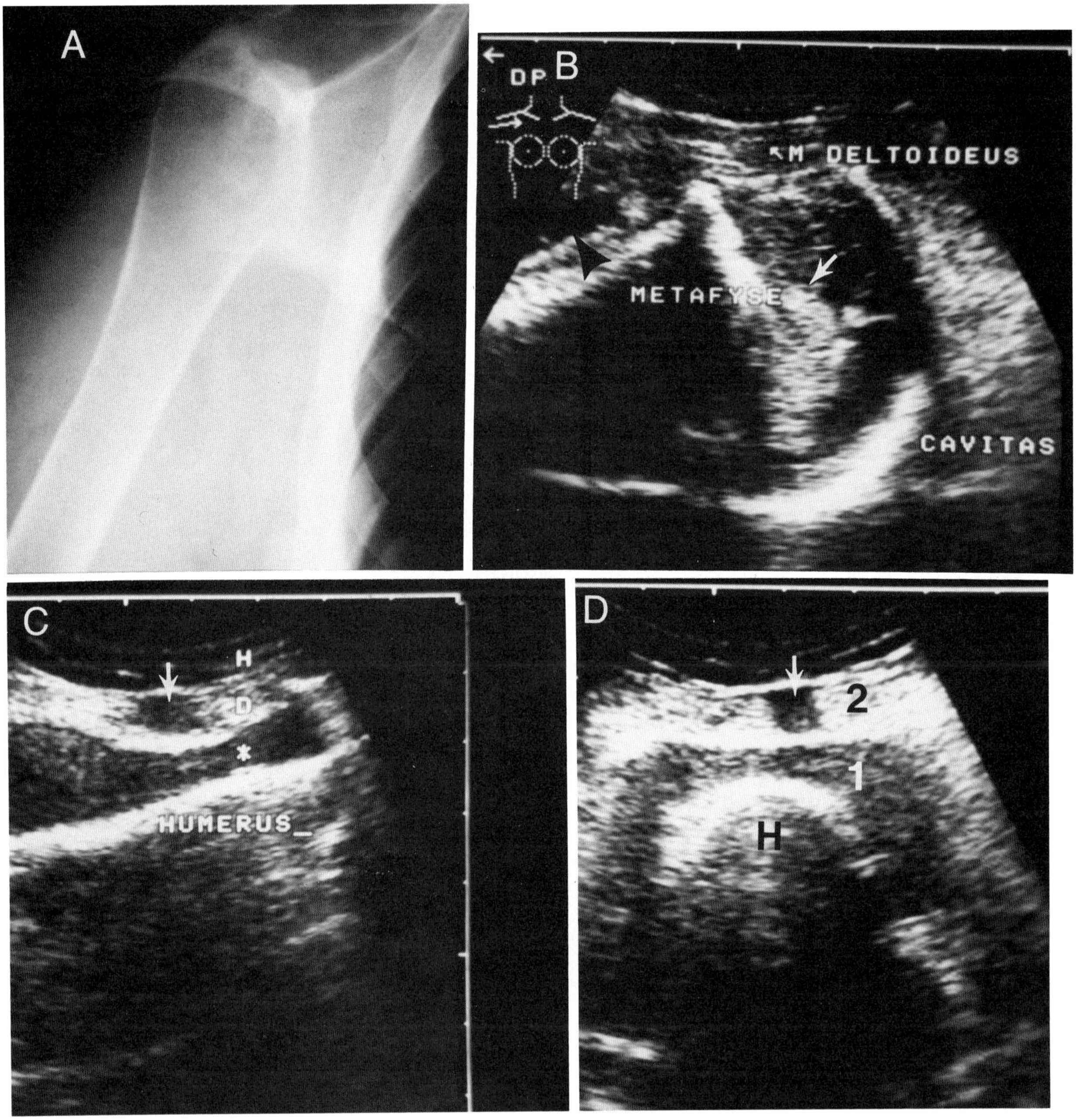


**FIG 10–17.**

**A,** rheumatoid shoulder: radiograph of the shoulder. The patient is a 50-year-old woman with a 14-year history of rheumatoid arthritis. She presented to the emergency room with acute loss of shoulder function. Radiograph of the shoulder demonstrates a markedly deformed joint. The humeral head projects over the glenoid on both anteroposterior and Grashey views, thus raising the possibility of dislocation. **B,** rheumatoid shoulder: sonogram of the shoulder. Same patient as in **A.** The transducer is positioned coronally over the lateral aspect of the shoulder. The humeral head is grossly deformed, almost completely eroded. Erosion of the glenoid had also occurred, producing a pseudarthrosis within the body of the scapula. The proximal humerus and neoglenoid are in gross anatomical alignment. Deltoid muscle covers the lateral aspect of the joint. The supraspinatus is absent. Pannus *(arrow)* and fluid are identified within the joint. Fluid is also noted in the subdeltoid bursa *(arrowhead).* **C,** rheumatoid shoulder: coronal sonogram of the shoulder. Same patient as in **A** and **B.** The transducer is moved inferior to that in **B.** A bulbous fluid-filled subacromial-subdeltoid bursa *(asterisk)* is noted lateral to the humerus. Pressure is applied with the transducer to demonstrate that the bursa is fluid filled. Rupture of the distal deltoid *(arrow)* is observed as a hypoechoic gap within the muscle. This accounts for the sudden loss of shoulder function. Muscle and tendon ruptures are more commonly seen in patients with rheumatoid arthritis, particularly those treated with corticosteroids. Abbreviations: deltoid *(D),* hypodermis *(H).* **D,** rheumatoid shoulder: transverse sonogram of the shoulder. Same patient as in **A–C.** The transducer is rotated to demonstrate the cross-section of the deltoid rupture *(arrow).* Two layers are identified. The superficial layer is not compressible and corresponds to the substance of the deltoid muscle *(2).* The deeper layer *(1)* was noted to be deformable, corresponding to the fluid filled bursa. Abbreviations: humerus *(H).*

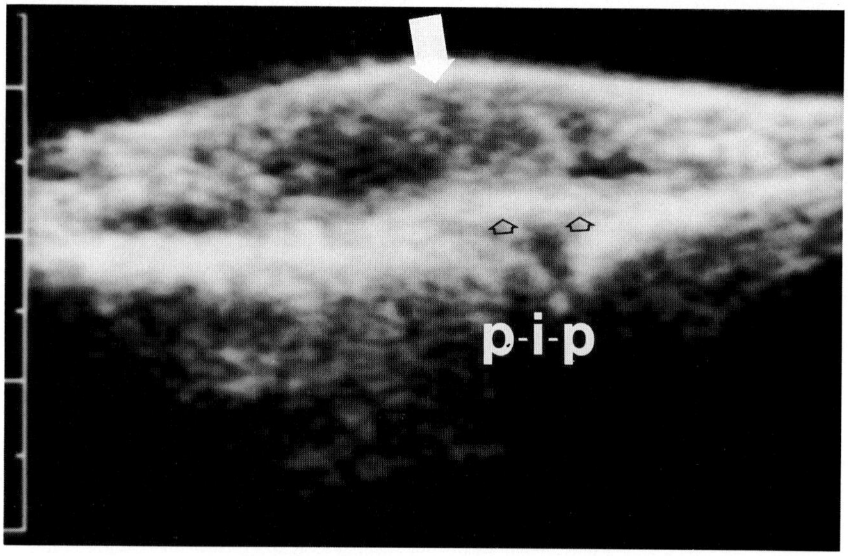


**FIG 10–18.**
Rheumatoid nodule at the finger-coronal sonogram. A 45-year-old woman with rheumatoid arthritis presents with a painful swelling over the proximal interphalangeal joint *(p-i-p)* of the left index finger. The clinical diagnosis of synovial cyst was made. This longitudinal image was obtained over the point of swelling. The joint capsule *(open arrows)* is demonstrated to be intact and nondisplaced. The nodule *(large arrow)* was composed of soft tissue without communication with the joint or tendon sheath. Surgical excision and histological analysis demonstrated this to be a rheumatoid nodule.

## REFERENCES

Adam R, Hendry GMA, Moss J, et al: Arthrosonography of the irritable hip in childhood. A review of 1 year's experience. *Br J Radiol* 1986; 59:205–208.

Aisen AM, McCune WJ, McGuire A, et al: Sonographic evaluation of the cartilage of the knee. *Radiology* 1984; 153:781–784.

Beecker HK: *Measurement of Subjective Responses.* New York, Oxford University Press, 1959.

Chandler GN, Jones DT, Wright V, et al: Charcot's arthropathy following intra-articular hydrocortisone. *Br Med J Clin Res* 1959; 1:952–953.

Cooperberg PL, Tsang L, Truelove L, et al: Gray scale ultrasound in the evaluation of rheumatoid arthritis of the knee. *Radiology* 1978; 126:759–763.

DeSilva M, Kyle V, Hazleman B, et al: Assessment of inflammation in the rheumatoid knee joint: Correlation between clinical, radioisotopic, and thermographic methods. *Ann Rheum Dis* 1986; 45:227.

Gielen JL, van Holsbeeck MT, Hauglustaine D, et al: Growing bone cysts in long-term hemodialysis. *Skeletal Radiol* 1990; 19:43–49.

Gray RG, Gottlieb NL: Intra-articular corticosteroids. An updated assessment. *Clin Orthop* 1983; 177:235.

Hammer M, Mielke H, Wagener P, Sonography and NMR imaging in rheumatoid gonarthritis. *Scand J Rheumatol* 1986; 15:157.

Marchal GJ, van Holsbeeck MT, Raes M, et al: Ultrasonography in transient synovitis of the hip in children. *Radiology* 1987; 162:825–828.

McGahan JP: *Controversies in Ultrasound.* New York, Churchill-Livingstone, 1987.

Miller WT, Restifo RA: Steroid arthropathy. *Radiology* 1966; 86:652.

Moore CP, Sarti DA, Lovie SS: Ultrasonographic demonstration of popliteal cysts in rheumatoid arthritis: a noninvasive technique. *Arthritis Rheum* 1975; 18:557–580.

Peck J: Ultrasound of the painful hip in children. *Br J Radiol* 1986; 50:293–294.

Resnick D, Niwayama G: *Diagnosis of Bone and Joint Disorders.* Philadelphia, WB Saunders Co, 1988.

Richardson ML, Selby B, Montana MA, et al: Ultrasonography of the knee. *Radiol Clin North Am* 1988; 26:63–75.

Ropes MW, Bennett GA, Cobb S, et al: Revision of the diagnostic criteria of rheumatoid arthritis. *Ann Rheum Dis* 1959; 18:49.

Smith RJ, Kaplan EB: Rheumatoid deformities at the metacarpophalangeal joints of the fingers. *J Bone Joint Surg (Am)* 1967; 49:31.

Swanson AB, Swanson GG: Pathogensis and pathomechanics of rheumatoid deformities in the hand and wrist. *Orthop Clin North Am* 1973; 4:1039.

van Holsbeeck M, van Holsbeeck K, Gevers G, et al: Staging and follow-up of rheumatoid arthritis of the knee. Comparison of sonography, thermography, and clinical assessment. *J Ultrasound Med* 1988; 7:561–566.

Worth WD, Hermann E, Meudt R, et al: Stellenwert der Arthrosonographie in die Beurteilung der Exsudativen und proliferativen Synovialitis. *Z Rheumatol* 1986; 45:263.

# 11

# Evaluation of Foreign Bodies

One of the most frequent requests made of musculoskeletal radiologists in the emergency setting is the identification and localization of foreign bodies. In an alert cooperative adult, this task is made easier. The patient can give an indication of the likelihood of the presence of a foreign body, the material involved, and the general location. However, many of these patients are not alert cooperative adults. Barefoot children possess a magnetic attraction to foreign bodies. All inquiries by the physician or parent about the accident are answered with screaming and tears. Equally challenging is the search for foreign bodies in patients involved in motor vehicle accidents. Often they are under the influence of alcohol, received intravenous (IV) analgesics, lost consciousness during the accident, or simply "hurt all over." Frequently foreign bodies in these patients go unrecognized, even after reassessment following emergency therapy. Unrecognized foreign bodies will result in chronic draining wounds, abscesses, and persistent pain. Surrounding infection can lead to devitalization of large amounts of tissue, joint destruction, and even limb loss (Gooding et al., 1987). Other complications seen less frequently are migration of foreign bodies, metallic synovitis, and erosion into vessels with concomitant embolization.

In the evaluation for foreign bodies, radiographs alone are not adequate (Gooding et al., 1987; Anderson et al., 1982). A retrospective study of foreign bodies showed that the average time between injury and detection was 7 months. Thirty-eight percent of retained foreign bodies were overlooked on initial examination (Anderson et al., 1982). Nonradiopaque foreign bodies and patients with orthopedic implants pose the greatest challenge. These are the cases in which ultrasound is the most beneficial. In our experience, sonography has increased the rate of success of identifying and localizing foreign bodies by approximately 20%.

## EXAMINATION TECHNIQUE

A 5-MHz linear array transducer is recommended for these examinations due to its utility for both superficial and deep foreign bodies. Examination of the superficial subcutaneous tissues requires a standoff pad. If a gel pad is unavailable or cannot be used because of a sterile field, a bag of IV fluid with the air removed will serve adequately. This will eliminate near-field artifacts and help to bring the object into the optimal focal zone of the transducer. The final piece of equipment needed for the quest is an indelible marking pen (Fig 11–1). It is extremely embarrassing when the mark indicating the location of the elusive foreign body is washed away by the surgeon preparing the skin with an antiseptic solution. In fact, some surgeons prefer a tattoo on the skin (Graf and Schuler, 1988).

Once the foreign body is identified, its relationship to adjacent structures must be carefully documented. Defining its position relative to neighboring anatomical landmarks will make the surgeons job much easier. In addition, location of nearby vessels, nerves and synovial membrane must be established, allowing the surgeon to select the most desirable approach (Fig 11–2). These structures have a characteristic sonographic appearance (Fornage, 1988; Weill, 1978; van Holsbeeck et al., 1988). The distance separating the foreign body and these vital structures can

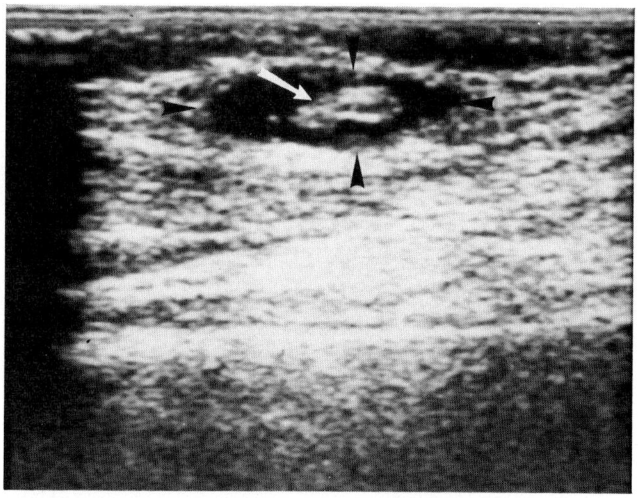

**FIG 11–1.**
Instrument breakage during arthroscopy: longitudinal sonogram of the knee. During therapeutic arthroscopy, the end of a curette was broken off in the subcutaneous tissues. Three radiographic examinations failed to demonstrate the fragment. This longitudinal image was obtained immediately superior to the suprapatellar bursa over the point of maximal tenderness. A hyperechoic foreign body *(white arrow)* is found within the subcutaneous tissues surrounded by fluid *(black arrowheads)*. The surrounding subcutaneous fat is increased in echogenicity, in addition to the increased through-transmission deep to the fluid collection. No increase in intra-articular fluid was observed, thus ruling out joint infection. The skin was marked, and an instrument fragment less than 5 mm in greatest dimension was surgically removed.

be easily measured using the postprocessing features available on state of the art ultrasound equipment. Occasionally, foreign bodies can penetrate tendons, fascia, or muscle. In these cases real-time sonography can provide a dynamic study, which better defines the structures involved and the location of the foreign body through a range of motion (Fornage and Schernberg, 1986).

## ULTRASOUND VS. RADIOGRAPHY

The standard imaging procedure used in the search for foreign bodies over the past 50 years has been conventional radiographs obtained in two projections. In the majority of cases the diagnosis of the presence or absence of a foreign body was correct using this method. In one large series of surgically

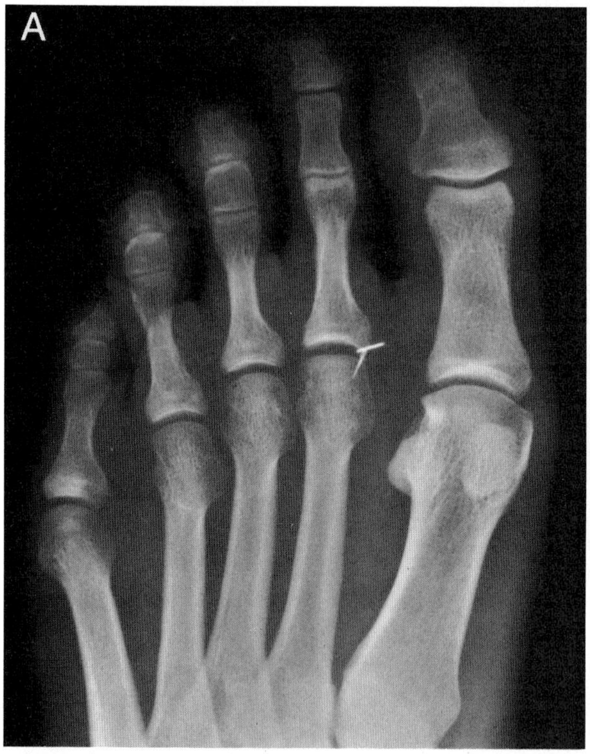


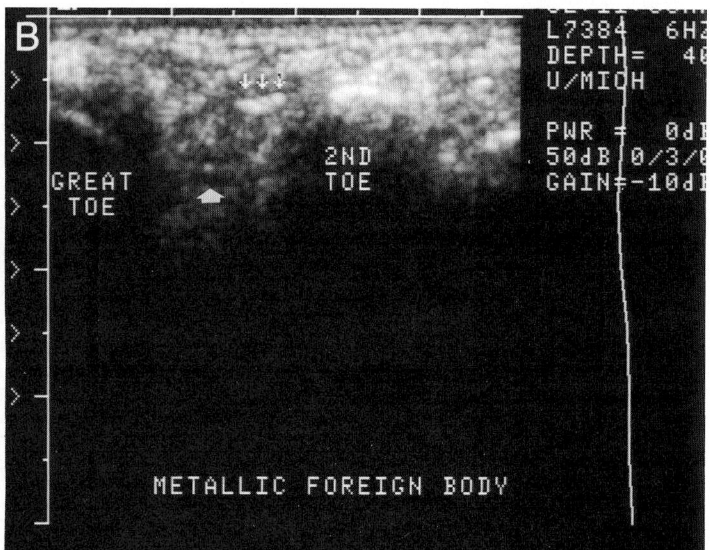


**FIG 11–2.**
**A,** localization of needle fragment in the foot: anteroposterior radiograph. This 28-year-old man stepped on a sewing needle lost in the carpet. The tip of the needle broke off during removal. The needle fragment is projected over the second metatarsophalangeal joint. Its relationship to the joint, vessels, and nerves is not well demonstrated on conventional radiographs. **B,** localization of needle fragment in the foot: transverse sonogram. A needle fragment slightly less than 1 cm long *(arrows)* is visualized as a hyperechoic line with minimal reverberation artifact. It was clearly demonstrated not to penetrate the joint capsule. The digital nerve and artery *(arrow)* are seen approximately 2 cm deep and slightly medial to the fragment. Again, the skin was marked, and surgical removal was successful.

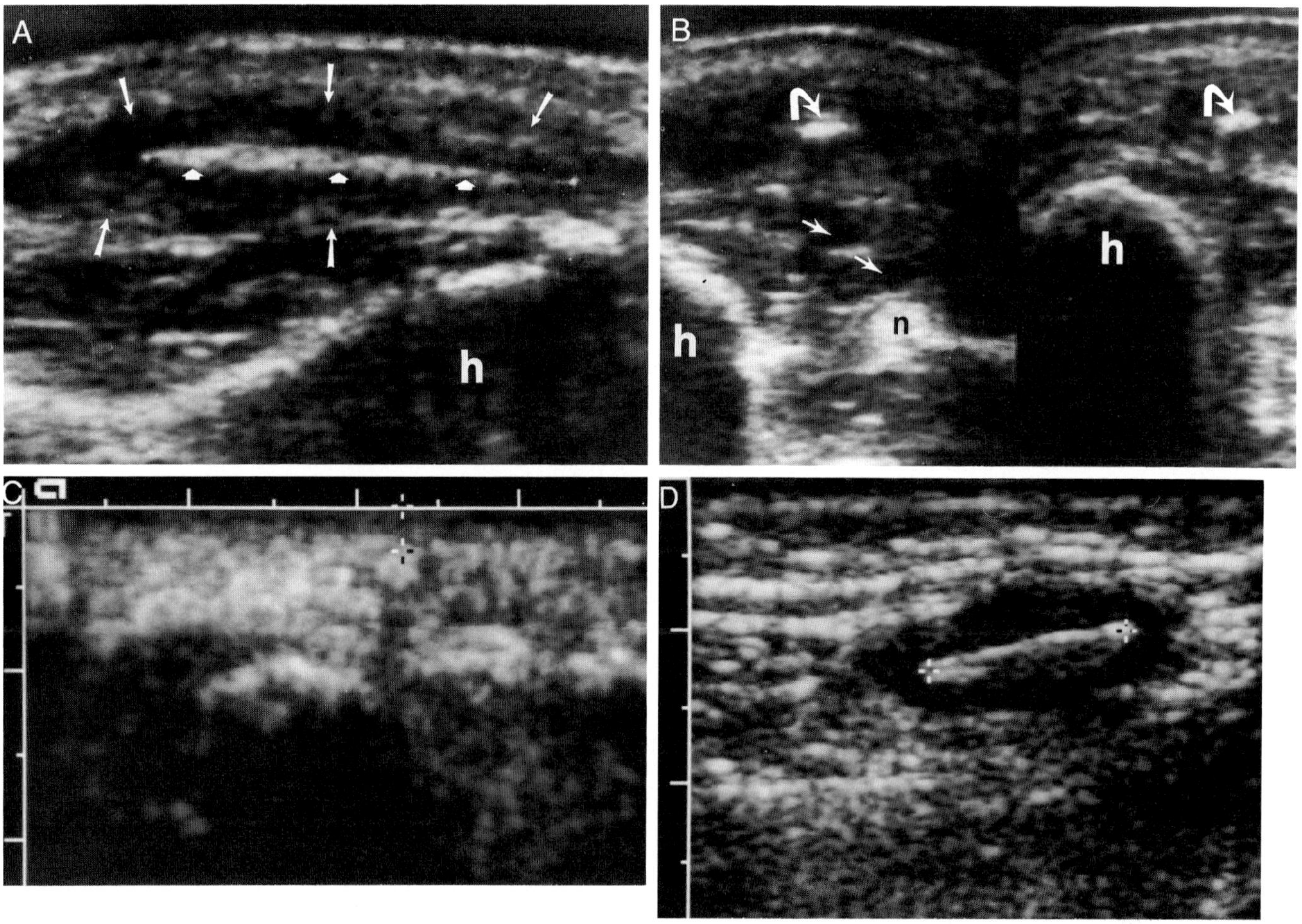


**FIG 11–3.**

**A,** wood splinter in the elbow: longitudinal sonogram. This 35-year-old wood worker punctured the skin with a wood splinter 10 days ago. He presents with swelling and erythema over the anteromedial aspect of the elbow. The splinter *(small arrows)* is identified as a hyperechoic line surrounded by a hypoechoic fluid collection *(large arrows).* Inflammatory collections are frequently seen around wooden foreign bodies. Initially the wood fragment is hyperechoic. As the inflammatory reaction evolves, the wood becomes progressively less echogenic. Eventually, the fragment will decompose completely in the inflammatory collection. Abbreviations: distal humerus *(h).* **B,** wood splinter in the elbow: transverse sonogram. Same patient as in **A.** A transverse image through the middle portion of the wood splinter *(curved arrow)* is shown on the *left,* distal portion on the *right.* Much more purulent material is seen surrounding the middle segment of the splinter. Its relationship to the brachial vessels *(arrows)* and median nerve *(n)* is clearly demonstrated. Abbreviations: humerus *(h).* **C,** toothpick within the heel pad of the foot: transverse sonogram. This 5-year-old boy on chemotherapy for leukemia stepped on a toothpick while playing yesterday. Two prior attempts at surgical localization and removal were unsuccessful. Radiographs were not helpful. Ultrasound demonstrated the toothpick fragment 0.5 cm below the skin (+). A mark was placed on the skin to indicate the location and orientation of the fragment. Subsequent surgical removal was simple. **D,** wood splinter at the knee: longitudinal sonogram. The patient is a 30-year-old female who presented with a history of chronic pain over the medial aspect of the right knee, which has been present for years. Intensity of the discomfort has varied over the years, but has increased over the past several months. This linear echogenic structure (+) surrounded by hypoechoic soft tissue was identified within the subcutaneous tissues of the medial right knee. Minimal shadowing is observed deep to this foreign body. Upon further questioning, the patient recalled an accident 15 years ago in which a piece of wood was imbedded beneath the skin, but had been removed by her father. At surgery a wooden foreign body surrounded by granualtion tissue was removed. These wooden fragments usually decompose in an intense inflammatory reaction over a period of days or weeks. This is the most unusual presentation of a wooden foreign body we have encountered.

removed foreign bodies, radiographs correctly identified metallic fragments in 100% of cases and glass in 96% of cases, but wood was correctly identified in only 15% of cases. These data would be considered quite good were it not considered that the most common foreign bodies in order of frequency are wood, glass, and metal fragments (Anderson et al., 1982). Nonradiopaque foreign bodies, such as wood, plastic, and certain types of glass, are better detected with xeroradiography or mammographic technique than with conventional radiographs (Gooding et al., 1987; Resnick and Niwayama, 1988). However, the availability of xeroradiography is rapidly decreasing, and mammographic technique can be used on only thin extremities. Conventional radiography remains the mainstay of foreign body imaging primarily due to the wide availability, low cost, and ease of examination (Anderson et al., 1982; Fornage and Schernberg, 1986; Gooding et al., 1987; Little et al., 1986).

Opthamologists were the first to apply ultrasound to the detection of foreign bodies. It was found to be extremely valuable in identification and localization of minute fragments in the eye (Hassani and Bard, 1978). The absence of ionizing radiation increased the attraction to this modality. Gynecologists started to use ultrasound to confirm the presence and location of intrauterine devices (Cochrane, 1985). This demonstrated that depth within the body did not hamper detection of foreign bodies with ultrasound.

Ultrasound offers a great deal more than radiography in the detection and localization of foreign bodies. Since the ability to detect an object on sonographic images is primarily a function of the difference between the acoustic impedance of the object and the surrounding tissue, objects that are radiolucent or radiopaque can be imaged quite easily. An in vivo study by Little et al. (1986) demonstrated wood fragments (Fig 11–3) to be visualized best, followed in order by glass, plastic, and metal. Glass and metal, though intrinsically more reflective than other materials, result in considerable reverberation artifact. This artifact can be both an aid to diagnosis and a hindrance to precise localization. Comet tail artifact can help to identify the foreign body, but it makes size measurement extremely difficult. Figures 11–4 through 11–9 demonstrate the sonographic appearance of various materials. Note that the diameter of the metal needles in Figure 11–7 appear the same regardless of their actual size. Compare this with the sonographic appearance of the plastic IV catheters (Fig 11–9). In vivo studies have shown sonography to be accurate in the identification and localization of foreign bodies with diameters as small as 0.5 mm. The smallest-diameter needle shown in Figure 11–7 is 0.3 mm, and it is clearly visualized.

Secondary signs associated with foreign bodies can aid in making the diagnosis. Surrounding hematoma or, in chronic cases, purulent material can help to outline a foreign body by providing a hypoechoic halo (see Fig 11–3). Also, in the acute setting, air may be entrapped along the path of insertion of the foreign body. This provides a trail of echogenic foci with shadowing that can be followed to the foreign body.

Wooden fragments are hyperechoic with acoustic shadowing deep to the object in an acute setting. This makes identification and localization quite simple. However, unlike many other types of foreign bodies, wood fragments tend to decompose quite rapidly due to the surrounding inflammatory reaction. Over the course of several days, the wooden fragment will become progressively less echogenic and finally disappear within the inflammatory collection (Graf and Schuler, 1988). These disappearing foreign bodies have misled clinicians into searching for sources of septic emboli when an accurate clinical history is not available. Plastic objects are slightly less echogenic

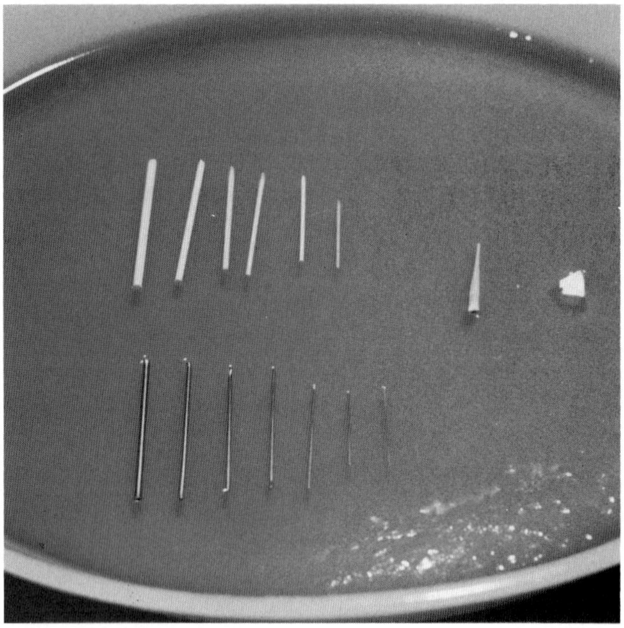

**FIG 11–4.**
Foreign body phantom. Metal needles and plastic catheters ranging in size from 14 to 30 gauge were embedded within gelatin in a glass dish. A fragment of toothpick and a small glass fragment were also included in the phantom.

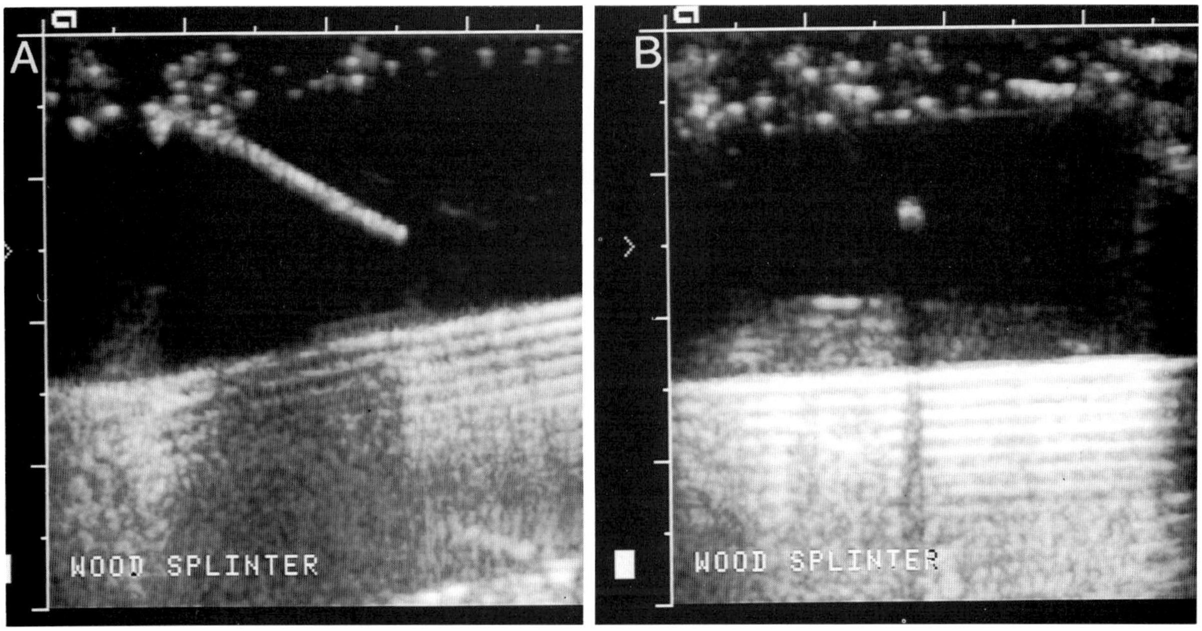


**FIG 11–5.**
Wooden splinter: longitudinal **(A)** and transverse **(B)** sonograms. The appearance of this wooden fragment is similar to that of the in vivo fragments in Figure 11–3. It appears hyperechoic without reverberation artifact. Incomplete acoustic shadowing is noted deep to this in vitro specimen. As a wooden fragment decomposes within an inflammatory collection, acoustic shadowing will no longer be observed.

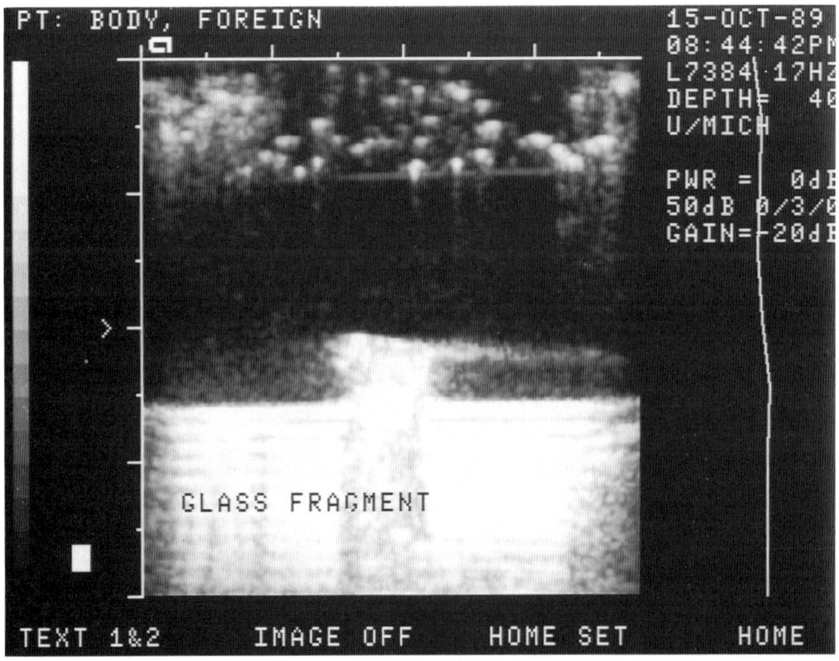


**FIG 11–6.**
Glass fragment. A striking comet tail artifact is seen deep to this glass fragment. This often helps in identifying glass fragments within the soft tissues. In this example, tiny echogenic foci within a layer immediately beneath the transducer face are gas bubbles in the coupling gel.

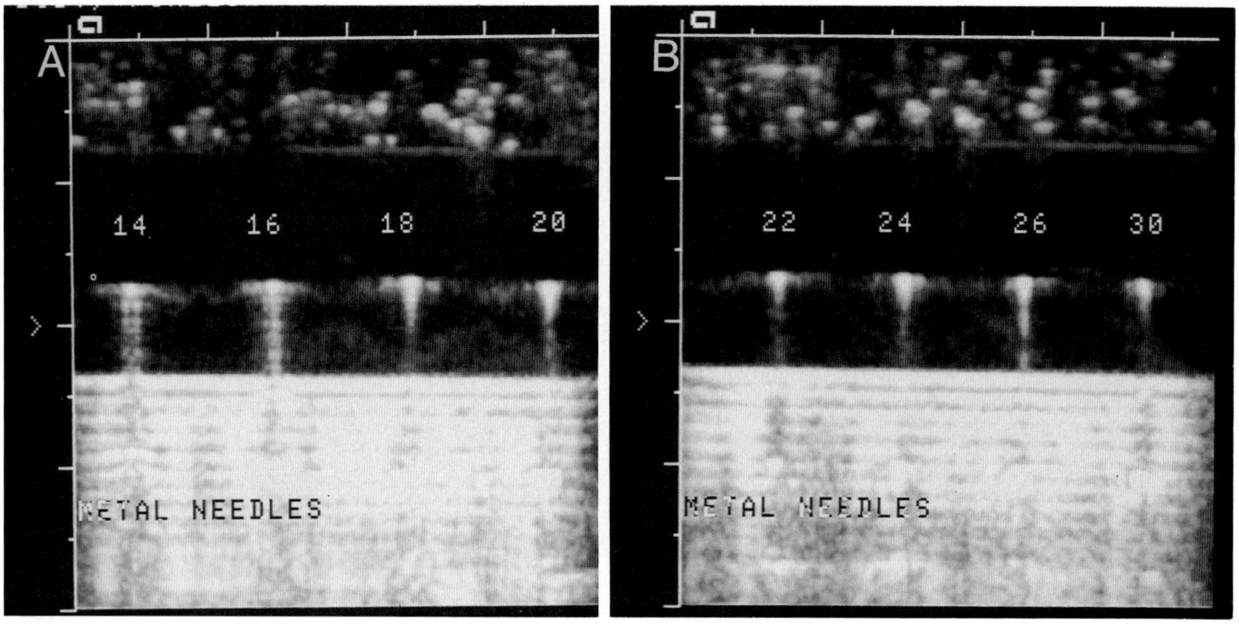


**FIG 11–7.**
**A and B,** series of metal needles: transverse sonogram. These transverse images demonstrate the appearance of metallic foreign bodies of varying diameters. The 14-gauge needle has an outer diameter of 2.1 mm, and the 30-gauge needle is 0.3 mm in diameter. Note that reverberation artifact makes the 30-gauge needle appear to be the same diameter as the 14-gauge needle. This aids in the detection of small metallic foreign bodies, but realistic measurement of the dimensions of the foreign body is impossible. Metallic foreign bodies produce a comet tail artifact similar to glass but less striking.

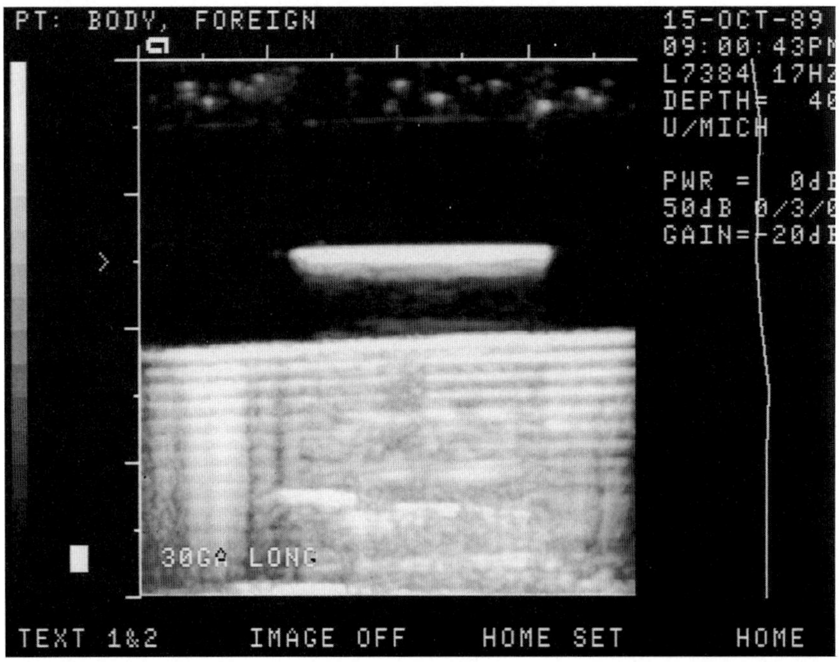


**FIG 11–8.**
A 30-gauge metal needle: longitudinal sonogram. The comet tail artifact is not as noticeable on longitudinal images due to beam width. This appearance is also noted in vivo, as demonstrated in Figure 11–2,**B.**

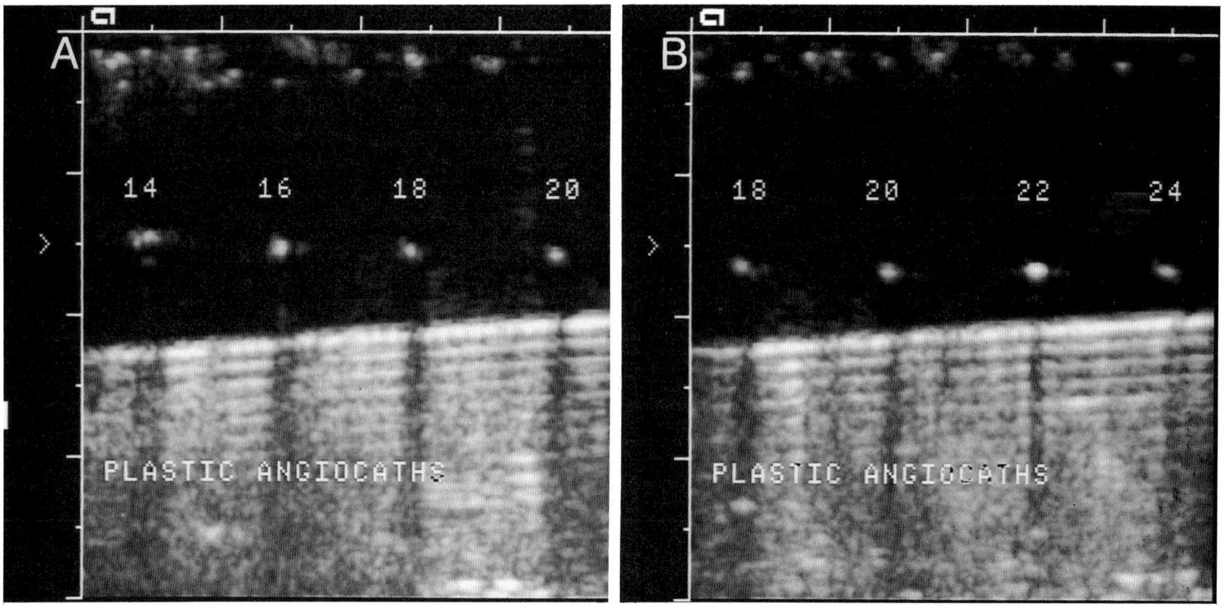


**FIG 11–9.**
**A and B,** series of plastic catheters: transverse sonograms. This series of plastic catheters ranges in size from 14 to 24 gauge. Less reverberation artifact is seen with plastic catheters, allowing relative comparison in size. Acoustic shadowing is present deep to the plastic catheters, but no comet tail is seen. Compare this with the appearance of the metal needles shown in Figure 11–7.

than intact wood but also demonstrate acoustic shadowing. These relatively inert materials do not decompose in the surrounding inflammatory reaction.

## DIAGNOSTIC APPROACH TO FOREIGN BODIES

Sonography has been demonstrated to be clearly superior to radiography in the evaluation of foreign bodies. However, it is not practical to use ultrasound in all cases. The algorithm outlined in Figure 11–10 provides a rational cost-effective approach to the diagnosis and localization of foreign bodies. When clinical history indicates that the foreign body is radiolucent, ultrasound is the study of choice. Otherwise, low-kilovoltage radiography should be the first step. If radiographs do not provide adequate information, ultrasound is then indicated for further evaluation. When surgical removal is unsuccessful following conventional radiography, real-time ultrasound guidance can be most helpful.

## ORTHOPEDIC IMPLANTS

Orthopedic implants comprise a special category of foreign bodies. Ultrasound can be very valuable in evaluation of these prostheses. Problems associated with orthopedic implants can be categorized as either acute or chronic, the acute problems occurring in the immediate postoperative period.

### Acute Complications

Postoperative hematoma is a serious complication because it predisposes to wound sepsis. This has been well established for hip prostheses, and this can reasonably be applied to other implants (Wilson et al., 1973). Clinical examination is of little value in the detection of these hematomas, unless they are massive or very superficially located (Magnussen et al., 1988). This is due in part to the amount of edema that accompanies this extensive surgery. Ultrasound is a fast accurate noninvasive examination for the diagnosis of postoperative hematomas (Fig 11–11). Localization and aspiration can be performed rapidly. In addition, a drain can be safely placed percutaneously under ultrasound guidance.

The distinction between superficial and deep hematomas is easily made using ultrasound. Hematomas will appear hypoechoic or anechoic depending on their stage of evolution or resolution. If the collection is separated from the prosthesis and bony structures by muscle or fascia, it is considered superficial.

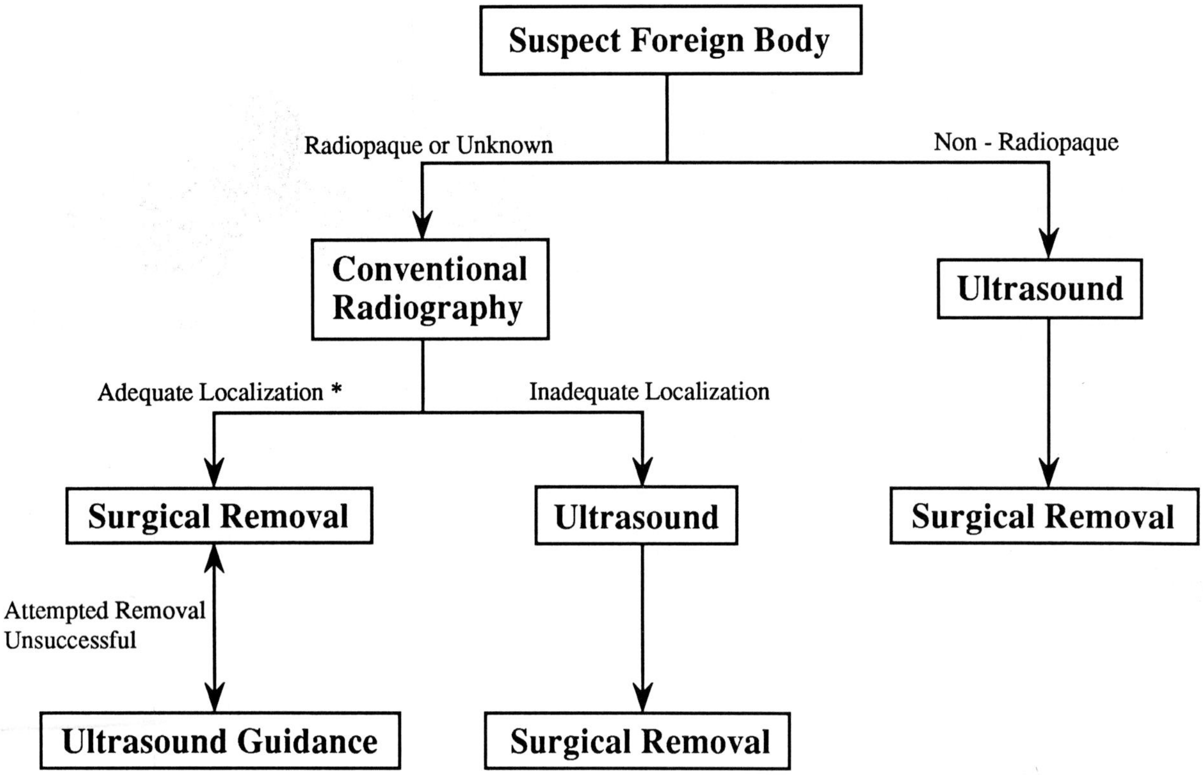


**FIG 11–10.**
Algorithm for evaluation of foreign bodies.
*If conventional radiographs indicate that the foreign body is in the vicinity of vital structures, ultrasound is indicated for further evaluation.

This distinction is important because most surgeons will take both superficial and deep cultures during the surgical procedure. If the hematoma is deep and the cultures were positive, surgical or percutaneous evacuation of the collection is imperative (Magnussen et al., 1988). Management of superficial hematomas varies considerably among clinicians, depending on the condition of the patient and other clinical factors.

## Chronic Complications

### *Infection*

Even following a successful uncomplicated postoperative period, orthopedic implants can become infected at any time (Figs 11–12 and 11–13). Since hip prostheses are by far the most widely employed, they are the most common site of an infected prosthesis. Whenever a patient with a prosthesis presents with fever of unknown origin or associated with pain in that extremity, infection is suspected. Surgeons usually perform blind or fluoroscopically guided aspiration of the joint (Resnick and Niwayama, 1988). When this yields joint fluid, the problem is solved, but a dry tap is not infrequent. Commonly fluid surrounding a prosthesis will be irregularly distributed and loculated. Ultrasound's ability to locate periprosthetic fluid collections and guide aspiration make it the perfect solution to the problem (Fig 11–14). Infected collections tend to be more echogenic than simple joint fluid (see Fig 11–13). Frequently debris floating within the infected fluid can be identified.

### *Loosening of the Prosthesis*

Modern orthopedic implants are composed of very inert materials. Foreign body granulomas around these implants are very rare. If present, they are usually found at the bone-cement interface (Harris et al., 1976). Granulomatous reaction in the surrounding soft tissues has not been reported. The significant complication resulting from foreign body granulomas at the bone-cement interface is loosening of the prosthesis. Motion of the prosthesis may result in pain and fracture of the adjacent bone. These fractures may be occult on conventional radiographs due to the radiopacity of the prosthesis. Real-time sonographic examination can easily demonstrate mobility of the prosthesis. Fractures can also

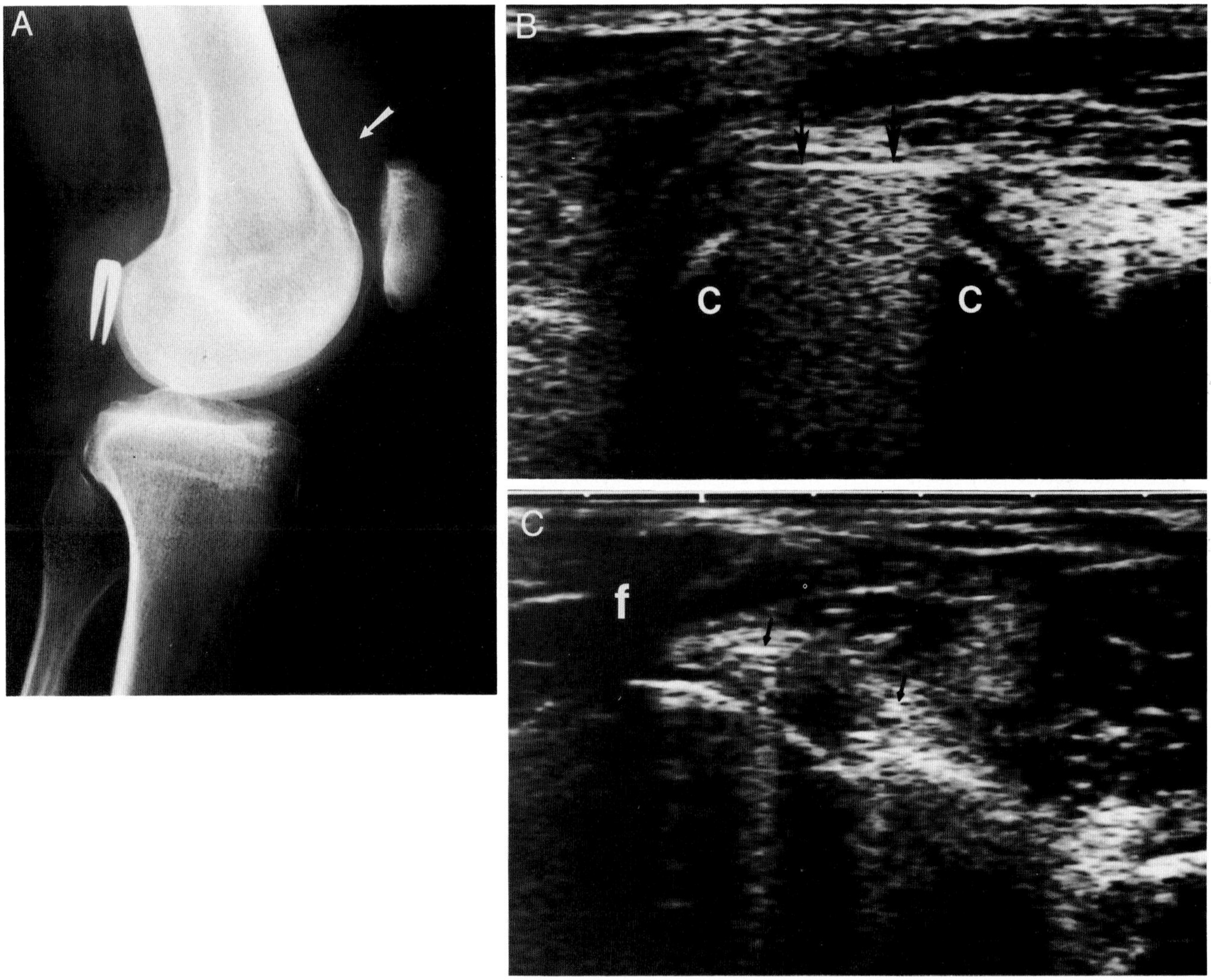


**FIG 11–11.**
**A,** intra-articular migration of a surgical staple: lateral knee radiograph. This 22-year-old man had ligament and posterior capsule repair 2½ years ago. Four days ago he reinjured the knee and immediately developed marked swelling. Joint aspiration at the time of initial presentation yielded frank blood. He was referred for imaging evaluation with a complaint of decreased range of motion and increased pain. This radiograph demonstrates a joint effusion in the suprapatellar bursa *(arrow)* and a metallic staple posterior to the femoral condyles. The staple is no longer fixed in bone. Position of the staple relative to the joint space is not clear from this study; therefore, ultrasound was performed. **B,** intra-articular migration of a surgical staple: longitudinal sonogram. Same patient as in **A.** The transducer is positioned over the lateral femoral condyle *(C).* A wide comet tail artifact is seen originating from the linear leg of the metal staple *(arrows).* **C,** intra-articular migration of a surgical staple: transverse sonogram. Same patient as in **A** and **B.** The transducer is rotated 90 degrees, demonstrating both legs of the staple with comet tail artifact *(arrows).* Fluid *(f)* clearly within the joint capsule surrounds the staple, indicating its intra-articular position. The staple was surgically removed, and the patient regained full range of motion.

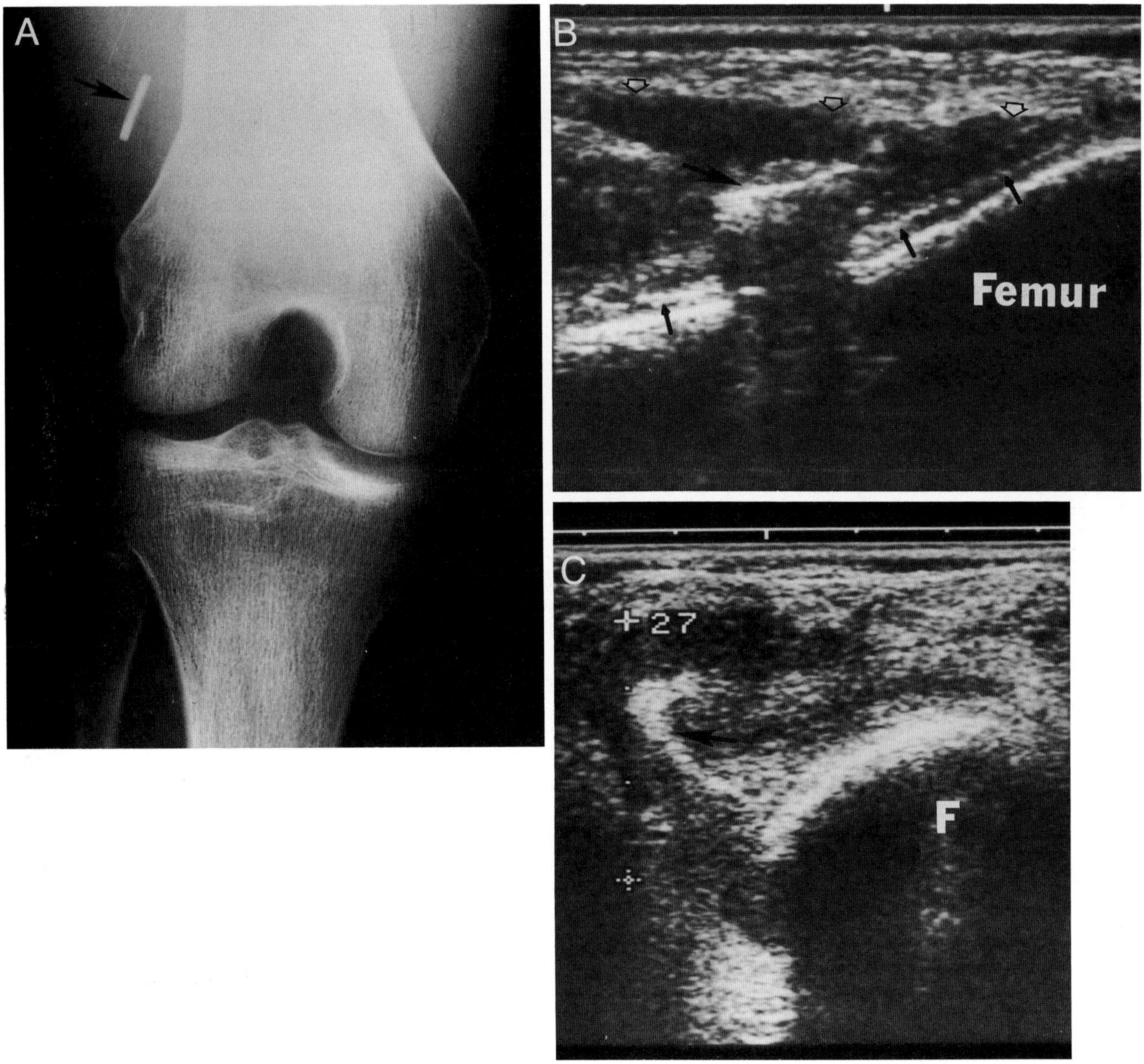


**FIG 11–12.**
**A,** painful knee following cruciate ligament reconstruction: anteroposterior radiograph. Four months after reconstructive surgery to repair torn cruciate ligaments, this 20-year-old football player presents with 2 weeks of pain on the lateral side of the right knee. A metallic foreign body *(arrow)* is identified lateral to the distal femoral metaphysis. **B,** painful knee following cruciate ligament reconstruction: coronal sonogram. Same patient as in **A.** The transducer is placed over the lateral aspect of the distal femoral metaphysis. A linear hyperechoic foreign body *(arrow)* is observed with associated comet tail artifact. Hypoechoic fluid *(open arrows)* surrounds the foreign body. Note the periosteal thickening *(solid black arrows)* adjacent to the foreign body. This periosteal reaction was not observable on the conventional radiographs **A. C,** painful knee following cruciate ligament reconstruction: transverse sonogram. Same patient as in **A** and **B.** The metallic foreign body is seen as a curvilinear hyperechoic line with subtle comet tail artifact *(arrow)* surrounded by a hypoechoic fluid collection 27 mm in diameter. Given the duration of the pain, finding fluid with definite internal echos is highly suspicious for purulent material. No fluid was identified within the knee joint, excluding joint sepsis. At surgery, purulent material was found surrounding the foreign body. Abbreviations: femur *(F).*

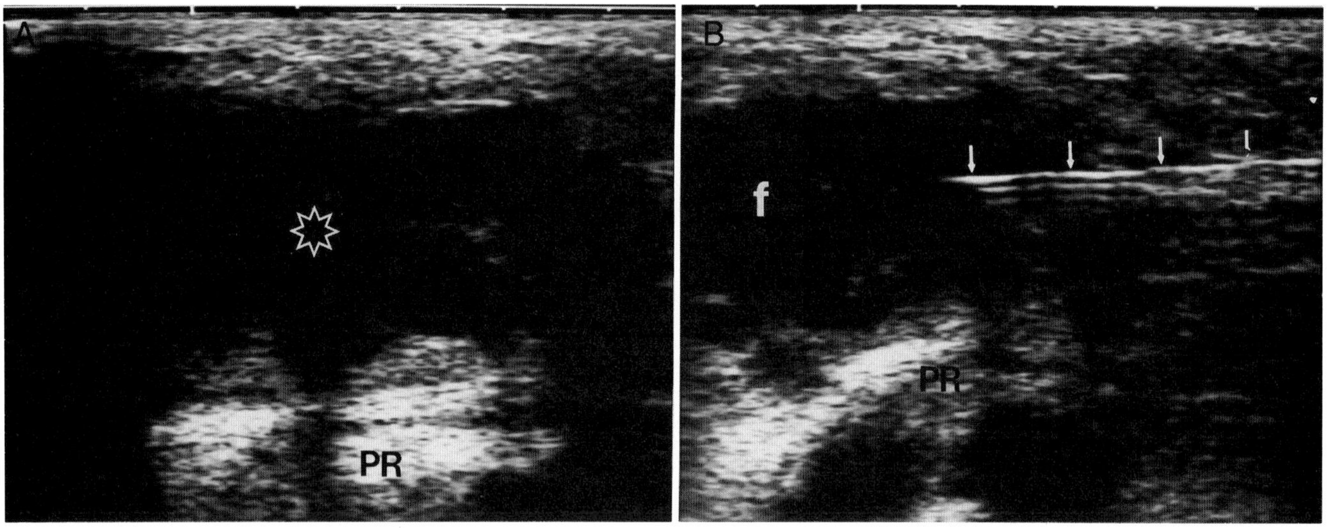


**FIG 11–13.**
**A,** infected hip prosthesis: longitudinal sonogram. The patient is a 90-year-old woman who received a left total hip arthroplasty 2 years ago. She was observed to have a deterioration in mental status accompanied by a slight fever of 2 days' duration. Chest radiographs were normal and urine cultures negative. Ultrasound of the hip was requested to evaluate the prosthesis for infection. A large fluid collection *(★)* containing echogenic debris is identified adjacent to the femoral prosthesis *(PR)*. This is clearly indicative of infection. **B,** infected hip prosthesis: transverse sonogram. Same patient as in **A.** The fluid collection *(f)* surrounds the anterior and medial borders of the prosthesis *(PR)*. Debris is again noted within the collection. A needle *(arrows)* was placed into this collection to confirm the diagnosis. A total of 80 mL of pus was removed following percutaneous placement of a drainage catheter. *Staphylococcus aureus* was the offending organism.

be detected as cortical discontinuity with adjacent subperiosteal hemorrhage or periosteal new bone formation. In addition, the relationship of the acetabular and femoral components can be observed through a range of motion in real time.

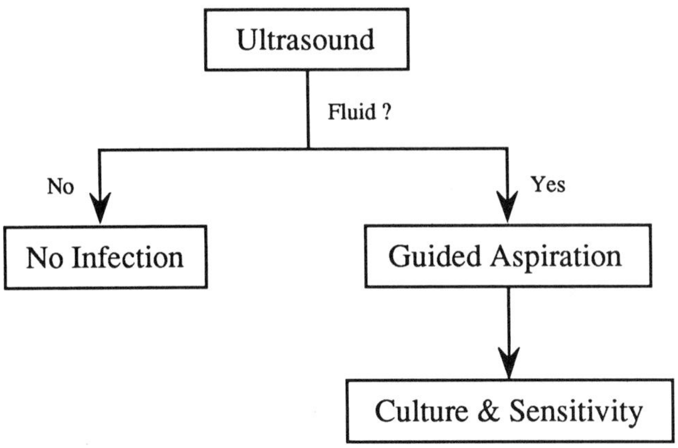


**FIG 11–14.**
Algorithm to evaluate joints for infection. The clinical question of joint infection most commonly arises in evaluation of hips following arthroplasty. Ultrasound's high sensitivity in detecting joint effusion makes it the best method to evaluate joints for infection. Effusions as small as 1 mL can be detected. If increased intra-articular fluid is detected, aspiration can be performed during the same session.

## SUMMARY

Ultrasound is more accurate than radiography in the detection and localization of all types of foreign bodies. In difficult cases, real-time sonographic examination can facilitate surgical removal of the foreign body. Sonography can also be very useful in the evaluation of orthopedic implants, both postoperatively and in long-term follow-up.

## REFERENCES

Anderson MA, Newmeyer WL, Kilgore ES: Diagnosis and treatment of retained foreign bodies in the hand. *Am J Surg* 1982; 144:63.

Cochrane WJ: Ultrasound and the intrauterine device, in Sanders RC, James AE: *Principles and Practice of Ultrasonography in Obstetrics and Gynecology*. Norwalk, Conn, Appleton-Century-Crofts, 1985.

Fornage BD, Schernberg FL: Sonographic diagnosis of foreign bodies of the distal extremities. *AJR* 1986; 147:567–569.

Fornage BD: Peripheral nerves of the extremities: Imaging with ultrasound. *Radiology* 1988; 167:179–182.

Gooding GAW, Hardiman T, Sumers M, et al: Sonography of the hand and foot in foreign body detection. *J Ultrasound Med* 1987; 6:441–447.

Graf R, Schuler P: *Sonographie am Stütz und Bewegungsapparat bei Erwachsenen und Kindern*. Weinheim, Edition Medizin VCH. 1988.

Harris WH, Schiller AL, Scholler JM, et al: Extensive localized bone resorption in the femur following total hip replacement. *J Bone Joint Surg (Am)* 1976; 58:612–618.

Hassani SN, Bard RL: Real-time opthalmic ultrasonography. *Radiology* 1978; 127:213.

Little CM, Parker MG, Callowich MC, et al: The ultrasonic detection of soft tissue foreign bodies. *Invest Radiol* 1986; 21:275–277.

Magnussen PA, Crozier AE, Gregg PJ: Detecting haematomas by ultrasound: Brief report. *J Bone Joint Surg (Br)* 1988; 70:150.

Resnick D, Niwayama G: *Diagnosis of Bone and Joint Disorders.* Philadelphia, WB Saunders Co, 1988.

van Holsbeeck M, van Holsbeeck K, Gevers G, et al: Staging and follow-up of rheumatoid arthritis of the knee. Comparison of sonography, thermography, and clinical assessment. *J Ultrasound Med* 1988; 7:561–566.

Weill FW: *Ultrasonography of Digestive Diseases.* St Louis, CV Mosby Co, 1978.

Wilson PD, et al: The problem of infection in endoprosthetic surgery of the hip joint. *Clin Orthop* 1973; 96:213–221.

# 12

# Sonography of Pain Syndromes Following Arthroscopy

Arthroscopy has proved its value as a diagnostic and therapeutic tool with regard to knee pathology. Refinements in technique over the last 15 years have expanded its role in therapy and enabled its use in other joints. Its application remains primarily in the knee, but arthroscopy is used increasingly in the shoulder, ankle, elbow, wrist, and hip (Table 12–1). All surgical procedures have an inherent risk and possibility for complication. As arthroscopy has become more sophisticated, the risk associated with these procedures has increased. Early studies by the Arthroscopy Association of North America reported an overall complication rate of 0.6%, but this reflected only "first-generation" procedures. A subsequent study that included "second-generation" procedures reported an overall complication rate of 1.8%, with the highest incidence of complications reported for lateral retinacular release (8.0%) (Fig 12–1). Other reports have stated complication rates as high as 20% for subcutaneous lateral release (Metcalf, 1982). The range of complications includes hemarthrosis, loose bodies, infection, thromboembolic disease, instrument failure, neuromuscular deficit, and vascular injury. Many of these are well evaluated with ultrasound.

## PAIN SYNDROMES FOLLOWING THERAPEUTIC ARTHROSCOPY

The vast majority of patients experience considerable improvement or complete resolution of their symptoms following arthroscopic surgery. However, a small group continue to experience pain following an arthroscopic procedure that was considered a surgical success. In some of these patients, the character of the discomfort differs from that experienced prior to surgery. Often ultrasound is helpful in evaluating these patients.

Therapeutic arthroscopy is more invasive than diagnostic arthroscopy, requiring at least two access portals and frequently three (Fig 12–2). One portal is used for viewing, and the others are used for grasping and cutting. Some orthopedic surgeons prefer a transtendinous approach for one of the portals, the Swedish technique. In this approach, access is obtained to the joint through the patellar tendon approximately 1 cm below the patellar apex. This provides access equidistant from the anterolateral and anteromedial portals. Rigidity of the tendon helps to avoid crowding of the portals. Damage to the tendon is minimized by using a sharp obturator to penetrate into the joint. Some patients will develop pain and tenderness over the anterior aspect of the knee centered over the transtendinous portal.

Patellar tendonitis (Fig 12–3) is the most frequent sonographic finding in these patients. Thickening and decreased echogenicity of the tendon may be observed for a period of up to 2 years following arthroscopy (van Holsbeeck and Introcaso, 1989). Sometimes a cleft will be observed in the tendon. These clefts are associated with significant weakening of the tendon and should be regarded as incomplete tears (Fig 12–4). In all cases, Hoffa's fat pad be-

**TABLE 12–1.**
Approximate Distribution of Arthroscopic Procedures by Site

| Site | % |
|---|---|
| Knee | 84.7 |
| Shoulder | 12.3 |
| Ankle | 1.3 |
| Elbow | 0.9 |
| Wrist | 0.7 |

comes hyperechoic relative to the tendon. Often a hypoechoic tract will be noted traversing the subcutaneous fat, the tendon, and Hoffa's fat pad. This tract corresponds to the arthroscopic portal. The fat pad is under compression in both full extension and flexion. Occasionally, fat will be identified herniating through the defect in the patellar tendon. When present, the incarcerated fat will prevent normal healing of the tendon.

Hoffa's fat pad is a highly vascular fibroadipose structure that extends medially and laterally to cover the anterior horns of the menisci. All of the commonly used anterior access portals pass through Hoffa's fat pad. After surgical trauma, the patient may experience chronic pain originating from the point of penetration through the fat pad. Several reports of surgical exploration of the fat pad in these patients have demonstrated foci of granulomatous reaction (Dandy, 1987; Lindenbaum, 1981). Following resection of these granulomatous lesions, symptoms resolved. Fibrosis of the fat pad at the site of surgical trauma may also occur, resulting in chronic pain. The combination of fibrosis and hyperplasia producing a sclerotic region within the fat pad and anterior retinaculum may also produce pain. In severe cases, limitation of motion will ensue (Paulos et al., 1987). Herniation of a portion of the fat pad through the extensor fascia has also been observed (Lindenbaum, 1981). Again, surgical repair is required.

Sonography will also demonstrate neuromas resulting from surgical trauma at the access portal. Cutaneous sensation over the anterior aspect of the knee is provided by the infrapatellar division of the saphenous nerve. Saphenous nerve damage is most frequently a complication of meniscal repair procedures (Fig 12–5), but small anterior branches may be damaged in any surgical procedure. Conservative therapy that includes injection of local anesthetic and steroids is often successful. Surgical excision is rarely indicated.

Cyst formation at the arthroscopic portal is not uncommon. Usually they are lined by synovium and therefore referred to as *synovial cysts.* If the wall of the cyst is predominantly fibrotic, it is called a *ganglion* (Figs 12–6 and 12–7). These cysts typically have a pedunculated appearance with a narrow base originating at the access portal (van Holsbeeck and Introcaso, 1989). Fluid within these cysts is markedly hy-

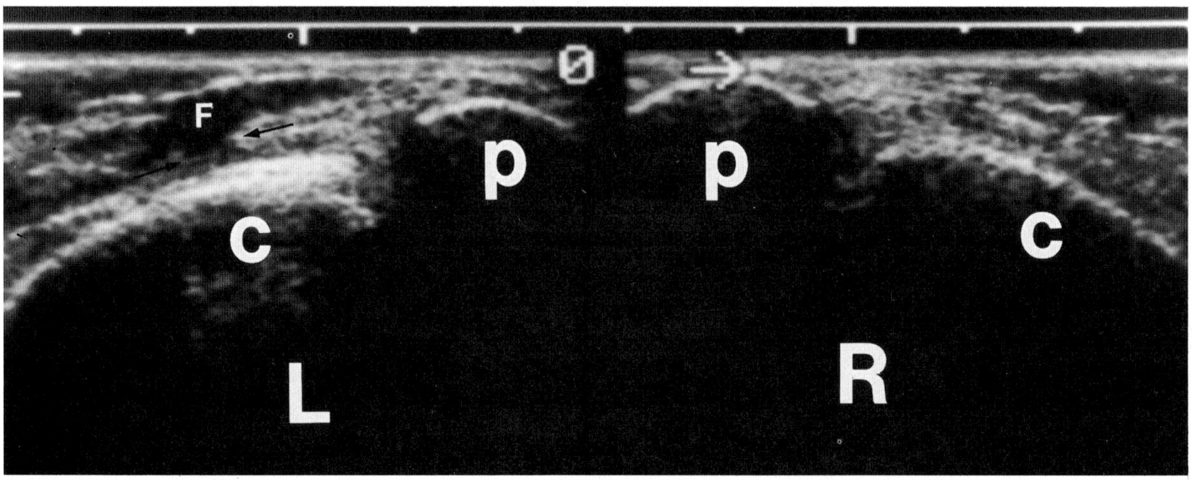


**FIG 12–1.**
Painful ganglion: transverse sonogram. The patient is a 19-year-old woman who underwent an arthroscopic lateral retinacular release 7 months ago. She presented here with pain and swelling lateral to the left patella. Side to side comparison is provided in this split screen image. On the *left* is a transverse image of the symptomatic lateral aspect of the left patellofemoral joint. The asymptomatic side is on the *right.* A fluid-filled cavity *(F)* is identified lateral to the patella *(p)* in direct communication with the joint through an interruption in the lateral retinaculum *(arrows).* Extravasation of synovial fluid is a common complication seen following arthroscopic procedures. Rarely, these collections may form fistulas to the skin. Abbreviations: lateral femoral condyle *(C).*

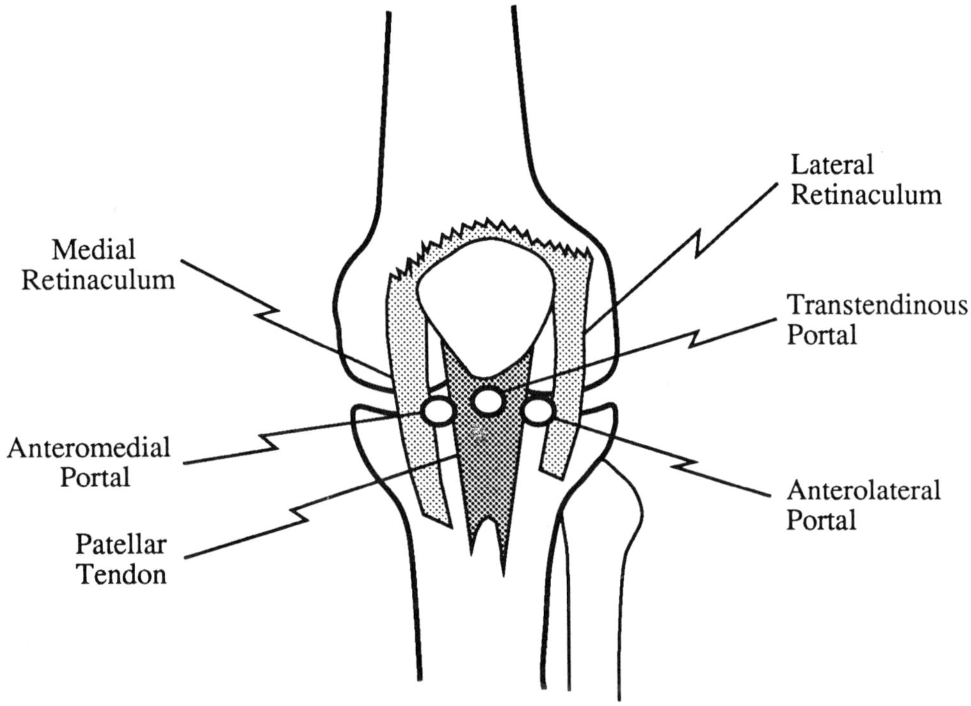


**FIG 12–2.**
Arthroscopic portals for therapeutic arthroscopy.

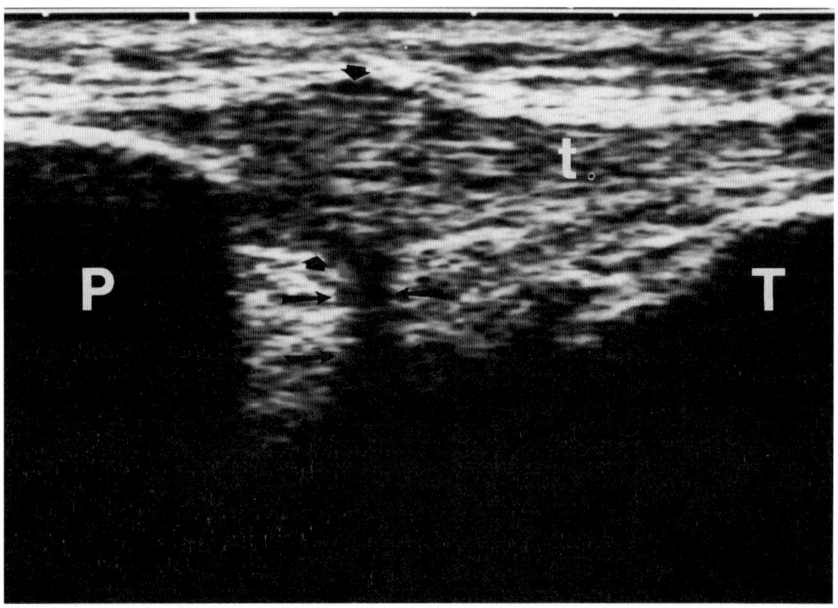


**FIG 12–3.**
Patellar tendinitis after arthroscopy: longitudinal sonogram. One and one half years after arthroscopy this 32-year-old male tennis player continued to have pain. This pain differed in character from the pain for which arthroscopy was performed and forced him to stop athletic activity. Clinical examination revealed pain, swelling, and tenderness just below the apex of the patella. A segment of the patellar tendon *(t)* is markedly thickened *(small arrows)* and hypoechoic, indicative of patellar tendonitis. Immediately deep to the involved tendon segment is a hypoechoic tract traversing Hoffa's fat pad *(large arrows)*. This is the characteristic appearance of an arthroscopic portal through the fat pad. The focal nature of the tendon abnormality associated with the transtendinous portal clearly indicates that this is a complication of the arthroscopic procedure. Abbreviations: patella *(P)*, tibia *(T)*.

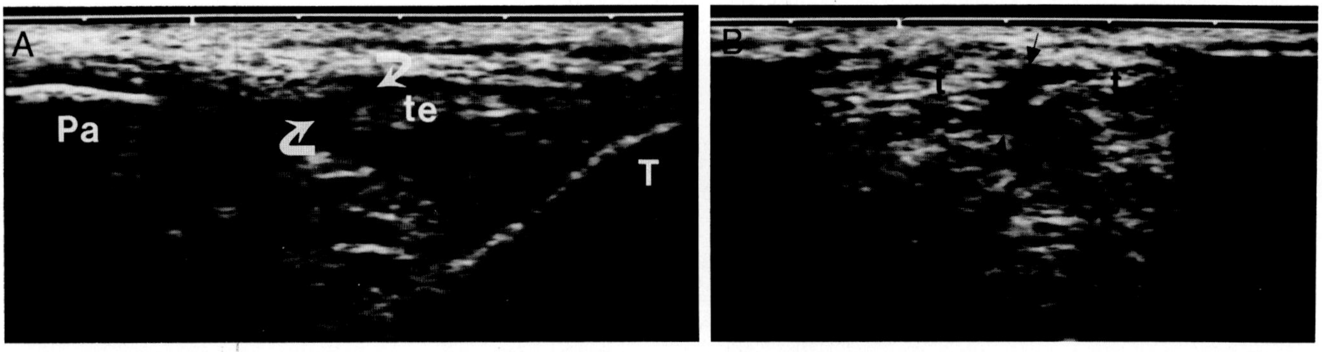


**FIG 12–4.**
**A,** incomplete patellar tendon rupture: longitudinal sonogram. This 41-year-old man experienced pain and focal tenderness beneath the left patellar apex 8 months following therapeutic arthroscopy. He also exhibited weakened extension of the lower leg. The longitudinal image of the left patellar tendon *(te)* demonstrates a gap *(curved arrows)* in the patellar tendon at the site of the transtendinous portal. Abbreviations: patellar apex *(Pa),* tibia *(T).* **B,** incomplete patellar tendon rupture: transverse sonogram. Same patient as in **A.** The transducer is positioned transversely over the point of maximal tenderness. A hypoechoic gap *(arrows)* is again noted in the tendon *(t).* Normal-appearing tendon is present on both medial and lateral sides of this partial rupture.

poechoic or anechoic. Rarely, meniscal fragments or other tissue fragments may become entrapped in these cysts. This is most commonly seen when a fragment slips from the grasp of a surgical instrument while being removed from the joint.

Fluid escaping from the joint at the access portals is another complication, which is easily diagnosed with sonography. During the procedure, irrigation fluid may extravasate from the joint. This is rarely significant, but when extensive, it has been known to result in compartment syndrome requiring fasciotomy (Peek and Haynes, 1984; Fruensgaard and Holm, 1988). Sonography is highly valuable in the diagnosis of compartment syndrome. The diagnosis of compartment syndrome is discussed in detail in chapter 3. Synovial fluid leaking into the subcutaneous tissues can cause significant inflammation and discomfort, sometimes mimicking venous thrombosis. Ultrasound is extremely valuable in making the definitive diagnosis in these cases. When gas is used for arthroscopy subcutaneous emphysema, pneumoscrotum, pneumoperitoneum, and pneumopericardium may occur (Lotman, 1987; Dick et al., 1978; Henderson and Clark, 1982). Partial airway obstruction has also been reported (Shupak et al., 1984). Cases of gas extravasation are best evaluated with conventional radiographs.

Vascular complications following arthroscopy are predominantly cases of deep venous thrombosis (Cohen et al., 1973; Walker and Dillingham, 1983). The incidence of deep venous thrombosis is greatest in patients more than 50 years of age and when tourni-

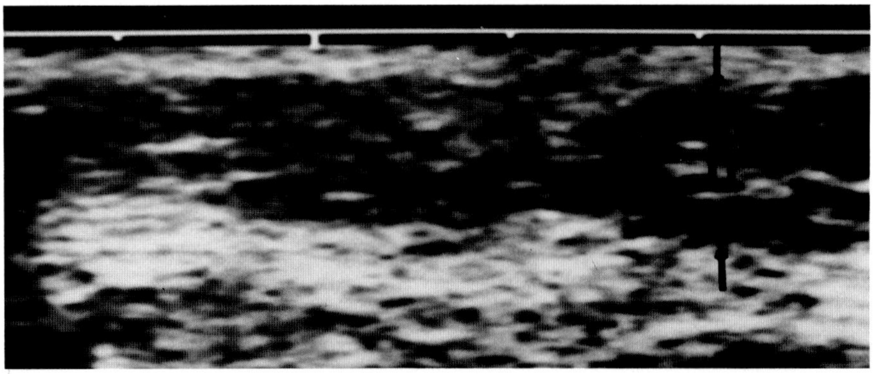

**FIG 12–5.**
Traumatic subcutaneous neuroma: transverse sonogram. This 45-year-old woman presented with pain and extreme tenderness at the medial aspect of the right knee 1½ years after arthroscopic repair of a meniscal tear. The transducer is positioned over the scar of the painful anterior medial portal. A crescentic anechoic fluid collection *(arrows)* borders a hyperechoic mass *(n)* continuous with the infrapatellar nerve. Surgical exploration revealed a neuroma of the infrapatellar nerve.

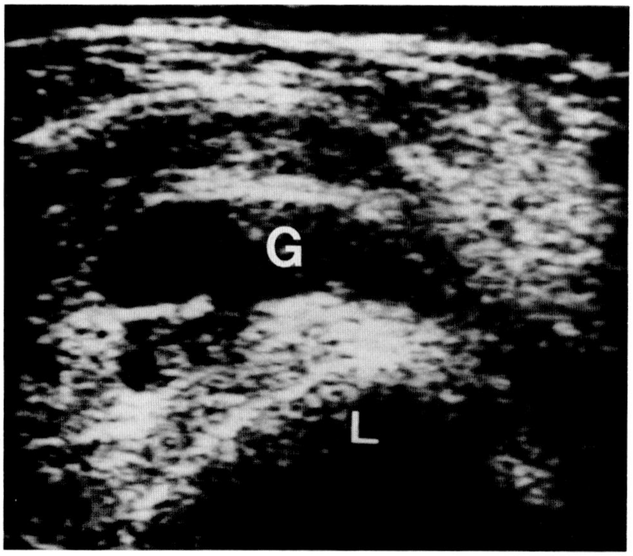


**FIG 12–6.**
Periarticular ganglion after arthroscopy: transverse sonogram. This large ganglion *(G)* was demonstrated at the anterior lateral arthroscopic portal in this 50-year-old woman. Arthroscopy was performed 2 years ago. Several months ago the patient noted a painful swelling over the lateral femoral condyle *(L)*. Surgery demonstrated a ganglion continuous with the joint at the site of the anterior lateral portal.

quet time exceeds 1 hour (Sherman et al., 1986). This has become an increasing problem as arthroscopic procedures have become more sophisticated and duration of procedures increases. Doppler techniques and compression ultrasound examination for deep venous thrombosis have demonstrated their value (Lensing et al., 1989). Absence of a Doppler signal alone may not be adequate to make the diagnosis of deep venous thrombosis. In cases of very slow flow, a Doppler signal may not be detected due to filter threshold. Compression examination should be used to confirm the absence of a Doppler signal. If a vein is not compressible when pressure is applied with the transducer, then thrombus must be present.

Arterial pseudoaneurysm is a very rare complication of arthroscopy (Sherman et al., 1986; Kvist and Ksaergaard, 1979). They occur secondary to perforation of the posterior joint capsule by sharp instruments, adjacent to the popliteal vessels. Some of these accidents have resulted in death or amputation (Sherman et al., 1986). If a diagnostic test is required, the correct diagnosis is quickly established with sonography. Ultrasound images will demonstrate a popliteal mass continuous with the artery. Doppler signal may be absent within the aneurysm due to thrombosis. Mural thrombus will be seen as hypoechoic material layering peripherally or completely filling the pseudoaneurysm. Emergent excision and graft placement are indicated because their natural history is inevitable rupture.

Instrument breakage (see Fig 11–1) is a complication that has decreased in incidence recently, perhaps due in part to refinement of instrument design (Sprague 1989). It is most often associated with partial lateral meniscectomy procedures. When fragments are located in the peripheral recesses of the joint or subcutaneous tissues, ultrasound can be of significant benefit. The approach to these situations is identical to that used for any foreign body, as outlined in Figure 11–10. However, if instrument fragments are located deep within the joint, conventional radiography is the examination of choice. Instrument fragments tend to have sharp edges, causing them to remain fixed in location when deep within the joint. This differs from tissue fragments that may become lost within the joint during arthroscopy. Meniscal fragments or pieces of osteophyte lost within the joint will migrate to the peripheral synovial recesses. Once sequestered within the synovial recesses, they are easily detected with sonography (see Fig 7–43).

Hemarthrosis is the most common complication following both therapeutic and diagnostic arthroscopy. To be considered a complication, bleeding into the joint must be of sufficient magnitude to require intervention (Fig 12–8). Significant hemarthrosis requires evacuation for two reasons: formation of adhesions and infection. Often untreated hemarthrosis will result in formation of adhesions within the joint during the process of resolution. Adhesions may limit the range of motion of the joint and be the cause of chronic pain. A second surgical procedure is required to lyse these adhesions.

Capsular rupture is one cause of hemarthrosis. Often these lesions go unrecognized. Involvement of the medial collateral ligament is a severe form of capsular rupture that greatly affects joint stability. Rupture of the medial collateral ligament during an arthroscopic procedure is rare but may result from excessive valgus stress applied in an attempt to better visualize the medial compartment (Fig 12–9). Peripheral extension of meniscectomy incision into the medial collateral ligament can also result in rupture.

Infection is the second most common complication of arthroscopic procedures. The risk of infection is considerably increased when hemarthrosis is present. Therefore, evacuation of significant intra-articular hemorrhage is recommended. The role of sonography in these cases is the detection and characterization of abnormal intra-articular fluid col-

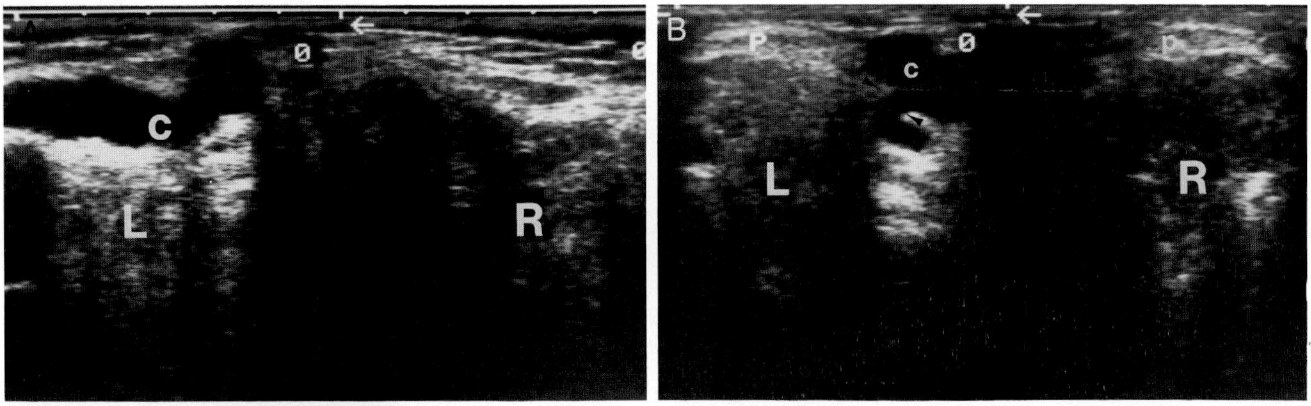


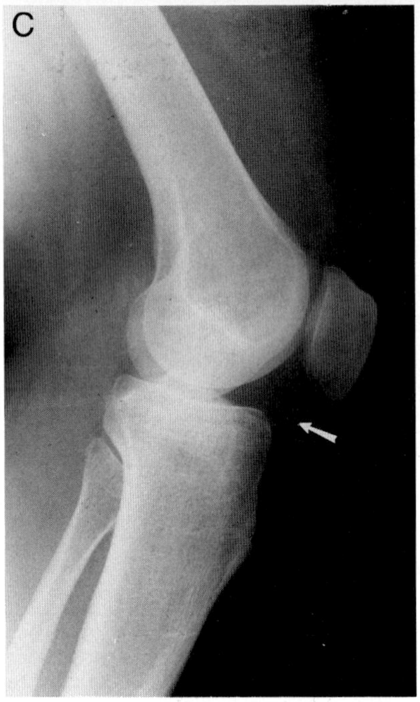


**FIG 12–7.**
**A,** portal ganglion: oblique sonogram. A 26-year-old woman presented with pain and swelling over the medial portal adjacent to the patellar tendon. The image on the *left* is an oblique image of the symptomatic left knee over the point of tenderness. A comparable image of the asymptomatic right knee is on the *right.* An anechoic fluid collection *(c)* is seen continuous with the arthroscopic portal. It extends laterally and superiorly through Hoffa's fat pad. Abbreviations: left *(L),* right *(R).* **B,** portal ganglion: transverse sonogram. Same patient as in **A.** The transducer is now positioned transversely over the point of greatest swelling. Again, a split screen comparison is provided. A point of communication with the joint through the retinaculum *(arrowheads)* is better demonstrated in this image. Abbreviations: patellar tendon *(P),* ganglion cyst *(c).* **C,** portal ganglion: lateral radiograph. Same patient as in **A** and **B.** This lateral radiograph is essentially normal. Minimal streaky density is seen in Hoffa's fat pad *(arrow).*

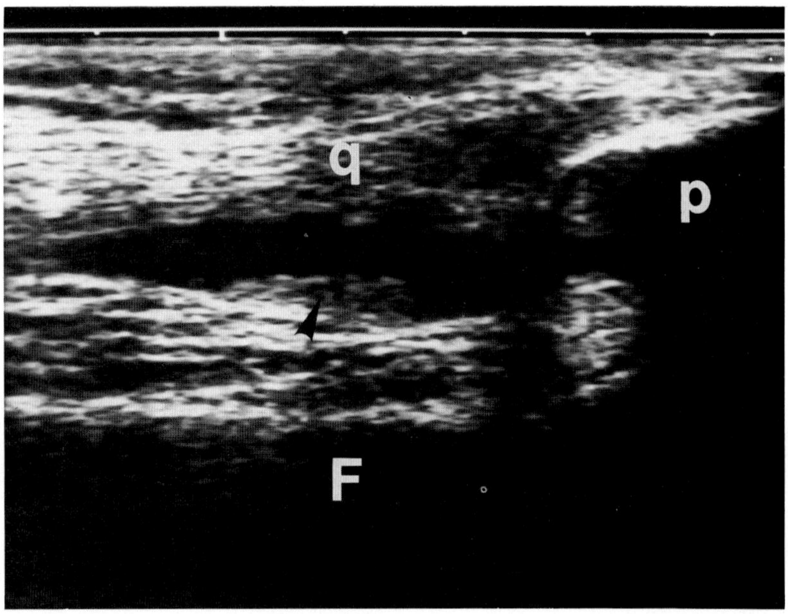


**FIG 12–8.**
Hemarthrosis following arthroscopy: longitudinal sonogram. This 30-year-old soccer player had arthroscopic medial meniscectomy 5 days ago. He was admitted for increased swelling of the knee and slight fever. A large effusion is noted in the suprapatellar bursa. Several blood clots *(arrowhead)* were noted moving freely within the joint. Joint aspiration was performed and yielded frank blood. Cultures were negative. Abbreviations: femur *(F),* patella *(p),* quadriceps *(q).*

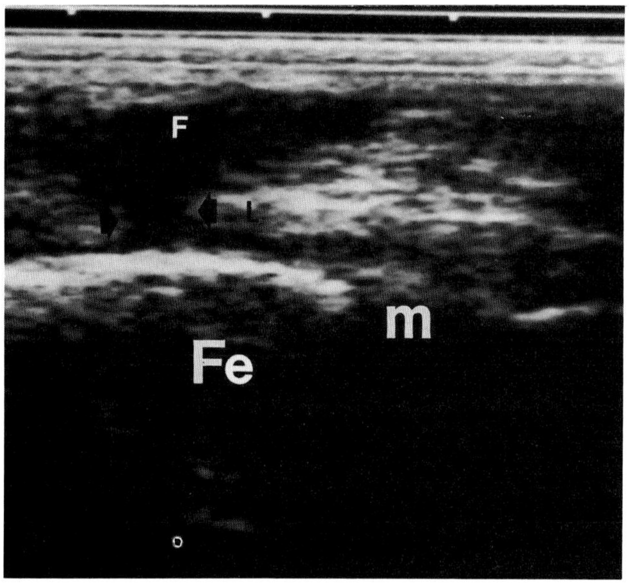


**FIG 12–9.**
Medial collateral ligament rupture secondary to arthroscopy: coronal sonogram. Four days after an arthroscopic medial meniscectomy, this 35-year-old man presented with a complaint of progressively increasing swelling over the medial aspect of the knee. Proximal is at the observer's left. A disruption *(arrows)* is noted in the medial collateral ligament *(L)*. Fluid *(F)* extends through the gap and pool beneath the skin in the subcutaneous tissue. Acute ruptures of the medial collateral ligament may be due to excessive valgus stress applied during the procedure or overaggressive incision. Abbreviations: femur *(Fe)*, meniscus *(m)*.

lections. Ultrasound is exquisitely sensitive in detecting increased intra-articular fluid, far more sensitive than aspiration (van Holsbeeck et al., 1988). If needed, ultrasound guidance can be provided for aspiration to obtain fluid for culture.

## PAIN FOLLOWING DIAGNOSTIC ARTHROSCOPY

The spectrum of complications related to diagnostic arthroscopy is identical to that of therapeutic arthroscopy. They differ only in incidence. Fewer complications are seen following diagnostic procedures because they tend to be less invasive and tourniquet times are considerably shorter. As stated earlier, the two most frequent complications are hemarthrosis and infection.

Pain that persists following a normal diagnostic arthroscopic procedure is usually not attributable to a complication of the procedure. It is more likely that the cause of the pain lies outside of the diagnostic range of the arthroscope. Arthroscopy is excellent for the evaluation of intra-articular pathology, but it cannot evaluate the periarticular structures. Examination of the posterior horns and the peripheral areas of the menisci is limited and requires extensive manipulation through multiple ports (Weissman and Sledge, 1986). This increases the risk of complications.

Ultrasound and magnetic resonance imaging have great utility in the evaluation of the periarticular tissues. Cost, availability, and capability for dynamic examination make ultrasound the favored examination for periarticular structures. Sonography can evaluate posterior and peripheral meniscal lesions far better than arthroscopy. Meniscal cysts associated with small degenerative meniscal tears are particularly well evaluated by ultrasound (Coral et al., 1989). These lesions are completely overlooked by arthroscopy. Sonography has demonstrated that these lesions occur far more frequently than previously thought. Peripheral meniscal lesions are commonly associated with collateral ligament and capsular injuries. These lesions are easily evaluated in the same ultrasound examination. In acute injury, small hematomas within the substance of the collateral ligaments can be demonstrated. Chronic intrasubstance injury of the medial collateral ligament (type I and II lesions) are also well demonstrated. These lesions are discussed in detail in chapter 6. Overall, sonography is excellent for the evaluation of lesions that lie outside of the diagnostic range of arthroscopy.

## REFERENCES

Cohen SH, et al: Thrombophlebitis following knee surgery. *J Bone Joint Surg [Am]* 1973; 55A:106–112.

Coral A, van Holsbeeck M, Adler R: Imaging of meniscal cysts of the knee in three cases. *Skeletal Radiol* 1989; 18:451–455.

Dandy DJ: *Arthroscopic Management of the Knee*, ed 2. New York, Churchill-Livingstone, 1987.

Dick W, et al: Complications of arthroscopy. A review of 3714 cases. *Arch Orthop Trauma Surg* 1978; 92:69–73.

Fruensgaard S, Holm A: Compartment syndrome complicating arthroscopic surgery: Brief report. *J Bone Joint Surg [Br]* 1988, 70:146–147.

Henderson CE, Hopson CN: Pneumoscrotum as a complication of arthroscopy. *J Bone Joint Surg* 1982; 8:1238–1240.

Kvist E, Ksaergaard E: Vascular injury complicating meniscectomy. Report of a case. *Acta Chir Scand* 1979; 145:191–193.

Lensing AWA, Prandoni P, Brandjes D, et al: Detection of deep-vein thrombosis by real-time B-mode ultrasonography. *N Engl J Med* 1989; 320:342–345.

Lindenbaum BL: Complications of knee joint arthroscopy. *Clin Orthop* 1981; 160:158.

Lotman DB: Pneumoperitoneum and acidosis during arthroscopy with $CO_2$. *Arthroscopy* 1987; 3:185–186.

Metcalf RW: An arthroscopic method for lateral release of subluxating or dislocating patella. *Clin Orthop* 1982; 167:9.

Paulos LE, et al: Infrapatellar contracture syndrome, an unrecognized cause of knee stiffness with patellar entrapment and patella infera. *Am J Sports Med* 1987; 15:331–341.

Peek RD, Haynes DW: Compartment syndrome as a complication of arthroscopy, a case report and a study of interstitial pressure. *Am J Sports Med* 1984; 12:464–468.

Sherman OH, Fox JM, Snyder SJ, et al: Arthroscopy—"no problem surgery." *J Bone Joint Surg [Am]* 1986; 68:256–265.

Shupak RC, Shuster H, Funch R: Airway emergency in a patient during $CO_2$ arthroscopy: Report of a case. *Anesthesiology* 1984; 60:161–172.

Sprague NF: *Complications in Arthroscopy.* New York, Raven Press, 1989.

van Holsbeeck M, van Holsbeeck K, Gevers G, et al: Staging and follow-up of rheumatoid arthritis of the knee. *J Ultrasound Med* 1988; 7:561–566.

van Holsbeeck M, Introcaso J: Ultrasound follow-up of knee arthroscopy. *Orthop Today* 1989; 9:10.

Walker RH, Dillingham M: Thrombophlebitis following arthroscopic surgery of the knee. *Contemp Orthop* 1983; 29–33.

Weissman BNW, Sledge CB: *Orthopedic Radiology.* Philadelphia, WB Saunders Co, 1986.

# 13

# Sonography of the Shoulder

In the 1960s, clinicians called the shoulder the "forgotten joint" (Golding, 1962). At that time, only plain films were available for the evaluation of a painful shoulder. Plain films are normal in more than 90% of patients with chronic shoulder pain. Tendon calcifications are present in some of these patients; others have ossified tendon insertions or cystic lesions around the anatomical neck. The significance of all these plain film findings is still debated in literature (Depalma, 1983; Refior et al., 1987). There is, as yet, no proof of the relationship of these findings to certain types of shoulder disease, such as hydroxyapatite deposition disease, impingement, and rotator cuff disease. More recently, finding ossification in the coracoacromial ligament, acromioclavicular osteoarthritis, and narrowing of the subacromial space have been cited as signs of rotator cuff disease (Petersson and Gentz, 1983; Resnick and Niwayama, 1988; Gielen et al., 1990).

Both arthrography and arthroscopy of the shoulder were introduced during the 1970s. Arthroscopy of the shoulder is, without doubt, the most complete examination of the rotator cuff (Ogilvie Harris and Wiley, 1986). The arthroscopist examines the undersurface of the rotator cuff through the joint. Then, a new portal is made to examine the superior aspect of the cuff through the subdeltoid bursa. Not only are rotator cuff tears seen, but tears of the labrum are also visualized. The arthroscopist has a clear view of the synovial cavity and can therefore diagnose synovial disease at an early stage. Arthroscopic surgery may be performed during the same session.

A disadvantage of arthroscopy is the invasive character of the procedure. Numerous complications such as neurovascular complications, damage to the labrum, considerable muscle injury, infection, and hemarthrosis have been described (Jeffries et al., 1987; Lindenbaum, 1981). Chronic draining fistulas and leak of synovial fluid are often cited as the most common complications (Henderson and Hopson, 1982). They are considered minor by the arthroscopist but are often regarded as a significant esthetic handicap by the patient.

Not all shoulder pathology is detectable with an arthroscope (Jobe and Jobe, 1983). Rotator cuff lesions communicating with the joint or subacromial-subdeltoid bursa are detected arthroscopically (Refior et al., 1987). Rotator cuff edema and fibrosis, generally regarded as the precursor of a rotator cuff tear, is not an arthroscopic diagnosis (Neer, 1983). Orthopedic surgeons refer to these lesions as *impingement syndrome.* No clinical tests to distinguish rotator cuff edema and fibrosis from rotator cuff tear are available.

Arthrography of the shoulder is less invasive than arthroscopy (Ahovuo, 1984). This technique detects complete tears of the rotator cuff and tears involving the undersurface of the supraspinatus tendon. Detection of incomplete tears of the subdeltoid bursal surface of the rotator cuff requires injection of the subdeltoid bursa (Neer, 1983). This type of tear is rare, less than 3% of all rotator cuff tears. The majority of the radiologists, therefore, never perform an injection of the subdeltoid bursa. Arthrography is not a good technique for the evaluation of edema and intrasubstance tears of the rotator cuff. Like arthroscopy, arthrography is not the ideal technique for routine screening of painful shoulders.

Impingement is the most common pathology causing chronic shoulder pain (Neer, 1983). It is caused by compression of the anterior cuff against the anterior acromial edge and coracoacromial ligament. The initial manifestations are edema and petechial hemorrhage, which progress to tendonitis and fibrosis. Incomplete and complete tears of the rotator cuff are the end point of the disease spectrum called impingement. Unfortunately, there are no clinical tests that will differentiate impingement from rotator cuff tear. Duration of symptoms may influence the clinical diagnosis, depending on the patient's history. The majority of patients will experience 10 to 15 years of chronic shoulder pain before impingement progresses to rotator cuff tear.

Treatment for the initial stages of impingement differs considerably from the treatment for rotator cuff tears. Edema of the cuff is managed with conservative measures for as long as possible. This usually consists of avoidance of painful elevation of the arm, oral anti-inflammatory medication, and injections of anti-inflammatory drugs beneath the acromion. Surgery is considered only when the pain persists for more than 18 months, despite conservative treatment. Surgery is then limited to resection of capsular structures and, in some patients, extended by an anterior acromioplasty. Rotator cuff tear is treated differently. Surgery is indicated in all complete tears. It is important to perform repair early before the defect becomes too big and then irreparable.

In a patient with impingement and rotator cuff edema, one would like to avoid arthroscopy. The arthroscopic procedure could possibly damage the cuff, resulting in tears due to scope introduction and patient manipulation. Therefore, we prefer to use a noninvasive technique. The ideal technique should be readily available for screening a large population with chronic shoulder pain and be able to distinguish tears from edema of the cuff in the majority of the cases. Ultrasound is the ideal technique. It is noninvasive and widely available for screening on a large scale (Crass, 1987; Mack et al., 1985). Reported sensitivity and specificity are more than 90%.

## WORKUP OF SHOULDER DISEASE

Whenever a new technique is introduced in diagnostic imaging several questions come up. Does it bring anything new? Does it replace an old technique? Finally, how do we integrate this new technique into the diagnostic workup?

The estimated frequency of rotator cuff tears in people more than 60 years old is approximately 30% (Cofield, 1985). With such a large potential group of patients, we prefer ultrasound over more invasive, expensive, and time-consuming methods, such as arthrography, arthroscopy, and magnetic resonance imaging (MRI). The cost effectiveness of the technique and its noninvasive character are strong factors favoring ultrasound as a screening method.

Every examination of the shoulder starts with plain films. The most detailed examination of the bony architecture is still the conventional radiograph. Substantial narrowing of the subacromial space with an acromiohumeral distance equal to or less than 5 mm is due to large rotator cuff tears. In these cases, no further investigations are necessary (Fig 13–1) (Cofield, 1985). Radiographs can also exclude referred pain to the shoulder due to a lesion in adjacent structures or a neurovascular syndrome secondary to a Pancoast tumor (Bateman, 1983).

We recommend an ultrasound examination in patients with chronic shoulder symptoms, normal plain films, and positive clinical test results indicating cuff disease. The best clinical indicators of rotator cuff disease are a painful arc, positive impingement test result, positive supraspinatus test result, or limited abduction of the shoulder. Arthrography or arthroscopy should then be limited to patients with normal plain films and normal ultrasound examination of the shoulder but with an unequivocal clinical examination (limited abduction). About 3% of the patients referred to us for ultrasound of the rotator cuff fall into this category.

On rare occasions, further investigation is necessary after a negative arthrogram. Superficial tears of the rotator cuff can be shown by injection of the subacromial-subdeltoid bursa. Bursography can be performed under sonographic or fluoroscopic guidance (van Holsbeeck and Introcaso, 1989; Neer, 1983).

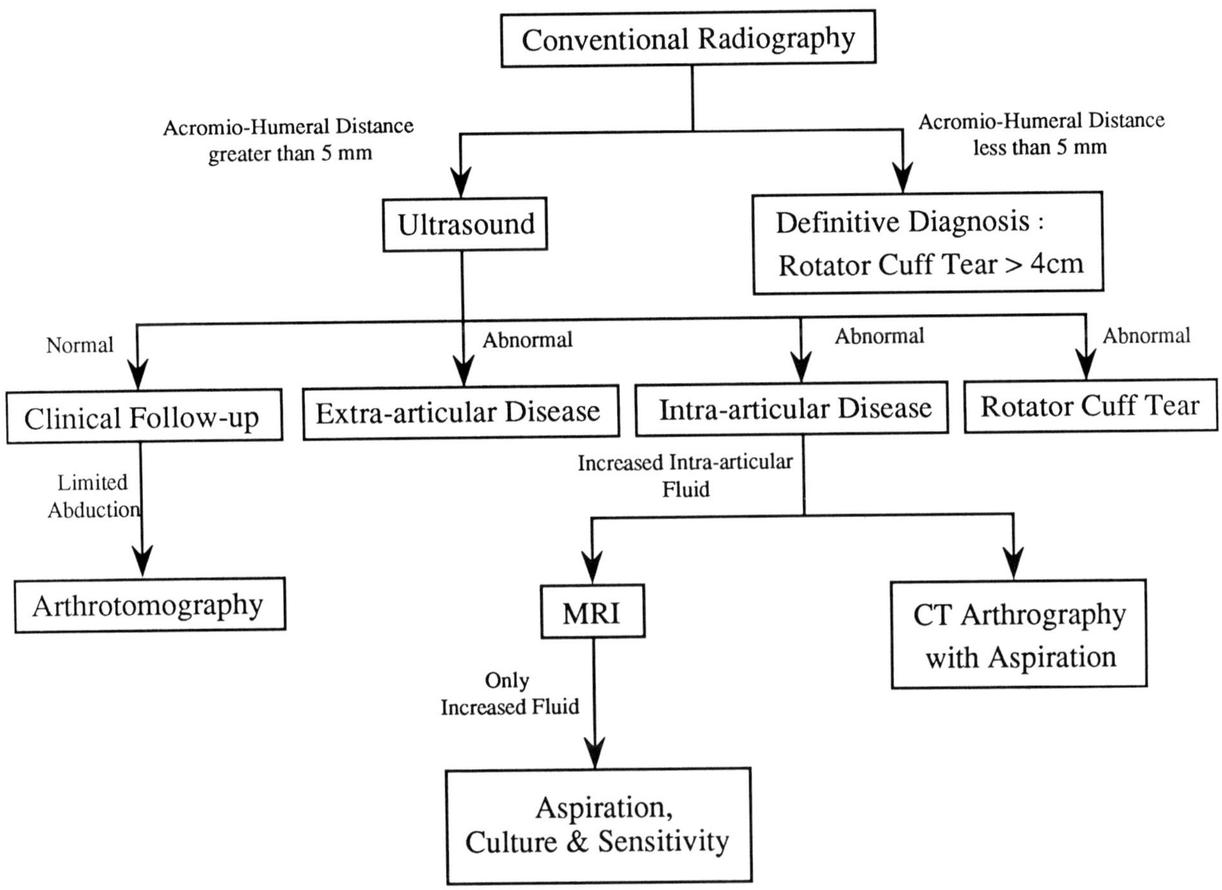


**FIG 13–1.**
Algorithm for evaluation of a painful shoulder.

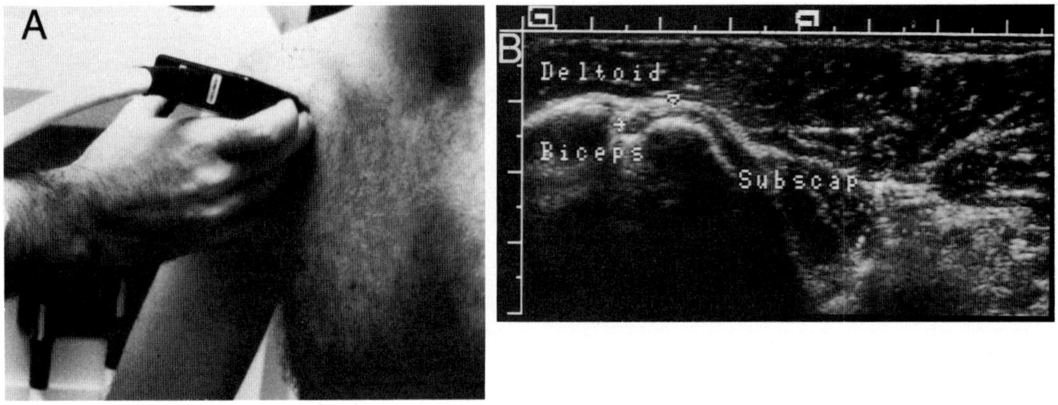


**FIG 13–2.**
**A,** sonography of the shoulder: the anterior rotator cuff. The examination is started by identifying the long biceps tendon in the bicipital groove on transverse images. **B,** sonography of the shoulder: transverse sonogram of the anterior rotator cuff. The anterior rotator cuff is located medial to the long biceps tendon *(arrow)*. The subscapularis tendon *(arrowhead)* covers the bicipital groove anteriorly.

## EXAMINATION OF THE SHOULDER

High-frequency linear array transducers (5, 7.5, 10 MHz) are essential for the evaluation of the rotator cuff. Linear-array transducers provide the wide field of view needed superficially; radial arrays are less desirable. We do not recommend the use of sector scanners (Mack, 1988). The wedge-shaped image produced with sector scanners severely limits the width of the superficial portion of the image.

We use the method proposed by Mack (1988). The long biceps tendon is considered the most important landmark in shoulder sonography, as well as in shoulder anatomy and surgery. This tendon helps us to distinguish the subscapularis muscle and tendon from the supraspinatus tendon. The anterior supraspinatus tendon is by far the tendon most often involved in rotator cuff tear due to impingement.

The patient is ideally positioned on a rotating chair with the observer facing the lateral side of the shoulder of the patient. We begin the examination by locating the long biceps tendon within the bicipital groove overlying the anterior surface of the humerus with the shoulder held in neutral position (Fig 13–2). The biceps tendon has a circular appearance in the transverse plane. Medial to the biceps tendon we identify the subscapularis muscle covering the anterior surface of the humeral head. Its tendon inserts

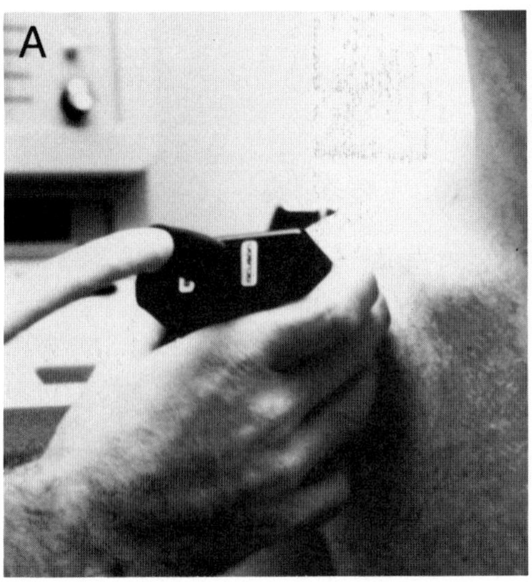


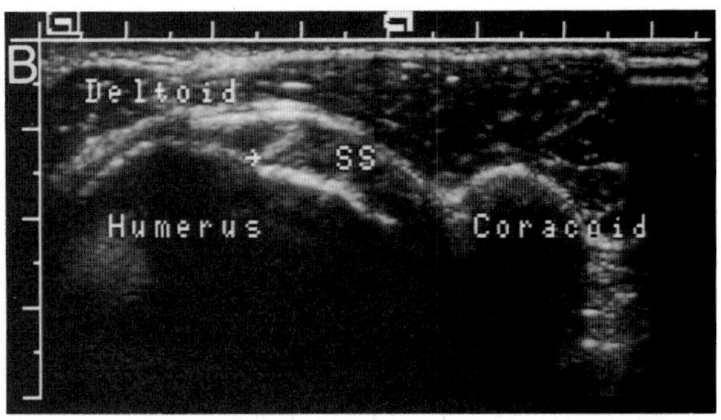


**FIG 13–3.**
**A,** sonography of the shoulder: the anterior rotator cuff. The transducer is then moved superiorly over the bicipital groove until its superior margin is immediately beneath the acromion. **B,** sonography of the shoulder: transverse sonogram of the anterior rotator cuff. The proximal bicipital groove is shallow, and the long biceps tendon *(arrow)* differs in shape due to its orientation. Abbreviations: subscapular muscle *(SS)*.

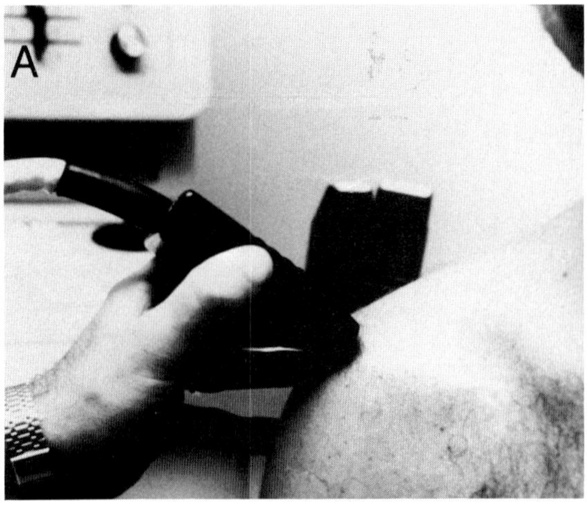


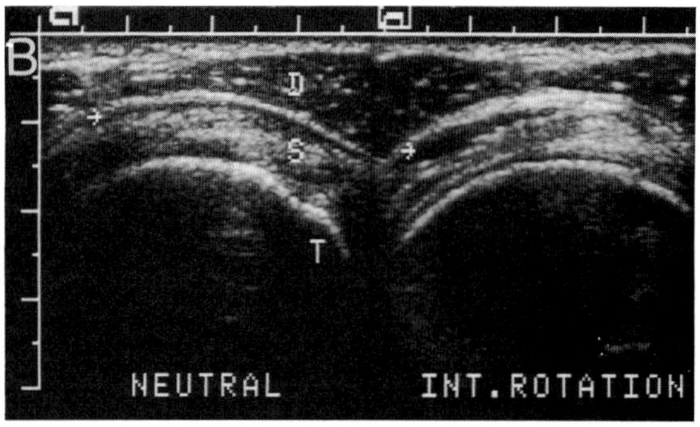


**FIG 13–4.**
**A,** sonography of the shoulder: lateral rotator cuff. The transducer is then moved laterally around the acromial edge. **B,** sonography of the shoulder: transverse sonograms of the lateral rotator cuff. Two soft tissue layers are recognized with this approach. The most superficial layer is the deltoid muscle *(D);* the deep layer is the supraspinatus tendon *(S).* The subacromial-subdeltoid bursa is noted as a hypoechoic layer in between the supraspinatus tendon and the deltoid. Internal rotation of the patient's shoulder during the examination results in displacement of fluid into the superior aspect of the bursa *(arrows).* Abbreviations: greater tuberosity *(T).*

on the lesser tuberosity with some fibers continuing across the bicipital groove to form the transverse humeral ligament. The most anterior extent of the subscapularis is rarely involved in rotator cuff tears. Dislocation of the biceps tendon usually occurs following separation of the subscapularis tendon from the lesser tuberosity while fibers forming the transverse ligament remain intact (see chapter 4) (Petersson, 1986). Therefore, the biceps tendon will usually be located posterior to the subscapularis muscle

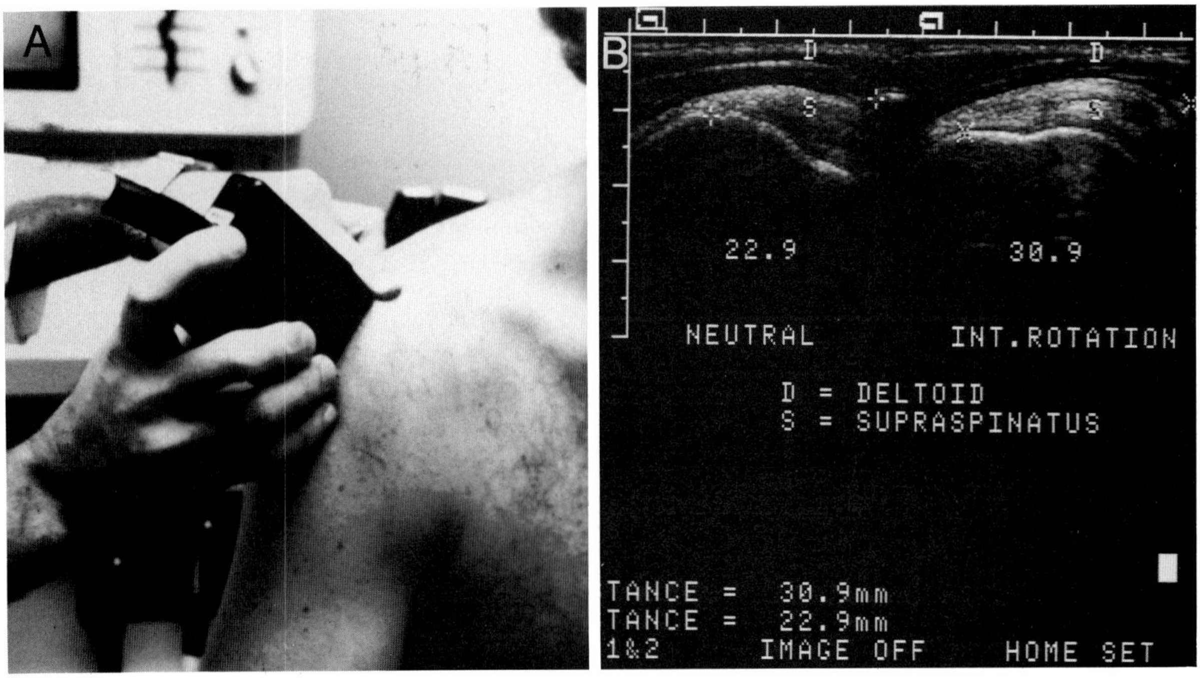


**FIG 13–5.**
**A,** sonography of the shoulder: lateral rotator cuff. The transducer is rotated 90 degrees if an abnormality is found in the transverse plane. The examiner ensures that the proximal end of the transducer touches the lateral edge of the acromion by marking it with a finger. **B,** sonography of the shoulder: coronal sonograms of the lateral rotator cuff. The supraspinatus tendon *(S)* is triangular shaped. The patient's shoulder is moved from a neutral position to internal rotation to view more of the supraspinatus tendon. This maneuver is also important in testing shoulder mobility. Ultrasound can examine the 3 distal cm of the supraspinatus tendon. The rest of the cuff remains hidden under the acromion. Abbreviations: deltoid *(D).*

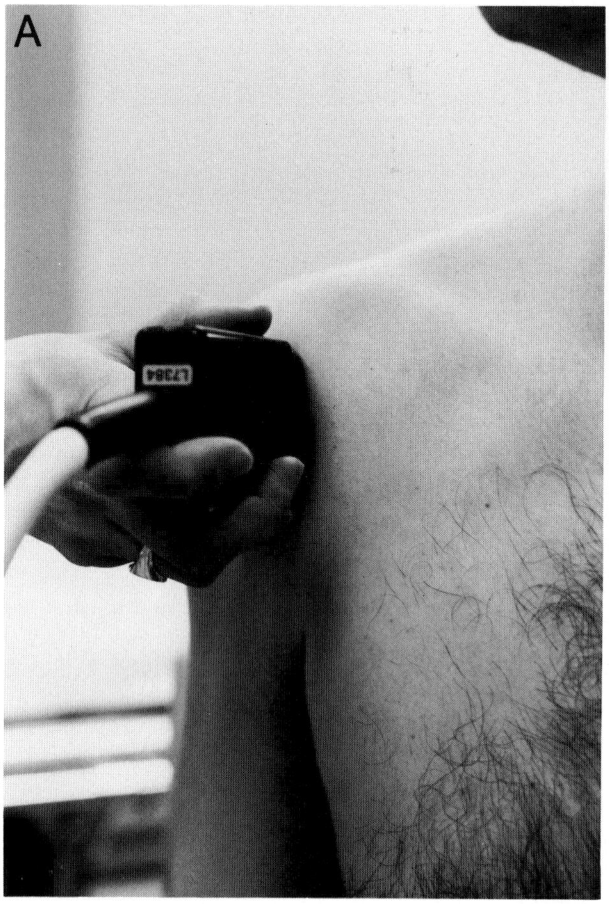


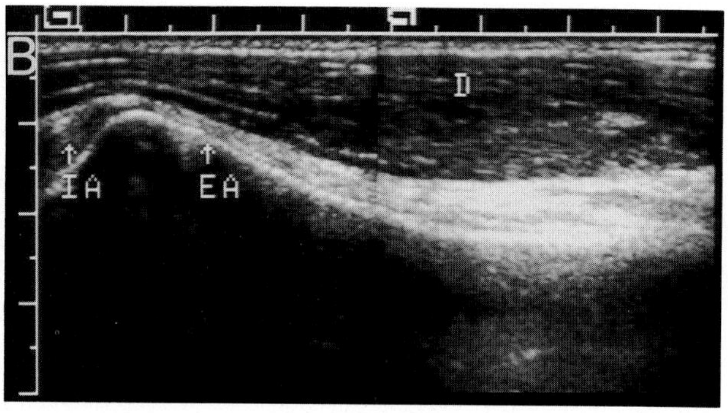


**FIG 13–6.**
Sonography of the shoulder: biceps tendon. **A,** the biceps tendon is examined in its long axis with the transducer placed over the bicipital groove. **B,** sonogram of the biceps tendon. The intra-articular *(IA)* and extra-articular *(EA)* segments of the long biceps tendon can be seen. The long biceps tendon appears hyperechoic if the transducer is positioned parallel to the long axis of the tendon. *D* = deltoid.

in cases of tendon dislocation. Once the biceps tendon and subscapularis (Fig 13–3) are identified, the transducer is moved laterally (Fig 13–4) to visualize the supraspinatus tendon adjacent to the lateral edge of the acromion. It may be helpful to mark the lateral edge of the acromion with the left hand while holding the transducer with the right hand. Scanning with palpation of the acromial edge ensures that the procedure is not performed too inferior relative to the supraspinatus insertion. An examination performed to low on the humerus results in a false impression of the absence of the supraspinatus tendon layer, thus giving a false positive examination. The supraspinatus tendon extends only a few centimeters inferior to the lateral acromion. It is therefore important to bring the transducer medially to confirm the absence of cuff material. The ideal transducer position is immediately beneath the lateral border of the acromion. The shoulder of the patient is then moved into internal rotation. This maneuver uncovers more of the supraspinatus tendon. The examiner has to keep in mind that sonography is able to screen about 95% of the rotator cuff substance, but the acromion covers the remaining 5%. Two percent of rotator cuff tears will be missed by sonographic examination because they are obscured from view by the acromion.

A normal supraspinatus tendon is found as a tendon layer deep to the deltoid muscle. The normal tendon is slightly more reflective than the deltoid. In older individuals, the supraspinatus tendon can be decreased in echogenicity, resulting in reflectivity less than or equivalent to the deltoid muscle. The normal tendon has a very homogeneous appearance; any inhomogeneity is suspicious for rotator cuff injury. Inhomogeneity of the rotator cuff can also be due to artifact. We can distinguish true lesions from artifact by changing the position of our transducer. The examiner rotates the transducer 90 degrees and tries to locate the abnormality in the coronal plane (Figs 13–5 and 13–6). A focal area of hyporeflectivity or hyperreflectivity found only in one plane is likely due to anisotropy. The supraspinatus tendon, which appears as a curvilinear band on transverse images, is a triangular structure on coronal images. An imaginary line between the acromion and the humeral

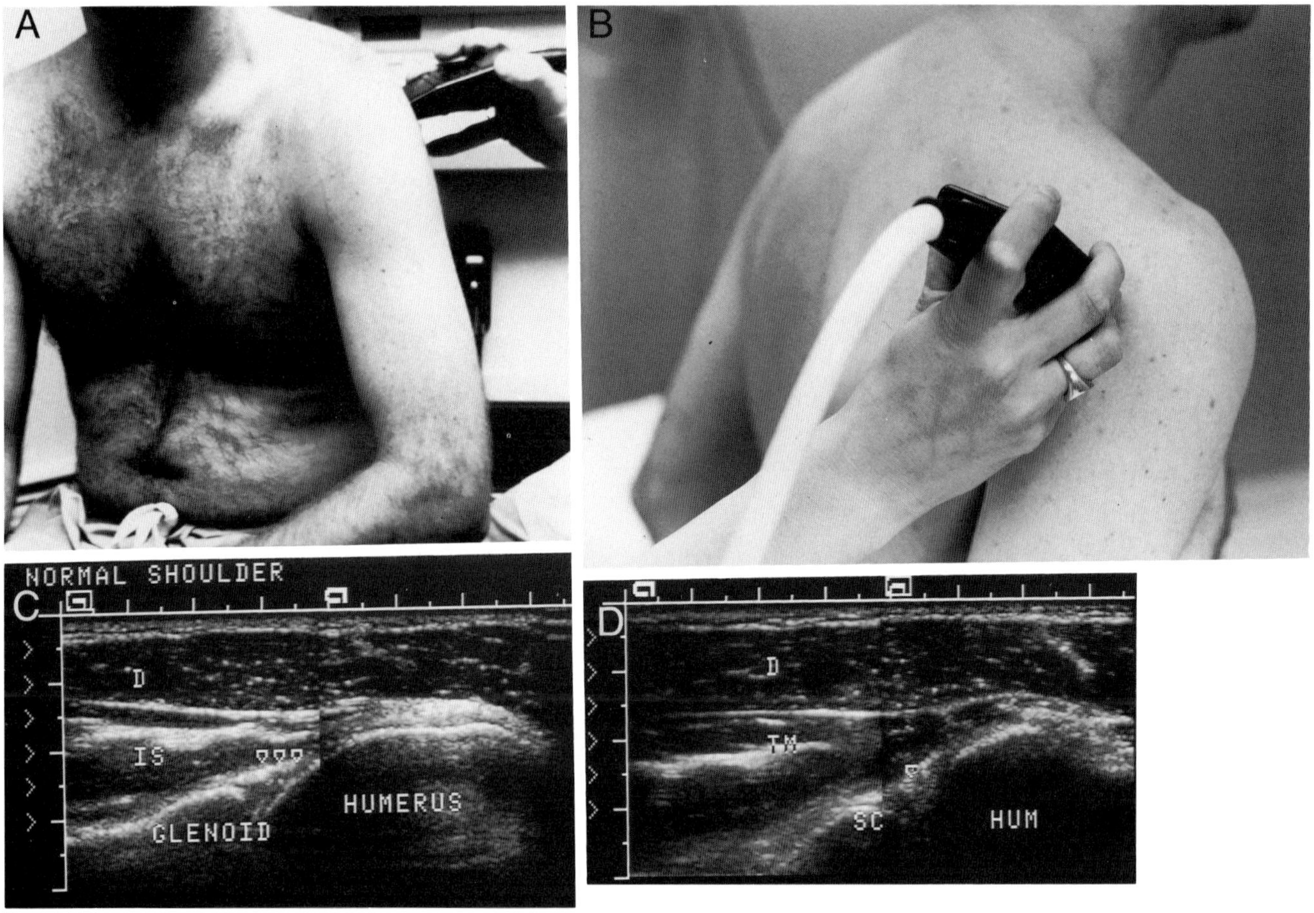


**FIG 13–7.**
**A** and **B,** sonography of the shoulder: posterior rotator cuff. Scanning of the posterior rotator cuff is performed in the transverse plane. The transducer is moved over the posterior aspect of the glenohumeral joint. This is the best approach to detect small intra-articular effusions. **C** and **D,** sonography of the shoulder: transverse sonogram of the posterior rotator cuff. **C,** the infraspinatus muscle and tendon *(IS)* are triangular. The hyperechoic fibrocartilage of the glenoid labrum *(arrowheads)* is well demonstrated. Muscle of the infraspinatus appears hypoechoic relative to the tendon. **D,** the teres minor *(TM)* is detected inferior to the infraspinatus. This muscle and tendon cover the inferior glenohumeral joint. *D* = deltoid; *SC* = scapula.

head forms the base of the triangle, and the apex is the insertion of the supraspinatus on the greater tuberosity.

After examination of the supraspinatus is completed, the posterior cuff is evaluated. This is best examined in the transverse plane (Fig 13–7). The transducer is placed over the posterior aspect of the shoulder at the level of the middle of the glenohumeral joint. In transverse images, the infraspinatus muscle and tendon are triangular shaped. The posterior labrum is seen as a hyperechoic triangle separating the infraspinatus tendon from the glenoid. Moving the transducer 2 to 3 cm inferiorly will demonstrate the teres minor. This muscle and tendon have a more rectangular appearance and form the posteroinferior aspect of the rotator cuff. The poste-

**TABLE 13–1.**
Diagnostic and Suggestive Signs for Rotator Cuff Tear

| |
|---|
| Signs diagnostic for a rotator cuff tear |
| Complete absence of the tendon |
| Focal atrophy of the tendon |
| Hypoechoic cleft in the tendon |
| Distended subdeltoid bursa with joint communication |
| Signs suggestive of a rotator cuff tear |
| Inhomogeneous areas in the cuff (hypoechoic or hyperechoic) |
| Increased amount of fluid in the biceps tendon sheath and subacromial-subdeltoid bursa |
| Small joint effusion |

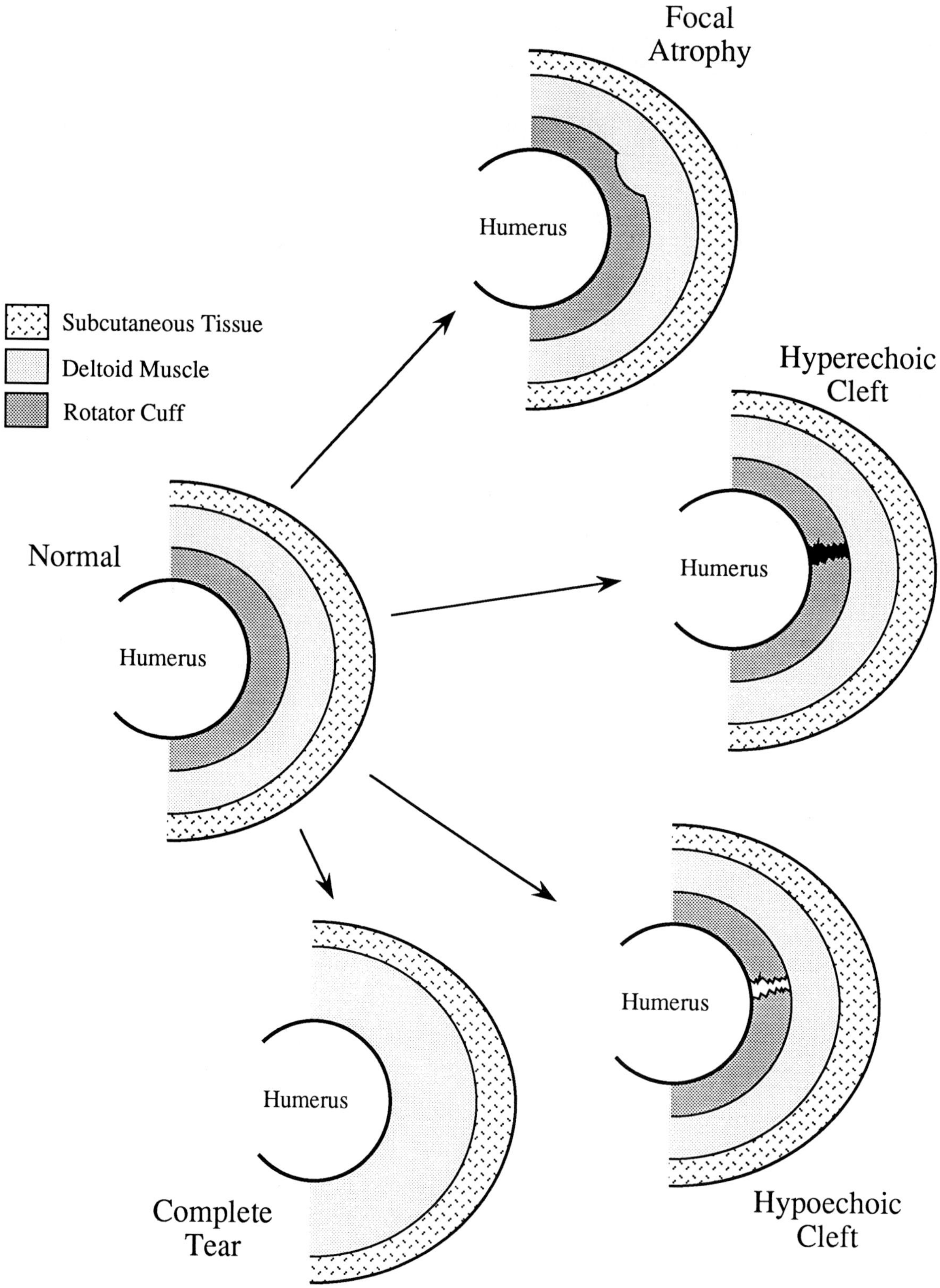


**FIG 13–8.**
Spectrum of rotator cuff tears.

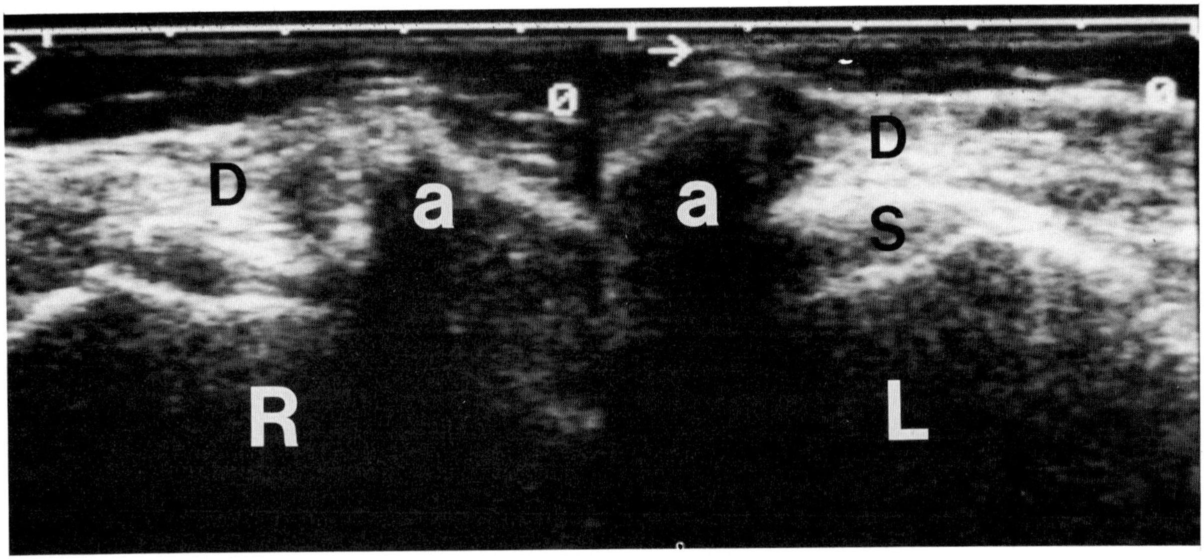


**FIG 13–9.**
Large rotator cuff tear: coronal sonograms. Complete absence of the supraspinatus is observed in the right shoulder *(R)*, whereas a normal supraspinatus *(S)* is seen on the left *(L)*. Abbreviations: acromion *(a)*, deltoid *(D)*.

rior portion of the cuff is torn only in the most severe rotator cuff injuries, but the examination is not complete without scanning this region. Small joint effusions are best detected from this posterior approach.

## SHOULDER PATHOLOGY

Rotator cuff tears and rotator cuff calcifications both cause inhomogeneities in the sonographic architecture of the cuff. Punctate echogenic foci with or without acoustic shadowing are consistent with calcification (Mack, 1988). The minority of rotator cuff tears are hyperechoic in appearance, and unlike calcifications, tears tend to be more linear. When evaluating the rotator cuff, we must distinguish between sonographic signs clearly indicative of rotator cuff tear (Fig 13–8) and signs suggestive of rotator cuff tear (Table 13–1) (Mack, 1988; Crass, 1987).

Large rotator cuff tears (Fig 13–9) show up sono-

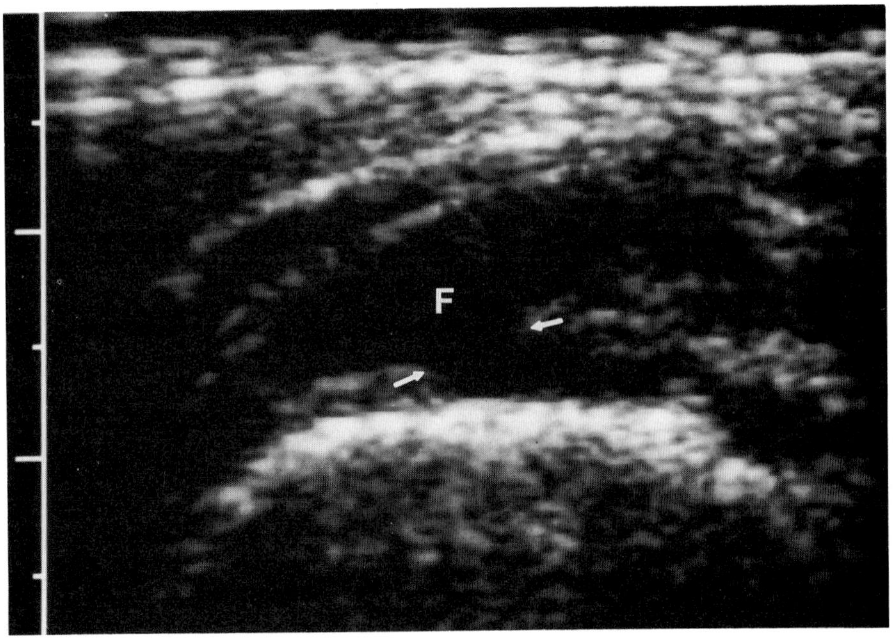


**FIG 13–10.**
Large rotator cuff tear: coronal sonogram. Fluid *(F)* replaces the torn cuff. The torn ends are indicated by *arrows*.

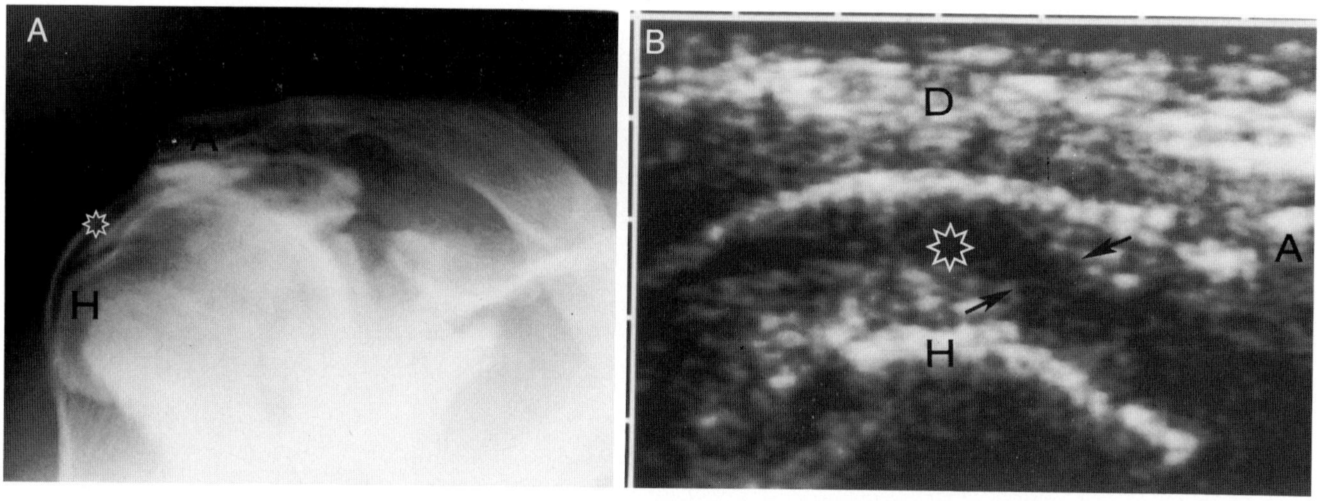


**FIG 13–11.**
**A,** large rotator cuff tear: shoulder arthrogram. The subacromial-subdeltoid bursa *(star)* fills with contrast material in this case of extensive rotator cuff injury. Acromion *(A)*, greater tuberosity *(H)*. **B,** large rotator cuff tear: coronal sonogram. Torn end of the rotator cuff *(arrows)* allows direct communication of the subacromial-subdeltoid bursa *(star)* with the joint space beneath the acromion *(A)*. Abbreviations: deltoid *(D)*, proximal humerus *(H)*.

graphically as complete absence of the supraspinatus tendon layer. Only the deltoid muscle layer is detected when the ultrasound transducer is placed adjacent to the acromion. In chronic disease, space between the acromion and the humerus is markedly decreased. This is best demonstrated with coronal images. The supraspinatus tendon may be replaced by a hypoechoic fluid layer (Fig 13–10), and often there is a small amount of fluid in the subacromial-subdeltoid bursa (Figs 13–11 and 13–12). This fluid is easily distinguished from the tendon by compression (Figs 13–13 and 13–14). The tendon is relatively rigid and deforms only minimally with compression. This simple test allows us to distinguish a hypoechoic tendon from fluid replacing a torn tendon.

Large tears in patients with arthritis, particularly those with rheumatoid arthritis, are often associated with significant distention of the subacromial-subdeltoid bursa. The synovium of the bursa is very irregular, and the lumen of the bursa is filled with

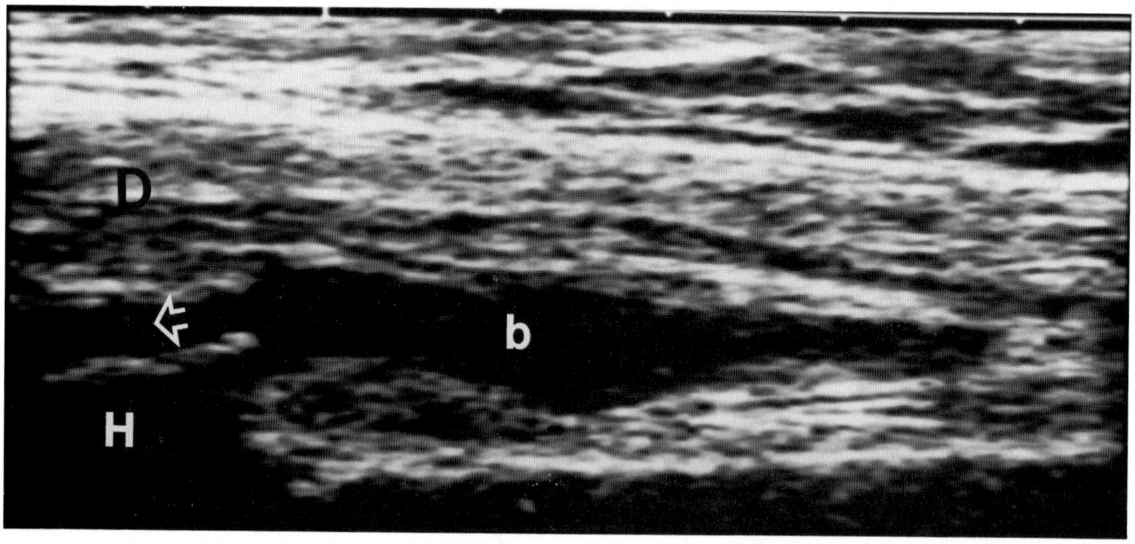


**FIG 13–12.**
Large rotator cuff tear: coronal sonograms. Communication of the fluid-filled subacromial-subdeltoid bursa *(b)* with the joint space is indicated by an *arrow*. Abbreviations: deltoid *(D)*, humeral head *(H)*.

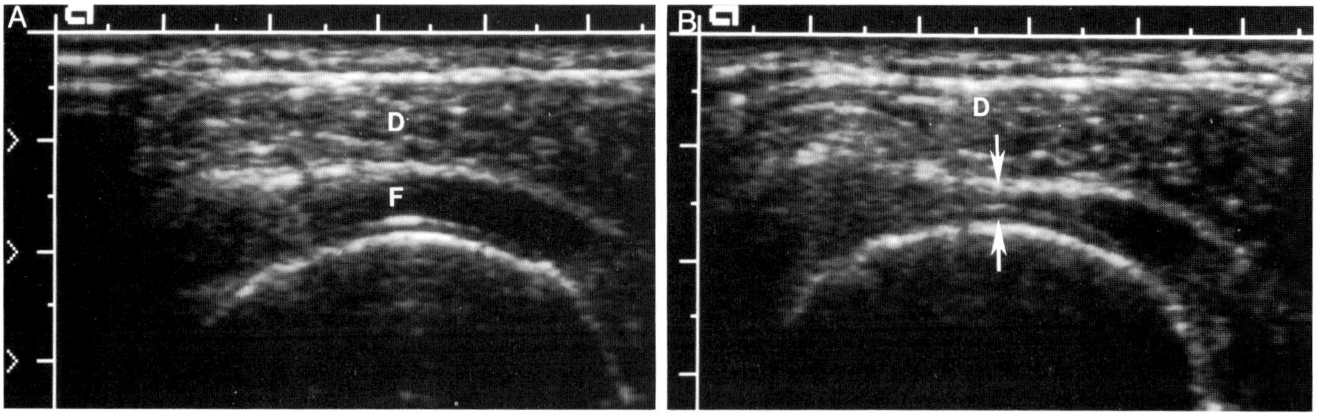


**FIG 13–13.**
**A,** large rotator cuff tear: transverse sonogram. A hypoechoic layer *(F)* is identified deep to the deltoid *(D)*. **B,** large rotator cuff tear: transverse sonogram with compression. To determine whether this hypoechoic layer was an abnormal tendon or a complete tear replaced by fluid, compression was applied. This image demonstrates displacement of fluid *(arrows)*.

echogenic material. These signs are nonspecific and can be seen in chronic infection. Ultrasound guided aspiration is always indicated in large shoulder effusions if infection is of clinical concern.

Small tears, 1 cm or less, are often harder to diagnose. If there is fluid present within the tear, the lesion will appear hypoechoic (Fig 13–15). However, when no fluid is present, the lesion will be observed as a linear echogenic band. The diagnosis of an echogenic band as a complete rotator cuff tear is made only when it extends through the entire thickness of the cuff (Fig 13–16). Small tears can manifest as focal atrophy of the cuff as well. Indentations (Fig 13–17) in the normal smooth curvilinear contour of the supraspinatus tendon must be regarded as retraction of tendon bundles, indicative of partial rotator cuff tear.

Many reports have appeared in the literature discussing shoulder sonography. One presents disappointing results (Brandt et al., 1989). All types of sonography are highly technique dependent. Eight percent of their examinations were performed using

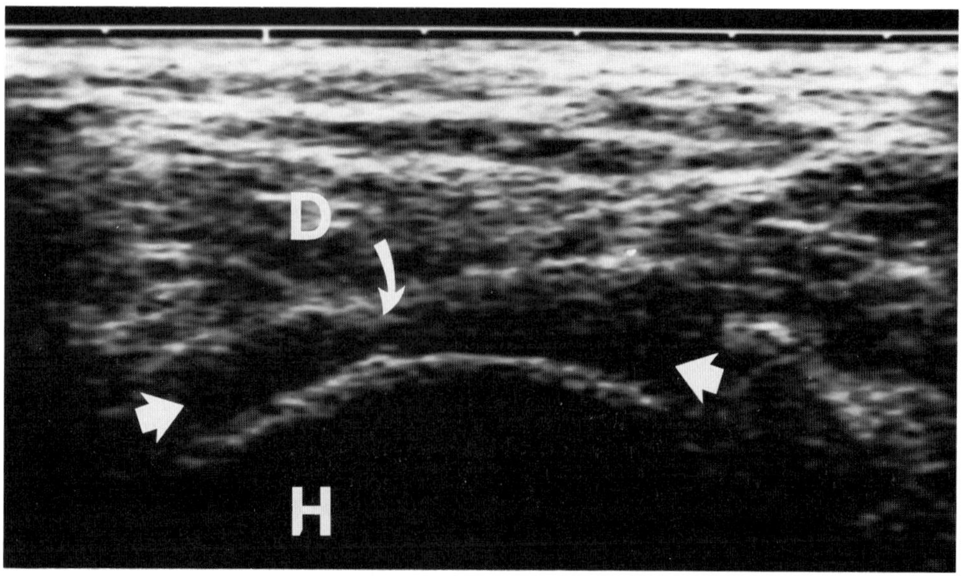


**FIG 13–14.**
Large rotator cuff tear: transverse sonogram. The entire supraspinatus tendon is absent and replaced by a thin fluid layer *(arrows)*. Slight pressure on the transducer deforms the fluid layer, which replaces the torn rotator cuff. Abbreviations: deltoid *(D)*, humerus *(H)*.

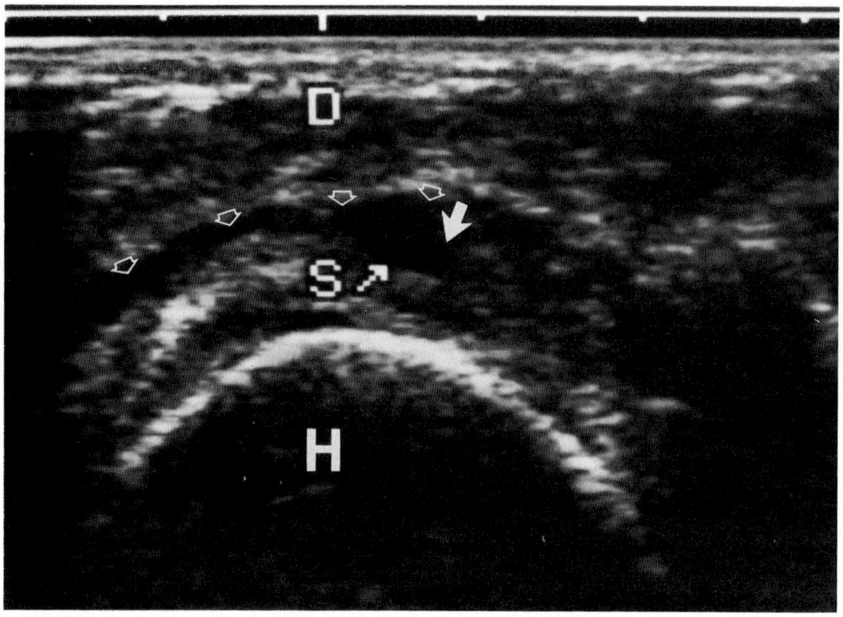


**FIG 13–15.**
Small rotator cuff tear: transverse sonogram. A hypoechoic tear is through the entire thickness of the supraspinatus tendon *(S)*. The tear is filled with anechoic fluid *(large arrow)*, which also mildly distends the subacromial-subdeltoid bursa *(open arrows)*. Abbreviations: deltoid *(D)*, humerus *(H)*.

mechanical sector scanners, which are generally unacceptable for shoulder sonography. In addition to the use of proper equipment, method of examination is important in obtaining reliable results. The real-time capability of ultrasound should be exploited to its fullest. Movement of the shoulder into full extension, internal and external rotation will demonstrate the greatest portion of the cuff, minimizing the portion obscured by bony structures. Recognition of imaging artifacts is essential. Failure to identify artifacts encountered when imaging a sphere and anisotropy of tendons will lead to misdiagnosis. This accounts for the difficulty encountered by some in the diagnosis of hyperechoic rotator cuff tears. Abnormalities should be confirmed in two imaging planes if there is question of imaging artifact. As in all areas of medicine, attention to detail significantly improves performance.

Ultrasound shows us more than just the rotator cuff. Tears of the surrounding muscles, including the deltoid, pectoral, brachial, and biceps muscles can be observed during the same examination. Lesions in any of these muscles may account for shoulder pain. Sonographic screening of these muscles is quickly

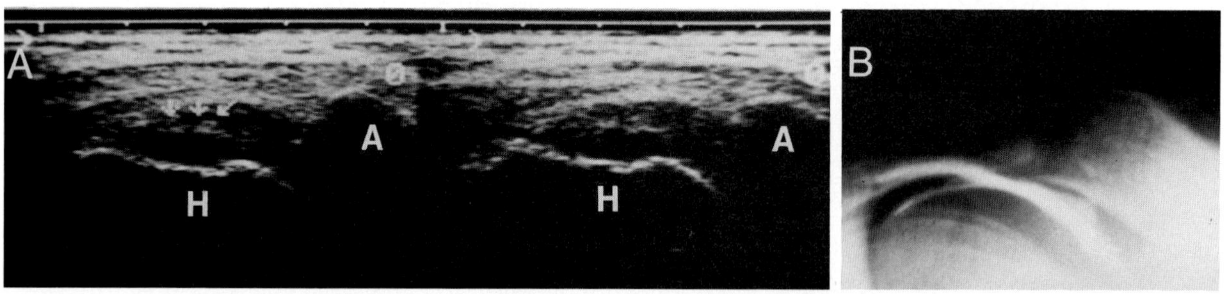


**FIG 13–16.**
**A,** small rotator cuff tear: coronal sonograms. This split screen displays two coronal images of the supraspinatus tendon, one slightly anterior to the other. A linear hyperechoic band is noted within the substance of the supraspinatus tendon on both images. Abbreviations: acromion *(A)*, humerus *(H)*. **B,** small rotator cuff tear: arthrogram with scapular Y view. A fine linear tear connects the subacromial-subdeltoid bursa with the joint space. Hyperechoic tears are generally quite small when seen on arthrography performed after sonographic examination. Abbreviations: humeral head *(H)*.

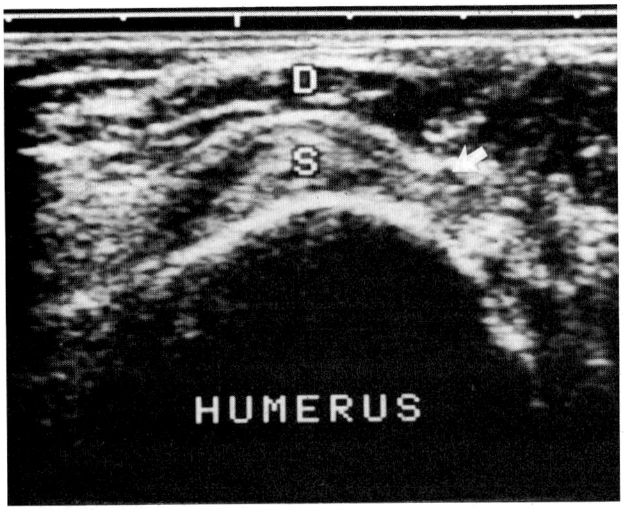


**FIG 13–17.**
Small rotator cuff tear: transverse sonogram. A region of focal atrophy *(arrow)* is noted in the outer surface of the supraspinatus tendon *(S)*. Abbreviations: deltoid *(D)*.

and easily performed as part of the routine shoulder examination.

Biceps tendon pathology is common in shoulder impingement (Figs 13–18 through 13–21). Characteristic findings are fluid filling of the tendon sheath and edema of the tendon, resulting in a hypoechoic appearance relative to a normal tendon. If fluid is found within the biceps tendon sheath, biceps tendinitis is the first consideration, but this fluid may also be secondary to intra-articular pathology. The presence of fluid elsewhere in the joint would exclude isolated biceps tendonitis. Rotator cuff tears may cause fluid accumulation within the subacromial-subdeltoid bursa. Arthritis of any etiology is often associated with fluid in all the synovial recesses of the shoulder. The most accessible recess of the glenohumeral joint is the posterior synovial recess (Fig 13–22). Elevation of the infraspinatus tendon more than 2 mm from the posterior glenoid labrum is a definite sign of effusion (Fig 13–23 and 13–24). Further investigation is then indicated, especially if a large cuff tear cannot be demonstrated. When the etiology (Figs 13–25 and 13–26) of a fluid collection cannot be established with any other technique, aspiration with culture must be performed to exclude infection.

Calcifications around the shoulder are common (Fig 13–27). As mentioned previously, not all calcifications are associated with acoustic shadowing. Size, location, transducer frequency, and focal zone are probably the most significant factors in establishing an acoustic shadow. Ultrasound may be helpful by supplementing plain films in cases of calcifications in unusual locations. Calcifications in the subscapular tendon are rare but very difficult to demonstrate on plain films. These calcifications are quickly demonstrated with sonographic shoulder examination (Fig 13–28).

In addition to evaluation of chronic injury, ultrasound is often valuable in cases of acute trauma to the shoulder (Fig 13–29). Ultrasound is quite helpful

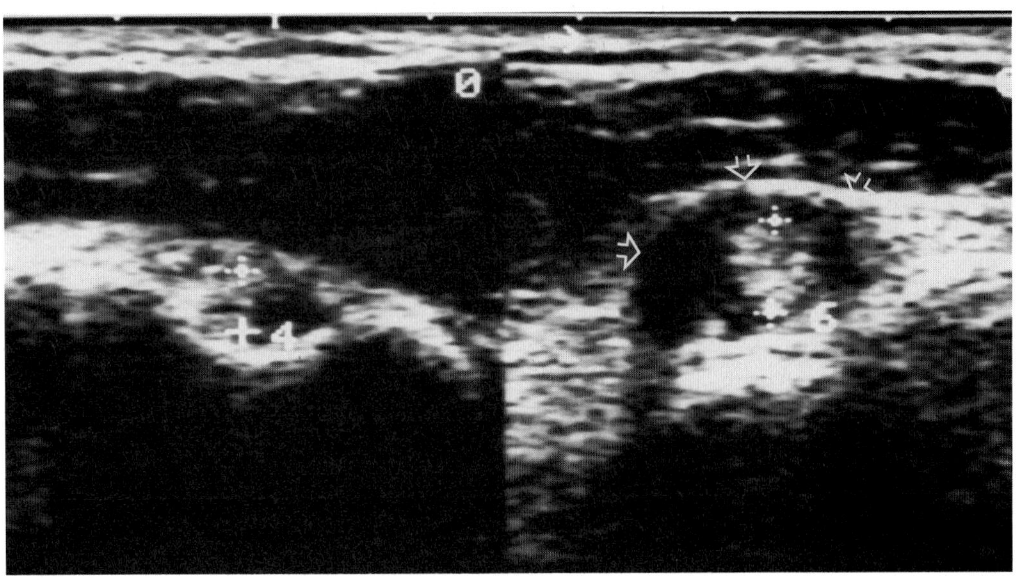

**FIG 13–18.**
Impingement syndrome: transverse sonogram. In this split screen image, an enlarged biceps tendon (+) surrounded by a fluid-filled sheath *(open arrows)* is noted on the *right*. The normal biceps tendon (+) is seen on the *left*.

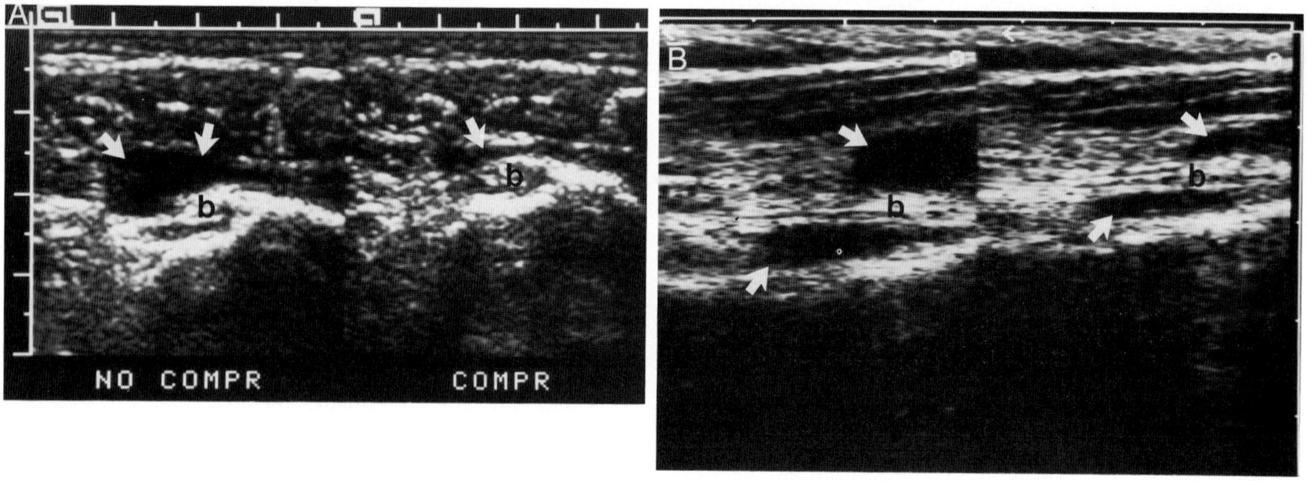


**FIG 13–19.**
**A,** impingement syndrome: transverse sonograms. Biceps tendonitis may manifest solely as fluid *(arrows)* within the tendon sheath. If there is any question of the presence of fluid, compression can be applied. Abbreviations: biceps tendon *(b)*. **B,** impingement syndrome: longitudinal sonograms. At its inferior extent, the biceps tendon sheath appears rounded when distended with fluid. Compression was applied in the image on the *right*. Abbreviations: biceps tendon *(b)*.

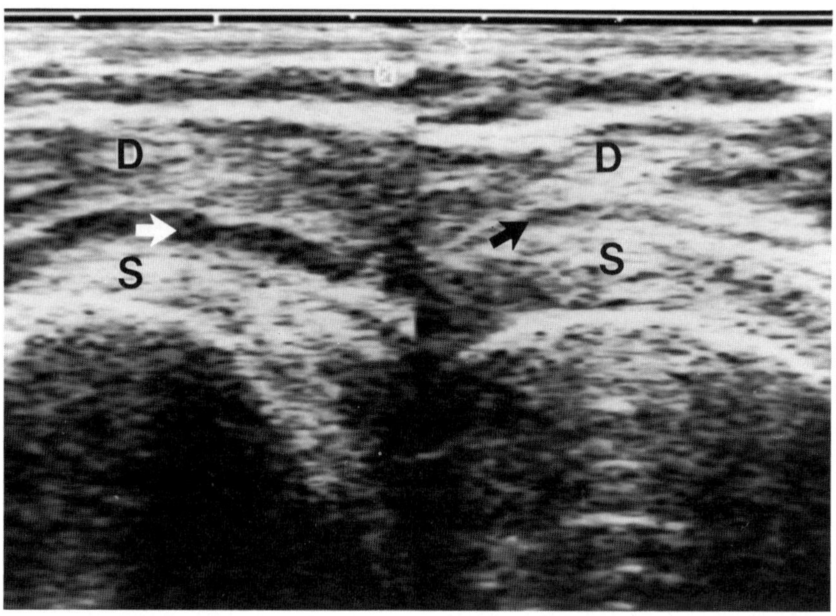


**FIG 13–20.**
Impingement syndrome: transverse sonograms. Fluid or hemorrhage may be seen within the subacromial-subdeltoid bursa *(arrows)* in impingement syndrome. There is no communication with the joint. Again, compression is applied in the image on the *right*. Abbreviations: deltoid *(D)*, supraspinatus *(S)*.

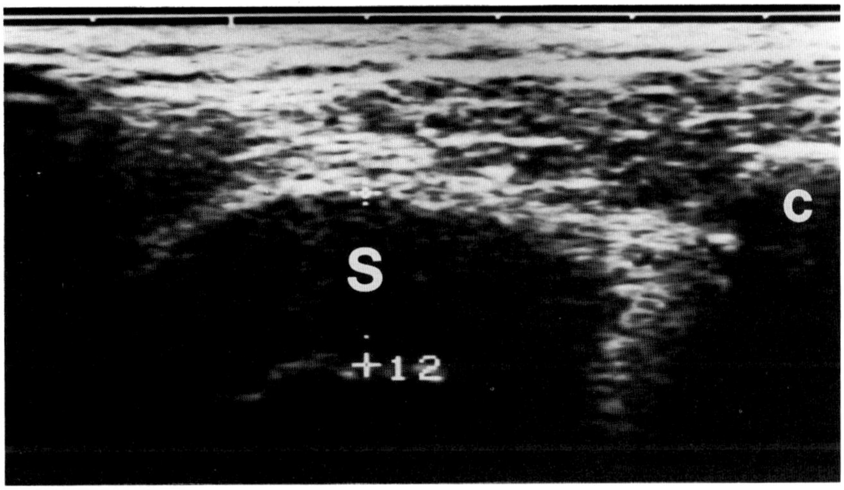


**FIG 13–21.**
Impingement syndrome: transverse sonogram in extreme internal rotation. The supraspinatus tendon *(S)* may become markedly thickened and hypoechoic in the early phase of impingement. Abbreviations: coracoid *(C)*.

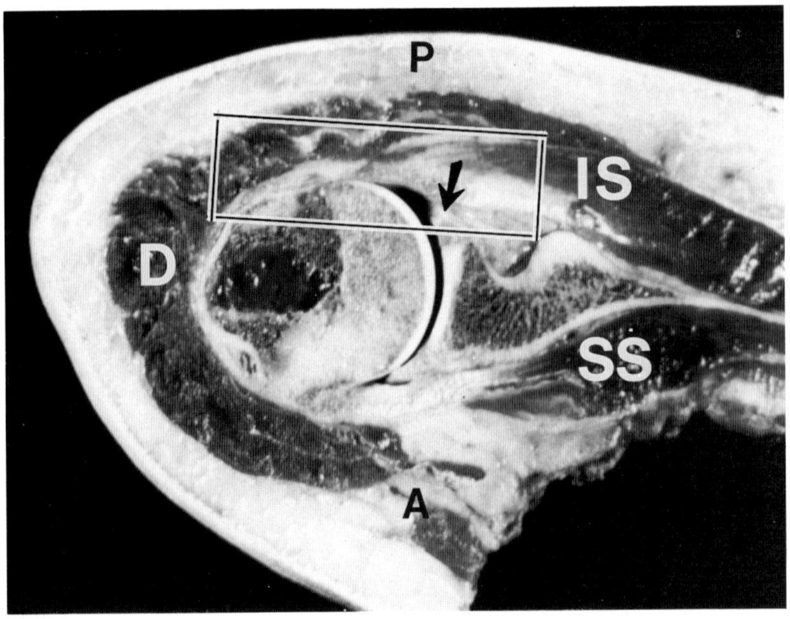


**FIG 13–22.**
Anatomy of the posterior synovial recess. The rectangular region indicated the best sonographic window for detection of fluid within the shoulder joint. Normally, no more than 2 mm of fluid is seen beneath the infraspinatus tendon. Abbreviations: deltoid *(D)*, glenoid labrum *(arrow)*, infraspinatus muscle *(IS)*, subscapularis *(SS)*, anterior *(A)*, posterior *(P)*.

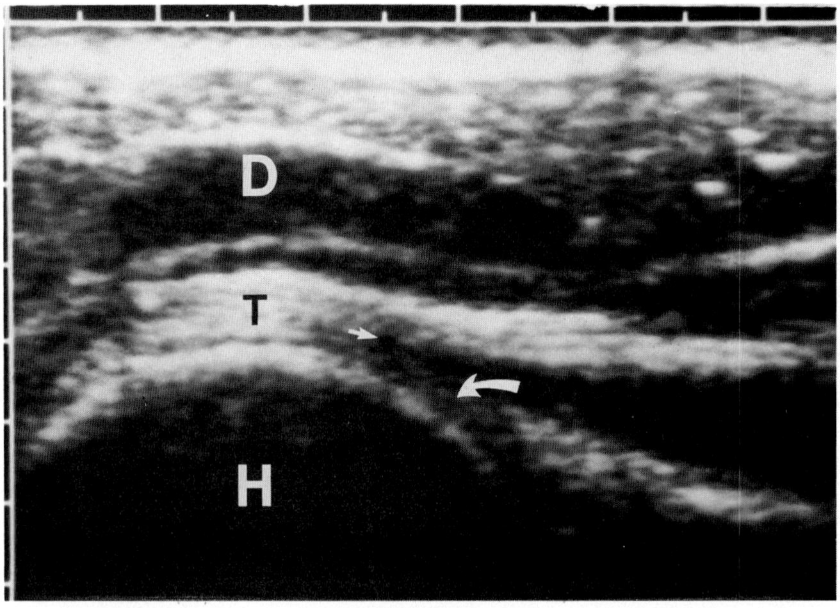


**FIG 13–23.**
Normal posterior synovial recess: transverse sonogram. Minimal fluid *(small arrow)* is seen between the infraspinatus tendon *(T)* and the glenoid labrum *(curved arrow)*. Abbreviations: deltoid *(D)*, humerus *(H)*.

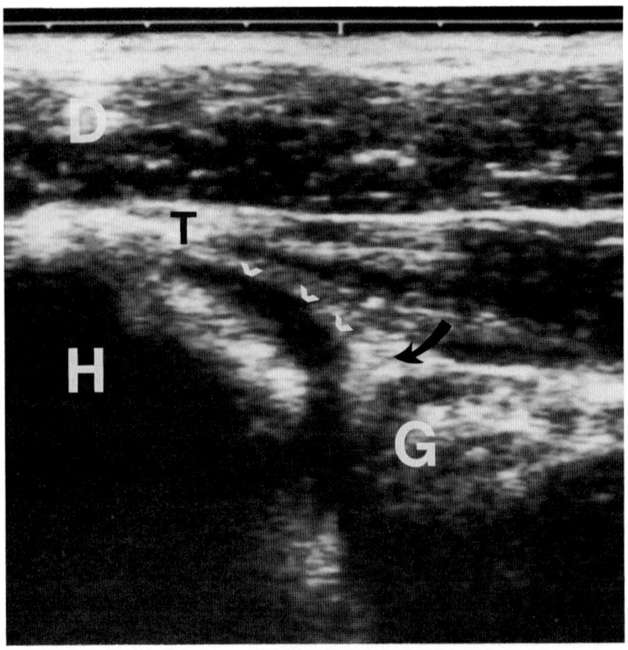


**FIG 13–24.**
Shoulder effusion: transverse sonogram. A fluid layer 5 mm thick *(arrowheads)* is seen between the glenoid labrum *(curved arrow)* and the infraspinatus tendon *(T)*. Abbreviations: deltoid *(D)*, humeral head *(H)*.

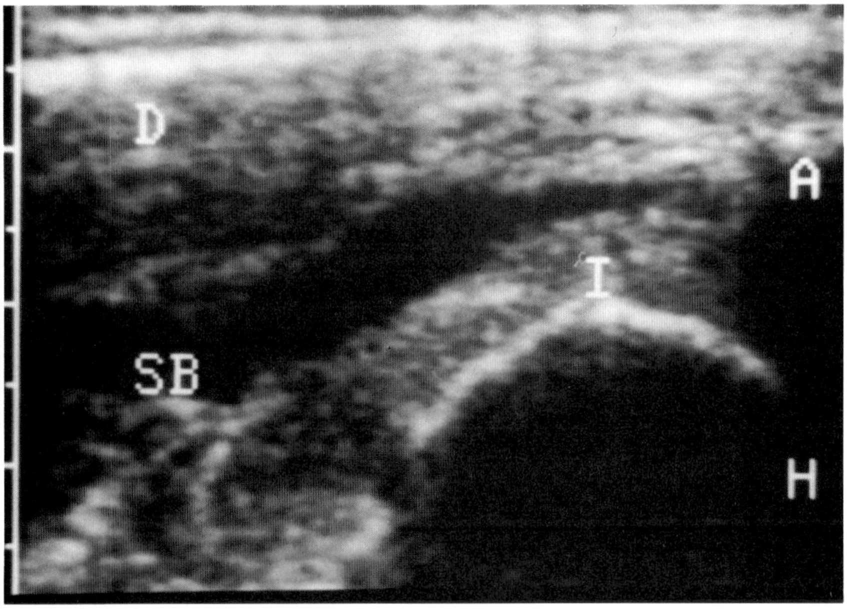


**FIG 13–25.**
Large effusion within the subacromial-subdeltoid bursa: transverse sonogram. This bursal fluid can easily be differentiated from joint effusion by its location dorsal to the infraspinatus *(I)*. Abbreviations: acromion *(A)*, deltoid *(D)*, humerus *(H)*, subacromial-subdeltoid bursa *(SB)*.

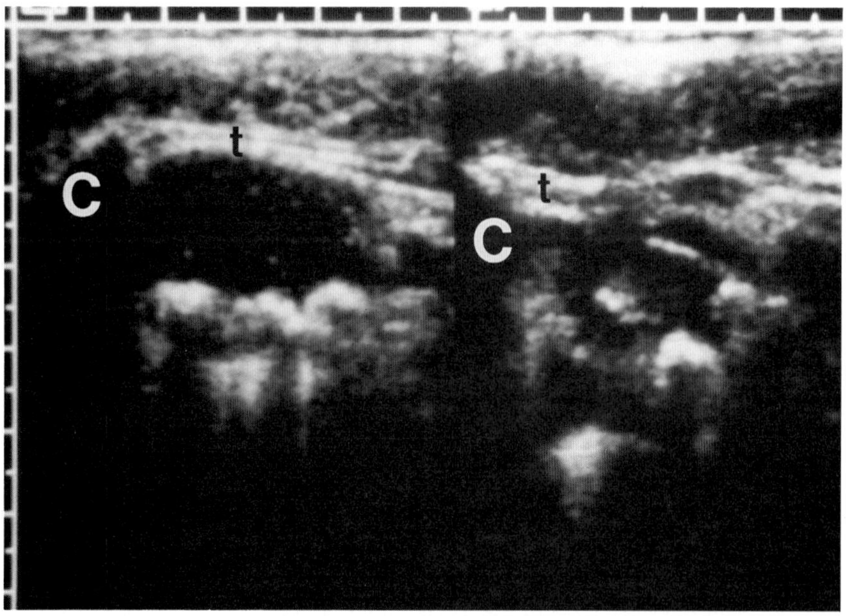


**FIG 13–26.**
Large effusion in the subcoracoid recess: longitudinal sonograms. The movement of numerous intra-articular loose bodies is demonstrated by imaging the patient while supine *(left)* and then upright *(right)*. The subcoracoid recess is noted deep to the short biceps tendon *(t)* and inferior to the coracoid process *(c)*.

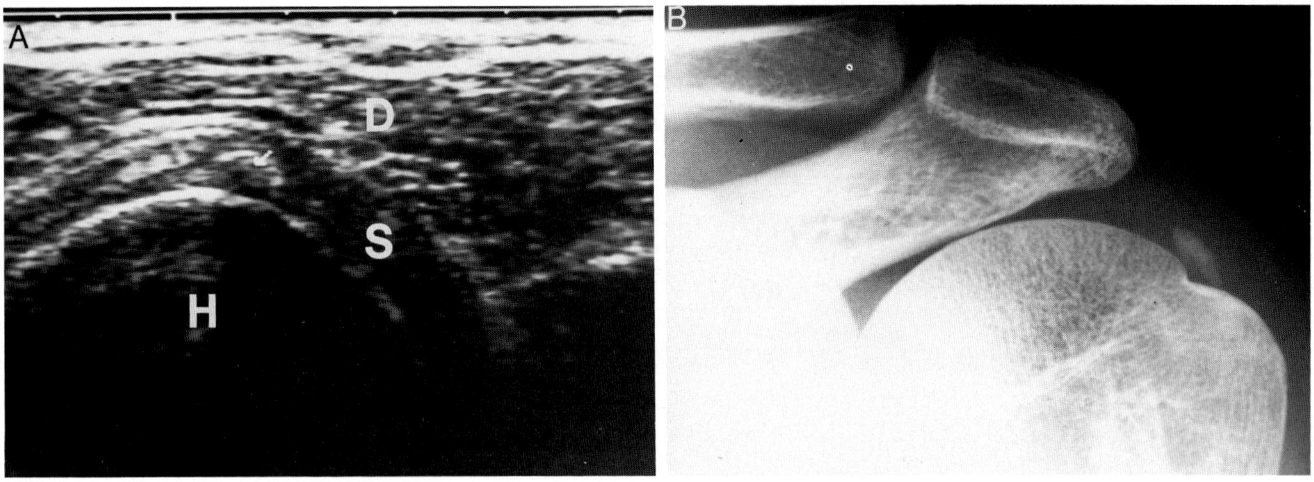


**FIG 13–27.**
**A,** supraspinatus tendon calcification: transverse sonogram. A 7-mm calcification *(arrow)* with incomplete acoustic shadowing is seen within the supraspinatus tendon. Abbreviations: deltoid *(D),* humerus *(H),* supraspinatus *(S).* **B,** supraspinatus tendon calcification: anteroposterior radiograph. Same patient as in **A.** Calcification is again noted in the region of the supraspinatus tendon.

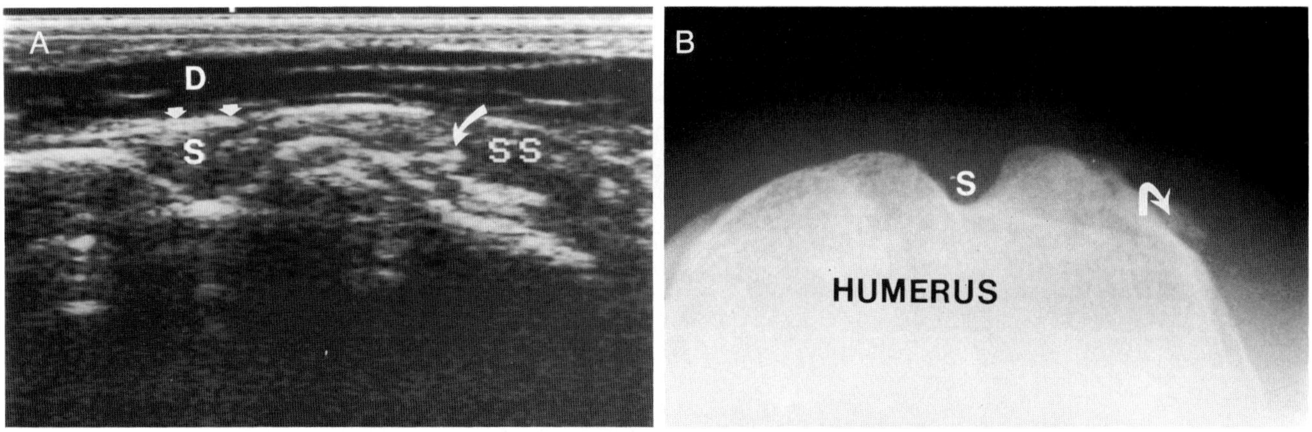


**FIG 13–28.**
**A,** subscapularis calcification: transverse sonogram. The patient is a 26-year-old woman with acute onset of severe pain in the right shoulder. Initial radiographic examination was normal. This transverse image of the subscapularis tendon *(SS)* revealed a small calcification *(curved arrow)* within the tendon. Abbreviations: bicipital groove *(S),* deltoid *(D),* transverse ligament *(small arrows).* **B,** subscapularis calcification: bicipital groove view. Same patient as in **A.** A tangential view was the only projection capable of demonstrating this calcification *(curved arrow)* due to its position so far medial to the bicipital groove *(S).*

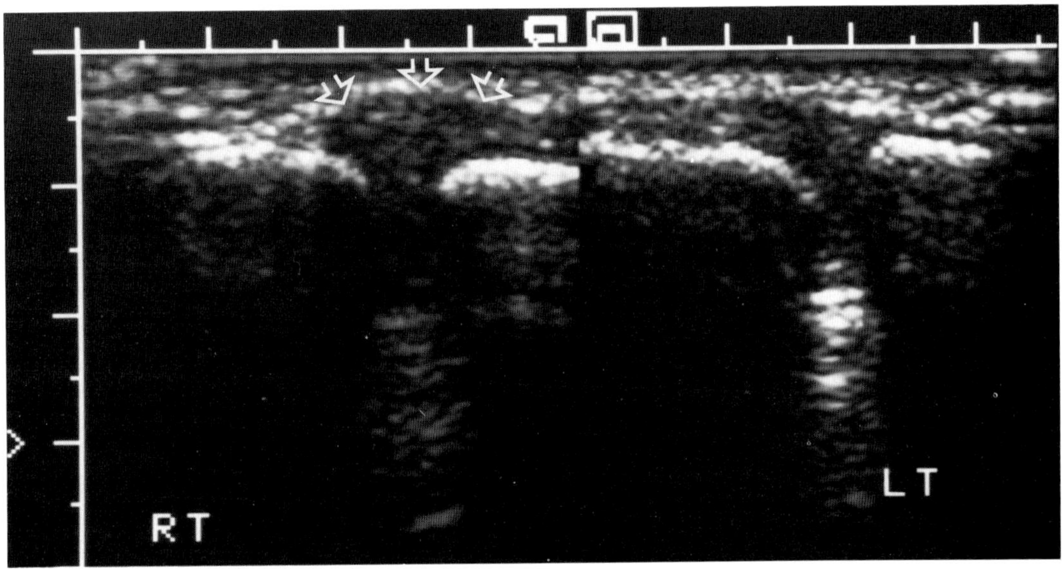


**FIG 13–29.**
Acromioclavicular disruption: coronal sonograms. This split screen provides side to side comparison of the acromioclavicular joints. Conventional radiographs with weighted views demonstrated no dislocation. Distention of the joint capsule *(open arrows)* of the right acromioclavicular joint is observed sonographically.

to examine for Hill-Sachs defects following anterior shoulder dislocation. Another application is the search for rotator cuff tears after greater tuberosity fractures. These tears are often linear and hyperechoic. Biceps tendon dislocation is uncommon as an isolated pathology, but biceps tendon dislocation is not unusual in large rotator cuff tears. It is a very difficult diagnosis to make clinically and on arthrograms. Many of these dislocations are intermittent. Therefore, it will often not be demonstrated on an MRI study. An ultrasound study in neutral position and in external rotation of the shoulder is the ideal technique to show the empty bicipital groove and the dislocated biceps tendon. Biceps tendon dislocation is invariably to the medial side, most commonly under the subscapular tendon (Petersson, 1986).

## CONCLUSION

Sonography is an exciting new technique for the evaluation of shoulder disease. Learning the technique demands a thorough knowledge of the anatomy of the rotator cuff. It is important to note that a "learning curve" exists with rotator cuff sonography (Mack et al., 1985). The novice should begin by performing sonographic examinations immediately following single contrast arthrography. Ultrasound is the ideal screening tool for rotator cuff pathology due to its noninvasive nature and widespread availability. It can easily be performed on the day of the patient's initial orthopedic consultation. We highly recommend ultrasound examination of the shoulder when plain radiographs are normal.

## REFERENCES

Ahovuo J: Single and double contrast arthrography in lesions of the glenohumeral joint. *Eur J Radiol* 1984; 4:237–240.

Bateman JE: Neurologic painful conditions affecting the shoulder. *Clin Orthop* 1983; 173:44–54.

Brandt TD, Cardone BW, Grant TH, et al: Rotator cuff sonography: A reassessment. *Radiology* 1989; 173:323–327.

Cofield RH: Rotator cuff disease of the shoulder. *J Bone Joint Surg [Am]* 1985; 67:974–979.

Crass JR: Ultrasonography of rotator cuff tears: A review of 400 diagnostic studies [abstract]. Presented at the American Roentgen Ray Society Annual Meeting, Miami Beach, Fl, 1987.

Depalma AF: Biologic aging of the shoulder, in AF De Palma (ed): *Surgery of the Shoulder* Philadelphia, JB Lippincott Co, 1983.

Gielen J, van Holsbeeck M, Hauglustaine D, et al: Growing bone cysts in long-term hemodialysis. *Skeletal Radiol* 1990; 19:43–49.

Golding FC: The shoulder, the forgotten joint. *Br J Radiol* 1962; 35(41):149–158.

Henderson CE, Hopson CN: Pneumoscrotum as a complication of arthroscopy. *J Bone Joint Surg [Am]* 1982; 64:1238–1240.

Jeffries JT, Gainor BJ, Allen WC, et al: Injury to the popliteal artery as a complication of arthroscopic surgery. *J Bone Joint Surg [Am]* 1987; 69:783–785.

Jobe FW, Jobe CM: Painful athletic injuries of the shoulder. *Clin Orthop* 1983; 173:117–124.

Lindenbaum BL: Complications of knee joint arthroscopy. *Clin Orthop* 1981; 160:158.

Mack LA, Matsen FA, Kilcoyne RF, et al: Sonographic evaluation of the rotator cuff. *Radiology* 1985; 157:205–209.

Mack LA: Sonographic evaluation of the rotator cuff. *Radiol Clin North Am* 1988; 26(1):161–177.

Neer CS: Impingement lesions. *Clin Orthop* 1983; 173:70–73.

Ogilvie Harris DJ, Wiley AM: Arthroscopic surgery of the shoulder. *J Bone Joint Surg [Am]* 1986; 68:201–207.

Petersson CJ, Gentz CF: Ruptures of the supraspinatus tendon. *Clin Orthop* 1983; 174:143–148.

Petersson CJ: Spontaneous medial dislocation of the tendon of the long biceps brachii. *Clin Orthop* 1986; 211:224–227.

Refior HF, Krödel A, Melzer C: Examinations of the pathology of the rotator cuff. *Arch Orthop Traumatic Surg* 1987; 106:301–308.

Resnick D, Niwayama G: *Diagnosis of Bone and Joint Disorders,* ed 2. Philadelphia, WB Saunders Co, 1988.

van Holsbeeck M, Introcaso J: Sonography of the postoperative shoulder. *AJR* 1989; 152:202.

# 14

# Sonography of the Elbow, Wrist, and Hand

Clinical examination of the elbow, wrist, and hand is much easier than examination of the shoulder. These joints are much more accessible due to their superficial nature. The majority of injuries in these regions involve bone and are therefore usually well evaluated with conventional radiographs. However, ultrasound is valuable in the diagnosis of muscle, tendon, and ligament pathology. When radiographic examination is normal, ultrasound is indicated to evaluate persistent pain and swelling.

## THE ELBOW

The most common type of soft tissue pathology of the elbow is epicondylitis. Tendonitis of the proximal radial extensors (tennis elbow) and proximal ulnar flexors (golfer's elbow) are often confused with joint pathology due to their close proximity. Lateral epicondylitis is encountered more frequently. Biceps and triceps tendonitis may also be mistaken for joint pathology. These entities are clearly best evaluated with ultrasound because calcifications are seen only in advanced disease. At the same time, the examiner can also evaluate for evidence of intra-articular pathology. Joint effusion, although nonspecific, is a clear indicator of joint pathology. The most frequently observed joint pathology in the absence of radiographic findings is cartilaginous loose bodies. Often these loose bodies will be found in the anterior joint recess. The algorithm presented in Figure 14–1 provides a rational approach for imaging patients with chronic elbow pain.

Examination of the elbow is initiated by evaluation of the anterior synovial recess. If no fluid is found within the anterior synovial recess, intra-articular pathology is unlikely. Figure 14–2,A demonstrates optimal patient and transducer positioning for imaging the anterior synovial recess. The patient sits facing the examiner with the arm lying slightly flexed on the examination table. Imaging is performed with the transducer in the long axis of the arm. Normally, a fluid film no greater than 1 mm is identified within the joint (Fig 14–2,B). If a joint effusion is present and loose bodies are not identified, the posterior joint recess should be examined with the arm in 90 degrees of flexion. Examination of the medial and lateral epicondyles is performed next (Fig 14–3). Careful side to side comparison of dimension and echogenicity of the tendons is most important. The patient places the palmar surfaces of the hands together with the arms extended on the examination table. The tendon origins appear as hyperechoic triangles. Right-left comparison is absolutely necessary to detect slight differences in dimension. Tendinitis is characterized by tendon swelling and inhomogeneity, as elsewhere in the extremities. Examination of the elbow is completed by evaluation of the biceps and triceps tendons. Figures 14–4 through 14–7 demonstrate a variety of pathology in patients who presented with generalized elbow pain and swelling.

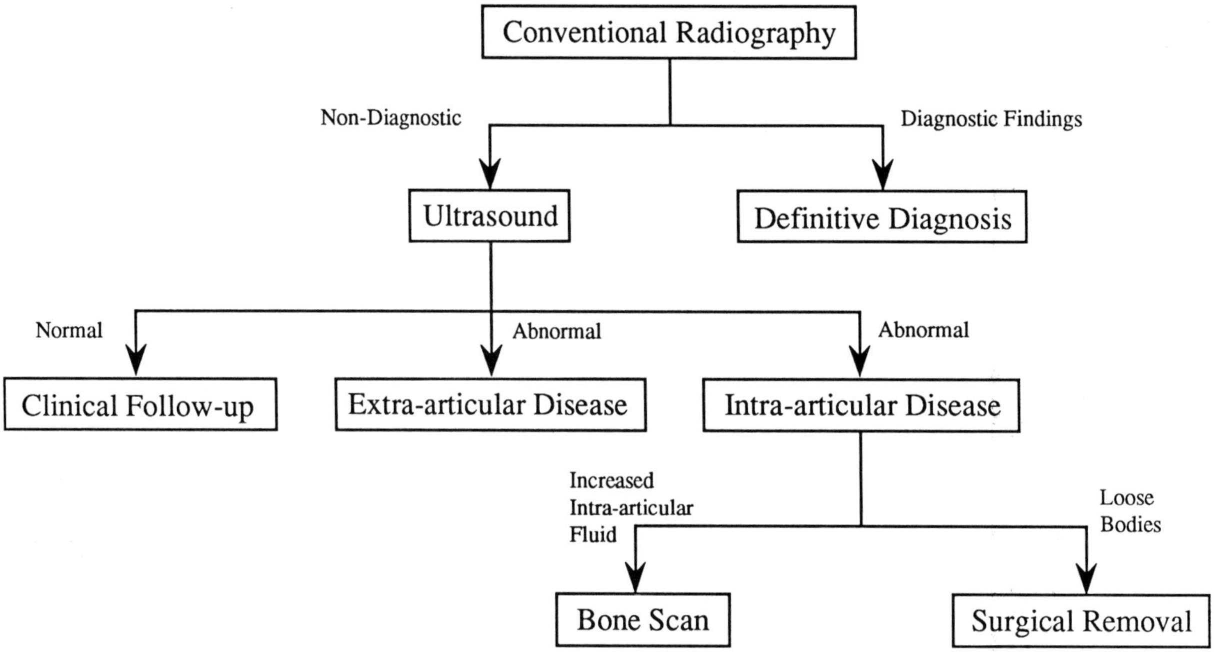


**FIG 14–1.**
Algorithm for diagnostic imaging of the painful or swollen elbow.

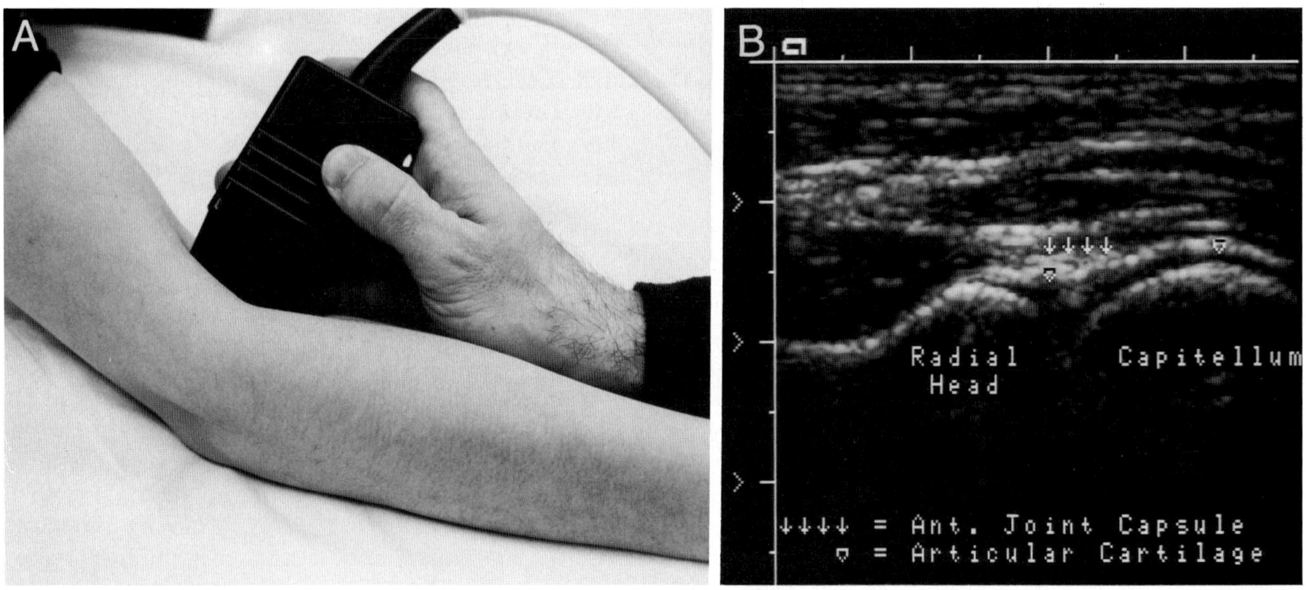


**FIG 14–2.**
**A,** sonography of the elbow: anterior synovial recess, transducer position. **B,** sonography of the elbow: longitudinal sonogram of the anterior synovial recess. The anterior joint capsule appears hyperechoic relative to the cartilage *(arrowheads)* and thin film of joint fluid surrounding the subchondral bone.

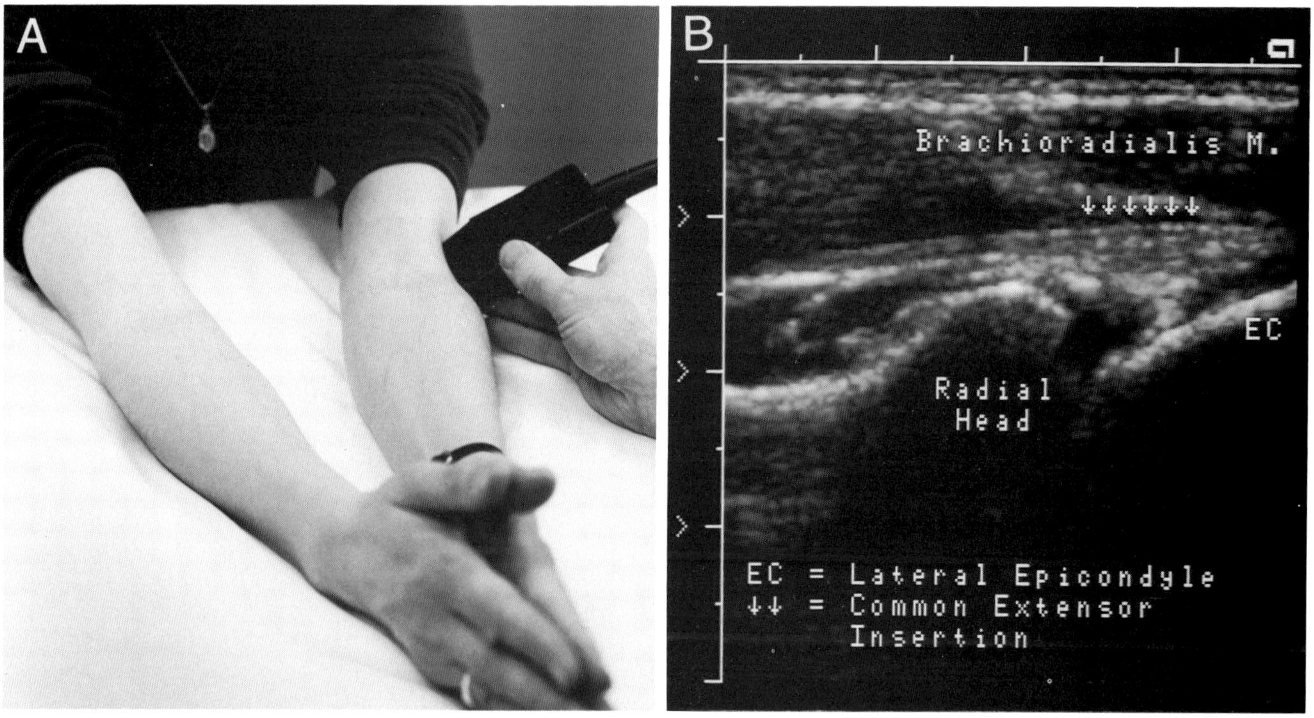


**FIG 14–3.**
**A,** sonography of the elbow: common extensor tendon insertion. Positioning of the forearms in this manner is helpful for side to side comparison of the elbow tendons. **B,** sonography of the elbow: coronal sonogram of the common extensor tendon insertion. The common tendon insertion appears as a hyperechoic triangle.

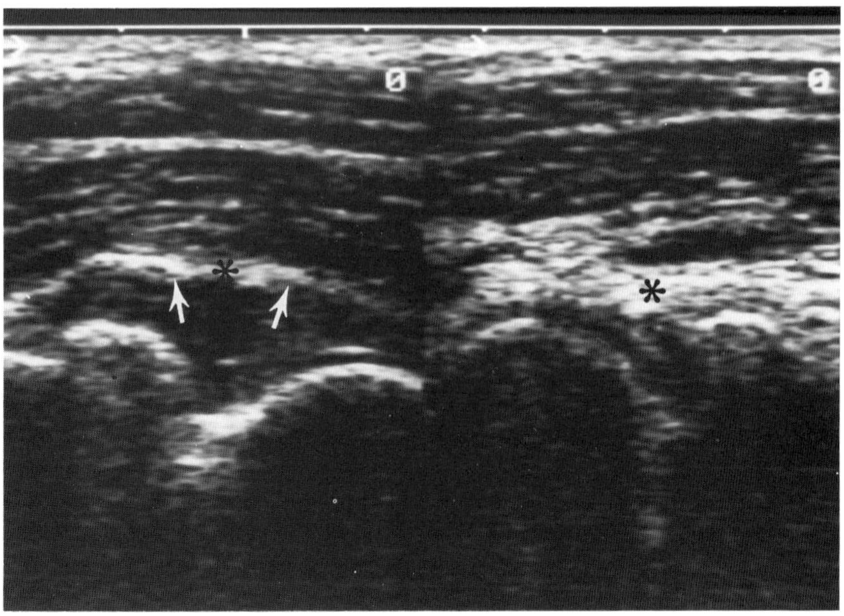

**FIG 14–4.**
Elbow effusion: longitudinal sonogram. Joint effusion in a patient with a hairline fracture of the proximal radius. Compare the displaced capsule *(arrows)* on the *left* with the normal capsule *(asterisk)* on the *right.*

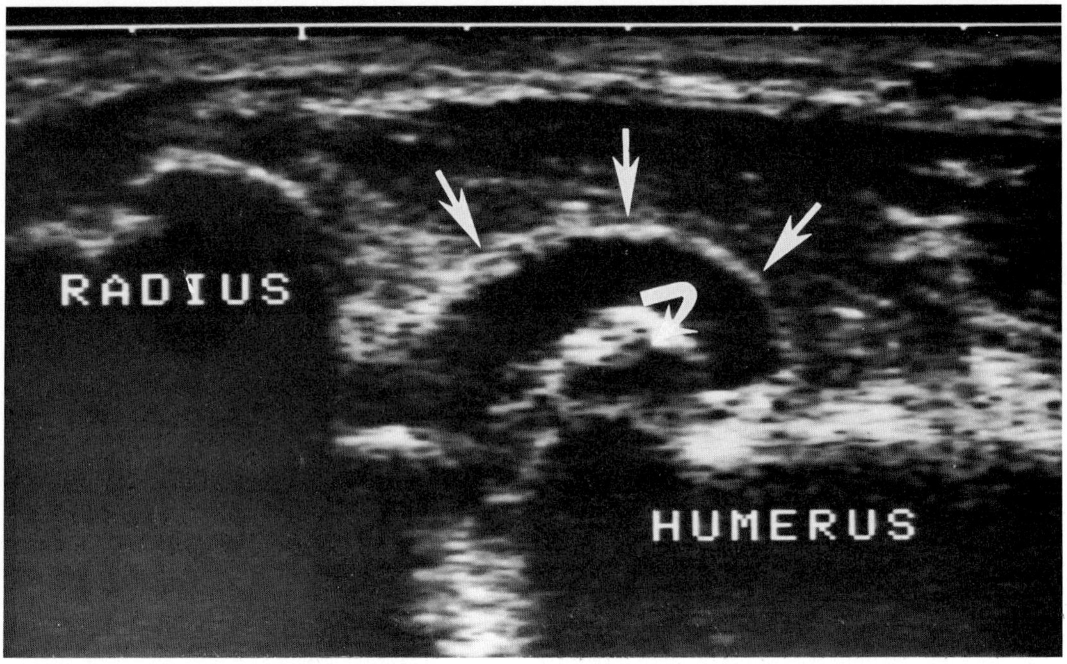


**FIG 14–5.**
Elbow effusion with intra-articular loose bodies: longitudinal sonogram. Joint effusion with elevation of the anterior joint capsule *(arrows)* due to an intra-articular loose body *(curved arrow)*.

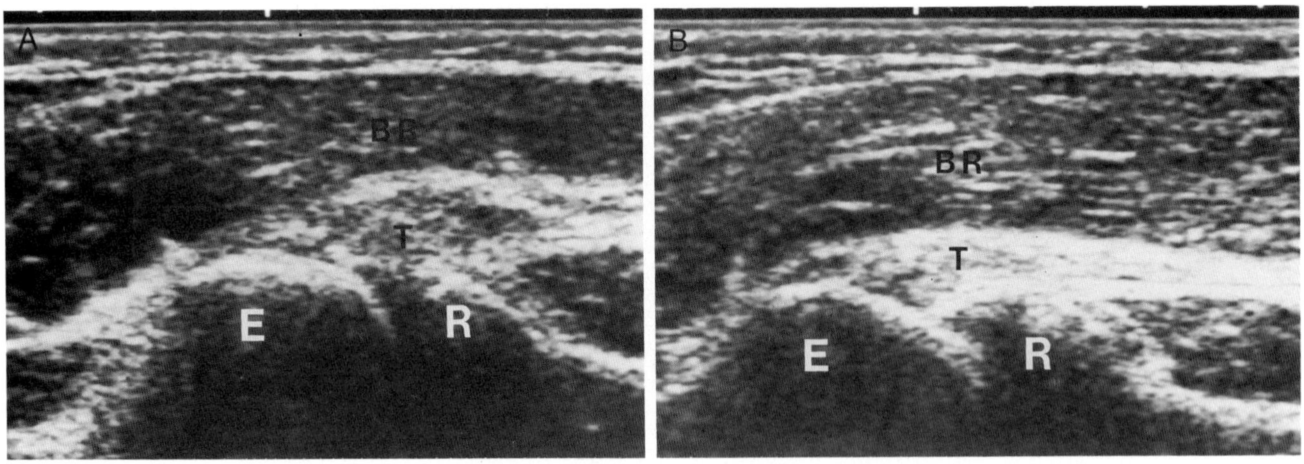


**FIG 14–6.**
**A,** radial epicondylitis: coronal sonogram. The tendon insertion *(T)* of the radial extensor muscles is swollen and inhomogeneous. Abbreviations: brachioradialis muscle *(BR)*, lateral epicondyle *(E)*, proximal radius *(R)*. **B,** normal contralateral side.

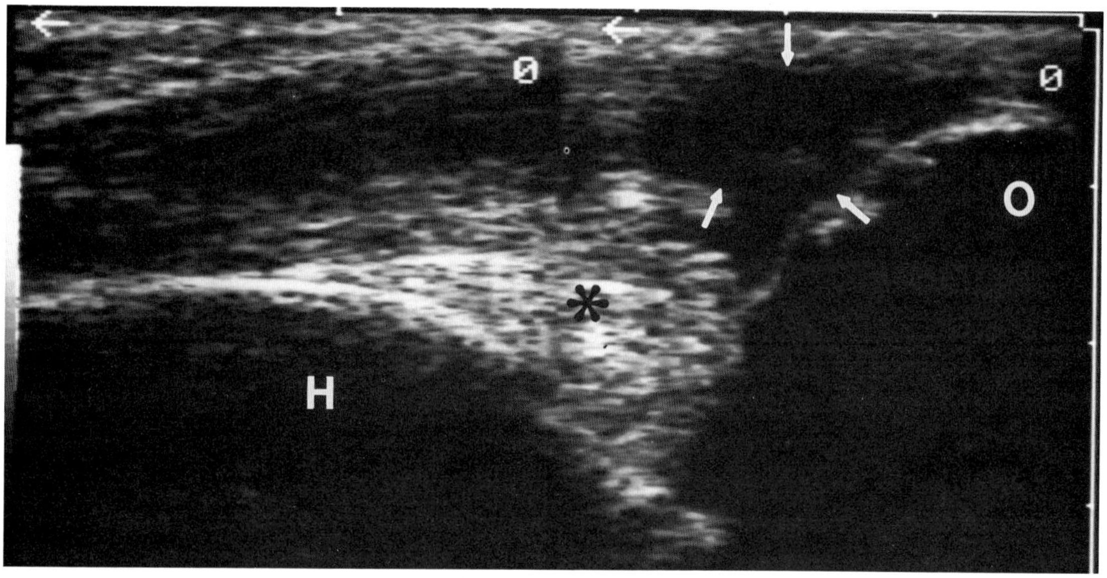


**FIG 14–7.**
Triceps tendon rupture: longitudinal sonogram. The triceps tendon has been avulsed and is retracted from the olecranon *(O)*. The space vacated *(arrows)* is filled with fluid. Fat of the olecranon fossa *(asterisk)* of the distal humerus *(H)* appears hyperechoic.

## THE WRIST AND HAND

The principle indications for ultrasound examination of the wrist and hand are evaluation of tendon pathology and identification of the origin of swelling. Tendon ruptures are easily detected on longitudinal images. Real-time examination of the tendons of the hand is valuable in the diagnosis of tendon dislocation and entrapment. Tenosynovitis of the flexor tendons of the wrist can result in carpal tunnel syndrome. Although this is one of the less common causes of carpal tunnel syndrome, it is important to identify due to the greater likelihood of success with conservative therapy. Diagnosis of synovial cysts and identification of their origin is possible with ultrasound. Surgical management of a cyst arising from a tendon sheath differs from that of a cyst originating from a joint. Often these cysts are very hard and may be mistaken for bony hypertrophy on physical exam. The triangular fibrocartilage of the wrist may also be examined. Tears of the triangular fibrocartilage result from rotational trauma during pronation and supination. Sonographic evaluation of the joints of the carpus is limited due to the very tightly fitting joint cap-

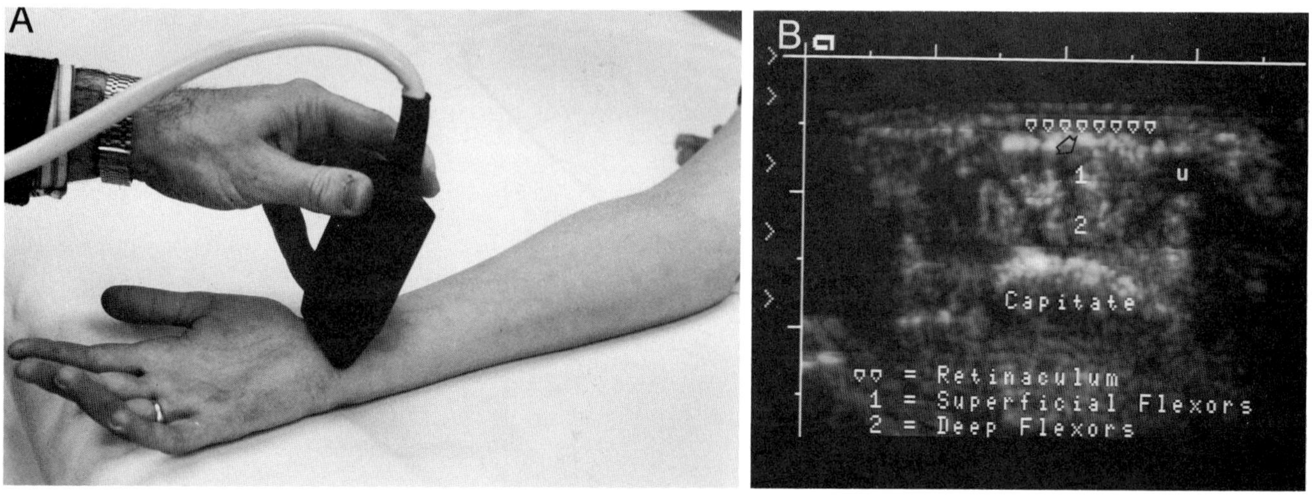


**FIG 14–8.**
**A,** sonography of the wrist: carpal tunnel examination. **B,** carpal tunnel sonogram. Median nerve, *open arrow*.

sule. Evaluation of these joints is best performed by radiographic examination and arthrography. Examinations of the wrist and hand performed for diffuse pain are generally low-yield studies. This is in marked contrast to sonographic examination of the shoulder, where investigation of nonspecific complaints will usually establish the cause. Sonographic examination of the wrist is most beneficial in investigation of a specific clinical question.

Examination of the hand and wrist is ideally performed using 10-MHz linear array transducers. These are currently available on only the best state of the art machines. A 7.5-MHz linear array will provide adequate visualization of hand and wrist anatomy. Often, a standoff pad must be utilized to visualize these superficial structures. An abundance of sonographic gel facilitates transverse imaging of the tendons of the fingers as a group. Transverse imaging of the tendons is performed first. This is essential for the identification of swelling of the tendons and tendon sheaths. Measurement of the transverse diameters of tendons allows easy identification of tendon swelling. When involvement of multiple adjacent tendons is suspected, comparison with the asymptomatic side is diagnostic. Evaluation of the echogenicity of tendons of the wrist, as elsewhere, must be performed with the beam perpendicular to the tendon surface. The anisotropic properties of tendons are most noticeable in the wrist (Fig 2–6). Figure 14–8 is a transverse image through the carpal tunnel. The flexor retinaculum is seen as a hyperechoic line, indicated by the row of arrowheads. Two rows of flexor tendons, superficial (1) and deep (2), are identified within the carpal tunnel. The median nerve is an ovoid structure in the middle of the carpal tunnel immediately beneath the flexor retinaculum (open arrow). It is slightly less echogenic than the surrounding tendons. Flexion and extension of the fingers during real-time examination will remove any doubt about which structure is the median nerve. The triangular fibrocartilage is then located by placing the index finger of the examiner's free hand over the tip of the ulnar styloid. The transducer is positioned immediately proximal to the examiner's index finger. If the cartilage of the ulnar head is seen, the transducer is slowly moved distally until the triangular fibrocartilage is visualized (Fig 14–9). After evaluation of the tendons and triangular fibrocartilage on transverse images, continuity of the tendons is examined with longitudinal images (Fig 14–10). Again, dynamic real-time examination easily differentiates tendons from nerves. Extensor tendons are examined the same way as flexor tendons.

Figures 14–11 through 14–13 demonstrate a spectrum of pathology seen in patients who presented with pain and swelling over the palmar surface of the wrist. Pathology seen on the dorsal aspect of the wrist and hand is seen in Figures 14–14 through 14–16. Extensor tendon dysfunction is seen in Figures 14–17 through 14–19.

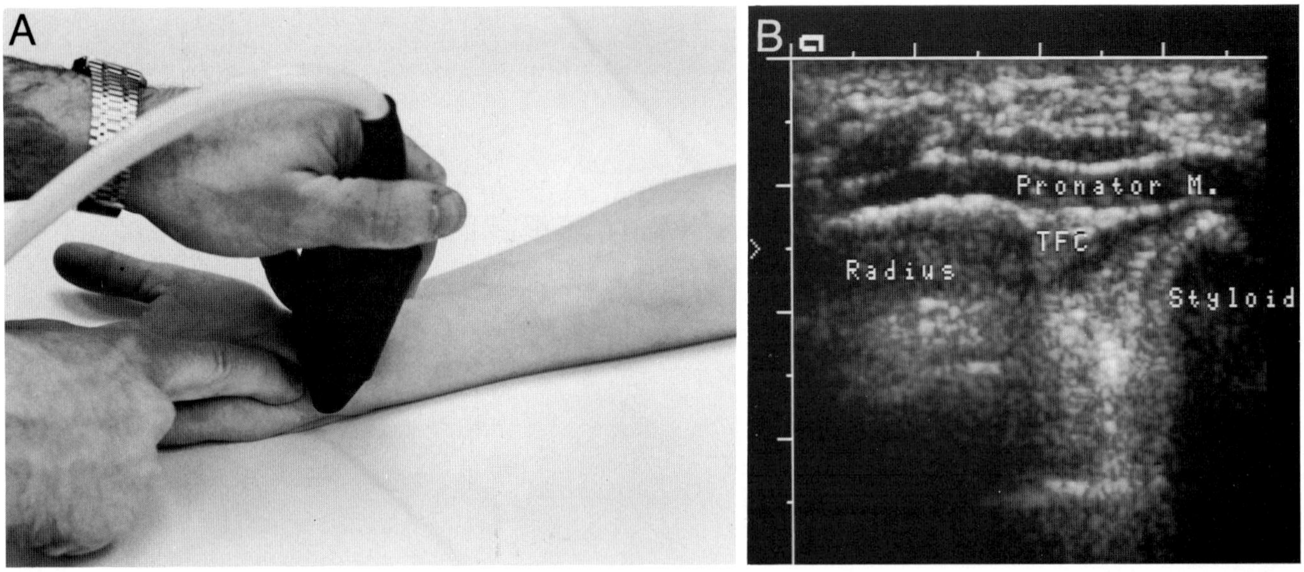


**FIG 14–9.**
**A,** sonography of the wrist: the triangular fibrocartilage. **B,** the triangular fibrocartilage *(TFC)* appears hyperechoic relative to the pronator quadratus muscle.

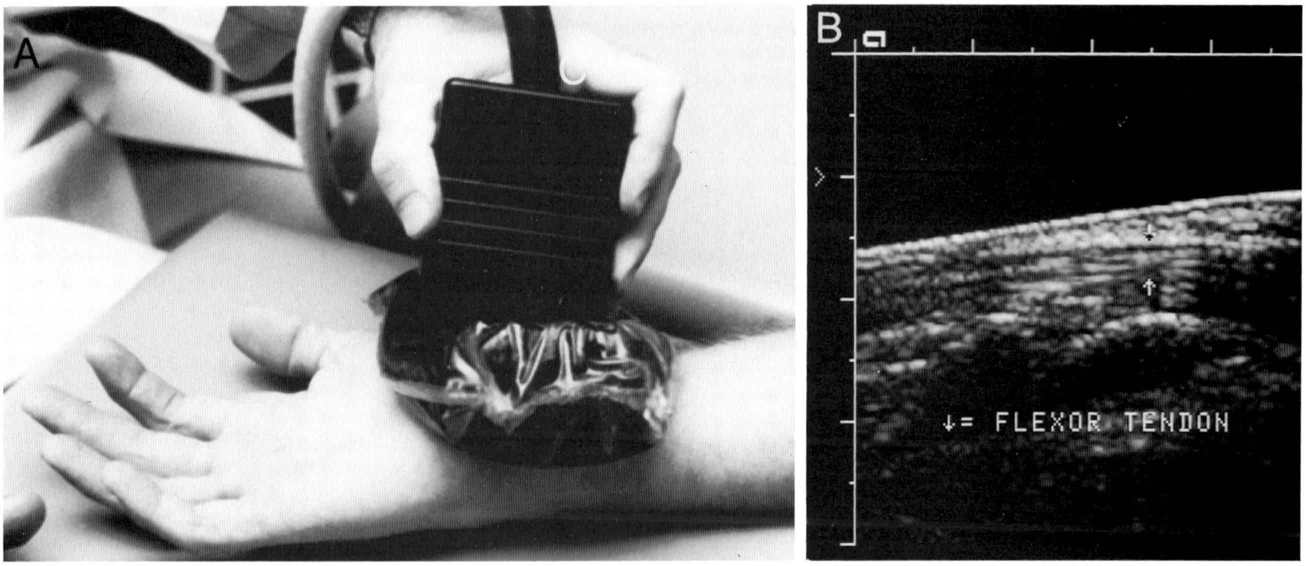


**FIG 14–10.**
**A,** sonography of the wrist: tendon continuity. **B,** the fibrillar structure of a tendon can be seen only in the longitudinal axis. This longitudinal image demonstrates a synovial sheath containing a fluid film less than 1 mm thick.

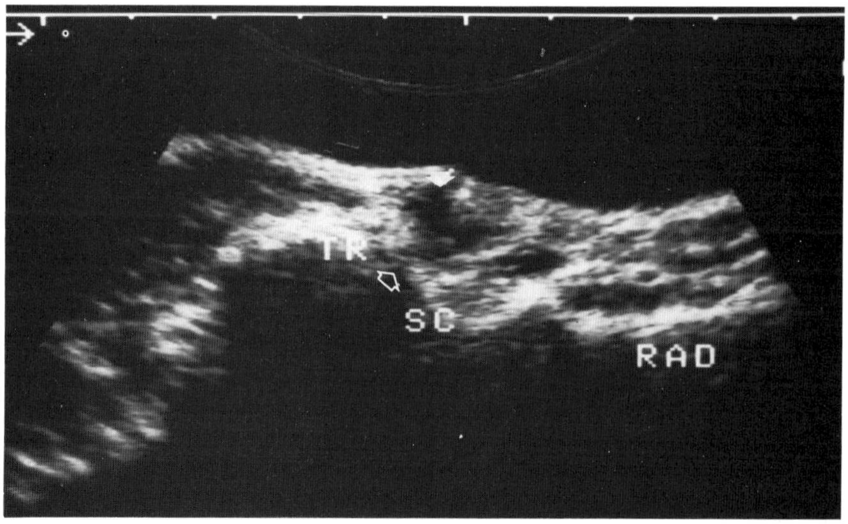


**FIG 14–11.**
Wrist cyst: longitudinal sonogram. This wrist cyst *(white arrow)* communicates *(open arrow)* with the midcarpal joint between the trapezium *(TR)* and scaphoid *(SC)*. Distal radius *(RAD)*. The origin of a synovial cyst is important due to the different surgical approaches utilized.

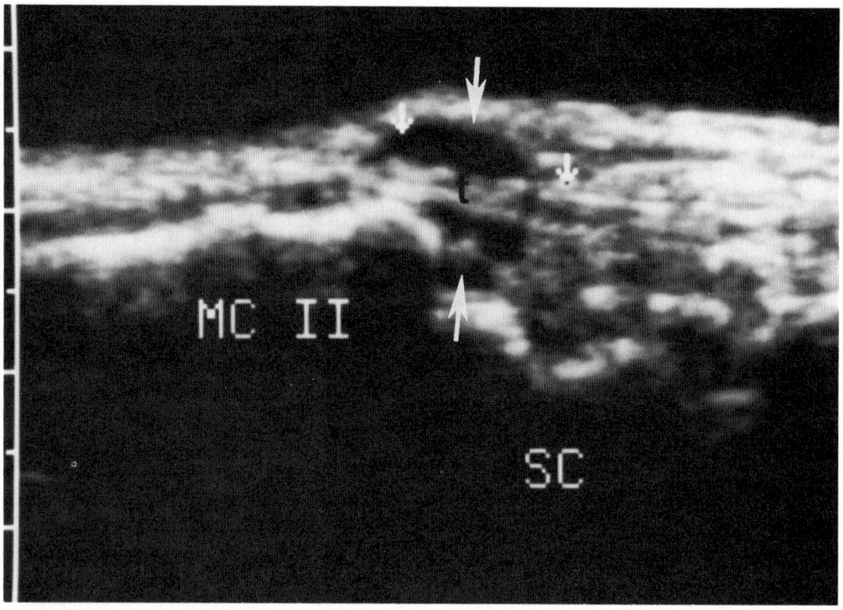


**FIG 14–12.**
Tendon cyst: longitudinal sonogram. The flexor tendon *(t)* of the index finger traverses *(small arrows)* a cyst arising from the tendon sheath *(large arrows)*. Abbreviations: scaphoid *(SC)*, second metacarpal *(MC II)*.

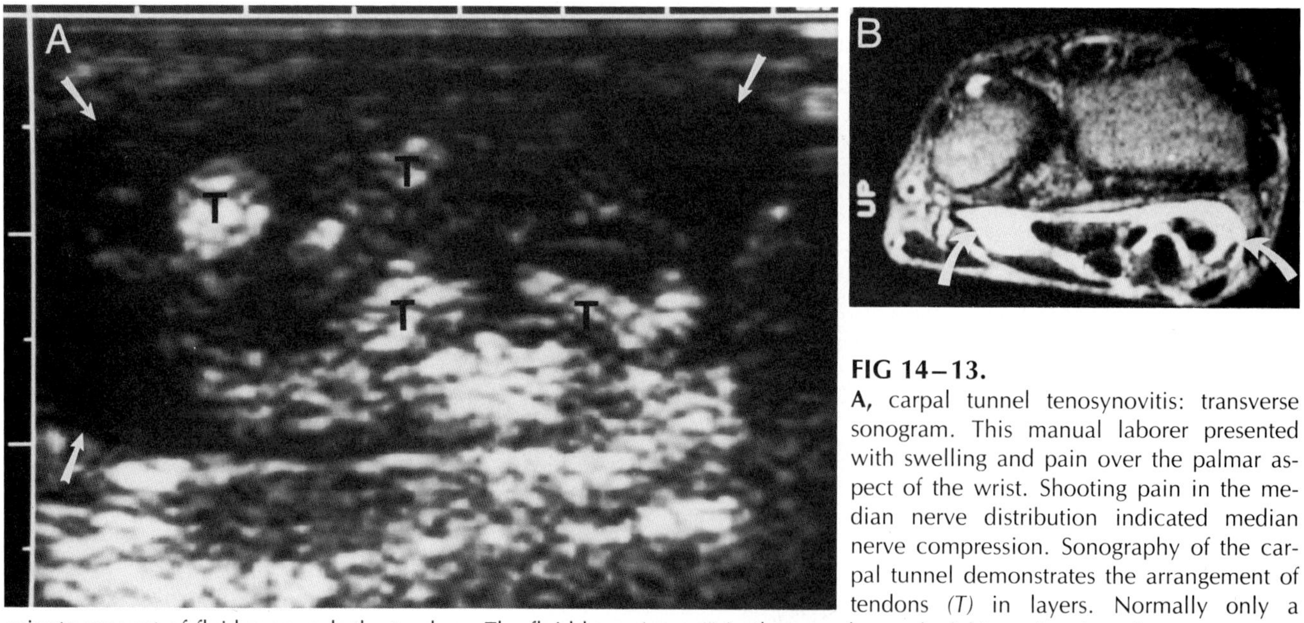


**FIG 14–13.**
**A,** carpal tunnel tenosynovitis: transverse sonogram. This manual laborer presented with swelling and pain over the palmar aspect of the wrist. Shooting pain in the median nerve distribution indicated median nerve compression. Sonography of the carpal tunnel demonstrates the arrangement of tendons *(T)* in layers. Normally only a minute amount of fluid surrounds the tendons. The fluid layer *(arrows)* in the carpal tunnel of this patient is as large as 1 cm in some areas. **B,** carpal tunnel tenosynovitis: axial magnetic resonance imaging study of the wrist. Same patient as in **A.** This T2-weighted axial spin-echo image of the wrist demonstrates the flexor tendons with their fluid-filled tendon sheath. The fluid *(curved arrows)* appears as bright signal surrounding the tendons with low-signal intensity.

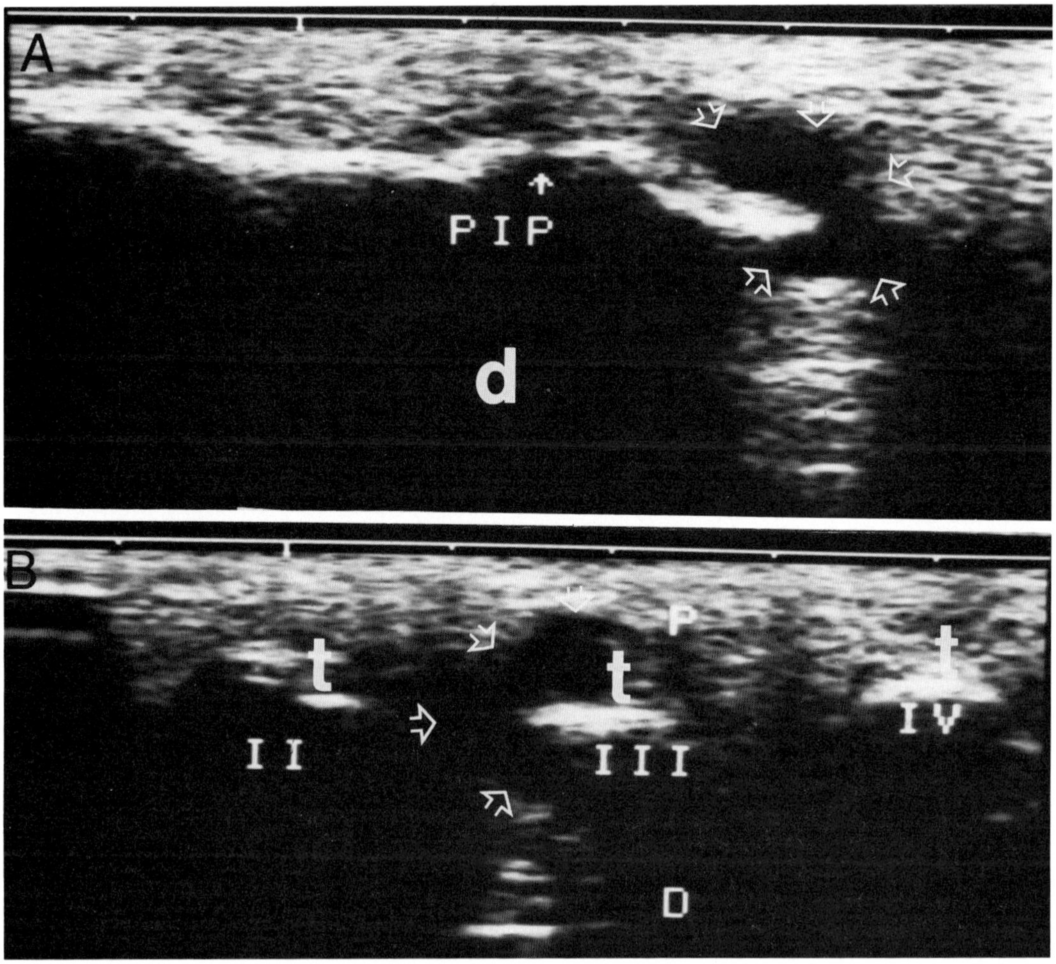


**FIG 14–14.**
**A,** tendon cyst: longitudinal sonogram. The clinician palpated a mass over the dorsal aspect *(d)* of the middle finger close to the proximal interphalangeal joint *(PIP)*. This longitudinal image of the palmar aspect of the finger demostrates a cyst *(open arrows)* extending more palmar than dorsal *(d)*. A communication between the cyst and the joint could not be demonstrated. **B,** tendon cyst: transverse sonogram. This transverse image of the fingers demonstrates a cyst *(arrowhead)* originating from the flexor tendon sheath of the middle finger. Abbreviations: flexor tendons *(t)*; *II, III, IV:* second, third, and fourth fingers; palmar aspect *(P)*; dorsal aspect *(D)*.

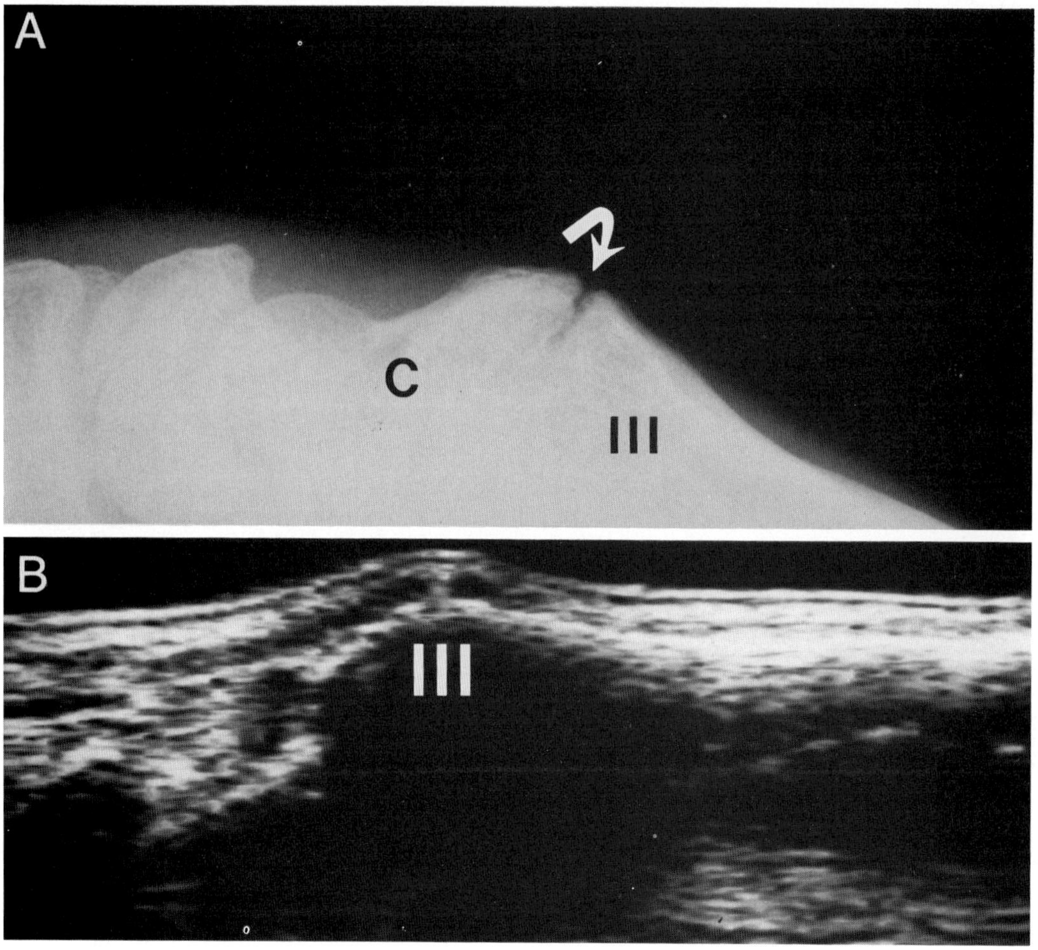


**FIG 14–15.**
**A,** carpal boss: lateral radiograph of the wrist. The styloid process of the third metacarpal *(III)* and the adjacent capitate *(C)* appear hypertrophic. This abnormality, clinically palpated as a hard swelling, is called a carpal boss *(curved arrow)*. **B,** carpal boss: longitudinal sonogram of the wrist. Ultrasound shows that the palpable swelling is due to the bony prominence of the third *(III)* metacarpal.

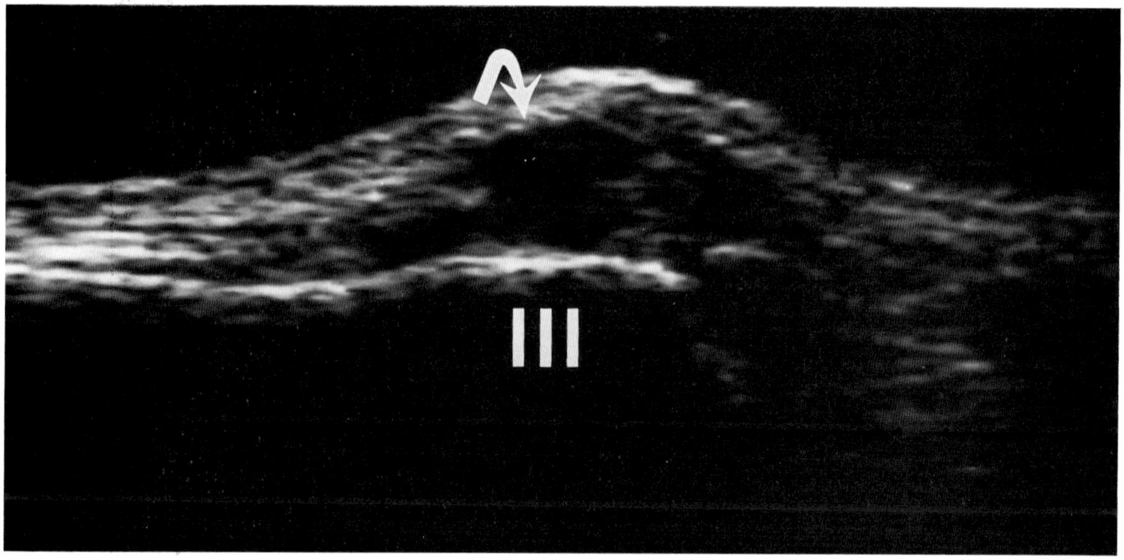


**FIG 14–16.**
Dorsal wrist cyst: longitudinal sonogram. A synovial cyst *(curved arrow)* over the third metacarpal *(III)* may clinically simulate a carpal boss. These cysts may feel very hard on clinical examination.

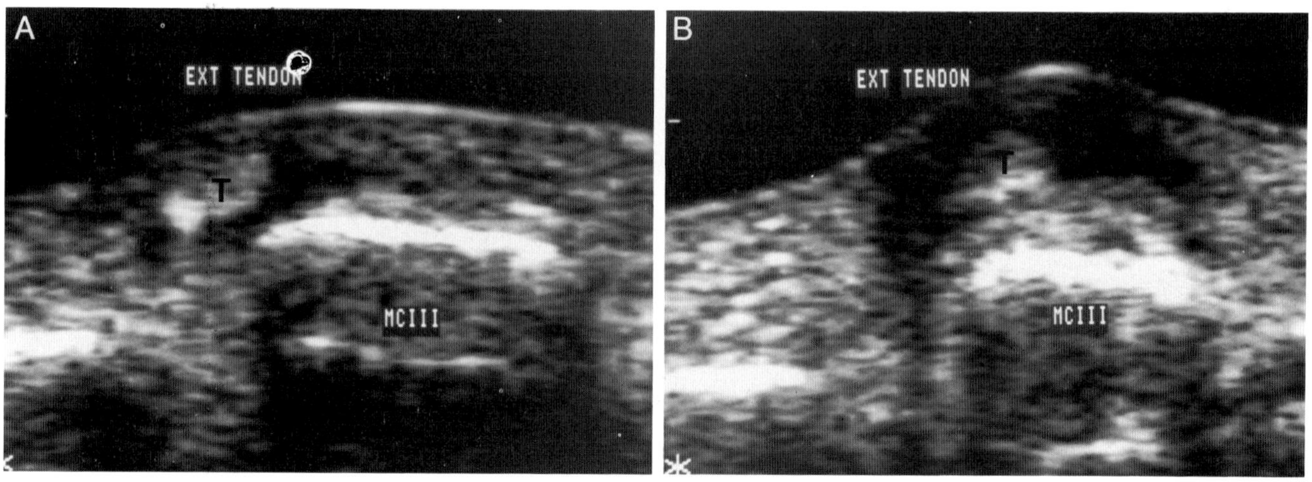


**FIG 14–17.**
Tendon subluxation: transverse sonogram of the middle finger. **A,** the extensor tendon *(T)* is laterally subluxed during flexion of the finger. Note the position of the tendon relative to the underlying bone of the distal third metacarpal *(MC III)*. **B,** extension of the finger partially reduces subluxation of the tendon *(T)*.

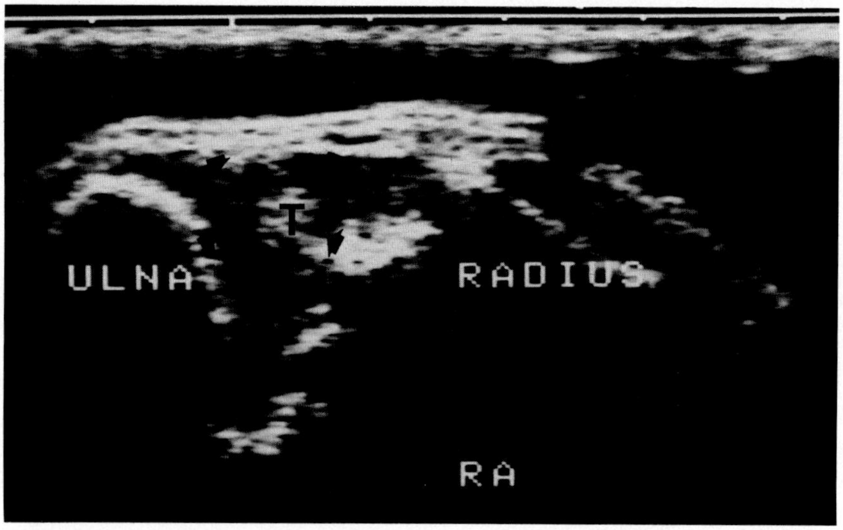


**FIG 14–18.**
Extensor tendon entrapment: transverse sonogram of the wrist. Extension deficit of the fifth finger in this patient with rheumatoid arthritis is explained by entrapment of the extensor digiti minimi tendon *(T)* between the subluxed distal radius and the ulna. Note that the tendon is surrounded by inflamed synovia *(arrows)*.

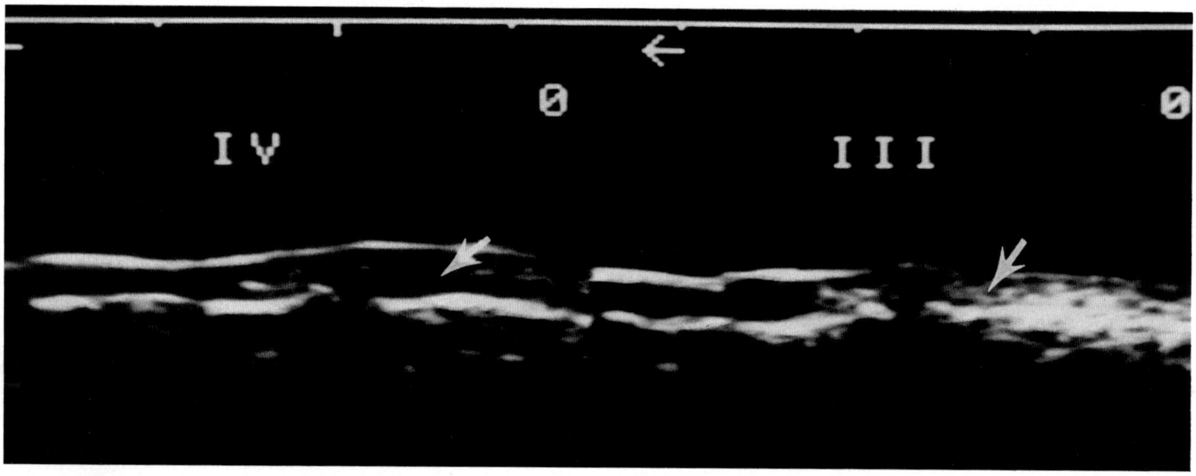


**FIG 14–19.**
Extensor tendon rupture: longitudinal sonogram of the finger. The extension deficit of the fourth finger *(IV)* in this patient was due to rupture of the distal insertion of the extensor tendon *(arrow)*. A normal hyperechoic tendon insertion *(arrow)* of the fourth digit is seen at the observer's right.

# 15

# Sonography of the Hip, Knee, and Ankle

Sonographic examination of the knee is the most frequently requested study of the lower extremity. The knee is a bicondylar joint stabilized by soft tissue structures: ligaments, tendons, menisci, and the joint capsule. These structures are all easily injured by trauma, particularly sports-related trauma. Ultrasound is the only fast noninvasive technique available for the examination of soft tissue injury at the time the patient presents for clinical evaluation. Normal plain film examination following injury is a definite indication for sonographic examination.

Trauma to the hip results in bony injury almost exclusively. This is due to the protection and stabilization inherent in its ball-in-socket configuration. Often the trauma is the result of a motor vehicle accident. Conventional radiography and computed tomography (CT) are the best means for evaluating these injuries. However, when radiographs and CT are normal, ultrasound examination may prove beneficial when clinical signs and symptoms are indicative of injury.

Injury to the ankle, like the hip, also predominantly involves bone injury. Ligaments stabilizing the ankle mortise are quite strong. Therefore, trauma to the ankle will frequently result in fracture. When no fracture is identified on conventional radiographs, the cause of swelling may be established by sonographic examination of the ankle ligaments.

## THE HIP

A large number of patients with pathology of the lumbar spine, pelvis, or abdomen will experience referred pain to the hip. The value of ultrasound in examination of the hip is its ability to determine if the pain is due to intra-articular or periarticular pathology (Fig 15–1). A joint effusion, although nonspecific, is a clear indication of joint pathology. If no joint effusion is found, careful examination of ligaments and muscles surrounding the joint may identify the cause of the pain.

Ultrasound is extremely sensitive in the detection of joint effusion. Increases in joint fluid as small as 1 mL are easily detected in the hip. Joint effusions may be seen in infection, inflammatory arthritis, osteonecrosis, osteoarthritis, trauma, and tumoral diseases.

In children with hip effusion diagnosed sonographically following normal radiographic examination, 90% will be the result of transient synovitis. A repeat sonographic examination performed in 2 to 3 weeks will demonstrate resolution of the effusion when it is the result of transient synovitis. However, if infection is of concern based on either clinical or sonographic criteria, joint aspiration should be performed immediately. Sonographic guidance is often valuable in directing aspiration. In the small number of children with persistence of joint effusion beyond 14 days, the investigation must be continued aggressively. Repeat examination with conventional radiographs is clearly indicated. If these films are normal, thin-section CT is indicated to exclude intra-articular osteoid osteoma. Magnetic resonance imaging (MRI) is the final study in the workup when CT examination is normal. Legg-Calvé-Perthes disease is often manifested by pain and joint effusion in the early stages, prior to radiographic changes. Magnetic resonance imaging is extremely sensitive in making the diagnosis of this form of osteonecrosis.

Joint effusion in adults is not unusual. Osteoar-

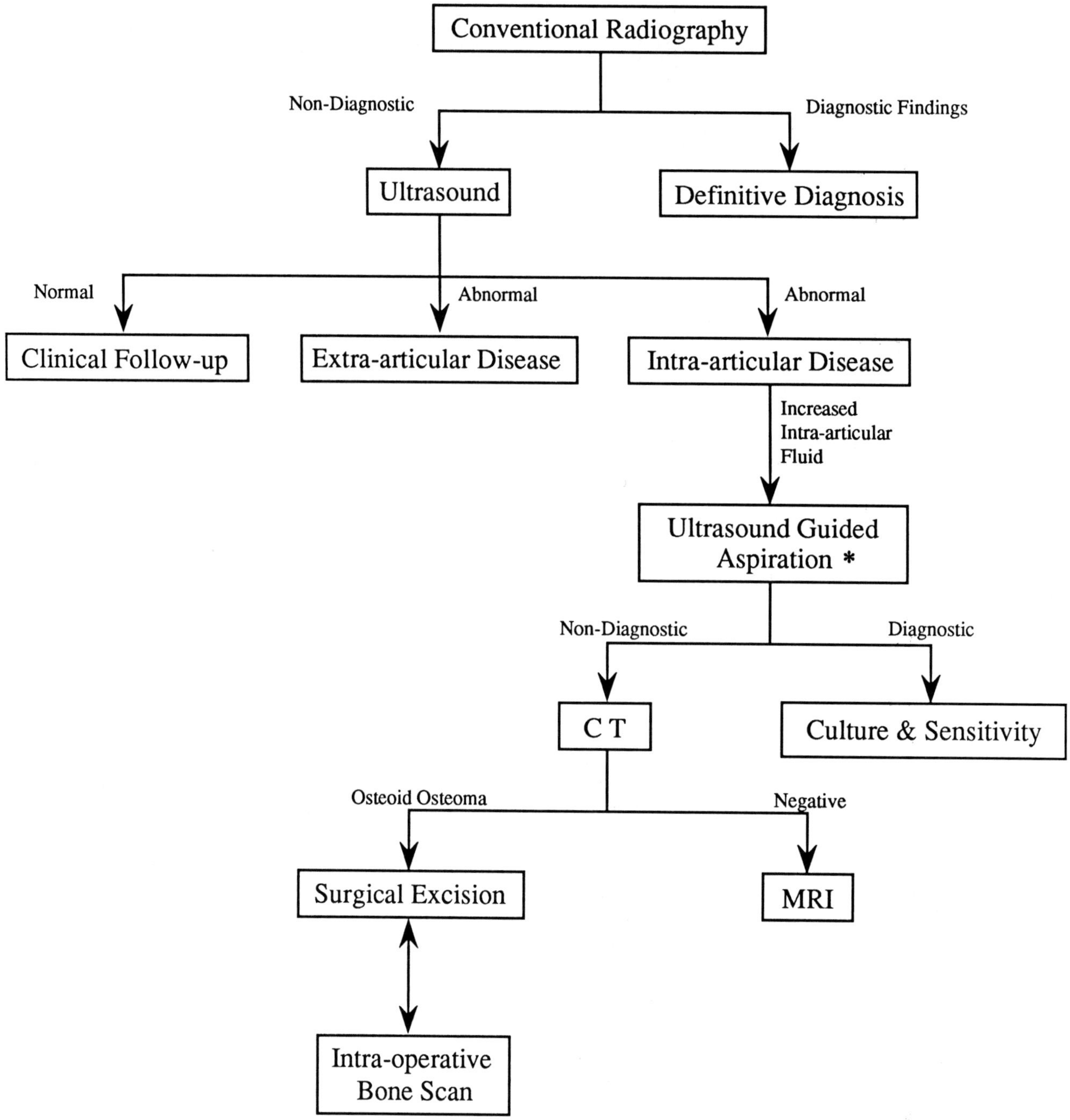


**FIG 15–1.**
Algorithm for diagnostic imaging of the painful hip.
*This step should be omitted if septic arthritis is not a clinical concern. A repeat ultrasound examination in 2 to 3 weeks is indicated before proceeding through the remainder of the algorithm if transient synovitis is suspected.

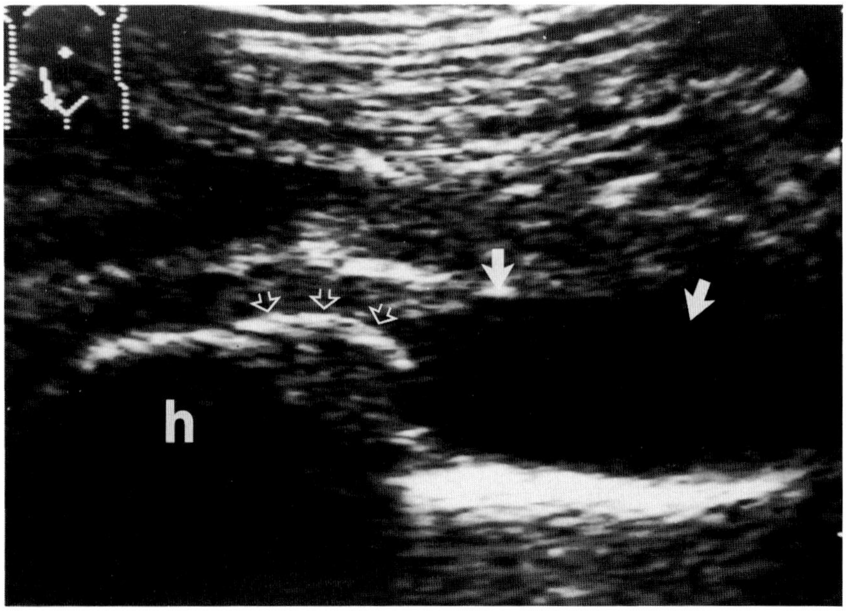


**FIG 15–2.**
Hip effusion: longitudinal sonogram. The sensitivity of sonography in detecting hip effusions is extremely high. An increase in joint fluid as small as 1 mL is easily detected sonographically. In this case, a hip effusion is detected in a patient with osteoarthritis of the hip (h). The specificity of joint effusion is low. Hip effusions are seen in a variety of articular and osseous diseases. Symbols: hip effusion *(arrows);* hip osteophyte *(open arrows).*

thritis (Fig 15–2) and osteonecrosis are the most common causes. In these cases, conventional radiography will often help to make the diagnosis. Synovial diseases will also result in increased intra-articular fluid but will also demonstrate synovial proliferation. Rheumatoid arthritis is the most common synovial disease involving the hip in adults. Pigmented villonodular synovitis and synovial chondromatosis are relatively rare diseases. Calcifications in the synovium will often be evident on radiographs in cases

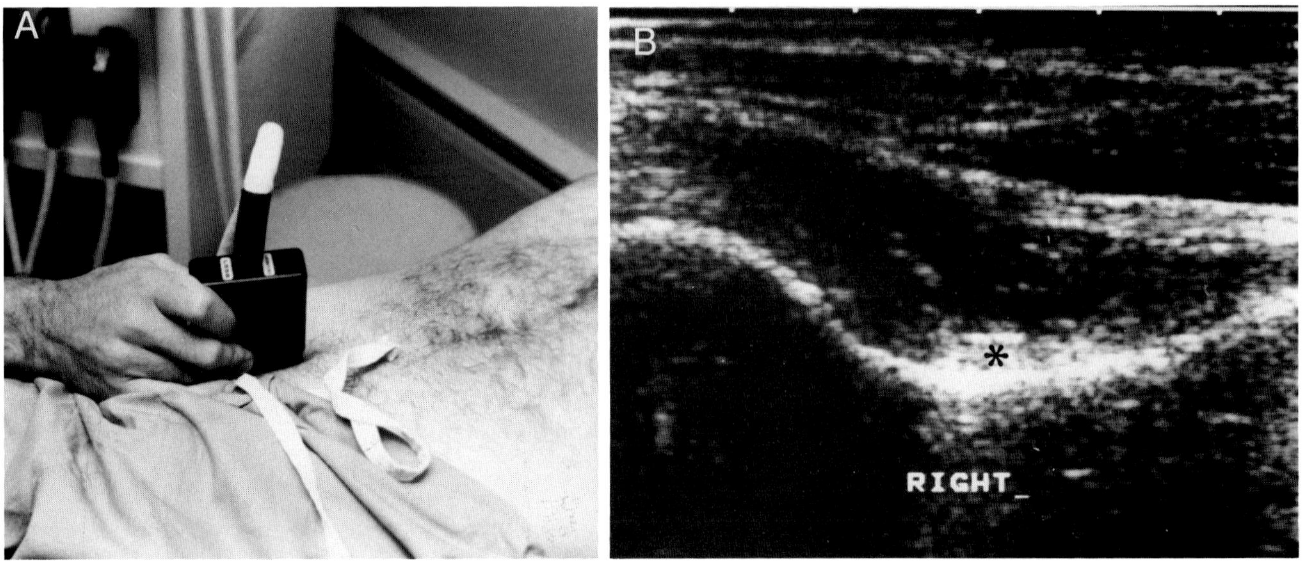


**FIG 15–3.**
**A,** Sonography of the hip: anterior synovial recess. **B,** the iliofemoral ligament *(asterisk),* which forms the anterior joint capsule, is adjacent to the femoral neck.

of synovial osteochondromatosis. Magnetic resonance imaging and, less frequently, arthrography may be indicated for further investigation of synovial disease.

Aspiration of fluid collections surrounding hip prostheses is always indicated when detected sonographically. These surgically introduced foreign bodies often serve as a nidus for infection. In the immediate postoperative period, ultrasound is very valuable in the detection and aspiration of hematomas that will predispose to infection. Following surgery, these patients are always difficult to evaluate clinically due to the degree of pain and swelling associated with these extensive procedures.

Sonographic examination of the hip is limited to an anterior approach, except in children. Examination for congenital hip dysplasia is described in detail in chapter 8. In adults, the anterior synovial recess of the hip is well visualized with the transducer aligned along the length of the femoral neck (Fig 15–3). This is the largest recess not protected by bone. The iliofemoral ligament is identified as a hyperechoic band tightly approximating the cortical surface of the femoral neck. Any increase in intra-articular fluid displaces the iliofemoral ligament anteriorly and is seen as a hypoechoic or anechoic layer separating the ligament from the femoral neck (Fig 15–4 ). After the anterior synovial recess is imaged, the surrounding muscles, tendons, and bursae are examined.

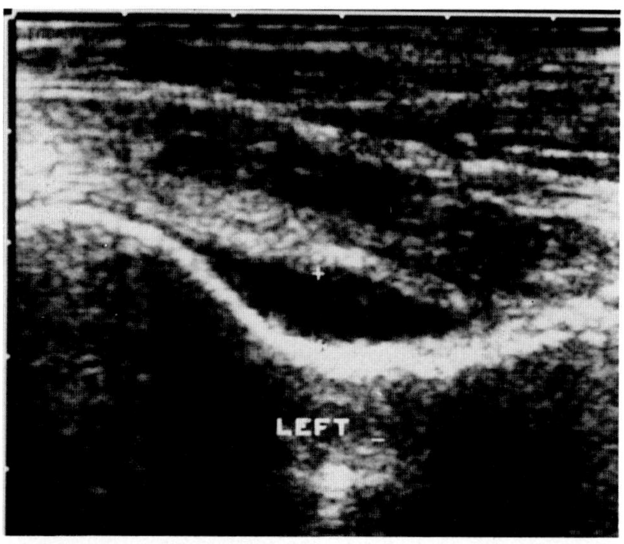


**FIG 15–4.**
Sonography of the hip: infected hip. A dialysis patient with fever and hip pain. The hyperechoic ligament is displaced anteriorly. Fluid and debris (pus) fill the anterior synovial recess (+).

## THE KNEE

Intra-articular and extra-articular disease are equally common causes of knee pain. Pain and swelling of the knee associated with a normal radiographic examination are clear indications for sonographic examination (Fig 15–5). Arthrography and arthroscopy are invasive techniques that evaluate only intra-articular pathology. Sonography has the ability to demonstrate noninvasively both intra-articular and extra-articular disease. This diagnostic modality can provide information about joint function during real-time examination, which is unavailable with other techniques. In addition the risk, discomfort, high cost, and delay associated with other methods of diagnosis are spared.

Tendonitis is a common cause of knee pain. Most frequently involved are the quadriceps, biceps, and patellar and tripod tendons. Swelling and decreased echogenicity of the involved tendon is observed. Calcifications are seen in association with tendonitis in the minority of cases. Longitudinal images distinguish tendon rupture from tendonitis by demonstrating discontinuities in the fibrillar tendon structure.

Bursitis may also be the cause of knee pain. Traumatic, septic, hemorrhagic, and chemical etiologies may be responsible for bursitis. Rupture of Baker's cyst is almost always associated with pain and swelling of the knee, which may extend to involve the entire lower leg. Ultrasound is also valuable in evaluating ligamentous injuries, synovial cysts, ganglia, muscle tears, aneurysms, and venous thrombosis.

Examination of intra-articular pathology can demonstrate synovial thickening, loose bodies, cartilage defects, and meniscal and cruciate ligament tears. Several limitations must be kept in mind. Peripheral meniscal tears are more easily diagnosed than tears of the inner aspect of the meniscus. Imaging of the anterior cruciate is performed with the knee in 60 degrees of flexion, which is rarely possible after acute injury. The only absolute limitation of the intra-articular examination is that the cartilage surface of the patella and tibial plateau cannot be examined due to the lack of an acoustic window. Whenever ultrasound demonstrates a joint effusion but cannot establish its cause, other imaging modalities or arthroscopy should be employed (see Fig 15–5). Technical advances, such as new transducers and the further development of transmission ultrasound, will reduce or eliminate these limitations. However, a normal sonographic examination demonstrating no effusion indicates that the source of pain lies outside of the joint.

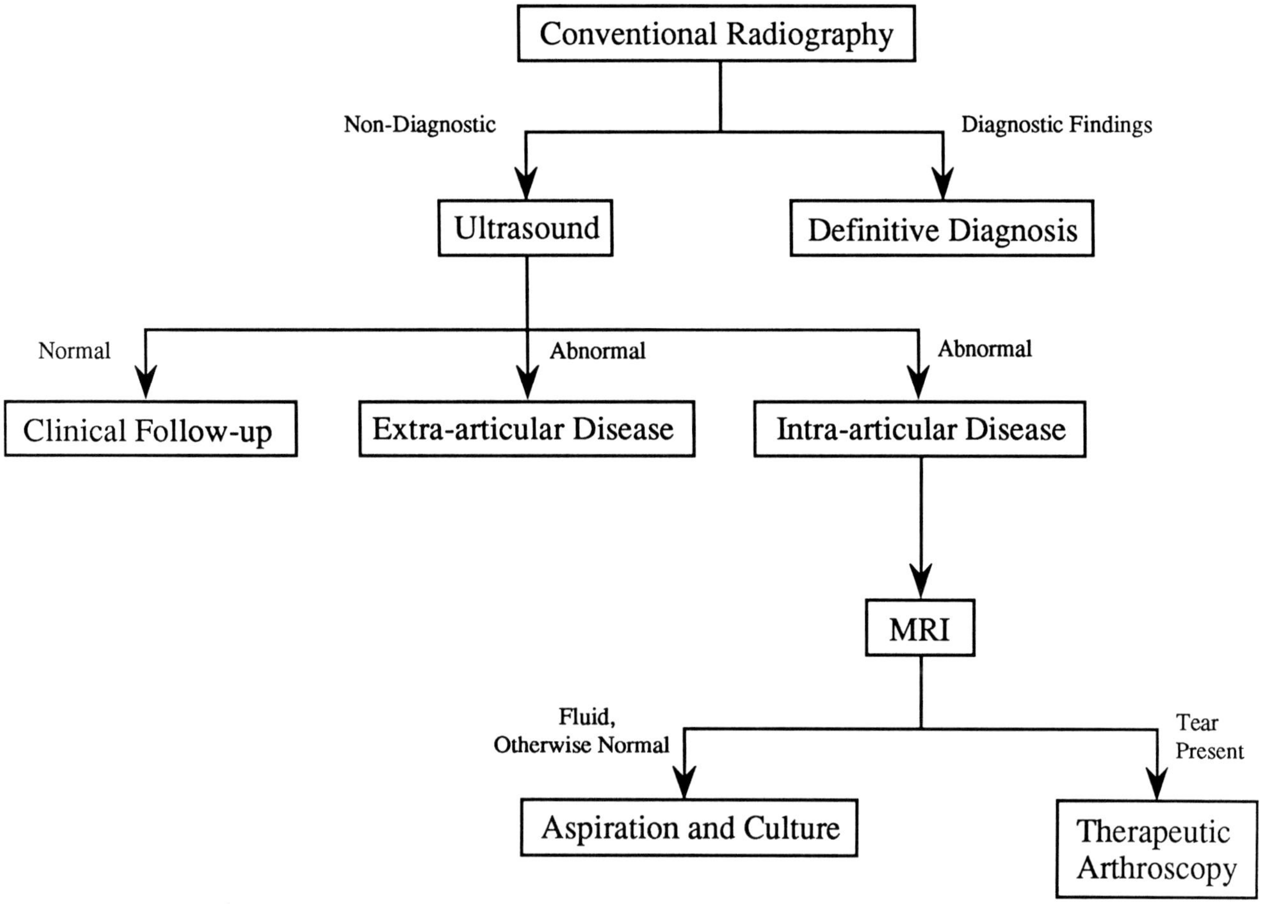


**FIG 15–5.**
Algorithm for diagnostic imaging of the painful or swollen knee.

Sonographic examination of the knee begins with evaluation of the suprapatellar bursa and quadriceps tendon. The transducer is placed in the midline of the knee in the long axis of the femur immediately superior to the patella, its inferior extent touching the patellar base. A fluid film no more than 2 mm thick separates the leaves of the bursa, which is identified immediately deep to the quadriceps tendon. Two sets of longitudinal images are obtained. The first is a compression image to evaluate synovial thickness (Fig 15–6). Firm pressure is applied with the transducer to express excess fluid. This yields a reproducible measure of synovial thickness, which can be used for comparison in follow-up exams. In the second set of images, an assistant applies manual compression of the lateral synovial recesses while the examiner obtains longitudinal images of the suprapatellar bursa. This forces all mobile joint fluid into the suprapatellar bursa. In a normal joint, compression of the lateral joint recesses should not separate the synovial leaves of the suprapatellar bursa (Fig 15–7). Thickness, continuity, and echogenicity of the quadriceps tendon may be compared with the asymptomatic side if a question of subtle pathology exists.

Images of the patellar tendon are obtained next. The transducer is positioned over the long axis of the patellar tendon with its superior aspect touching the patellar apex (Fig 15–8). A normal patellar tendon is uniform in thickness throughout its length. Further information may be obtained about focal abnormalities in the tendon by rotating the transducer 90 degrees and imaging in the transverse plane. Care must be taken to keep the beam perpendicular to the tendon surface to avoid artifactual changes in tendon echogenicity due to anisotropy.

Following examination of the anterior aspect of the knee, the medial collateral ligament is evaluated with the patient in the lateral decubitus position. This ligament is best evaluated on coronal images with the knee extended (Fig 15–9). After the medial collateral ligament is examined, the menisci are imaged, first with the transducer aligned in the long

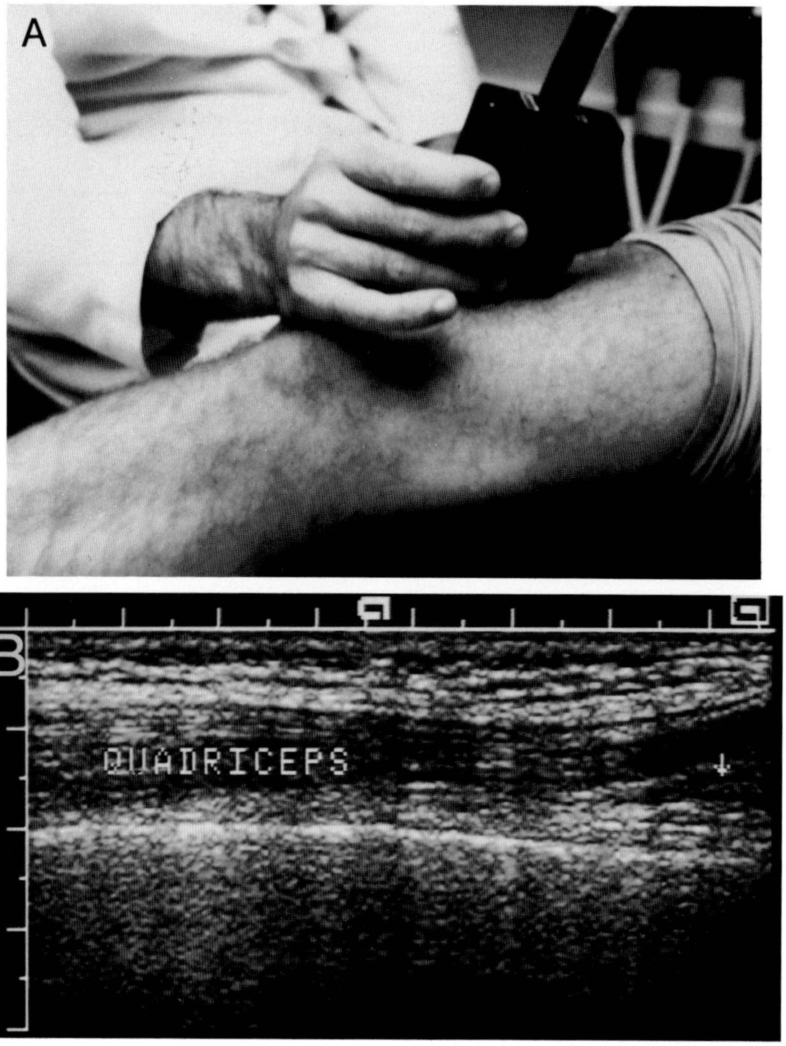


**FIG 15–6.**
Sonography of the knee: synovial measurement of the suprapatellar bursa *(arrow)*.

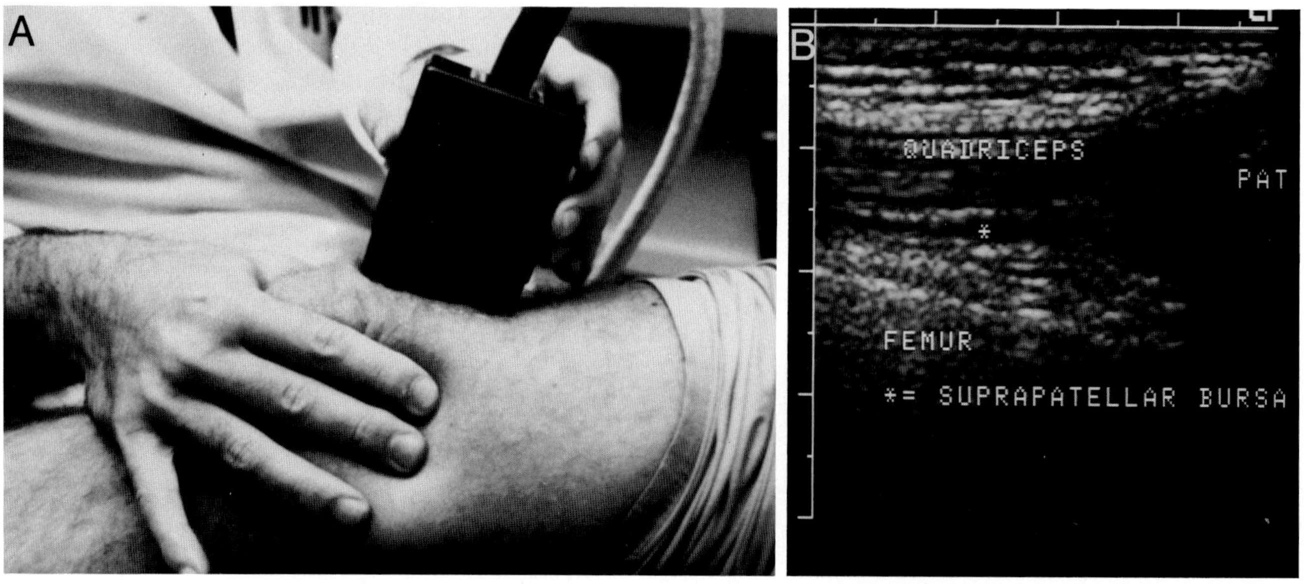


**FIG 15–7.**
Sonography of the knee: fluid measurement of the suprapatellar bursa.

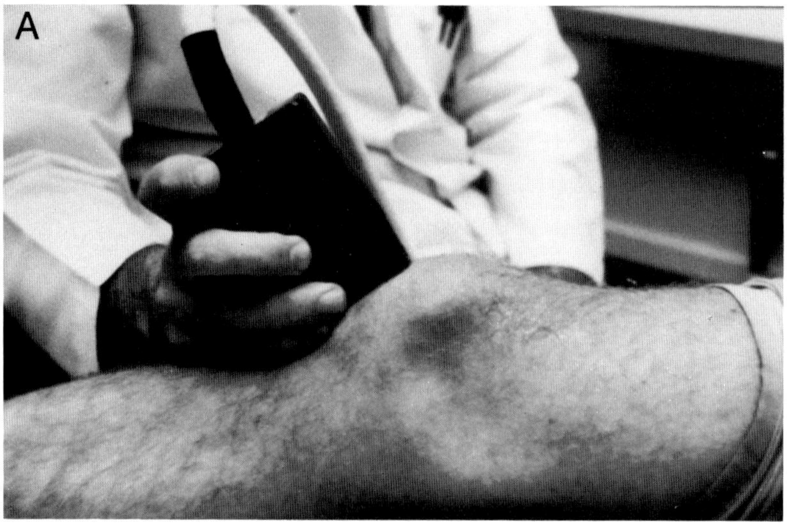

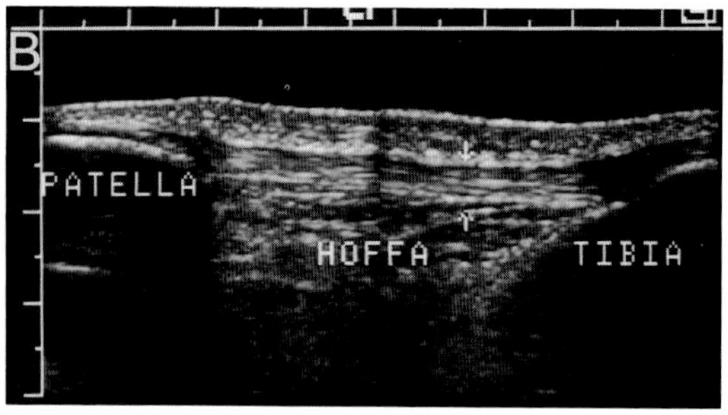


**FIG 15–8.**
**A,** sonography of the knee: patellar tendon *(arrows)*. **B,** longitudinal sonogram.

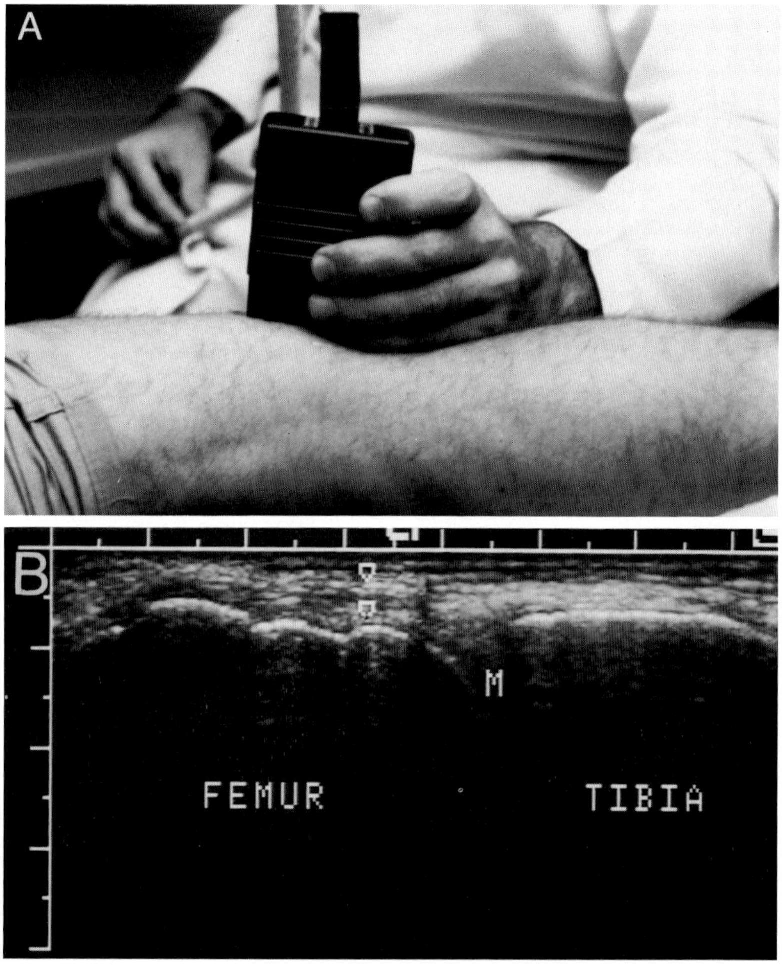


**FIG 15–9.**
**A,** Sonography of the knee: medial collateral ligament. **B,** coronal sonogram. Hyperechoic layers of the medial collateral ligament are indicated by *arrowheads*. Medial meniscus, *M*.

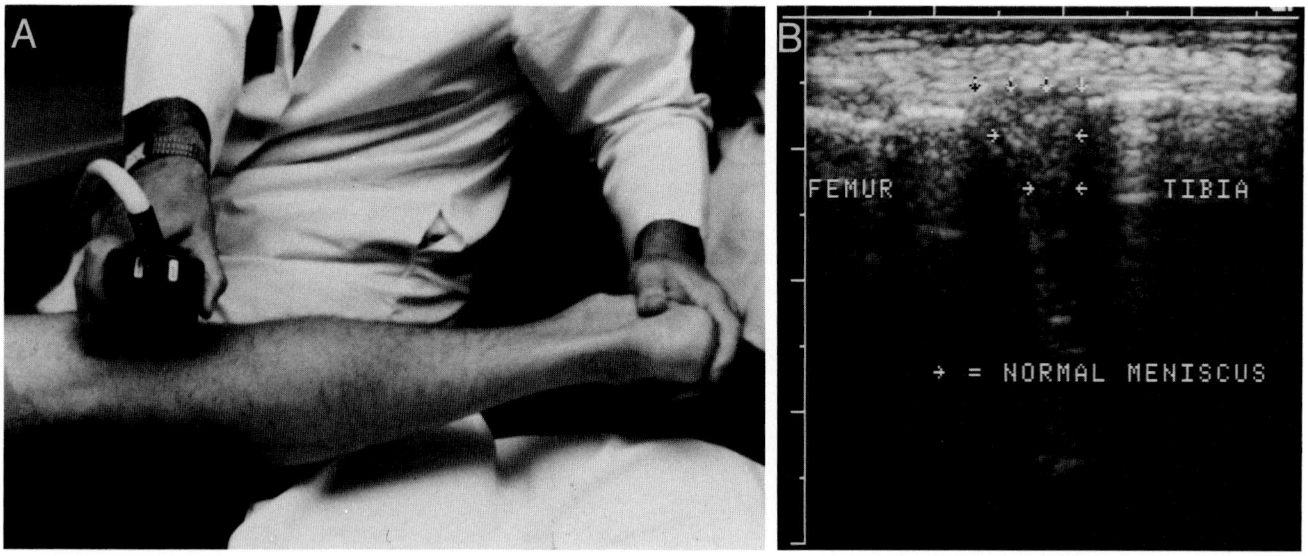


**FIG 15–10.**
Sonography of the knee: coronal examination **(A)** and coronal sonogram **(B)** of the medial meniscus.

axis of the leg for localization. Once they are localized on longitudinal images (Fig 15–10), the transducer is rotated 90 degrees to yield transverse images (Fig 15–11). Menisci are composed of fibrocartilage that has a homogeneous hyperechoic appearance. When the menisci are imaged, the gain setting must be increased to visualize the innermost portion of the meniscus. A cushion is placed beneath the knee and valgus, or varus stress is applied dependent on which meniscus is being examined. When large tears are present, a vacuum phenomenon may be observed sonographically, but is unusual due to accompanying joint effusion.

Examination of the popliteal fossa and lateral collateral ligament is performed with the patient prone. The medial aspect of the popliteal fossa is examined with both longitudinal and transverse images to check for Baker's cyst between the gastrocnemius and semimembranosus tendons (Fig 15–12). Vessels of the popliteal fossa are evaluated for patency and

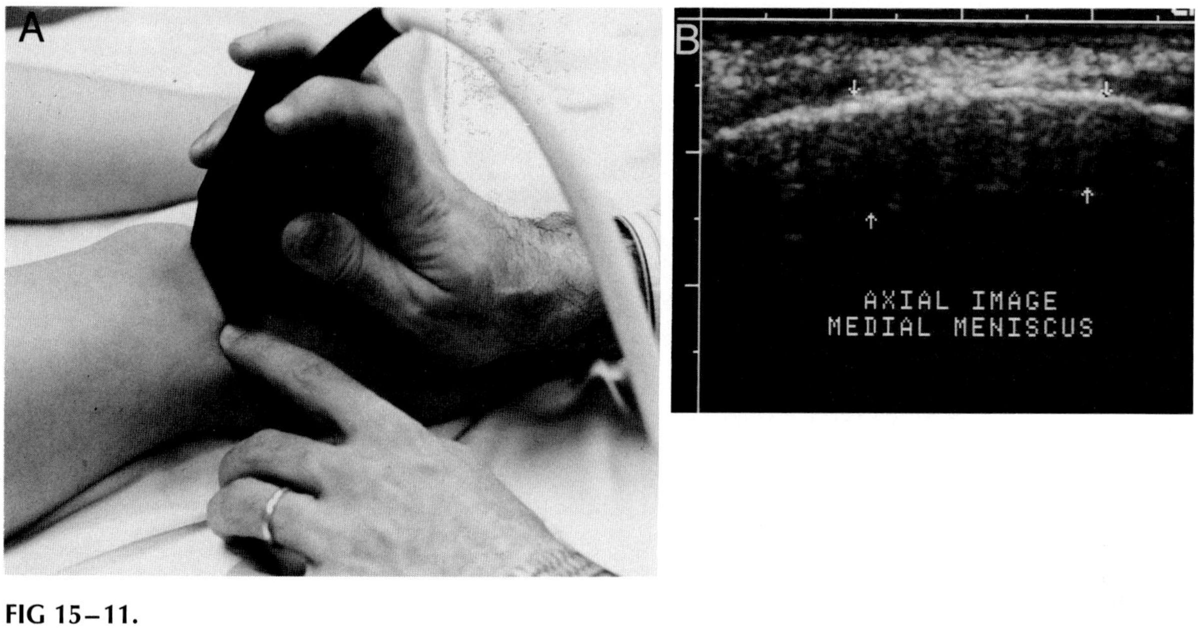


**FIG 15–11.**
Sonography of the knee: transverse examination **(A)** and transverse sonogram **(B)** of the medial meniscus.

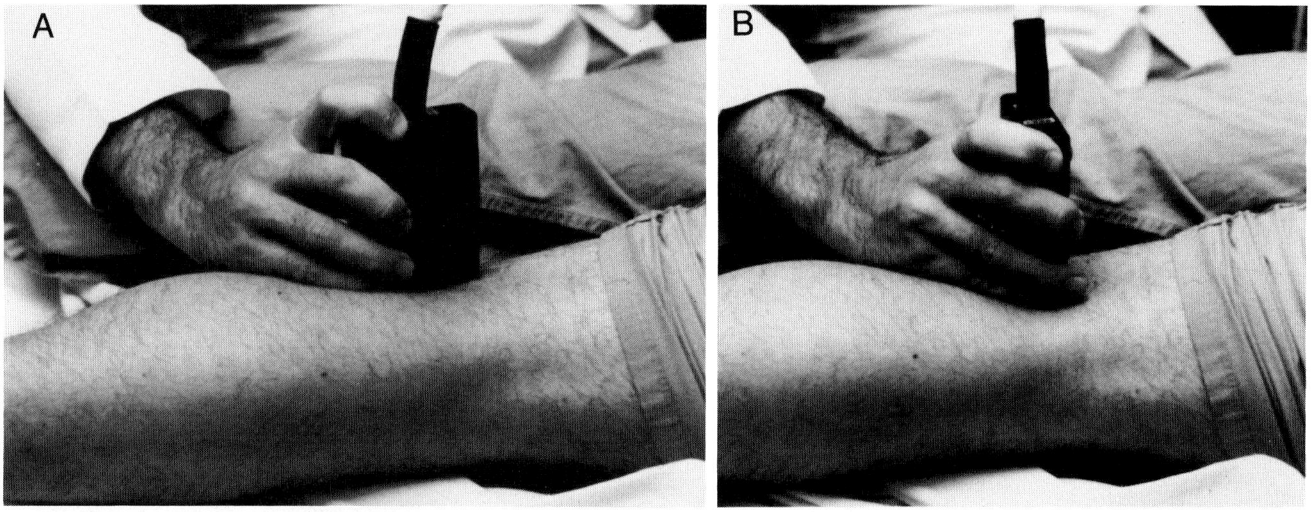


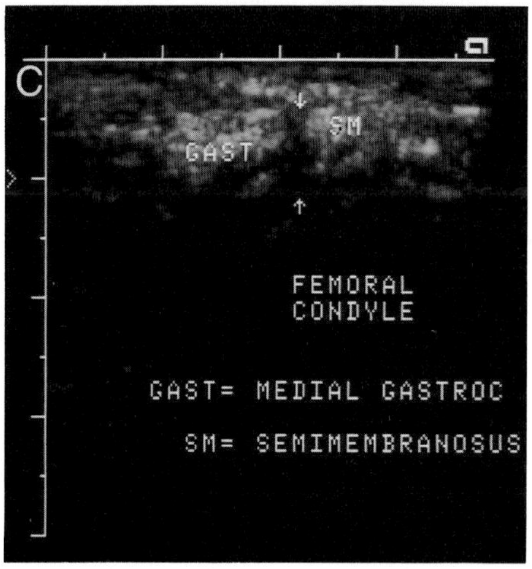


**FIG 15–12.**
Sonography of the knee: longitudinal **(A)** and transverse **(B)** examination and transverse sonogram **(C)** of the medial popliteal fossa. The normal gastrocnemius-semimembranosus bursa is indicated by *arrows*.

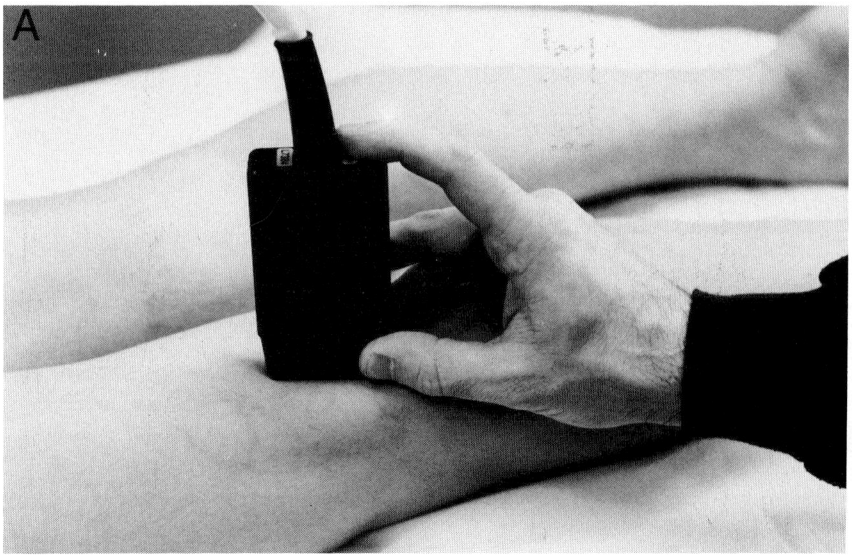


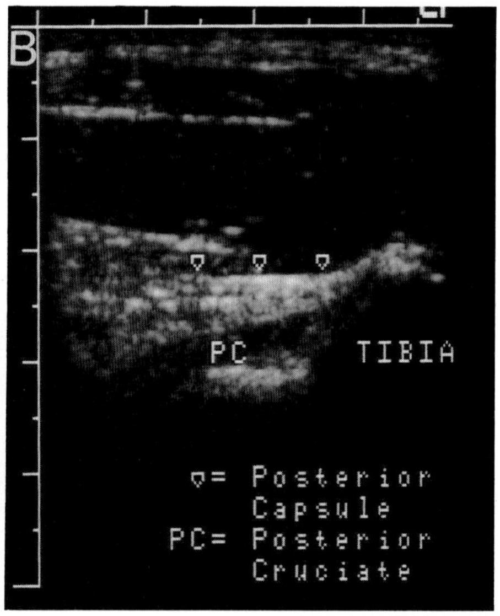


**FIG 15–13.**
Sonography of the knee: longitudinal examination **(A)** and longitudinal sonogram **(B)** of the posterior cruciate ligament.

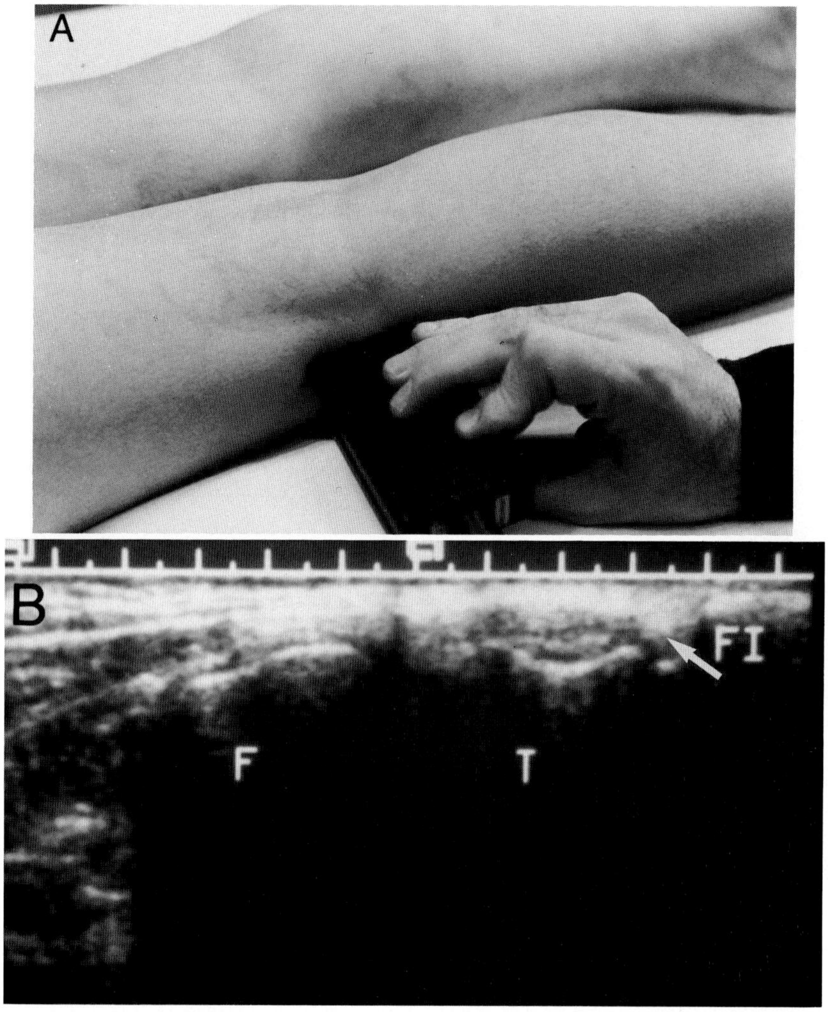


**FIG 15–14.**
Sonography of the knee: coronal examination **(A)** and coronal sonogram **(B)** of the lateral collateral ligament and long biceps tendon. Abbreviations: femur *(F)*, tibia *(T)*, fibular head *(FI)*.

aneurysm. Conventional Doppler, color-flow Doppler, and compression imaging are all valuable in evaluation of the popliteal vessels.

While the patient is prone, the posterior cruciate and lateral collateral ligaments are examined. Visualization of the entire length of the posterior cruciate ligament requires the transducer be obliqued slightly with the superior aspect deviated to the medial (Fig 15–13). The lateral collateral ligament and the distal biceps tendon are seen as one structure sonographically (Fig 15–14). At the level of the joint, the biceps tendon is in close approximation with the lateral aspect of the lateral collateral ligament. Longitudinal images are most valuable for evaluation of these two structures. Images in Figures 15–15 and 15–16 demonstrate disease processes involving the lateral aspect of the knee.

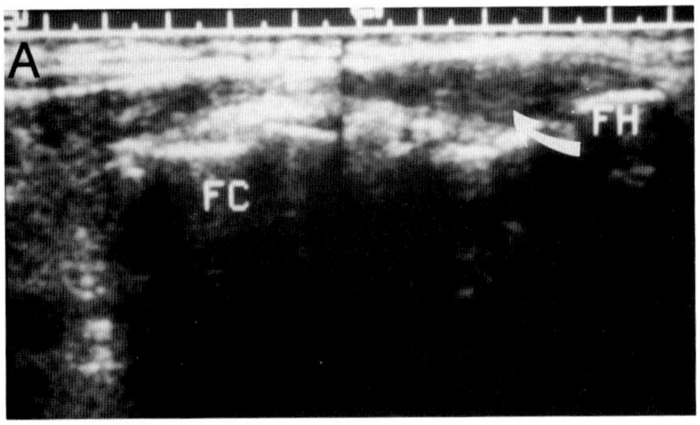


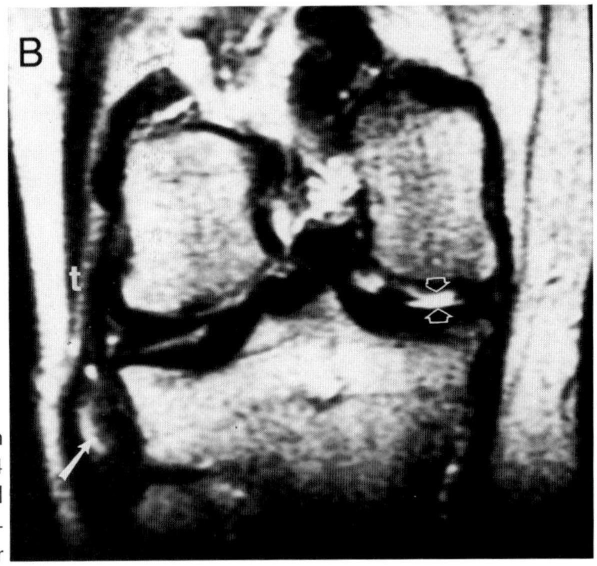


**FIG 15–15.**
**A,** biceps tendonitis: coronal sonogram. This 44-year-old man presented with painful swelling over the fibular head *(FH)* 4 months after a football injury. An arthrogram revealed a medial meniscal tear, but pain over the lateral aspect of the knee persisted. Ultrasound demonstrates swelling *(curved arrow)* over the fibular head secondary to biceps tendonitis. Compare this image with Figure 14–13,**B**. Abbreviations: lateral femoral condyle *(FC)*. **B,** biceps tendonitis: coronal MR image. This coronal T2-weighted spin-echo examination of the knee confirms the edema *(arrow)* of the distal biceps tendon *(t)*. The distal tendon joins the distal fibular collateral ligament. A bucket handle tear *(open arrows)* is noted in the medial femorotibial compartment.

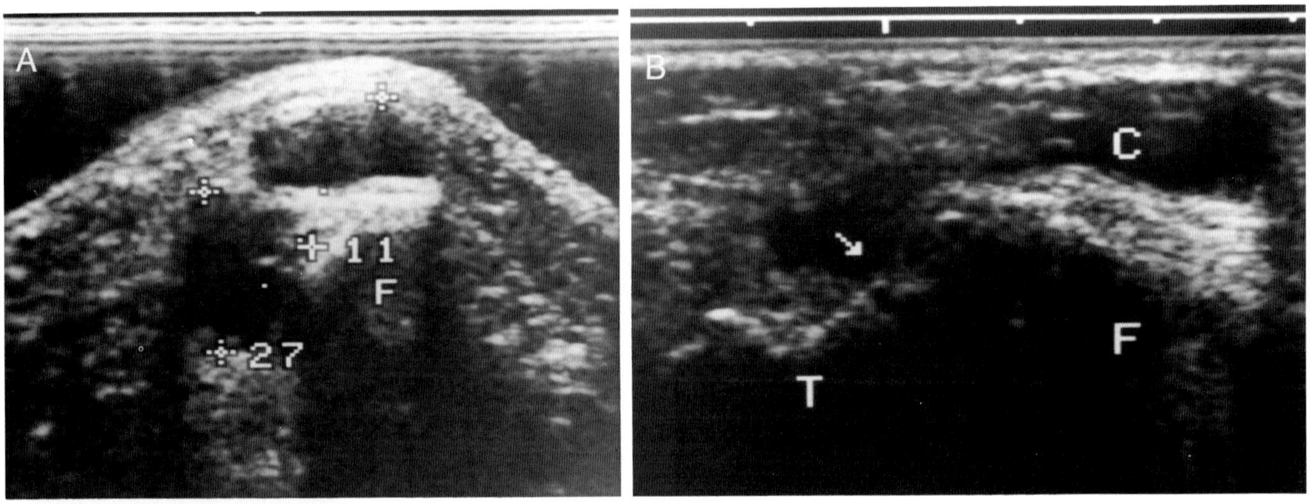


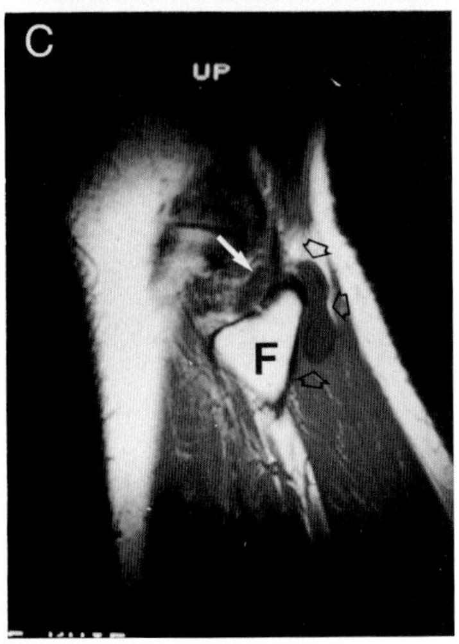


**FIG 15–16.**
**A,** synovial cyst: transverse sonogram. A synovial cyst surrounds the fibular head *(F)*. **B,** synovial cyst: longitudinal sonogram. This longitudinal sonogram demonstrates communication *(arrow)* between the cyst and the tibiofibular syndesmosis. Proximal fibula *(F)*, proximal tibia *(T)*. **C,** synovial cyst: sagittal MR image. This sagittal T1-weighted image demonstrates low signal of the fluid-filled cyst *(open arrows)* overlying the posterior proximal fibula. Communication with the joint is seen more anteriorly *(arrow)*.

## THE ANKLE

Acute trauma to the ankle often results in fractures of the malleoli or avulsions. Ligamentous injury on the side opposite the fracture is common. However, in a significant number of cases, conventional radiographs demonstrate swelling without fracture. This is suggestive but not diagnostic of ligamentous injury. Ultrasound can quickly demonstrate ligament tears of the ankle. The anterior fibulotalar ligament is most frequently involved.

Chronic ankle pain (Fig 15–17) is most commonly the result of extra-articular disease. Numerous tendons adjacent to the joint can be the source of ankle pain. Rupture of the plantar fascia and retrocalcaneal bursitis may also result in chronic pain. In the rare cases when intra-articular pathology is responsible for ankle pain, ultrasound is able to demonstrate effusions and loose bodies in the anterior synovial recess.

Ultrasound is poor for evaluation of chondral fractures. Arthrotomography, CT arthrotomography, MRI, and arthroscopy all have been advocated for the detection of chondral fractures, but a comparative study has not been performed. Clearly arthroscopy is the most invasive and therefore should be used sparingly.

Examination of the ankle is initiated by imaging the anterior synovial recess (Fig 15–18). The foot is dorsiflexed and the transducer positioned over the joint space aligned along the length of the tibia. Normally the hyperechoic joint capsule fits tightly against the anterior tibial margin and cartilage of the tibial dome. Joint effusion will displace and bow the joint capsule anteriorly with hypoechoic fluid filling the gap.

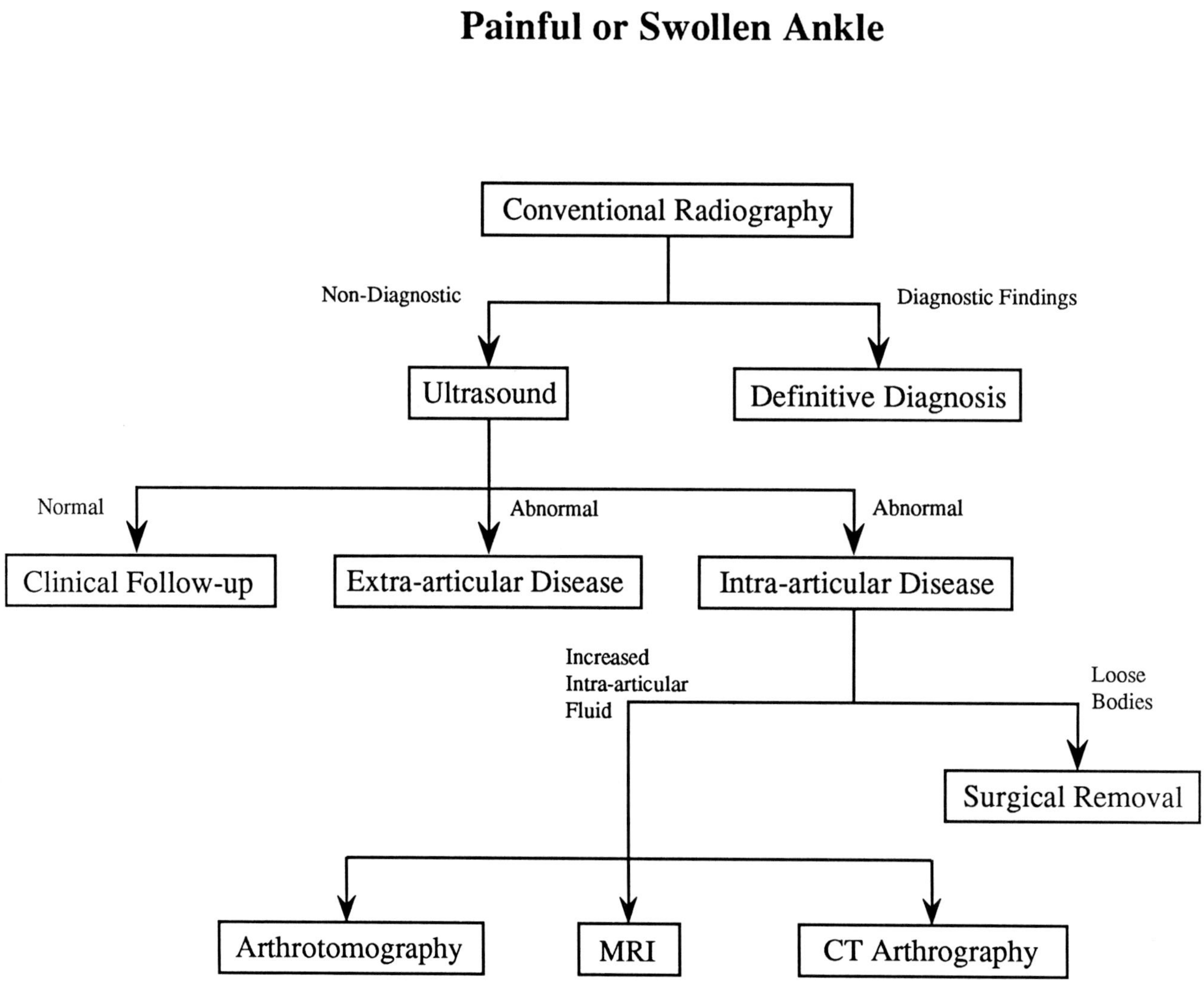


**FIG 15–17.**
Algorithm for diagnostic imaging of the painful or swollen ankle.

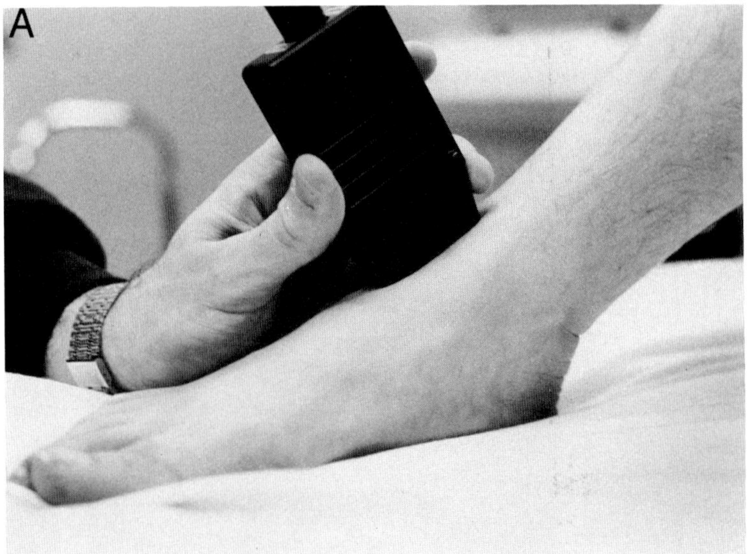


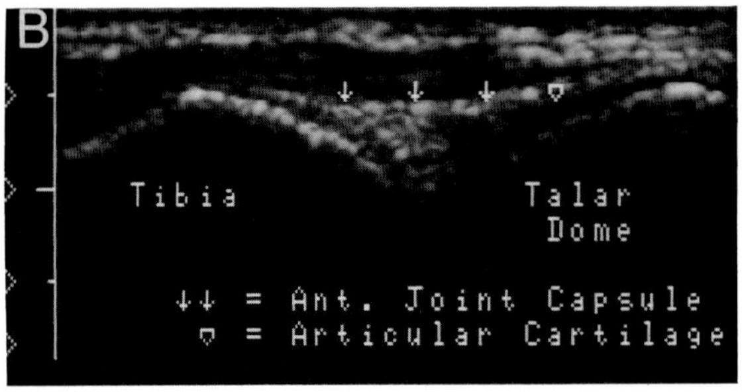


**FIG 15–18.**
**A,** sonography of the ankle: anterior joint capsule. **B,** longitudinal sonogram.

Ligaments of the medial and lateral aspects of the ankle are then examined. In acute trauma, swelling is usually severe and a standoff pad is rarely necessary. However; a standoff pad is obligatory in cases where swelling is minimal, as in chronic injury. The ligament of particular interest is the anterior fibulotalar ligament (Fig 15–19) due to its high incidence of rupture.

The sonographic evaluation is completed by examination of the tendons and bursae. In the supine position, the peroneus, anterior tibial, and posterior tibial tendons are evaluated. Examination of the Achilles tendon (Fig 15–20 and 15–21), subcutaneous Achilles tendon bursa, and retrocalcaneal bursa are performed with the patient prone. Figures 15–22 and 15–23 demonstrate pathology of the dorsum of the foot and anterior ankle. Disease involving the posterior ankle is presented in Figures 15–24 and 15–25. Pathology of the plantar surface of the foot is shown in Figures 15–26 and 15–27.

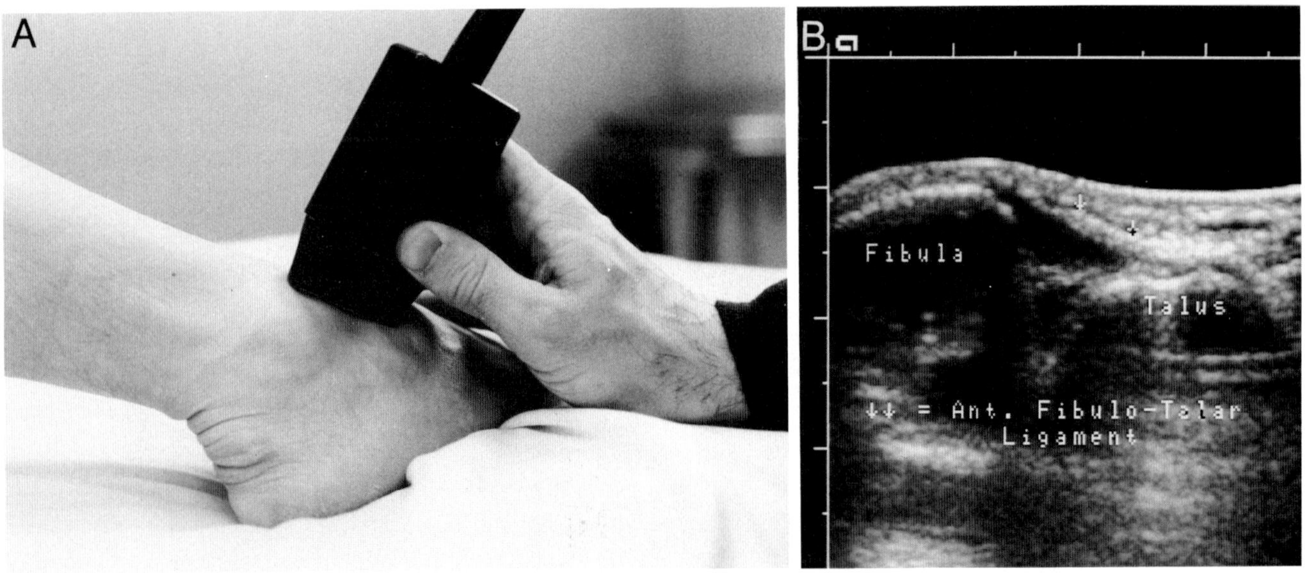


**FIG 15–19.**
Sonography of the ankle: anterior fibulotalar ligament.

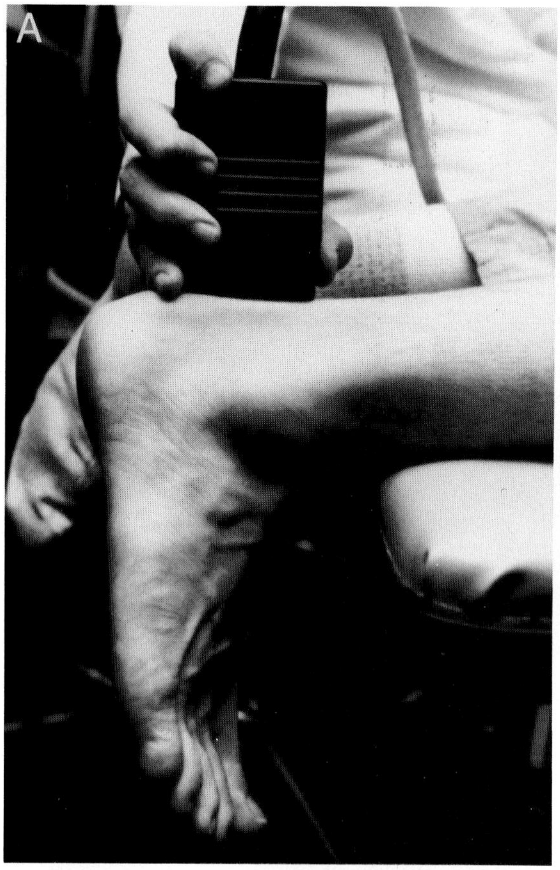


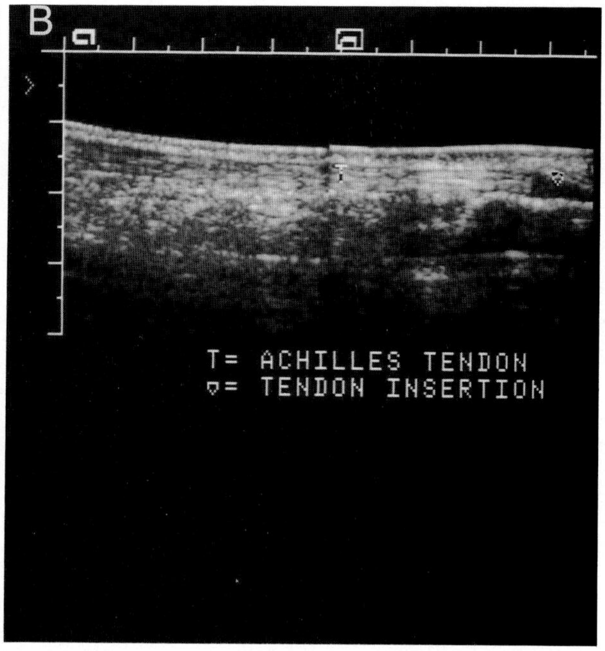


**FIG 15–20.**
Sonography of the ankle: longitudinal examination **(A)** and longitudinal sonogram **(B)** of the Achilles tendon.

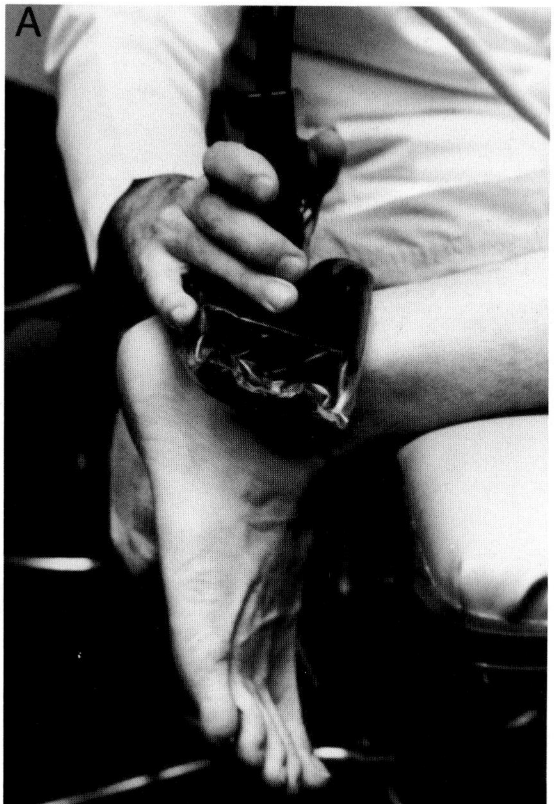


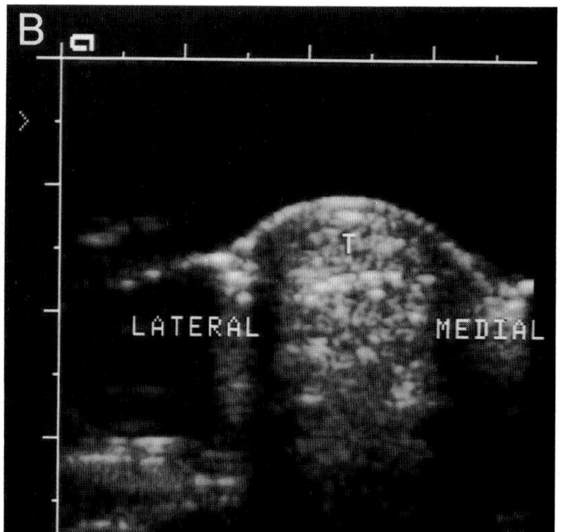


**FIG 15–21.**
Sonography of the ankle: transverse examination **(A)** and transverse sonogram **(B)** of the Achilles tendon *(T)*.

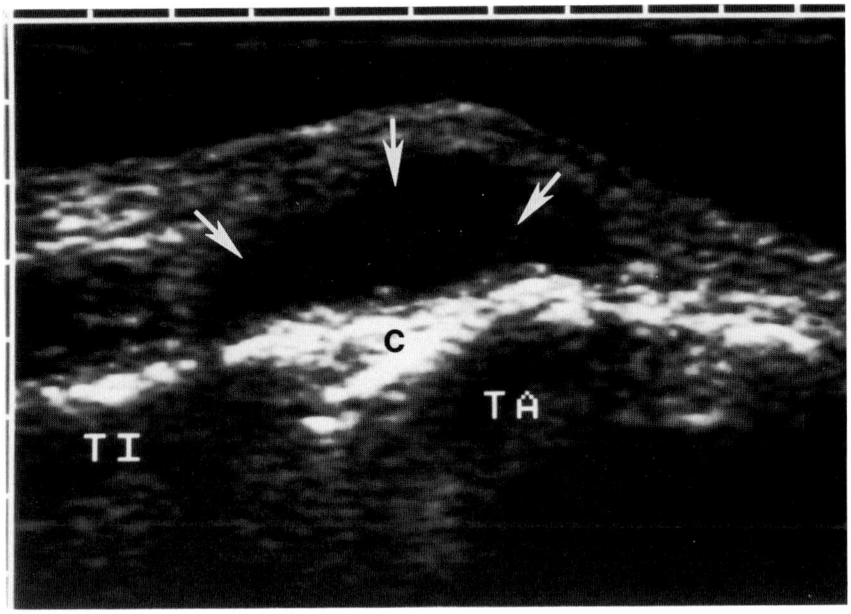


**FIG 15–22.**
Synovial ankle cyst: longitudinal sonogram. A hypoechoic fluid collection *(arrows)* is noted anterior to the ankle joint. The anterior joint capsule *(c)* of the tibiotalar joint is not displaced, unlike an ankle effusion. Abbreviations: distal tibia *(TI)*, talus *(TA)*.

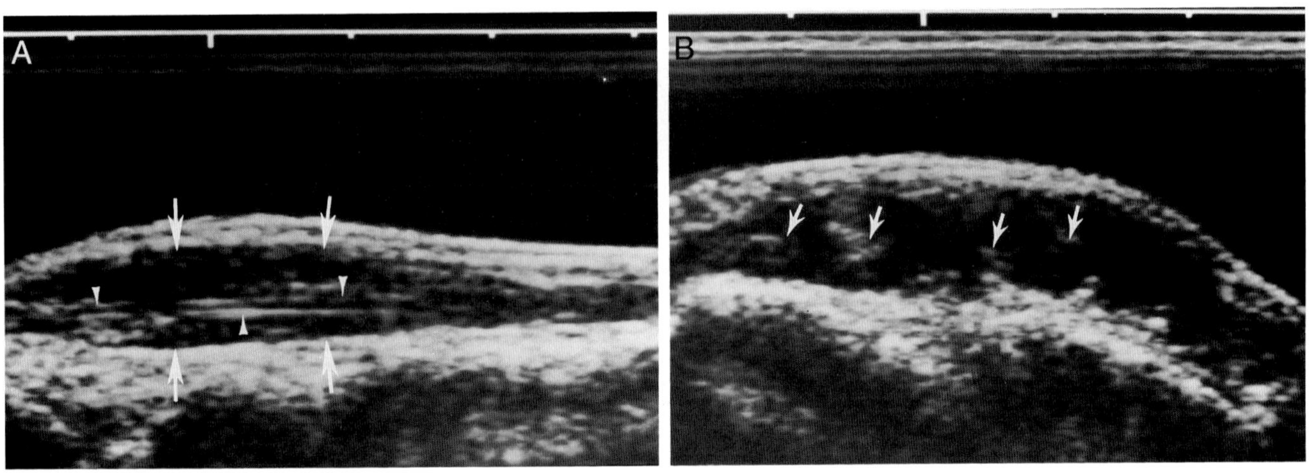


**FIG 15–23.**
**A,** synovial tendon cyst: longitudinal sonogram. A fusiform hypoechoic mass *(arrows)* surrounds the extensor tendons *(arrowheads)* at the dorsum of the foot. **B,** synovial tendon cyst: transverse sonogram. Several extensor tendons *(arrows)* are recognized within the hypoechoic synovial cyst. Clinically, a sarcoma was suspected. At surgery a large synovial cyst filled with mucoid material was found surrounding the tendon.

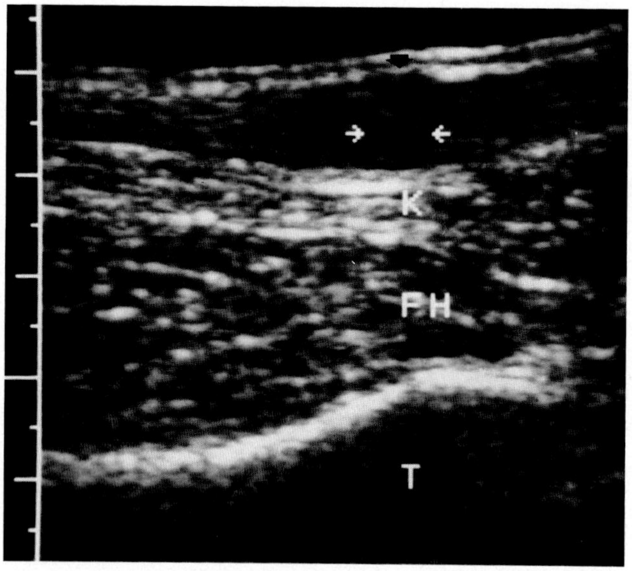


**FIG 15–24.**
Achilles tendon rupture: longitudinal sonogram. Following rupture, the tendon can appear swollen and hypoechoic, as in tendonitis. Careful examination of the entire tendon is necessary to detect the hypoechoic cleft *(white arrows),* which is characteristic of tendon rupture. Interruptions in the epitendineum *(black arrow)* are seen in some patients. Abbreviations: flexor hallucis longus muscle *(FH),* Kager's fat pad *(K),* distal tibia *(T).*

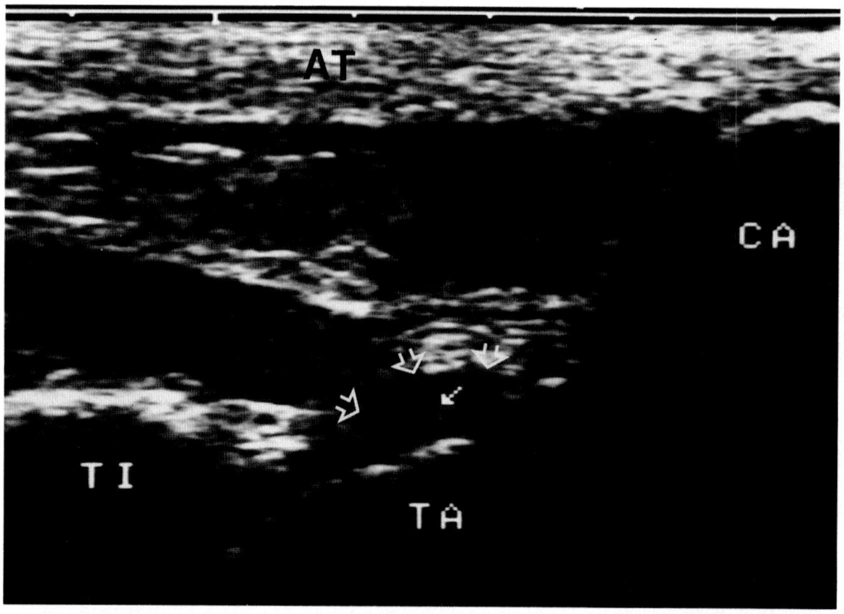


**FIG 15–25.**
Ankle effusion and intra-articular loose bodies: longitudinal sonogram. This patient was referred with the clinical diagnosis of Achilles tendonitis. The Achilles appears normal, but small loose bodies *(arrow)* were observed in a tibiotalar joint effusion *(open arrows).* Abbreviations: calcaneus *(CA),* distal tibia *(TI),* talus (TA).

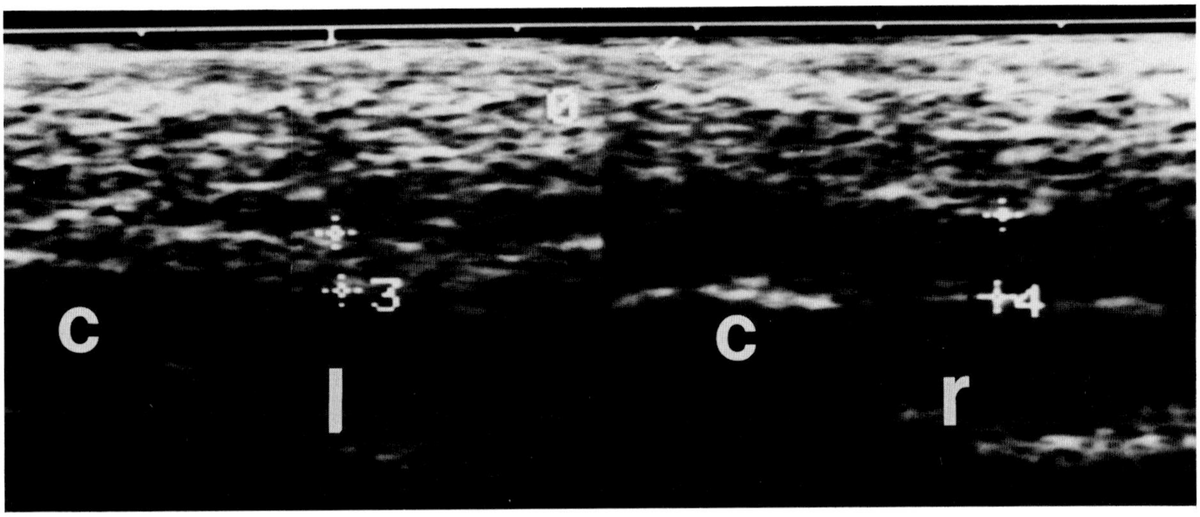


**FIG 15–26.**
Plantar fasciitis: longitudinal sonograms. The plantar fascia of the right foot *(r)* is thickened and hypoechoic. Compare the abnormal right fascia with the normal fascia of the left foot *(l)*. Abbreviations: calcaneal tuberosity *(C)*.

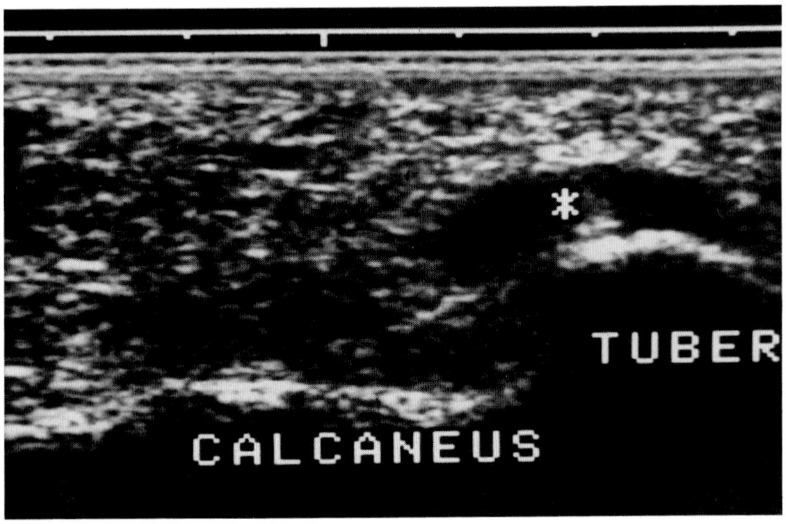


**FIG 15–27.**
Calcaneal bursitis: longitudinal sonogram. Calcaneal bursitis presents as an oval-shaped hypoechoic swelling *(asterisk)* medial to the plantar fascia and overlying the calcaneal tuberosity.

# APPENDIX

## Table of Normal Values

## ACHILLES TENDON MEASUREMENTS

### Transverse Image

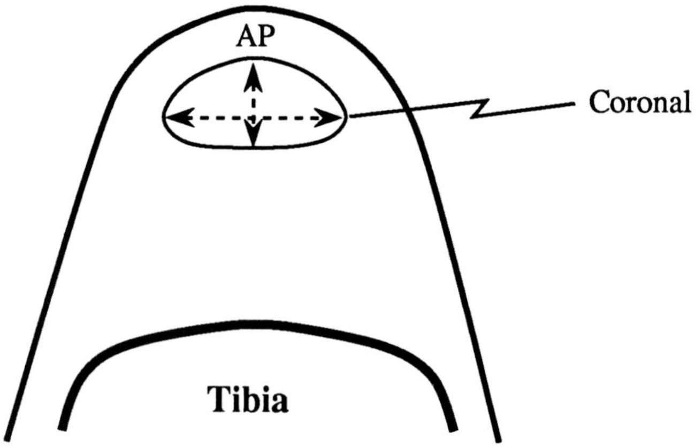


| | Sedentary Males (n=24) | Sedentary Females (n=21) | Male Athletes (n=52) | Female Athletes (n=43) |
|---|---|---|---|---|
| AP right tendon | 6.2 mm (SD±0.8) | 5.5 mm (SD±0.7) | 5.7 mm (SD±0.8) | 5.4 mm (SD±0.7) |
| AP left tendon | 6.1 mm (SD±0.8) | 5.7 mm (SD±1.1) | 5.9 mm (SD±1.0) | 5.5 mm (SD±0.7) |
| Coronal right tendon | 9.0 mm (SD±1.0) | 9.2 mm (SD±0.9) | 13.0 mm (SD±2.8) | 12.3 mm (SD±1.4) |
| Coronal left tendon | 8.8 mm (SD±0.8) | 8.7 mm (SD±1.0) | 12.8 mm (SD±2.4) | 12.3 mm (SD±1.5) |

## BICEPS TENDON MEASUREMENT

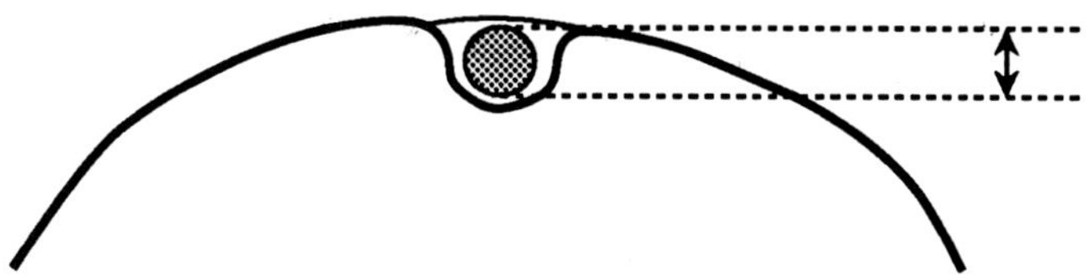

| | Sedentary Males (n=22) | Sedentary Females (n=21) | Male Athletes (n =24) | Female Athletes (n=29) |
|---|---|---|---|---|
| Biceps, right shoulder | 3.2 mm (SD±0.5) | 2.7 mm (SD±0.3) | 3.5 mm (SD±0.4) | 3.0 mm (SD±0.3) |
| Biceps, left shoulder | 3.3 mm (SD±0.4) | 3.0 mm (SD±0.3) | 3.7 mm (SD±0.5) | 3.2 mm (SD±0.3) |

## MCL MEASUREMENTS

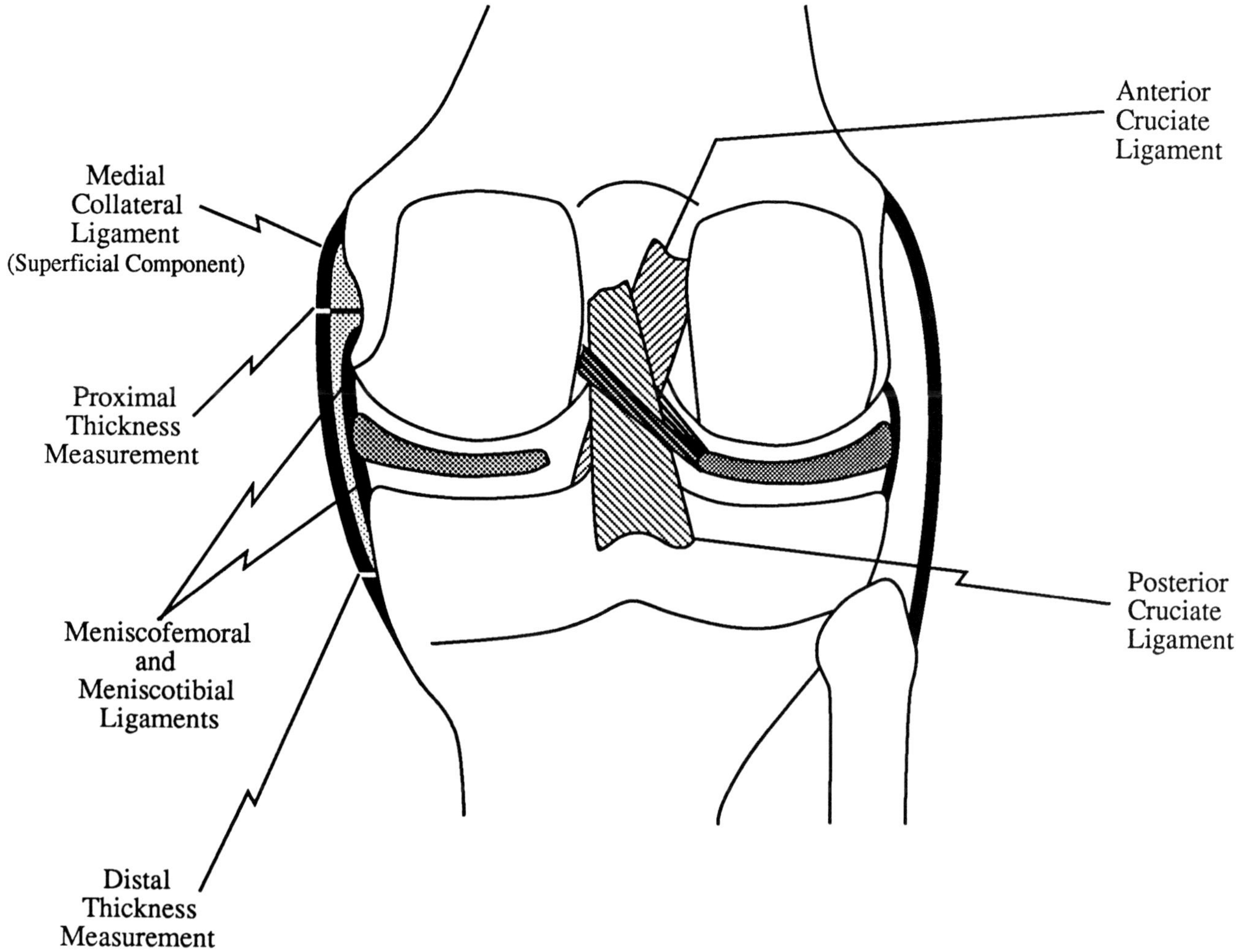


| | Sedentary Males (n=24) | Sedentary Females (n=21) | Male Athletes (n=52) | Female Athletes (n=43) |
|---|---|---|---|---|
| Right proximal MCL | 3.8 mm (SD±0.5) | 3.6 mm (SD±0.7) | 3.8 mm (SD±0.5) | 3.8 mm (SD±0.5) |
| Left proximal MCL | 3.7 mm (SD±0.6) | 3.3 mm (SD±0.7) | 3.8 mm (SD±0.5) | 3.6 mm (SD±0.5) |
| Right distal MCL | 2.3 mm (SD±0.3) | 2.3 mm (SD±0.4) | 2.0 mm (SD±0.4) | 2.1 mm (SD±0.3) |
| Left distal MCL | 2.3 mm (SD±0.4) | 2.3 mm (SD±0.3) | 2.0 mm (SD±0.4) | 2.1 mm (SD±0.3) |

## QUADRICEPS AND PATELLAR TENDON MEASUREMENTS

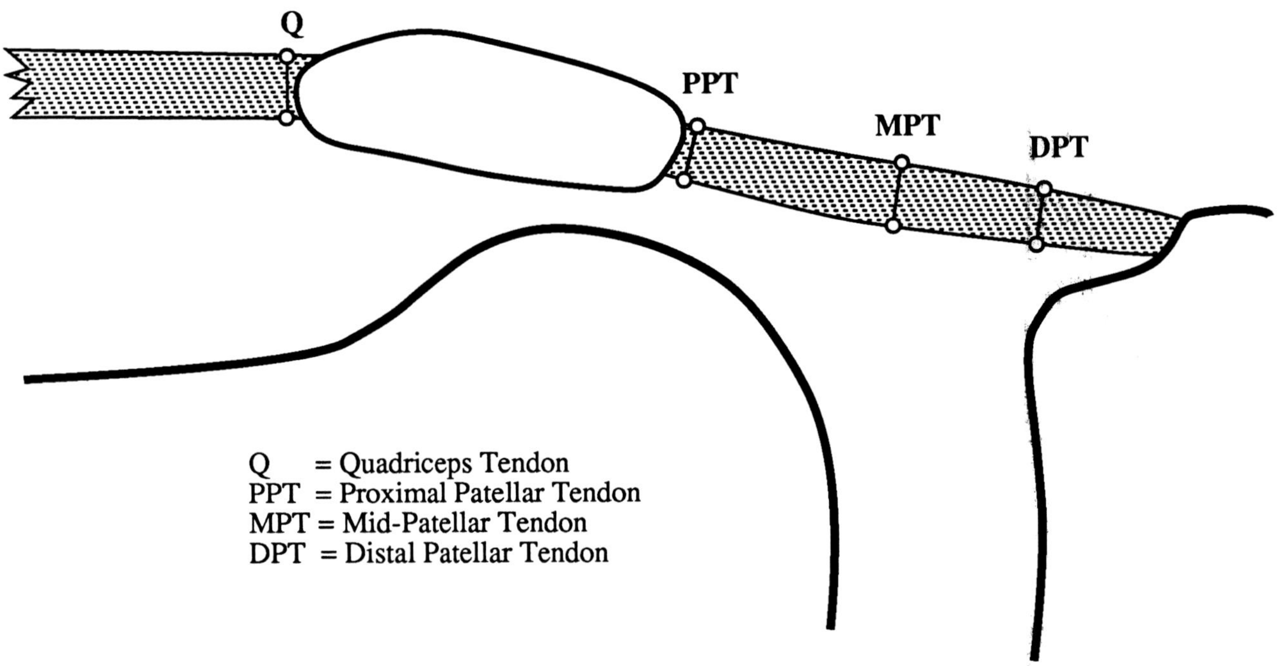


| | Sedentary Males (n=24) | Sedentary Females (n=21) | Male Athletes (n=52) | Female Athletes (n=43) |
|---|---|---|---|---|
| Right Q | 5.1 mm (SD±0.6) | 4.9 mm (SD±0.6) | 4.9 mm (SD±0.8) | 5.1 mm (SD±0.7) |
| Left Q | 5.2 mm (SD±0.8) | 4.8 mm (SD±0.5) | 5.2 mm (SD±0.9) | 5.1 mm (SD±0.6) |
| Right PPT | 3.0 mm (SD±0.4) | 2.6 mm (SD±0.4) | 3.4 mm (SD±0.9) | 3.3 mm (SD±0.8) |
| Left PPT | 3.1 mm (SD±0.3) | 2.7 mm (SD±0.4) | 3.5 mm (SD±0.7) | 3.1 mm (SD±0.4) |
| Right MPT | 3.1 mm (SD±0.4) | 2.9 mm (SD±0.5) | 3.6 mm (SD±0.7) | 3.3 mm (SD±0.8) |
| Left MPT | 3.2 mm (SD±0.4) | 2.9 mm (SD±0.3) | 3.5 mm (SD±0.6) | 3.1 mm (SD±0.4) |
| Right DPT | 3.1 mm (SD±0.4) | 2.6 mm (SD±0.4) | 3.4 mm (SD±0.9) | 3.3 mm (SD±0.8) |
| Left DPT | 3.1 mm (SD±0.4) | 2.7 mm (SD±0.4) | 3.5 mm (SD±0.7) | 3.1 mm (SD±0.4) |

## SUPRAPATELLAR BURSA MEASUREMENTS

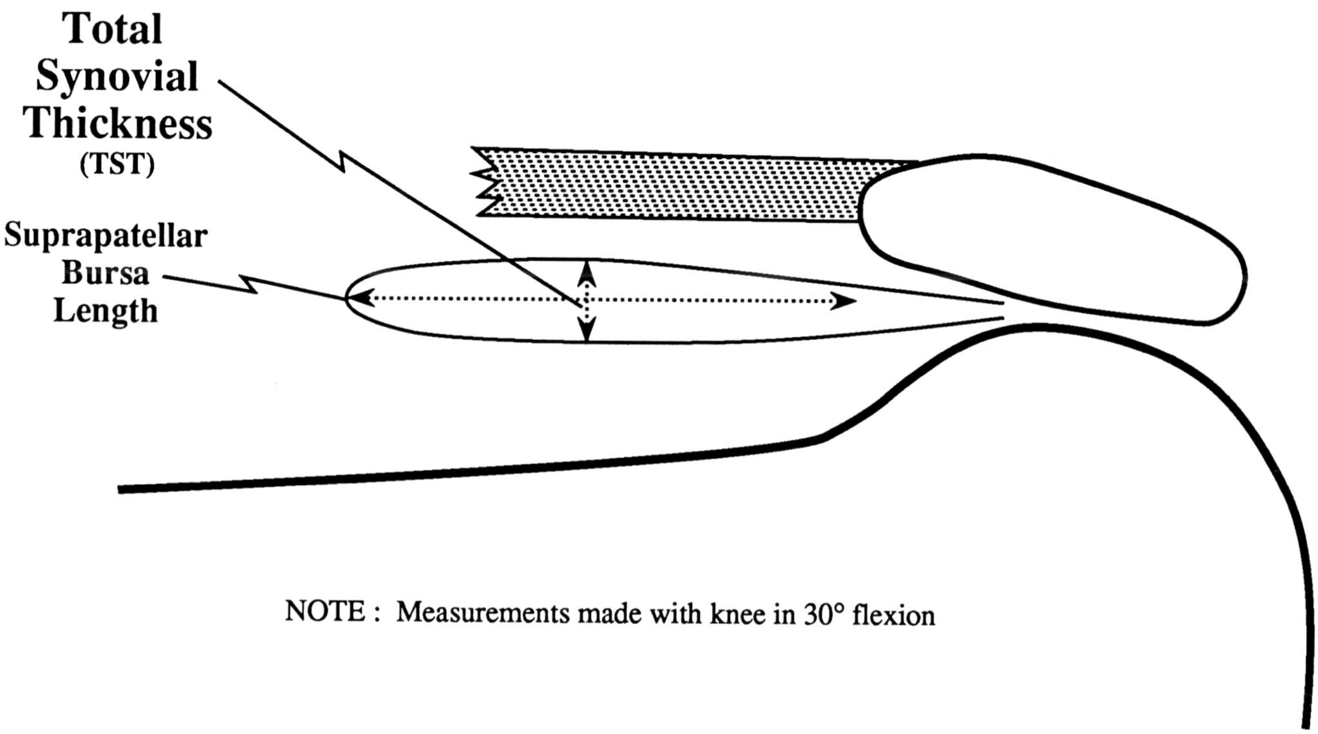


| | Sedentary Male n=24 | Sedentary Female n=21 | Male Athlete n=52 | Female Athlete n=43 |
|---|---|---|---|---|
| TST, right knee | 2.5 mm (SD±0.5) | 2.3 mm (SD±0.5) | 2.5 mm (SD±0.7) | 2.3 mm (SD±0.6) |
| TST, left knee | 2.5 mm (SD±0.4) | 2.3 mm (SD±0.5) | 2.6 mm (SD±0.6) | 2.2 mm (SD±0.6) |
| Bursa length, right | 22.3 mm (SD±3.8) | 21.2 mm (SD±4.3) | 21.2 mm (SD±4.2) | 19.5 mm (SD±5.0) |
| Bursa length, left | 21.8 mm (SD±4.2) | 20.6 mm (SD±5.3) | 23.9 mm (SD±4.8) | 20.8 mm (SD±4.9) |

# Index

## F



## G



## H



## I



## J



## K

## T